High-temperature Metamorphism and Crustal Anatexis

TITLES OF RELATED INTEREST

Boninites
A. J. Crawford (ed.)

Carbonatites
K. Bell (ed.)

Cathodoluminescence of geological materials
D. J. Marshall

Chemical fundamentals of geology
R. Gill

Crystal structures and cation sites of the rock-forming minerals
J. R. Smyth & D. L. Bish

The dark side of the Earth
R. Muir Wood

Deformation processes in minerals, ceramics and rocks
D. J. Barber & P. G. Meredith (eds)

Geology and mineral resources of West Africa
J. B. Wright

Heavy minerals in colour
M. Mange & H. Maurer

Igneous petrogenesis
B. M. Wilson

Image interpretation in geology
S. Drury

The inaccessible Earth
G. C. Brown & A. E. Mussett

The interpretation of igneous rocks
K. G. Cox *et al.*

Introduction to X-ray spectrometry
K. L. Williams

Komatiites
N. Arndt & E. Nisbet (eds)

Mathematics in geology
J. Ferguson

Metamorphism and metamorphic belts
A. K. Miyashiro

Petrology of the igneous rocks
F. Hatch *et al.*

Petrology of the metamorphic rocks
R. Mason

A Practical introduction to optical mineralogy
C. D. Gribble & A. J. Hall

Rheology of the Earth
G. Ranalli

Rutley's elements of mineralogy
C. D. Gribble

Simulating the Earth
J. Holloway & B. Wood

Statistical methods in geology
R. F. Cheeney

Volcanic successions
R. Cas & J. V. Wright

The young Earth
E. G. Nisbet

High-temperature Metamorphism and Crustal Anatexis

Edited by

J. R. Ashworth

University of Birmingham

M. Brown

Kingston Polytechnic

Published in association with
The Mineralogical Society of Great Britain and Ireland

London
UNWIN HYMAN
Boston Sydney Wellington

Published by the Academic Division of
Unwin Hyman Ltd
15/17 Broadwick Street, London W1V 1FP, UK

Unwin Hyman Inc.,
955 Massachusetts Avenue, Cambridge, Mass. 02139, USA

Allen & Unwin (Australia) Ltd,
8 Napier Street, North Sydney, NSW 2060, Australia

Allen & Unwin (New Zealand) Ltd in association with the
Port Nicholson Press Ltd,
Compusales Building, 75 Ghuznee Street, Wellington 1, New Zealand

First published in 1990

British Library Cataloguing in Publication Data

High-temperature metamorphism and crustal anatexis.
1. Lithosphere. Evolution
I. Ashworth, J. R. II. Brown, M. III. Series
551.14

ISBN 0-04-445721-9

Library of Congress Cataloging-in-Publication Data

High-temperature metamorphism and crustal anatexis/[edited by] J. R. Ashworth, M. Brown.
p. cm. – (The Mineralogical Society series; 2)
Includes bibliographical references.
ISBN 0–04–445721–9 (alk. paper)
1. Metamorphism. 2. Earth–Crust. I. Ashworth, J. R.
II. Brown, M. III. Series.
QE475.A2H54 1990
552'.4–dc20 90–11928
CIP

Typeset in 10 on 12 point Times by Mathematical Composition Setters Ltd, Salisbury
and printed in Great Britain by Cambridge University Press.

Preface

This second volume in the new series produced by the Mineralogical Society is concerned with the study of rocks from the deep continental crust. It is, we hope, timely to summarize recent petrological advances contributing to this field of active interest. Based mainly on review papers read at a conference, the chapters have subsequently been revised and expanded, while the editors have produced an introductory overview as Chapter 1. The conference was the Winter Meeting of the Mineralogical Society on 15 December 1988, at which Prof. R. C. Newton delivered the 20th Hallimond Lecture of the Society (which forms the basis of Ch. 7). The editors are grateful to all who contributed to the smooth running of the meeting at Kingston Polytechnic, and in the ensuing preparation of the volume: in particular, we sincerely thank all of the following for their labours as referees: A. J. Baker, L. M. Barron, M. J. Bickle, A. D. Chambers, J. D. Clemens, J. S. Daly, G. T. R. Droop, C. R. L. Friend, E. S. Grew, S. L. Harley, R. S. Harmon, N. B. W. Harris, B. Harte, T. J. B. Holland, N. F. C. Hudson, W. S. MacKenzie, W. Perkins, H. R. Rollinson, J. W. Sheraton, D. J. Waters, R. H. Worden and B. W. D. Yardley.

John R. Ashworth

Acknowledgements

We are grateful to the following individuals and organizations who have kindly given permission for the reproduction of copyright material (see figure captions and References at the end of each chapter for full details of sources):

Figure 2.2a after Hensen & Essene 1971, by permission of Springer-Verlag; Figure 4.4 after Johannes 1989, by permission of Springer-Verlag; Figures 6.2, 6.3 and 6.4 after Bickle *et al.* 1988, by permission of Springer-Verlag; Figures 7.4 and 7.5 after Peterson & Newton 1989, by permission of the University of Chicago Press; Figure 9.3 after Schumacher *et al.* 1989, by permission of the Geological Society of London; Figure 11.3 after Clarke *et al.* 1990, by permission of Blackwell Scientific Publications Ltd; Figures 11.25 and 11.26 after Clarke *et al.* 1989, by permission of Blackwell Scientific Publications Ltd; Figures 12.1, 12.4 and 12.9 after Harley 1989a, by permission of Cambridge University Press; Figure 13.3 after Barnicoat *et al.* 1987, by permission of Scottish Academic Press (Journals) Ltd.

Contents

List of tables

List of contributors

J. R. Ashworth, School of Earth Sciences, University of Birmingham, P.O. Box 363, Birmingham B15 2TT, UK

A. J. Brearley, Institute of Meteoritics, Department of Geology, University of New Mexico, Albuquerque, New Mexico 87131, USA

M. Brown, School of Geological Sciences, Kingston Polytechnic, Penrhyn Road, Kingston upon Thames, Surrey KT1 2EE, UK

I. Cartwright, Department of Earth Studies, University College of Wales, Llandinam Building, Aberystwyth, Dyfed SY23 3DB, UK

G. L. Clarke, School of Earth Sciences, Macquarie University, Sydney, NSW 2109, Australia

W. J. Collins, Department of Geology, University of Newcastle, Newcastle, NSW 2308, Australia

J. Downes, Department of Geology, Monash University, Clayton, Victoria 3168, Australia

S. L. Harley, Grant Institute of Geology, University of Edinburgh, West Mains Road, Edinburgh EH9 3JW, UK

B. J. Hensen, Department of Applied Geology, University of New South Wales, Box 1, Kensington, NSW 2033, Australia

K. T. Hollocher, Department of Geology, Union College, Schenectady, New York 12308, USA

F. Holtz, Institut für Mineralogie, Universität Hannover, Welfengarten 1, D-3000 Hannover 1, FRG

W. Johannes, Institut für Mineralogie, Universität Hannover, Welfengarten 1, D-3000 Hannover 1, FRG

R. C. Newton, Department of the Geophysical Sciences, University of Chicago, 5734 South Ellis Avenue, Chicago, Illinois 60637, USA

R. Powell, Department of Geology, University of Melbourne, Parkville, Victoria 3052, Australia

P. Raase, Mineralogisch–Petrographisches Institut, Universität Kiel, Olshausenstraße 40, 2300 Kiel, FRG

P. Robinson, Department of Geology and Geography, University of Massachusetts, Amherst, Massachusetts 01003, USA

D. C. Rubie, Bayerisches Geoinstitut, Universität Bayreuth, Postfach 10 12 51, D-8580 Bayreuth, FRG

V. Schenk, Mineralogisch–Petrographisches Institut, Universität Kiel, Olshausenstraße 40, 2300 Kiel, FRG

J. C. Schumacher, Mineralogisches Institut, Universität Kiel, Olshausenstraße 40, D-2300 Kiel, FRG

R. Schumacher, Mineralogisch–Petrographisches Institut, Universität Kiel, Olshausenstraße 40, 2300 Kiel, FRG

R. J. Tracy, Department of Geological Sciences, Virginia Polytechnic Institute and State University, Blacksburg, Virginia 24061, USA

J. W. Valley, Department of Geology and Geophysics, University of Wisconsin–Madison, Weeks Hall, 1215 W. Dayton Street, Madison, Wisconsin 53706, USA

R. H. Vernon, School of Earth Sciences, Macquarie University, Sydney, NSW 2109, Australia

P. W. Vitanage, Department of Geology, University of Peradeniya, Peradeniya, Sri Lanka

S. M. Wickham, Department of the Geophysical Sciences and Enrico Fermi Institute, University of Chicago, Chicago, Illinois 60637, USA

CHAPTER ONE

An overview of diverse responses to diverse processes at high crustal temperatures

J. R. Ashworth & M. Brown

1.1 Introduction

At the highest temperatures of regional metamorphism, crustal rocks undergo combinations of deformation, mineral reactions, partial melting, and geochemical transport processes. The scientific challenge goes beyond simple description of these processes, to the estimation of physical conditions (notably pressure, *P* and temperature, *T*), eventually leading to inferences about the tectonic régimes and heat sources responsible for the pressure–temperature–time (*P–T–t*) evolution of a region (from cold to hot followed by cooling and uplift to its present position at the Earth's surface). This volume comprises reviews which explain the current state of progress towards these goals using different methods and in several different high-grade terranes.

1.2 Processes and materials

High-grade processes act on assemblages of solid minerals, melts, and also fluids which are thought to be dominated usually by the components H_2O and CO_2. Of these substances, only the mineral assemblages remain in anything like their original, high-grade condition: a possible exception to this rule is that some fluid inclusions in the minerals may have been trapped under the high-grade conditions (Newton, Ch. 7, this volume), though there are ample possibilities for their later modification (Hollister 1988). The melts have crystallized, and their compositions can only be deduced indirectly; current experimental work (Johannes & Holtz, Ch. 4, this volume) is important here. Even the minerals themselves may have been modified by exsolution or by cation exchange processes, particularly Fe–Mg exchange in ferromagnesian minerals.

The aqueous fluid is sometimes called 'vapour' although it is really a supercritical fluid. H_2O-CO_2 fluids can be described in terms of the mole fractions of the components, X_{H_2O} and X_{CO_2}, or activities a_{H_2O} and a_{CO_2} which, like mole fractions, are unity in pure end-member fluids, but which differ from mole fractions in non-ideal mixtures so that an activity-composition model must be used (e.g. Johannes & Holtz, Ch. 4, this volume). It is also possible thermodynamically to define activities in a 'fictive' fluid which could have been at equilibrium with a given assemblage, even if the real conditions were fluid-absent (e.g. J. C. Schumacher *et al.* Section 9.8.2, this volume). The fluids can dissolve other components from the rocks, but the ways in which these interact to form species in solution are not well known at high P and T (Newton, Ch. 7, this volume). In principle, the presence of solute will affect the activity-composition relations for H_2O and CO_2; there is even the intriguing possibility that, at very high salinities, immiscibility between H_2O-rich and CO_2-rich fluids may extend up to temperatures approaching those of interest here (Bowers & Helgeson 1983).

Both melt and H_2O-CO_2 fluid may move through the rocks, with some difficulty in the case of water-undersaturated melt owing to high viscosity (Wickham 1987), although the volume expansion due to melting can produce the porosity necessary for flow. Partial melting of rocks within the Earth's crust (crustal anatexis) may produce effects ranging from (a) textural modification due to unsegregated melt (considered by Vernon *et al.*, Ch. 11, this volume), through (b) small-scale differentiation between melt-poor and melt-rich domains (these being the two complementary neosomes, respectively the melanosome or restite and the *leucosome*, of typical anatectic migmatites: e.g. Powell & Downes, Ch. 5; J. C. Schumacher *et al.*, Ch. 9, this volume), to (c) accumulation of large bodies of magma (melt ± suspended crystals) which tend to move upwards because of density contrasts with their surrounding rocks, producing intrusive anatectic granites (Wickham, Ch. 6; Newton, Ch. 7, this volume). H_2O-CO_2 fluids may be driven in various directions (Section 1.7), transporting their solute components (Newton, Ch. 7).

1.3 Mineral equilibria

The fundamental principle of metamorphic petrology is that rocks react towards equilibrium assemblages of minerals (± melt ± fluid). At high T, approach to equilibrium is often close. Both univariant and divariant equilibria are hypothetically important. Univariant ones, giving lines or curves in a two-variable diagram (usually the $P-T$ diagram) that represents the postulated stability relations, divide the diagrams into 'pigeon-holes' where different assemblages may be formed. The resulting line diagrams are often called 'petrogenetic grids' (Hensen & Harley, Ch. 2, this volume). Divariant equilibria occupy fields of the diagram rather than being restricted to lines, and therefore arguably have higher probability of being actually approximated by rocks: in the case where phases of variable composition undergo continuous ('sliding') reactions as a divariant field is traversed, model equilibrium compositions may be correlated with actual mineral analyses. Both kinds of equilibrium thus contribute to estimation of metamorphic *P–T* conditions (geothermobarometry). Further, reaction textures and zoning of the solid-solution minerals are indicators of changing conditions with time, and are often used as evidence for a segment of the *P–T–t* path across the *P–T* diagram (e.g. J. C. Schumacher *et al.*, Fig. 9.3; R. Schumacher *et al.*, Fig. 10.10, Vernon *et al.*, Fig. 11.26, this volume). There is disagreement over whether rocks that show reaction textures can be used for quantitative geothermobarometry. The problem is to determine which parts of the mineral assemblage, if any, were at equilibrium at any one point in the history of the rock. Harley & Hensen (Ch. 12) compile *P–T–t* paths from many different bits of information, and Cartwright (Fig. 13.2) connects several metamorphic events into a path for the most ancient part (Scourian complex) of the Lewisian of north-west Scotland. *P–T–t* paths are considered further in Section 1.8.

Modelling of any equilibria involving a fluid must include the component H_2O, as must melting, because the relevant melts inevitably contain H_2O (Johannes & Holtz, Ch. 4). On the other hand, most granulite facies minerals are anhydrous, and it is possible to consider their mutual equilibria independently of H_2O. Hensen & Harley (Ch. 2), modelling assemblages in metapelites, start from the simple three-component ($n = 3$) system $MgO–Al_2O_3–SiO_2$ (MAS). One of the modelled minerals, cordierite, can contain noticeable amounts of H_2O, but Hensen & Harley (Section 2.10) argue that this does not cause noticeable uncertainty in their treatment. With $n = 3$, the divariant assemblages have three phases and can be fully represented by the usual kind of triangular compositional diagram (Fig. 2.1). This compositional picture (chemography) then places constraints on what *P–T* grids can be postulated; a grid must have a self-consistent topology (set of stable invariant points interconnected by some of the univariant lines that radiate from each point), as is illustrated by Hensen & Harley for MAS. Because of the important Mg–Fe solid solution in natural minerals, they must progress to the system $FeO–MgO–Al_2O_3–SiO_2$ (FMAS). Considering seven phases altogether with $n = 4$ (an '$n + 3$ multisystem'), they have seven possible invariant points, but it is shown that these cannot all appear in a stable manner (some must be metastable). A purely theoretical treatment could become very complicated, so it is reasonable that Hensen & Harley apply constraints from natural assemblages and experiments at an early stage of the analysis (Section 2.4). On the other hand, the natural minerals contain minor amounts of additional components which may cause their stability relations to differ from those in the FMAS system employed for the experiments and theory. Using only a few constraints from natural and experimental assemblages, Hensen & Harley (Section 2.4) arrive at two alternative grids and face the paradox that, while only one of these should theoretically be stable, the full set of natural assemblages requires both. Their solution of the problem is that one grid is stable under oxidizing conditions (high oxygen fugacity, f_{O_2}), and the other at low f_{O_2}. The minerals are implicitly involved in redox equilibria and must contain Fe_2O_3 in addition to FeO (the system is really FMAS–O_2, FMASO), but in small amounts so that the FMAS representation of minerals remains a viable approximation. The Fe_2O_3 contents become explicit (are used to calculate slopes of lines) in the diagram using the chemical potential of this component, $\mu_{Fe_2O_3}$, on one axis (Fig. 2.10); it should be borne in mind that in electron-probe analyses Fe_2O_3 is estimated indirectly and roughly, adding to the uncertainties of the model equilibria.

With $n = 4$ or 5, the use of $n - 3$ chemical potentials as variables in a grid (while holding one or both of P and T constant) enables the remaining chemography to be fully and validly represented by triangles (e.g. Lal *et al.* 1987), one or two of the compositional dimensions describing variation between rocks having been incorporated into the construction of the grid itself. Alternatively, and with extensions to $n > 5$, chemographic diagrams projected from selected phases (which are constantly present in considered assemblages) are usually constructed, the simple case of projection from quartz in FMAS being illustrated by Hensen & Harley in Figure 2.3. Several other projections are exemplified in the chapter by J. C. Schumacher *et al.* (Figs 9.4 & 9–13). For metapelites at slightly lower grades, where biotite is important, the system commonly used for modelling is K_2O–FeO–MgO–Al_2O_3–SiO_2–H_2O (KFMASH). Here, the construction of unique triangular chemographic diagrams requires projection from three phases, for which Powell & Downes (Ch. 5) select quartz, K feldspar and sillimanite. By not projecting from H_2O, they are able to treat equilibria at $a_{H_2O} < 1$.

For assemblages containing more than two solid solution phases, the form of the P–T diagram depends on the proportions of the phases present, i.e. on bulk rock composition. The construction of a unique diagram for fixed bulk composition with three solid solution phases is discussed by Hensen (1971) and briefly in Chapter 2 by Hensen & Harley (see Figs 2.3c & d). Powell & Downes (Ch. 5) call such diagrams ‘pseudosections’, and use them to illustrate the important point that univariant reactions in end-member systems (KMASH and KFASH in the present case) become divariant fields, representing continuous reactions, in the compound system (KFMASH; Powell & Downes, Fig. 5.6). Because the diagrams differ for different bulk compositions, various rocks with similar but not identical compositions experience different reaction histories (Hensen & Harley, Fig. 2.3; Powell & Downes, Section 5.4). This is why it is important to collect samples from the full range of bulk rock compositions in any detailed study of an area which aims to establish the P–T–t path followed by those rocks (e.g. J. C. Schumacher *et al.*, Ch. 9). Powell & Downes (Ch. 5) also use T–a_{H_2O} diagrams (see Powell 1983), or T–a_{H_2O} pseudosections for specific bulk compositions, which are an underused tool in the understanding of reaction sequences in systems with internally buffered a_{H_2O}, and are particularly useful in the consideration of dehydration melting (e.g. Waters 1988; next section).

If the approximate orientation of a *P–T* grid with respect to its axes is known (because a few lines of well constrained slope have been used to generate it: Hensen & Harley, Ch. 2), then it can be used for 'pigeon-holing' rocks as, say, high-*T*, low-*P* etc., even without absolute calibration of the *P* and *T* axies. However, calibration must be attempted for proper *P–T–t* studies: Harley & Hensen (Ch. 12) use a calibration of part of their FMAS grid from Chapter 2, to infer very high temperatures and pressures for some Antarctic mineral assemblages (Fig. 12.4). Curvature of univariant lines is likely in principle, and in this grid one univariant boundary is strongly curved because of the inferred behaviour of solid solutions. Hensen & Harley (Ch. 2) even begin to draw experimentally constrained contours of mineral composition across the high-*P,T* divariant fields, although the uncertainties are such that two versions of the diagram are offered (Figs 2.13 & 14).

1.4 Melting equilibria

It is likely that rocks under granulite facies conditions will begin to melt, particularly if they contain hydrate minerals (Johannes & Holtz, Ch. 4). There may be a link between metasedimentary rocks and 'S-type granites' (those with mineralogical and geochemical features which indicate derivation from a sedimentary source material). Other melts may be trondhjemitic (crystallizing quartz + sodic plagioclase, without the major K-feldspar of granites), as has been argued in studies of some migmatite leucosomes (e.g. Evans & Speer 1984). Granitic to trondhjemitic leucogneisses in the Lewisian are interpreted by Cartwright (Ch. 13) as representing partial melts. Tait & Harley (1988) infer partial melting for the origin of plagioclase-orthopyroxene dominated leucosomes, of dioritic to tonalitic bulk composition, in mafic granulites. J. C. Schumacher *et al.* (Section 9.8.1) also mention tonalitic melts derived from metabasic rocks.

If the system contains important K_2O, probable hydrate minerals are micas. Powell & Downes (Ch. 5) treat melting equilibria in terms of the usual pelitic model system, KFMASH. Solid solutions are again useful in geothermobarometry. Since hydrate minerals are reservoirs of H_2O, a hydrous melt can be produced in the absence of aqueous fluid, given high enough *T* for a reaction in which the hydrate mineral breaks down as the melt is generated. Such reactions are *dehydration melting* reactions, also sometimes called '*vapour-absent*' reactions: they are certainly fluid-absent in KFMASH models, and are sometimes labelled (V), to indicate absence of this phase, on the KFMASH grids. In a system with other volatile components in addition to water, fluid-present dehydration melting is possible (with a_{H_2O} in the mixed fluid low enough to be at equilibrium with the melt), but dehydration melting in the crust may usually be truly fluid-absent (the rocks having lost their fluid through incipient fluid-consuming melting reactions at lower T; Clemens & Vielzeuf 1987).

Dehydration melting of muscovite produces some natural melts (e.g. in some migmatites discussed by J. C. Schumacher *et al.*, Ch. 9). Melting to a sufficient extent for large-scale melt segregation (next section) probably requires higher T and commonly involves biotite dehydration (Clemens & Vielzeuf 1987). A fairly comprehensive grid for KFMASH in the dehydration melting region of P–T space has been erected by Grant (1985). Modifications and extensions of parts of this for particular biotite reactions are proposed by Hensen & Harley (Figs 2.17–19) and Powell & Downes (Fig. 5.3). Vernon *et al.* (Fig. 11.26) throw in TiO_2 and Fe_2O_3 (producing KFMASHTO) to accommodate ilmenite and magnetite.

Uncertainty about the melt composition is a problem for such modelling. Even within KFMASH, subtle features of melt composition can affect the chemography so as to change the grid (Grant 1985). Also, real melts can take in many additional components. 'Fluxing' components such as B_2O_3 and F can noticeably lower the temperatures of melting (e.g. Pichavant 1987). In the presence of plagioclase, the component Na_2O can lower the temperatures somewhat (Johannes & Holtz, Fig. 4.2; Rubie & Brearley, Fig. 3.1). With CaO additionally partitioned between plagioclase and melt, but no additional Ca mineral present, there is also the complication that the univariant model reactions from KFMASH become divariant.

The solubility of CO_2 in granite-like melts has long been thought to be low, so that melting temperatures are much higher with CO_2 than with the melt-soluble H_2O as volatile component. Johannes & Holtz (Ch. 4) confirm this in the simplified granite system ('haplogranite' system, Qz–Ab–Or, with H_2O and CO_2). They also report that the solubility of FeO and MgO in such melts is small (Section 4.5). However, Newton (Ch. 7) argues that the joint presence of MgO and CO_2 profoundly affects melt compositions, and that in KMASH–CO_2 fluid-present melting with CO_2-rich fluids occurs at temperatures comparable to those with H_2O-rich ones. This surprising inference was initially based almost entirely on one set of experiments (Wendlandt 1981), but is now supported by new experimental work (Peterson & Newton 1989). Newton (Ch. 7) adds Na_2O and CaO to KMASH–CO_2: the system has an assumed compositional degeneracy (K : Al = 1 in all phases), and so effectively has $n = 7$ rather than $n = 8$. By fixing three quantities (P and X_{H_2O} in the fluid, and plagioclase composition), Newton constructs diagrams (Fig. 7.7) in which an invariant point is defined by six phases (quartz and three other minerals + melt + 'vapour'). In his model (Fig. 7.8), at constant X_{H_2O} in the fluid, an increase in temperature can produce either dehydration without melting (if $K_2O : Na_2O$ is low) or a vapour-consuming melting reaction (at high $K_2O : Na_2O$).

1.5 Amount and mobilization of melt

Small amounts of melt may fail to segregate from their source rocks. It has been argued that the origin of large granite bodies is related to the rather abrupt change in mechanical properties of partially molten rock at the 'rheologically critical melt percentage' (RCMP), perhaps 30–50% melt (Arzi 1978, van der Molen & Paterson 1979, Wickham 1987). The behaviour changes from that of a solid with interstitial liquid to that of melt with suspended crystals: the cohesion of the solid part is lost.

Melting can occur in an open system (Section 1.7) or a closed one. In the latter case, the fixed total H_2O content permits estimation of the amount of melt which can possibly be generated at given P,T, calculated from the minimum H_2O content of the granitic melt at that P, T: Johannes & Holtz (Ch. 4) review the evidence that high temperatures are required to produce large melt fractions in metasediments in closed systems (Fig. 4.6). Similarly, with igneous starting material (tonalite), Rutter & Wyllie (1988) find that, at 10 kbar, $T \approx 950^\circ C$ is required for approximately 35% melting if the H_2O content of the system is 0.8 wt% (initially held in biotite and hornblende). These experimental conditions approach those invoked for partial melting in the Scourian by Cartwright (Ch. 13), i.e. $1000 \pm 50^\circ C$ and >11 kbar.

According to Wickham (1987), at melt proportions below the RCMP, only small-scale segregation is likely. However, Clemens & Vielzeuf (1987) argue that some granites must have come from quartz-feldspar dominated source rocks at melt fractions below 30%, probably below 20%. Segregation may be aided by inhomogeneities caused or enhanced by deformation, e.g. fractures. Rubie & Brearley (Ch. 3) argue that the local increase in volume due to partial melting may promote cracking, particularly if melting occurs rapidly after the rock has been metastably heated above the equilibrium melting temperature. Chapter 3 is particularly concerned with the rapidity of the diffusion-controlled grain-boundary melting reaction that is expected, on experimental evidence, once the barrier causing the excess heating (overstepping of equilibrium T) has been overcome. The overstep makes heat immediately available for the endothermic reaction, which should proceed rapidly as T falls back towards the equilibrium value: Rubie & Brearley (Ch. 3) present models in which overstepping by a few tens of degrees is followed by production of tens of per cent melt in ~ 1 year. Whether the overstepping itself is likely during deep-seated metamorphism is a difficult question. As discussed in Chapter 3, one possible cause would be a kinetic barrier (nucleation of a new mineral produced by an incongruent melting reaction, or diffusion between reactant minerals that are not in direct contact); and another possibility is that, given an initially dry rock, melting is caused by an abrupt influx of H_2O (Section 1.7).

Although the scale of such processes, including fracture-assisted segregation, may often be small (migmatite-forming: Wickham 1987; Rubie & Brearley, Ch. 3), the interaction between deformation and melt segregation could be self-accelerating as the melt-enriched zones become mechanically weakened; geological evidence for some such effect on a large scale is adduced by Hollister & Crawford (1986). Both melt segregation (e.g. Strachan *et al.* 1989, p. 425, Fig. 3) and granite emplacement (Hutton 1988) are enhanced when anatexis occurs in association with major crustal deep-shear zones. Crustally derived leucogranites are commonly associated with shear zones (e.g. Strong & Hanmer 1981).

Powell & Downes (Ch. 5) discuss the rather common association between large garnets and leucosomes in migmatites (cf. Tracy & Robinson 1983, Fig. 4; Waters & Whales 1984 and references therein; and Waters 1988). One might speculate that the garnets act as mechanical inhomogeneities during deformation, but Powell & Downes (Ch. 5) prefer a chemical process; the complementary products of incongruent melting are spatially associated. It seems possible that the differentiation process is a diffusive segregation of garnet-forming components to leave a residual leucosome depleted in dark minerals, with the partial melt acting merely as a catalyst. However, Powell & Downes (Ch. 5) argue that melt-forming as well as garnet-forming constituents diffuse towards garnet from elsewhere in the rock; this appears to require a mechanism that inhibits melt production at a distance from garnet and localizes it around garnet (Waters 1988), even though the garnet is not contributing matter to the melt (garnet is actually growing) during the main melting reaction.

1.6 Melt–restite interaction and retrograde evolution

In a partial melting process, the residual rock and also any residual crystals suspended in melt constitute the *restite*. Wickham (1987) proposes a mechanism that effectively mixes the melt with the restite: kilometre-scale convective overturn of the whole mass once its effective viscosity falls drastically, as melt fractions reach 30–50%. (This is different from diapiric intrusion, which is not controlled by the viscosity of the diapir: see Wickham 1987.) This suggestion is consistent with the commonness of mafic minerals in granites, since Johannes & Holtz (Ch. 4) find that a restite-uncontaminated melt should be Fe,Mg-poor (leucogranitic); however, Newton (Ch. 7) would not agree if CO_2 is acting additionally.

Detailed interpretations of the mineralogy and geochemistry of intrusive granites in terms of mixing of melt with solids (largely restitic) are now well developed (e.g. Clark & Lyons 1986). On the other hand, it can be difficult to distinguish such processes from assimilation-fractional crystallization or from mixing of melts (Wall *et al.* 1987). In migmatites, Sawyer (1987) has made out a good case for fractional crystallization to explain the trace element geochemistry of a suite of leucosomes, which are structurally discordant to the country rocks (suggesting that the melt has become detached from its source). For larger granitic bodies, fractional crystallization models from the literature are rejected as explanations of geochemical data by Newton (Ch. 7) and Cartwright (Ch. 13) in their respective areas (southern India, Lewisian). Noting that rare earth element (REE) and other trace element data present the only substantial obstacles to an anatectic interpretation for the Lewisian granitic–trondhjemitic bodies, Cartwright (Ch. 13) accepts that interpretation, and rightly draws attention to the shortage of information on partition of these elements between minerals and melts (particularly at low a_{H_2O}). Trace elements such as REE may be controlled by accessory minerals (e.g. Barbey *et al.* 1989). Another idea discarded by Cartwright (Ch. 13) for the Lewisian is invoked by Barbey *et al.* (1989) for some migmatites: this is *disequilibrium melting*, for which equilibrium partition data would not be appropriate. It could, of course, affect major elements as well as trace elements. It has been studied experimentally for major elements in plagioclase (Johannes 1985), and something similar may happen in shallow-seated, rather rapid natural melting (e.g. Naslund 1986). Rubie & Brearley (Ch. 3) note that their rapid melting process would favour it; the rate-determining diffusion in the melt is so much faster than in the solid reactants that the latter would fail to adjust internally to new compositions at equilibrium with the melt produced.

In the 'extreme' model of disequilibrium melting (Allègre & Minster 1978), adopted by Barbey *et al.* (1989) for lack of more detailed knowledge, there is no fractionation at all; as a mineral is partly dissolved into the melt, the residual mineral does not change composition. It is clear that the result is indistinguishable from mechanical mixing between the melt and the restite mineral hosting a particular element (Barbey *et al.* 1989). Also, as pointed out by Pride & Muecke (1982) in discussing the disequilibrium melting hypothesis for the Lewisian, the outcome is very difficult to distinguish from that of equilibrium melting followed by *homogenization* of the minerals, throughout the crystallizing melt plus adjacent restite, during cooling. In both cases, the bulk composition (including the trace elements) of a part of the rock (e.g. leucosome or adjacent restitic melanosome) is determined solely by its mode (mineral proportions) plus the mineral compositions, which are homogeneous throughout. Given that the melt may not segregate cleanly from its restite, and that the retrograde reaction with adjacent restite may further influence the mode of the leucosome (Powell & Downes, Ch. 5), the recognition of criteria for equilibrium melting becomes very difficult. For major elements in plagioclase, the late homogenization effect is the usually accepted interpretation for the failure of anatectic leucosome–melanosome pairs to show the fractionation predicted by melt-crystal equilibrium (Johannes 1985, pp. 77–8). This could explain the plagioclase compositions in the migmatites of Barbey *et al.* (1989). If so, there is no record of whether melting was at equilibrium or disequilibrium, but the latter is intrinsically unlikely since equilibration was so effective during cooling. The end-stage equilibration effect is related to 'equilibration volume' by Powell & Downes (Ch. 5). In their migmatites, they infer that mineral homogenization occurred over distances ~20 cm, larger than the dimensions of leucosomes. The process may also be relevant in melt–restite mixtures behaving as granite magmas.

Incidentally, the difficulty of discriminating geochemically among migmatization processes extends to the distinction between anatectic and non-anatectic (subsolidus) differentiation. This problem has already been pointed out by Price & Taylor (1977). Barbey *et al.* (1989) note the mathematical equivalence between the latter process and 'extreme' disequilibrium melting. In summary, in migmatites there are good reasons why geochemistry may fail to distinguish among leucosomes, of similar modal composition, produced by (a) subsolidus differentiation, and by any of the following anatectic process: (b) disequilibrium melting, (c) melting plus admixture of restite minerals with melt, and (d) melting followed by subsolidus homogenization (equilibration of minerals). On the other hand, Sawyer & Barnes (1988) did report trace element differences between anatectic and non-anatectic leucosomes. Also, the minerals of Barbey *et al.* (1989) are not altogether homogeneous, as shown by backscattered electron images of zircon, an important trace element host. Future studies of zoning within accessory minerals, combined with bulk rock data for trace elements and an understanding of partition coefficients, may be illuminating.

In addition to mineral homogenization, another common effect in retrograde evolution is hydration reaction during melt crystallization (the reverse of dehydration melting). Replacement textures include symplectic pseudomorphing of cordierite (Vernon *et al.*, Ch. 11; cf. Ashworth & McLellan 1985, pp. 189–92). The reactions often fail to go to completion, and may fail even to start, suggesting either kinetic barriers or loss of H_2O from the cooling system (Waters 1988). Movement of H_2O from cooling melts to their country rocks may account for retrogression on a large scale (the Inverian event in the Lewisian is interpreted in these terms by Cartwright in Ch. 13). Powell & Downes (Ch. 5) note that their 'equilibration volume' concept, while accounting for homogeneity of garnet, tends to predict that the garnet should not exist (should have reacted back to biotite). Rather than escape of fluid H_2O, the authors suggest escape of some melt to reduce the amount of locally available H_2O. Late-crystallizing melt fractions may concentrate H_2O (Johannes & Holtz, Ch. 4), thus becoming less viscous and more liable to move. From a different viewpoint, the process favoured by Newton (Section 7.4.2) for H_2O removal would be a late influx of CO_2. Late fluid movements may also modify the geochemical patterns discussed in the preceding paragraphs. REE are usually thought to be unmoved by aqueous fluids (e.g. Barbey *et al.* 1989) but may be soluble in CO_2, as suggested by Tait & Harley (1988), who invoke leaching of REE at a stage postdating melt crystallization in a migmatite.

1.7 Fluid movements

Metamorphic histories can be profoundly affected by the flow of fluids through the rocks (*infiltration*). Possibly more important than the retrograde effects just discussed are influxes near the climax of some high-grade metamorphic episodes. Because the bulk H_2O content tends to limit the amount of melt produced (Johannes & Holtz, Ch. 4), influx can promote melting. It may also promote segregation, because water-rich melts are less viscous (Wickham 1987). Barr (1985, pp. 242–5) models a migmatized shear zone with deformational control of both H_2O influx and melt segregation. In dry rocks at some T above the H_2O-saturated solidus (defined by Johannes & Holtz, Ch. 4), the sudden arrival of water could trigger rapid melting (Rubie & Brearley, Ch. 3). A self-accelerating process is conceivable, in which melting assists deformation, which in turn provides channels for incoming H_2O (Rubie & Brearley, Ch. 3). On the other hand, the change of deformation style accompanying large-scale anatexis may prevent cracking and hence stop fluid flow (Wickham & Taylor 1985).

Wickham (Ch. 6) emphasizes the probable isotopic effects of fluid-induced metasomatism during metamorphism, prior to the onset of melting which generates granitic magmas. These effects are to be borne in mind when using isotopes in granites as indicators of magma sources. There is a considerable volume of literature on the inference of source rocks and mixing effects from isotopic systems in granitoids, e.g. classic work on the Peninsular Ranges Batholith of the western U.S.A. and Mexico (Hill *et al.* 1986), and work in the British Caledonides, pioneered by Harmon & Halliday (1980). In some cases, sources are obscured by later evolution (e.g. Hildreth & Moorbath 1988). Although not contradicting such work, Chapter 6 presents the argument that in the context of this book (i.e. crustal anatexis with no mantle melt contribution), metamorphic modification of the source rocks may greatly influence the isotope ratios. The data are still sparse, and rather few variables are measured (mainly $\delta^{18}O$ and $^{87}Sr/^{86}Sr$), but Wickham (Ch. 6) presents evidence for source-rock modification in several terranes. This does not prevent him from using isotopes in a general way as indicators of source-rock type: e.g. high $\delta^{18}O$ indicates an ultimately sedimentary protolith in Nevada and in the Hercynian of north-west Europe.

Consistent with the large melt proportions inferred for Pyrenean granite genesis, large H_2O influxes are inferred by Wickham (Ch. 6). Thorough melting is indicated at a temperature of only 750°C, possibly below that needed for biotite dehydration melting (Wickham 1987).

The fluids envisaged by Wickham (Ch. 6) may be of shallow origin, perhaps derived from sedimentary pore fluids. Upward fluid fluxes from deeper sources, causing more metasomatism than melting, have often been invoked to explain the geochemistry of ancient Precambrian (particularly Archaean) granulite facies regions. In this volume, Cartwright (Ch. 13) firmly rejects such interpretation of the Lewisian, while Newton (Ch. 7) just as stoutly upholds it for southern India. In both cases, the events are thought to be late Archaean (but see Aftalion *et al.* (1988) for a sceptical viewpoint on the dating in India).

The interpretation of small-scale field relations by Newton (Section 7.1.4) and his colleagues, as indicating fluid infiltration, has been queried by authors concerned with localized melting (e.g. Waters 1988). In Chapter 7, Newton is mainly concerned with large-scale distribution of rock types; anatectic granites were apparently generated in mid-crust but not at greater depths. His interpretation is a development of previous work by his school (Janardhan *et al.* 1979, 1982) which emphasized low X_{H_2O} in the fluids as an explanation of dehydration without melting (charnockitization). The fluids, arriving from a deep source, are supposed to be CO_2-rich, partly on fluid-inclusion evidence. Now (Ch. 7), because recent experiments indicate that melting temperatures may vary little between H_2O-rich and CO_2-rich systems (Section 1.4), emphasis is placed on K-metasomatism (already noted by Janardhan *et al.* 1982) as the explanation of the depth-correlated differences in petrology. The fluids leach K at depth and then deposit it to promote melting at shallower levels. They also transport heat. In the relevant diagrams, at constant X_{H_2O} (say, 0.5; Fig. 7.8), whether or not the rocks melt is controlled by a combination of T and potash activity (modelled by the K : Na ratio in the fluid). However, because the granites are granites, rather than resembling some mafic-enriched melt that might be suggested by Wendlandt's (1981) experiments with CO_2, the fluids associated with granitic melting were probably H_2O-rich. Once the granitic melt has segregated, the fluids can freeze it again by becoming more CO_2-rich with advancing time (provided that they do not also introduce Mg to modify the melt type): this explains the time sequence from granite to charnockite (Section 7.3.2) inferred from field observations. To Newton (Ch. 7), a crustal-scale shear zone is likely to be a conduit for metasomatizing fluids rather than melts, and such effects are not confined to the Archaean.

In contrast to the regions discussed by Wickham (Ch. 6) and Newton (Ch. 7), the Adirondack Mountains are regarded as representing granulite facies metamorphism without pervasive fluid flow (Cartwright & Valley, Ch. 8). Small-scale isotopic inhomogeneities are thought to have been preserved from the pre-metamorphic stage. Fluid movements were intermittent and channelized. A recent comparison of oxygen isotopes in southern India and the Adirondacks (Jiang *et al.* 1988) is consistent with the contrast drawn here. The sampling is not detailed enough to seek small inhomogeneities, but a general difference is found between the two regions: $\delta^{18}O$ values are lower in southern India, consistent with the expected effect of deep-sourced fluids, while the Adirondack values are consistent with pre-metamorphic processes (Jiang *et al.* 1988).

1.8 *P–T–t* paths and tectonic models

Many different *P–T–t* paths are recorded by high-temperature metamorphic terranes. The *P–T–t* path represents the actual track of a rock, or terrane, through *P–T* space with time, in contrast to the peak metamorphic mineral assemblage in a rock with an equilibrium texture, which represents one point on the *P–T–t* path frozen in at peak *T* (or, more correctly, maximum entropy). Additionally, it is not the same as the *metamorphic field gradient* (also known as the *piezothermic array*), which is the line in *P–T* space connecting successive peak metamorphic mineral assemblages from low to high grade along the present-day erosion surface (see discussion in England & Richardson 1977; also J. C. Schumacher *et al.*, Ch. 9). *P–T–t* paths can be generated in one orogenic cycle, or the information recorded in the rocks may be the result of two or more orogenic cycles, in which case the apparent *P–T–t* path determined from evidence in the rocks *may* have no real meaning, being a random composite of information partially preserved from more than one path. Additionally, single-cycle *P–T–t* paths are commonly represented as simple smooth curves, whereas in reality they are likely to have a more complex form because rates of burial and uplift, with or without magmatic heating, vary during the orogenic cycle.

Early work investigating *P–T–t* paths concentrated on the evolution of regions of thickened continental crust (England & Richardson 1977, England & Thompson 1984). The principal features of the *P–T–t* paths followed by rocks in such a tectonic setting are a period of thermal relaxation with increasing temperature followed by a period of cooling as the rock approaches the Earth's surface, generating a 'conventional' clockwise path in *P–T* space, in which peak *T* follows peak *P* after a significant time interval (see, for example, J. C. Schumacher *et al.*, Ch. 9; R. Schumacher *et al.*, Ch. 10; Harley & Hensen, Ch. 12). Since the work of Wells (1980), magmatic accretion has been proposed as a driving force of high-temperature metamorphism. The principal features of the *P–T–t* paths followed by rocks in such circumstances are heating with increasing pressure followed by cooling without significant exhumation, generating an anticlockwise path in *P–T* space (e.g. J. C. Schumacher *et al.*, Ch. 9; R. Schumacher *et al.*, Ch. 10; Harley & Hensen, Ch. 12) in which peaks *P* and *T* are coincident (Bohlen 1987, Bohlen & Mezger 1989). There are two particular kinds of incomplete *P–T–t* path, derived largely from evidence preserved from the retrograde rather than the prograde metamorphic history, which allow the division of many metamorphic terranes into two types; those which show near-isobaric cooling (IBC, Harley 1989) and those which show near-isothermal decompression (ITD, Harley 1989). IBC paths have been identified from many granulite terrains (see reviews by Bohlen 1987 and Harley 1989) but the tectonic setting in which they are generated, either crustal thickening or magmatic accretion (Ellis 1987, Bohlen 1987, Bohlen & Mezger 1989, Harley 1989) or, possibly, crustal or lithospheric extension (Sandiford

& Powell 1986), remains a matter of debate. Furthermore, since evidence for the prograde path is generally lacking, rocks which exhibit IBC paths can have followed either a clockwise or an anticlockwise path in *P–T* space. Where evidence for the prograde path is preserved (e.g. Warren 1983, Bohlen 1987, Hensen & Motoyoshi 1988), the rocks appear to have followed an anticlockwise path. Such a path also can explain the differences that are commonly observed between granulite xenoliths in volcanic suites and exposed granulite terranes (Bohlen & Mezger 1989). The ambiguity which results from evidence of the prograde path not being preserved is well illustrated in the final discussion of the Sri Lanka rocks studied by R. Schumacher *et al.* (Ch. 10). ITD paths occur as part of clockwise *P,T* evolution but have steeper dP/dT slopes than those generated by the erosion-controlled exhumation of the kind discussed by England & Richardson (1977) and England & Thompson (1984). The model of Albarède (1976) corresponds closely with many of the petrologically determined *P–T–t* paths (e.g. Hollister 1982, Brown & Earle 1983, Droop & Bucher-Nurminen 1984, Harris & Holland 1984). Such 'fast exposure paths' (Harley 1989) can be generated by rapid upward crustal movement; for example, by diapiric upward movement of the core to a migmatite belt (Jones & Brown 1990) or by melt-enhanced deformation or 'surges' (Hollister & Crawford 1986), or by tectonic thinning by extension of previously thickened crust (e.g. Sonder *et al.* 1987). One possible tectonic setting to generate a clockwise *P–T–t* path is a behind-arc basin along a continental margin. Here the extensional phase can generate the required high basal heat flux and high initial surface heat flow to allow eventual metamorphic temperatures in the granulite facies, and the subsequent moderate overthickening during basin inversion can generate the required clockwise *P–T–t* path. The exact mechanism by which an ITD path is generated along the high-temperature part of the clockwise loop will be determined by the degree of thinning and consequent heat flux, the degree of overthickening and the consequent erosional or tectonic thinning components, and the degree of crustal melting which will determine whether the core of the terrane might move upward diapirically. The thermal and tectonic settings of granulite facies terranes are not easy to generalize, simply because appropriate conditions can be attained in many different ways (cf. Harley & Hensen, Ch. 12). However, it is interesting to note that a behind-arc basin environment to develop 'interrupted' clockwise *P–T–t* paths (Jones & Brown 1990) is a complementary tectonic setting to an arc environment for anticlockwise ones (Bohlen 1987). These contrast with continent and arc collision settings which will generate 'conventional' clockwise *P–T–t* paths and, in extreme cases, some examples of IBC paths (Ellis 1987, Harley 1989). It is often only the serendipitous preservation of evidence of the prograde *P–T–t* path that makes it possible to distinguish between alternative tectonic settings (e.g. Hensen & Motoyoshi 1988, Brown 1988). None of the foregoing discussion helps to explain the unusually high-temperature metamorphism, at such low pressure, described by Vernon *et al.* (Ch. 11). Some kind of deep, intra-plate thermal anomaly in the mantle seems to be required, and may indicate some differences in the tectonics of the Proterozoic, at least for part of the continental lithosphere, in comparison with the Phanerozoic.

References

Aftalion, M., D. R. Bowes, B. Dash & T. J. Dempster 1988. Late Proterozoic charnockites in Orissa, India: a U–Pb and Rb–Sr isotopic study. *Journal of Geology* **96**, 663–76.

Albarède, F. 1976. Thermal models of post-tectonic decompression as exemplified by the Haut-Allier granulites (Massif Central, France). *Bulletin de la Société Géologique de France* **18**, 1023–32.

Allègre, C. J. & J. F. Minster 1978. Quantitative models of trace element behavior in magmatic processes. *Earth and Planetary Science Letters* **38**, 1–25.

Arzi, A. A. 1978. Critical phenomena in the rheology of partially melted rocks. *Tectonophysics* **44**, 173–84.

Ashworth, J. R. & E. L. McLellan 1985. Textures. In *Migmatites*, J. R. Ashworth (ed.) 180–203. Glasgow: Blackie.

Barbey, P., J.-M. Bertrand, S. Angoua & D. Dautel 1989. Petrology and U/Pb geochronology of the Telohat migmatites, Aleksod, Central Hoggar, Algeria. *Contributions to Mineralogy and Petrology* **101**, 207–19.

Barr, D. 1985. Migmatites in the Moines. In *Migmatites*, J. R. Ashworth (ed.), 204–64. Glasgow: Blackie.

Bohlen, S. R. 1987. Pressure–temperature–time paths and a tectonic model for the evolution of granulites. *Journal of Geology* **95**, 617–32.

Bohlen, S. R. & K. Mezger 1989. Origin of granulite terranes and the formation of the lowermost continental crust. *Science* **244**, 326–9.

Bowers, T. C. & H. C. Helgeson 1983. Calculation of the thermodynamic and geochemical consequences of non-ideal mixing in the system H_2O–CO_2–NaCl on phase relations in geologic systems: equation of state for H_2O–CO_2–NaCl fluids at high pressures and temperatures. *Geochimica et Cosmochimica Acta* **47**, 1247–75.

Brown, M. 1988. *P–T–t* paths and melting in garnet–cordierite gneisses. *Terra Cognita* **8**, 267.

Brown, M. & M. M. Earle 1983. Cordierite-bearing schists and gneisses form Timor, eastern Indonesia: *P–T* conditions of metamorphism and tectonic implications. *Journal of Metamorphic Geology* **1**, 183–203.

Clark, R. G. & J. B. Lyons 1986. Petrogenesis of the Kinsman intrusive suite: peraluminous granitoids of western New Hampshire. *Journal of Petrology* **27**, 1365–93.

Clemens J. D. & D. Vielzeuf 1987. Constraints on melting and magma production in the crust. *Earth and Planetary Science Letters* **86**, 287–306.

Droop, G. T. R. & K. Bucher-Nurminen 1984. Reaction textures and metamorphic evolution of sapphirine-bearing granulites from the Gruf Complex, Italian central Alps. *Journal of Petrology* **25**, 766–803.

Ellis, D. J. 1987. Origin and evolution of granulites in normal and thickened crusts. *Geology* **15**, 167–70.

England, P. C. & S. W. Richardson 1977. The influence of erosion upon the mineral facies of rocks from different metamorphic environments. *Journal of the Geological Society of London* **134**, 201–13.

England, P. C. & A. B. Thompson 1984. Pressure–temperature–time paths of regional metamorphism. I. Heat transfer during the evolution of regions of thickened continental crust. *Journal of Petrology* **25**, 894–928.

Evans, N. H. & J. A. Speer 1984. Low-pressure metamorphism and anatexis of Carolina Slate Belt phyllites in the contact aureole of the Lilesville pluton, North Carolina, U.S.A. *Contributions to Mineralogy and Petrology* **87**, 297–309.

Grant, J. A. 1985. Phase equilibria in partial melting of pelitic rocks. In *Migmatites*, J. R. Ashworth (ed.), 86–144. Glasgow: Blackie.

Harley, S. L. 1989. The origins of granulites: a metamorphic perspective. *Geological Magazine* **126**, 215–47.

Harmon, R. S. & A. N. Halliday 1980. Oxygen and strontium isotope relationships in the British late Caledonian granites. *Nature* **283**, 21–5.

Harris, N. B. W. & T. J. B. Holland 1984. The significance of cordierite–hypersthene assemblages from the Beitbridge region of the central Limpopo Belt: evidence for rapid decompression in the Archaean? *American Mineralogist* **69**, 1036–49.

Hensen, B. J. 1971. Theoretical phase relations involving cordierite and garnet in the system $MgO-FeO-Al_2O_3-SiO_2$. *Contributions to Mineralogy and Petrology* **33**, 191–214.

Hensen, B. J. & Y. Motoyoshi 1988. Sapphirine–quartz–orthopyroxene symplectites after cordierite in granulites from the Napier Complex, Antarctica: evidence for a counter-clockwise *P–T* path? *Terra Cognita* **8**, 263.

Hildreth, W. & S. Moorbath 1988. Crustal contributions to arc magmatism in the Andes of central Chile. *Contributions to Mineralogy and Petrology* **98**, 455–89.

Hill, R. I., L. T. Silver & H. P. Taylor Jr. 1986. Coupled Sr–O isotope variations as an indicator of source heterogeneity for the northern Peninsular Ranges Batholith. *Contributions to Mineralogy and Petrology* **92**, 351–61.

Hollister, L. S. 1982. Metamorphic evidence for rapid (2 mm/yr) uplift of a portion of the Central Gneiss Complex, Coast Mountains, B. C. *Canadian Mineralogist* **20**, 319–32.

Hollister, L. S. 1988. On the origin of CO_2-rich fluid inclusions in migmatites. *Journal of Metamorphic Geology* **6**, 467–74.

Hollister, L. S. & M. L. Crawford 1986. Melt-enhanced deformation: a major tectonic process. *Geology* **14**, 558–61.

Hutton, D. H. W. 1988. Granite emplacement mechanisms and tectonic controls: inferences from deformation studies. *Transactions of the Royal Society of Edinburgh: Earth Sciences* **79**, 245–55.

Janardhan, A. S., R. C. Newton & J. V. Smith 1979. Ancient crustal metamorphism at low P_{H_2O}: charnockite formation at Kabbaldurga, South India. *Nature* **278**, 511–14.

Janardhan, A. S., R. C. Newton & E. C. Hansen 1982. The transformation of amphibolite facies gneiss to charnockite in southern Karnataka and northern Tamil Nadu, India. *Contributions to Mineralogy and Petrology* **79**, 130–49.

Jiang, J., R. N. Clayton & R. C. Newton 1988. Fluids in granulite facies metamorphism: a comparative oxygen isotope study on the South India and Adirondack high-grade terrains. *Journal of Geology* **96**, 517–33.

Johannes, W. 1985. The significance of experimental studies for the formation of migmatites. In *Migmatites*, J. R. Ashworth (ed.), 36–85. Glasgow: Blackie.

Jones, K. A. & M. Brown 1990. A high temperature 'clockwise' *P–T* path and melting in the development of regional migmatites: An example from Southern Brittany, France. *Journal of Metamorphic Geology* **8**, (in press).

Lal, R. K., D. Ackermand & H. Upadhyay 1987. *P–T–X* relationships deduced from corona textures in sapphirine–spinel–quartz assemblages from Paderu, southern India. *Journal of Petrology* **28**, 1139–68.

van der Molen, I. & M. S. Paterson 1979. Experimental deformation of partially-melted granite. *Contributions to Mineralogy and Petrology* **70**, 299–318.

Naslund, H. R. 1986. Disequilibrium partial melting and rheomorphic layer formation in the contact aureole of the Basistoppen Sill, East Greenland. *Contributions to Mineralogy and Petrology* **93**, 359–67.

Peterson, J. W. & R. C. Newton 1989. CO_2-enhanced melting of biotite-bearing rocks at deep-crustal pressure–temperature conditions. *Nature* **340**, 378–80.

Pichavant, M. 1987. Effects of B and H_2O on liquidus phase relations in the haplogranite system at 1 kbar. *American Mineralogist* **72**, 1056–70.

Powell, R. 1983. Fluids and melting under upper amphibolite facies conditions. *Journal of the Geological Society of London* **140**, 629–33.

Price, R. C. & S. R. Taylor 1977. The rare earth element geochemistry of granite, gneiss and migmatite from the Western Metamorphic Belt of south-eastern Australia. *Contributions to Mineralogy and Petrology* **62**, 249–63.

Pride, C. & G. K. Muecke 1982. Geochemistry and origin of granitic rocks, Scourian Complex, NW Scotland. *Contributions to Mineralogy and Petrology* **80**, 379–85.

Rutter, M. J. & P. J. Wyllie 1988. Melting of vapour-absent tonalite at 10 kbar to simulate dehydration-melting in the deep crust. *Nature* **331**, 159–60.

Sandiford, M. A. & R. Powell 1986. Deep crustal metamorphism during continental extension: modern and ancient examples. *Earth and Planetary Science Letters* **79**, 151–8.

Sawyer, E. W. 1987. The role of partial melting and fractional crystallization in determining discordant migmatite leucosome compositions. *Journal of Petrology* **28**, 445–73.

Sawyer, E. W. & S.-J. Barnes 1988. Temporal and compositional difference between subsolidus and anatectic migmatite leucosomes from the Quetico metasedimentary belt, Canada. *Journal of Metamorphic Geology* **6**, 437–50.

Sonder, L. J., P. C. England, B. P. Wernicke & R. L. Christiansen 1987. A physical model for Cenozoic extension of western North America. In *Continental extensional tectonics*, M. P. Coward, J. F. Dewey & P. L. Hancock (eds), 187–201. Geological Society of London Special Publication 28.

Strachan, R. A., P. J. Treloar, M. Brown & R. S. D'Lemos 1989. Cadomian terrane tectonics and magmatism in the Armorican Massif. *Journal of the Geological Society of London* **146**, 423–6.

Strong, D. F. & S. K. Hanmer 1981. The leucogranites of Southern Brittany: origin by faulting, frictional heating, fluid flux and fractional melting. *Canadian Mineralogist* **19**, 163–76.

Tait, R. E. & S. L. Harley 1988. Local processes involved in the generation of migmatites within mafic granulites. *Transactions of the Royal Society of Edinburgh: Earth Sciences* **79**, 209–22.

Tracy, R. J. & P. Robinson 1983. Acadian migmatite types in pelitic rocks of Central Massachusetts. In *Migmatites, melting and metamorphism*, M. P. Atherton & C. D. Gribble (eds), 163–73. Nantwich, U.K.: Shiva.

Wall, V. J., J. D. Clemens & D. B. Clarke 1987. Models for granitoid evolution and source composition. *Journal of Geology* **95**, 731–49.

Warren, R. G. 1983. Metamorphic and tectonic evolution of granulites, Arunta Block, central Australia. *Nature* **305**, 300–3.

Waters, D. J. 1988. Partial melting and the formation of granulite facies assemblages in Namaqualand, South Africa. *Journal of Metamorphic Geology* **6**, 387–404.

Waters, D. J. & C. J. Whales 1984. Dehydration melting and the granulite transition in metapelites from southern Namaqualand, S. Africa. *Contributions to Mineralogy and Petrology* **88**, 269–75.

Wells, P. R. A. 1980. Thermal models for the magmatic accretion and subsequent metamorphism of continental crust. *Earth and Planetary Science Letters* **46**, 253–65.

Wendlandt, R. F. 1981. Influence of CO_2 on melting of model granulite facies assemblages: a model for the genesis of charnockites. *American Mineralogist* **66**, 1164–74.

Wickham, S. M. 1987. The segregation and emplacement of granitic magmas. *Journal of the Geological Society of London* **144**, 281–97.

Wickham, S. M. & H. P. Taylor Jr. 1985. Stable isotope evidence for large scale seawater infiltration in a regional metamorphic terrane; the Trois Seigneurs Massif, Pyrenees, France. *Contributions to Mineralogy and Petrology* **91**, 122–37.

CHAPTER TWO

Graphical analysis of *P–T–X* relations in granulite facies metapelites

B. J. Hensen & S. L. Harley

2.1 Introduction

The study of the complex mineral reactions in metamorphic rocks is facilitated by graphical representation in the form of chemographic projections and phase diagrams. In this chapter we discuss and develop the graphical analysis of phase relations in multicomponent systems containing Fe–Mg solid solutions. Particular emphasis is placed on the relevance of divariant and univariant equilibria for the interpretation of mineral reactions in high-grade metapelites.

Since the pioneering work of Goldschmidt, Eskola and others, Thompson (1955, 1957) and Korzhinskii (1959) have provided the main impetus for the graphical approach to the analysis of mineral assemblages in metamorphic rocks. They extended the use of chemographic projections and popularized Schreinemakers principles for the construction of phase diagrams (see also Niggli 1954). Albee (1965) produced the first petrogenetic grid with univariant reactions for the system K_2O–FeO–MgO–Al_2O_3–SiO_2–H_2O (KFMASH) and gave rules for the construction of such grids based on Schreinemakers principles. Hess (1969) and Hensen & Green (1973) noted the relationship of univariant reactions in the complex F–M grid to invariant points in the simple F- and M-end-member systems, and Harte & Hudson (1979) developed an extensive KFMASH grid in which F- and M-end-member reactions were combined with F–M multisystem reactions. The use of divariant reactions involving F–M solid solutions as potentially sensitive pressure (P) and temperature (T) indicators, first suggested by Ramberg (1964), was further developed by Hensen (1971). The relationship of divariant and univariant reactions in FMAS and the P–T-sensitive compositional variations of coexisting minerals were graphically represented by P–X_{Mg} ($X_{Mg} = Mg/(Mg + Fe^{2+})$) sections, P–T diagrams contoured with compositional isopleths, and AFM projections.

Wider application of this approach to KFMASH was made by A. B. Thompson (1976), who added a thermodynamic framework for calculation of the *P–T–X* relationships. The effect of compositional variation can also be portrayed in *P*(or *T*)–μ and μ–μ diagrams (e.g. Marakushev & Kudryavtsev 1965) which have application to both open (Korzhinskii 1959) and closed systems (Lal *et al.* 1984). The utility of these approaches to the study of high-grade metapelites is discussed herein with reference to the MAS, FMAS, and more complex chemical systems.

Most graphical analyses to date have assumed rectilinear behaviour of reaction boundaries and compositional isopleths in *P–T* space. However, there is increasing evidence for curvilinear reaction boundaries, particularly in the case of pyroxene-bearing equilibria (Hensen 1987). The topological implications of curvature in reaction boundaries are explored here with reference to the some important granulite facies equilibria.

Breakdown reactions of biotite in the amphibolite to granulite transition zone are sensitive to a_{H_2O}. At pressures greater than a few kilobars, final biotite breakdown probably takes place by dehydration melting (Holdaway & Lee 1977, Thompson 1982, Grant 1985). The univariant 'vapour absent' dehydration melting reactions for biotite, with other FMAS phases, are independent of a_{H_2O} and as such are ideally suited to construct a *P–T* framework for phase relationships in KFMASH. We propose a model *P–T* grid applicable to high-grade metapelites, incorporating the minerals cordierite (Crd), biotite (Bt), sapphirine (Spr), orthopyroxene (Opx), spinel (Spl), garnet (Grt), sillimanite (Sil), corundum (Crn), quartz (Qtz), K-feldspar (Kfs), and H_2O-undersaturated liquid (L). Pyrope (Prp) and enstatite (En) are considered in the system MAS. The ferromagnesian minerals listed above are given in the assumed order of decreasing X_{Mg}, based on observations of natural mineral assemblages (Hensen 1971).

2.2 Phase relations in a simple system: $MgO–Al_2O_3–SiO_2$ (MAS)

In order to gain a proper understanding of phase relations in complex systems it is necessary to consider a simple end-member system first. Chemographic relations in a three-component ($n = 3$) system are portrayed in triangular compatibility diagrams that allow the representation of any mineral in the system, usually in terms of its molar proportions.

The phase rule states: variance (number of independent variables) equals number of components minus number of phases plus two ($V = C - P + 2$). Thus considering five minerals (phases) in the system MAS we have one potential invariant point ($V = 3 - 5 + 2 = 0$) and can calculate five balanced univariant equations involving four phases each ($V = 3 - 4 + 2 = 1$). It is customary to designate each reaction by the absent phase in brackets, e.g. Crd = En + Spr + Qtz (Sil). The qualitative balancing of reaction equations

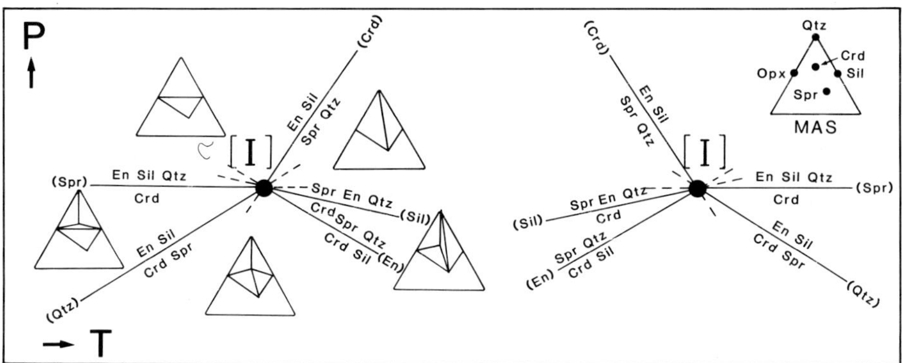

Figure 2.1 Schreinemakers bundles for an invariant point (I) in the simple system MAS considering $n+2$ (=5) phases. The sequence of tie line changes is shown in chemographic MAS triangles for each divariant field. The two mirror-image versions are both consistent with volume constraints. The left-hand version is qualitatively correct, but signs of the slopes of some of the reactions remain uncertain. Note the convention of labelling reactions with the absent phase.

can be done graphically (Fig. 2.1) but the calculation of reaction coefficients, the quantities of each of the reactants and products in moles, is done algebraically by simultaneously solving the equations for different elements or oxides and substituting, or by matrix algebra (Korzhinskii 1959) usually by computer (Greenwood 1975, Finger & Burt 1972, Andrew & Linde 1980). Once the reaction coefficients are known molar volume changes (ΔV) and entropy changes (ΔS) can be calculated. These are required to determine the slopes of the univariant reactions (i.e. $\mathrm{d}P/\mathrm{d}T = \Delta S/\Delta V$). Molar volumes of end-member minerals at 298 K and 1 bar are known fairly accurately (e.g. Robie *et al.* 1967), but the exact entropies of a large number of minerals are still in doubt. Nevertheless, it is often possible to deduce the high-temperature (high-entropy) side of a reaction by reference to experimental data or to evidence from field occurrences. Devolatilization reactions can usually be assumed to take place with increasing temperature, so that qualitative P–T grids can be readily constructed (Thompson 1955, Albee 1965). For solid–solid reactions at least, the high-pressure side of the reaction can be determined from the volume change, keeping in mind that high pressure favours high density.

If we assume that the univariant reactions intersect, they will do so in a single P–T, invariant, point. Simply applying only molar volume constraints, two topologically valid invariant points can be constructed, which are each other's mirror images (Fig. 2.1). The order in which the reactions are crossed in either clockwise or anticlockwise fashion remains unchanged. If the position of one of the reactions is known, the orientation of the reaction bundle is fixed. The mathematical procedure to establish the order of the reactions and their relative orientations can be found in Niggli (1954) and Zen (1966). In practice, once the reactions have been balanced, their disposition around an invariant point can be drawn quite easily by keeping a few rules in mind:

(a) The stability of a phase or combination of phases can never be extended by the addition of another phase: e.g. A + B + C must be in the stability field of A + B, A + C and B + C; and A + B within the field of A and B (see Fig. 2.1).
(b) The stability of a phase or combination of phases can never extend over a sector of greater than 180 degrees (Fig. 2.1).
(c) The stable portion of the A-absent reaction (A) must be outside the area of the diagram in which A can stably coexist with one or more other phases of the system.
(d) Only degenerate reactions, that have less than $n + 1$ phases, can be stable on both sides of an invariant point. No such reaction occurs in Figure 2.1, and all reactions become metastable on passing through the point (dashed lines are metastable extensions).

So far we have only considered five phases in the system MAS. If we add one more phase, such as pyrope (Prp), an $n + 3$ multisystem (Korzhinskii 1959) is created, with six possible invariant points and 15 possible non-degenerate reactions. A hypothetical, internally consistent diagram for this system is shown in Figure 2.2a. The invariant points are labelled, with the absent phase in square brackets. Given the position of two invariant points and the slopes of two reactions in each point, the rest of the grid is determined. Even though the absolute values for the reaction slopes are not known, it is possible to construct a topologically self-consistent version of the diagram. The grid is based on the assumption that the reactions are non-parallel straight lines that have intersections in the pressure–temperature range of interest. As we will see in the following, this assumption is invalid in this case and the diagram is entirely hypothetical, because some of the reactions curve away from each other and do not intersect. However, it can be employed to demonstrate a number of properties of such diagrams:

(a) The relative slopes of the (Sil) and (En) absent reactions are constrained by the position of the metastable invariant points [En] and [Sil]. This simple relationship, applying to *all* multigrids, allows one to test the internal consistency of a grid at a glance.
(b) Even though probably only invariant point [Prp] is actually stable in MAS, all reactions are stable in FMAS as divariant reactions and all invariant points as univariant reactions, as will be demonstrated below.

Figure 2.2 (a) Hypothetical $(n + 3)$ multisystems for six phases in MAS. The relative slopes of the (En) and (Sil) absent reactions are controlled by the positions of the metastable invariant points [En] and [Sil] (after Hensen & Essene 1971, Fig. 2, by permission of Springer-Verlag). (b) Metastable alternative (or 'residual') grid for the same system.

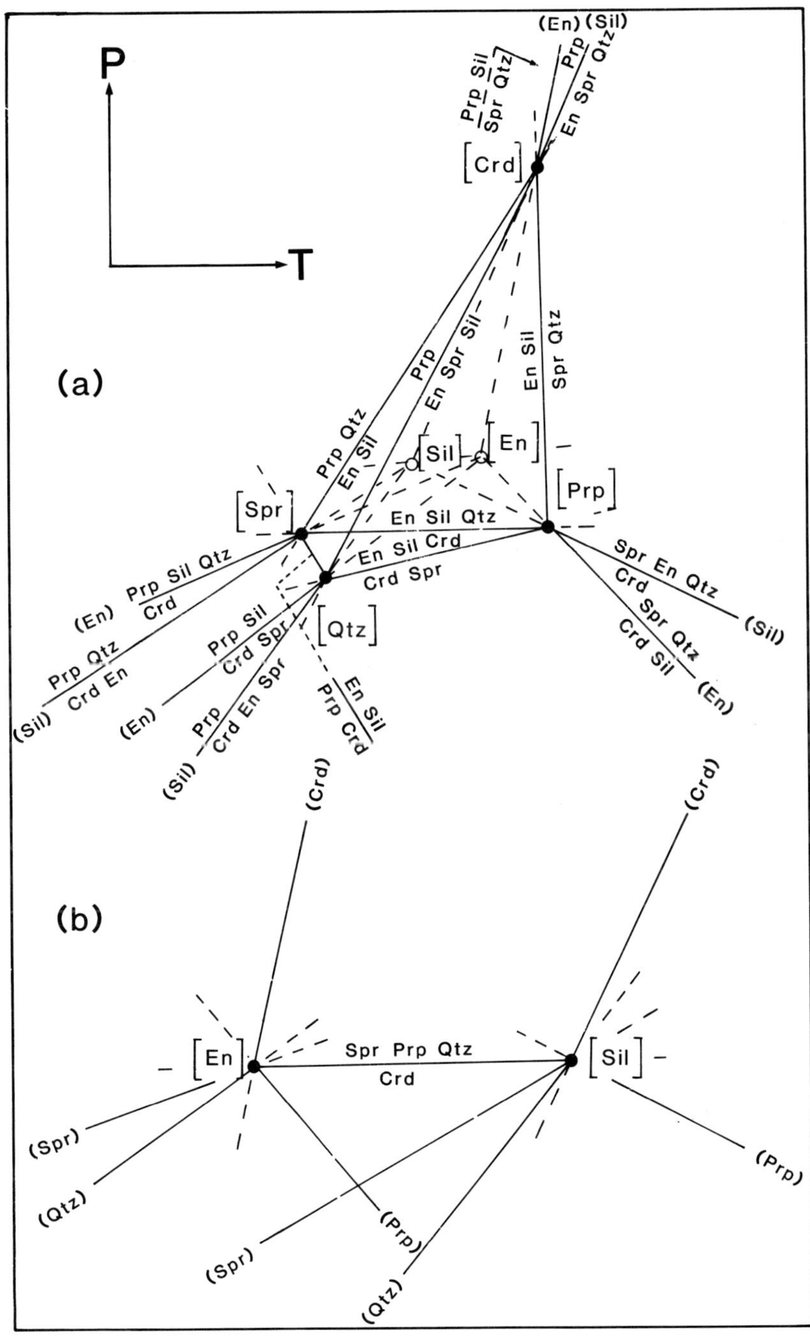

P
T
(a)
(b)
[Crd]
[Spr]
[Sil]
[En]
[Prp]
[Qtz]
(En) (Sil)
Prp Sil / Spr Qtz
Prp
En Spr Qtz
Prp Qtz
En Sil
Prp
En Spr Sil
En Sil
Spr Qtz
En Sil Qtz
En Sil Crd
Crd Spr
Prp Sil Qtz / Crd
(En)
Prp Qtz
Crd En
(Sil)
Prp Sil
Crd Spr
(En)
Prp
Crd En Spr
(Sil)
En Sil / Prp Crd
Spr En Qtz
(Sil)
Crd
Spr Qtz
Crd Sil
(En)
(Crd)
(Crd)
Spr Prp Qtz
Crd
(Spr)
(Qtz)
(Prp)
(Spr)
(Qtz)
(Prp)

(c) There is a no *a priori* reason to assume that the grid of Figure 2.2a is actually stable with regard to the alternative, residual grid (Korzhinskii 1959, Vielzeuf & Boivin 1984) in which stable and metastable invariant points are exchanged (Fig. 2.2b).

Most reactions have stable portions in both grids and only a small number of elements are unique to each grid, namely the univariant reactions linking the stable invariant points and the divariant assemblages that are unique to the triangular divariant fields between these. The absent phases in these assemblages correspond to combinations of the phases absent at the stable invariant points (Burt 1971, Hensen 1986). For example, in Figure 2.2a the divariant assemblage En–Sil–Qtz (Spr, Crd, Prp) is stable in the triangle with corners at points [Spr], [Crd], and [Prp]. In this case the metastable grid of Figure 2.2b does not contain any diagnostic divariant assemblage because it contains only two invariant points. It is possible to choose the correct stable version of a grid only if there is evidence for the stable occurrence (or, conversely, the complete absence) of certain divariant (or, more rarely, univariant) assemblages. In this particular case, the occurrence of the En–Sil tie line would be sufficient to identify Figure 2.2a as the stable grid, because En and Sil cannot coexist in a stable manner in Figure 2.2b, in which [En] and [Sil] are the only two invariant points.

2.3 The system FMAS

Before we consider the method of derivation and the topology of the FMAS grid, the properties of divariant (also called sliding or continuous) reactions must be explained. Taking the terminal* reaction Crd = En + Sil + Qtz in MAS (Fig. 2.2a between [Spr] and [Prp]) as an example, what is the result of the addition of FeO to the system? The reaction is no longer univariant because we have added another component without adding an extra phase. In metamorphic rocks coexisting ferromagnesian minerals rarely have the same $Mg/(Mg + Fe^{2+})$ ratio (X_{Mg}). For instance, cordierite is always more magnesian than coexisting orthopyroxene. So if we add FeO, it will partition preferentially into orthopyroxene, extending its stability field, as can be most elegantly illustrated with a P(or T)–X diagram (Fig. 2.3c). Note that if the partitioning relationships were reversed, the two-phase loop would have the opposite slope. For a fixed temperature, the compositions of coexisting orthopyroxene and cordierite, with sillimanite and quartz, are a unique function of pressure. This is what divariant means: if two variables are fixed, all other variables, in this case X_{Mg} of cordierite and orthopyroxene, and the Al_2O_3 content

* Reactions are known as 'terminal' if a phase is eliminated and 'non-terminal' if only a tie line between two phases is broken (Niggli 1954, Hess 1969).

of the latter, are determined. This relationship is also well shown in a P–T diagram contoured with compositional isopleths for both minerals (Fig. 2.3b). Obviously then, divariant equilibria are ideal geothermometers and barometers: analysis of two solid solutions yields both pressure and temperature, albeit with varying magnitudes of uncertainty.

The distribution of Mg and Fe^{2+} between coexisting minerals can be described by a distribution coefficient (K_D):

$$K_D(\text{Crd–Opx}) = \frac{(\text{Mg}/\text{Fe}^{2+})^{\text{Crd}}}{(\text{Mg}/\text{Fe}^{2+})^{\text{Opx}}} = \frac{X_{\text{Mg}}(\text{Crd}).(1 - X_{\text{Mg}}(\text{Opx}))}{X_{\text{Mg}}(\text{Opx}).(1 - X_{\text{Mg}}(\text{Crd}))}$$

The divariant reaction can be considered as a combination of a net-transfer reaction, Mg–cordierite = enstatite + sillimanite + quartz, and an Mg–Fe exchange reaction, Mg–cordierite + ferrosilite = Fe–cordierite + enstatite.

Invariant points in MAS become univariant reactions in FMAS. Such reactions involving five phases can be balanced if the compositions of the minerals at constant P and T are known. Qualitative balancing requires only the relative Mg-numbers of the participating minerals, and can be done with a chemographic projection from quartz onto the AFM plane for the silica-oversaturated portion of FMAS (Fig. 2.3a). The compositions of all solid solutions are fixed at any P (or T) along the univariant phase boundary and, consequently, solutions for reaction coefficients pertain to specific P–T conditions.

Once the balanced reaction is known, the sector in which it emanates from the MAS invariant point can be found (Fig. 2.3b). The reaction emanating from the invariant point [Prp] of Figure 2.2a and shown in Figure 2.3 is: Crd + Opx + Sil = Spr + Qtz. The sliding divariant reactions fit on to this boundary with reactions (Spr) and (Qtz) on one side, where the assemblage Crd–Opx–Sil is stable and (Crd), (Opx), and (Sil) on the other where Spr–Qtz is stable. As all ferromagnesian minerals decrease in X_{Mg} away from the MAS invariant point, the sense of displacement of all divariant reactions with decreasing X_{Mg} of a rock (X_{Mg} (bulk)) can be determined by using the rule that the mineral assemblage with the lower X_{Mg} will be stabilized in each case. The sense of displacement of reactions which are potentially balancing for M and F and thus not *a priori* predictable, such as the reaction (Qtz), follow from the other reactions. Reaction (Qtz) is not shown in Figure 2.3, which is only relevant to Qtz-present conditions.

Whereas on the univariant line the ferromagnesian minerals have unique compositions at given P and T (Fig. 2.3a; P_1, T_2), away from the boundary the Spr–Sil, Spr–Crd, and Spr–Opx tie lines expand to become trivariant fields (Fig. 2.3a; P_1, T_3). These, the divariant fields, and their topological relationship to the univariant boundary are displayed in P–T, P–X, and P–A/(A + F + M) diagrams in Figure 2.3b–d. The nature of P–T–X relations varies as a function of the A/(A + F + M) ratio as well as X_{Mg}. Note

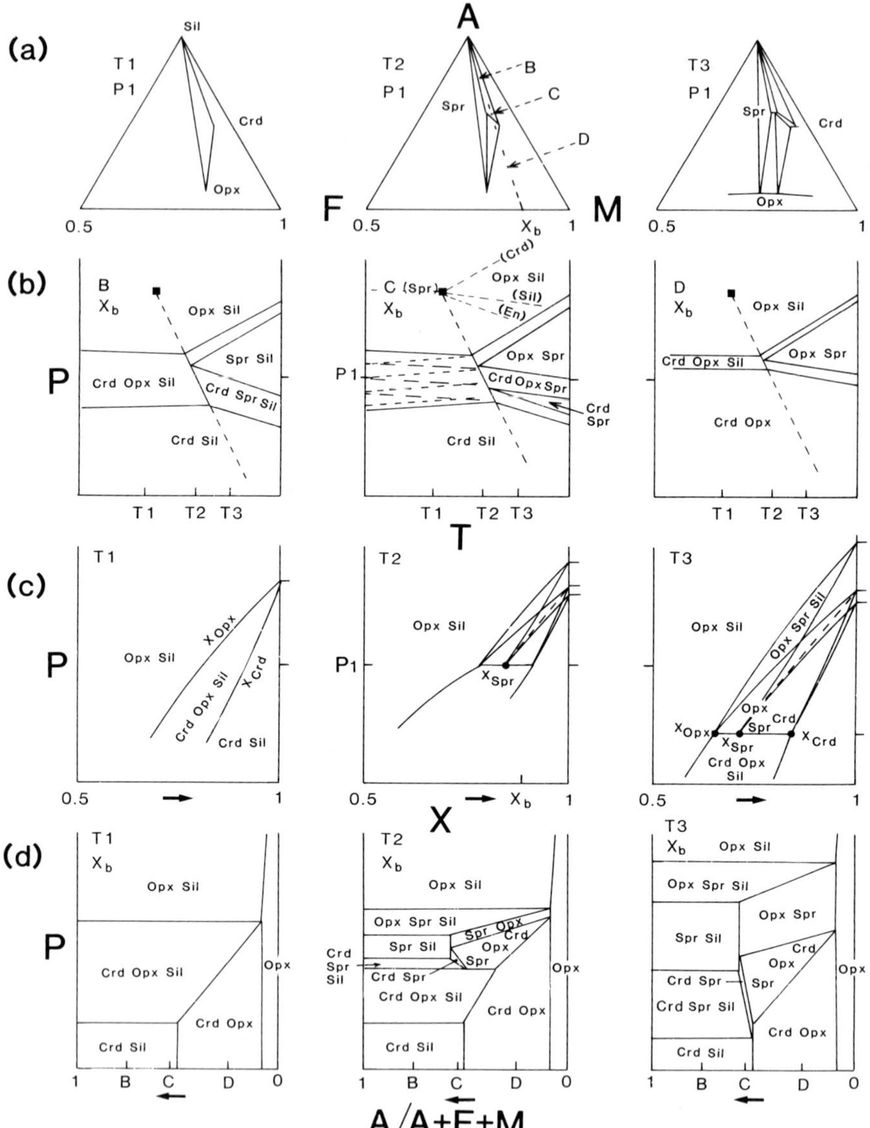

Figure 2.3 FMAS phase relations for quartz-present divariant reactions and the associated univariant reaction Crd + Opx + Sil = Spr + Qtz (Grt, Spl), which emanates from the end-member invariant point [Prp] in MAS (Fig. 2.2), indicated by a solid square in the *P–T* sections. (a) AFM triangles for three temperatures at a single pressure (P_1). Bulk compositions B, C, and D have the same Mg-number but different A/(A + F + M) ratios. At P_1, T_2, a point on the univariant boundary, phases of variable composition (Opx, Crd, and Spr) have unique compositions. (b) *P–T* diagrams for bulk compositions B, C, and D. Compositional (X_{Mg}) isopleths for Opx and Crd are shown in one divariant field. The isopleths for cordierite are parallel to the lower boundary and those for orthopyroxene to the upper boundary of the divariant field. Dashed lines

applies at low f_{O_2} (Hensen 1986). The topology of the invariant (inversion) point in P–T–f_{O_2} space shown in Figure 2.5c is discussed below in the section on $\mu_{Fe_2O_3}$–μ_{FeO} diagrams (Section 2.8).

2.5 Relationships between FMAS, MAS, and FAS

An invariant point [X] in FAS cannot have a stable MAS analogue if the univariant reaction in FMAS that emanates from [X] passes into a stable invariant point, which may belong to an FMAS multisystem. This follows from the Schreinemakers rule that each non-degenerate univariant reaction changes from stable to metastable on passing through an invariant point. If it is assumed that all univariant reactions emanating from an FMAS invariant point terminate in invariant points in FAS and MAS, then the points stable in MAS *must* be metastable in FAS and vice versa. Therefore the stable grid in one end-member system will have the metastable, or residual, topology of the other. While calling attention to this fact, Vielzeuf & Boivin (1984) have suggested that the geometrical relationships linking stable and metastable grids by inversion have *special* significance for the construction of multisystems involving Mg Fe solid solution. As demonstrated by the foregoing discussion of Figure 2.4, this suggestion does not bear critical scrutiny for the following reasons:

(a) Most importantly, it is *not* possible by graphical means to verify the stability of an invariant point in a multisystem (i.e. FMAS) by extrapolation from the topology of one or both of the end-member systems.
(b) End-member systems commonly do not have corresponding phases: e.g. sapphirine and orthopyroxene are not stable in FAS; pyrope and cordierite are not compatible in MAS. This renders most, and strictly speaking all, inversions meaningless for the FMAS system under consideration.
(c) It is impossible for all univariant reactions emanating from an invariant point in a *multisystem* (FMAS or KFMASH) to terminate in *stable* invariant points in the corresponding end-member systems, because non-

Figure 2.5 Complete P–T grids for FMAS(O), i.e. FMAS–O_2. (a) FMAS low-f_{O_2} grid, modified after Hensen (1971). Solid circles, stable invariant points; open circles, metastable invariant points. (b) High-f_{O_2} grid, after Hensen (1986). Solid squares, stable invariant points; open circles, metastable invariant points. (c) Inversion of the low-f_{O_2} to high-f_{O_2} grid through an invariant point in P–T–f_{O_2} space, projected onto the P–T plane. This P–T–f_{O_2} invariant point is indicated by the large solid circle. Other symbols as for (a) and (b). Dashed lines, stable portions of univariant reactions in the full system FMASO; stippled lines, metastable extensions. The topology of the univariant bundle is based on compositional data from granulite facies mineral assemblages (see Fig. 2.8a).

degenerate reactions connecting invariant points *within* the multisystem (e.g. Figs 2.4c & f) cannot be stable in *either* of the end member systems.

The claim that 'petrogenetic grids constructed from systems in which iron and magnesium are considered separately have multiple consequences that are hard to control' (Vielzeuf & Boivin 1984, p. 777; Vielzeuf 1983, p. 309) only applies when multisystems with incompatible mineral assemblages are considered; as, for instance, in a system in which the stability of muscovite plus quartz overlaps with cordierite plus garnet plus K-feldspar (Vielzeuf & Boivin 1984, Figs 5–9). Thus the problem is not inherent in the method of constructing the multigrid, but derives from the *inputs* used to construct it. Contrary to what is suggested by Vielzeuf & Boivin (1984), consideration of end-member phase relations has, in general, no predictive value for deriving the *stable* topology of multisystem invariant points. Information derived either from careful study of metamorphic assemblages or from experimental work on the end-member and multisystems is required for this purpose.

2.6 Variation of X_{Mg} along FMAS univariant boundaries

As pointed out above, the X_{Mg} of coexisting phases will, in general, show a steady change along the univariant boundary. For those reactions terminating in stable MAS (or FAS) invariant points, all phases will simultaneously become pure Mg- or Fe-end-members as the univariant reaction reaches its end-point. This is analogous to a divariant reaction becoming univariant on reaching the side of a P(or T)–X plot (e.g. Fig. 2.3c; Hensen 1971). The 'skeletal' diagrams of Thompson (1976) are a shorthand way in which to show the orientation of divariant boundaries with regard to univariants. The skeletal diagrams correspond to the reaction bundles in the FAS and MAS end-member systems. These are best suited to derive an internally consistent set of X_{Mg} vectors for a multigrid (see Fig. 2.6b; also Hensen 1986, Fig. 3; 1988, Fig. 3).

If the partitioning of Fe^{2+} and Mg between coexisting minerals is minor and the value of K_D is close to unity, inversion of the distribution relationship may occur. There is evidence for inversions involving garnet and staurolite (Ballevre *et al.*, in press) and garnet and liquid (Ellis 1986), and there is a suggestion of this behaviour for garnet–spinel (Hensen & Green 1972, p. 353). The latter inversion causes changes to some, but not all of the univariant reactions involving garnet and spinel, and does not greatly affect the overall topology of the diagram (Hensen, 1971, p. 212).

A K_D inversion on a univariant boundary creates a singular point at which the amount of one of the reactants reduces to zero, and on either side of which that same mineral changes side in the reaction. Such a point may be expressed as an inflection point on a curved reaction boundary.

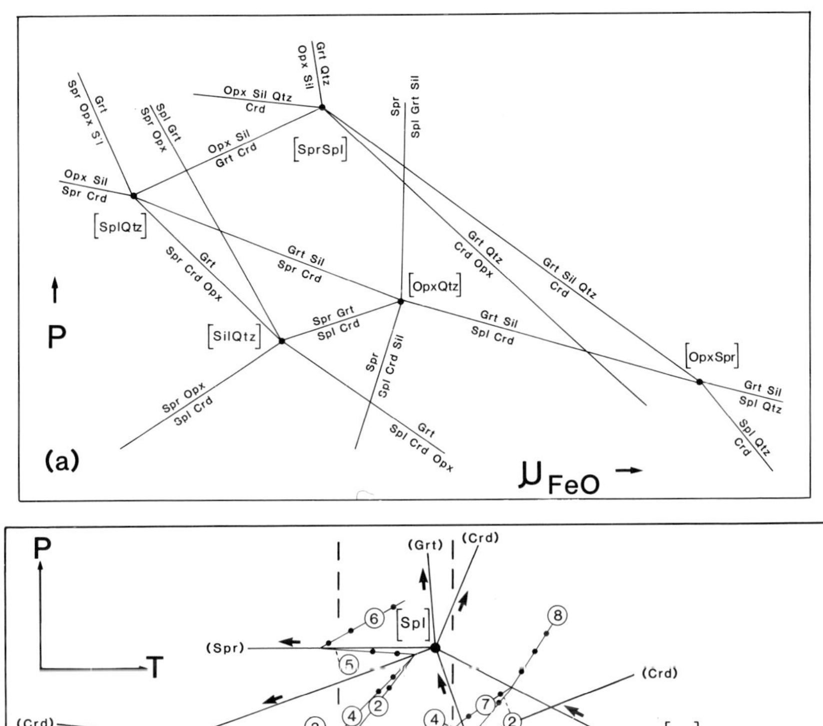

Figure 2.6 (a) P–μ_{FeO} section modified after Mohan *et al.* (1986) and Hensen (1988). (b) P–T diagram for FMAS, at low f_{O_2}, showing the position of P–μ sections of Figures 2.6a & 7a. Arrows indicate increasing X_{Mg} of coexisting phases. The relative positions of invariant points in the P–μ sections is related to the X_{Mg} vectors. This is demonstrated with X_{Mg} isopleths of garnet for the divariant assemblages (numbered lines with dots): 1, Grt = Crd + Opx + Spl; 2, Grt = Crd + Opx + Spr; 3, Grt + Spr = Crd + Spl; 4, Grt + Sil = Spr + Crd; 5, Crd + Grt = Opx + Sil; 6, Grt + Qtz = Opx + Sil; 7, Crd = Grt + Spr + Qtz; 8, Grt + Sil = Spr + Qtz.

2.7 Chemical potential diagrams: P–μ_{FeO}

Chemical potential diagrams such as P–μ_{FeO} and $\mu_{Fe_2O_3}$–μ_{FeO} plots have recently found favour as alternative or additional graphical devices to describe the behaviour of divariant reactions (Lal *et al.* 1984, 1987; Mohan *et al.* 1986). These diagrams are indispensable for the description of metasomatic processes, in which reactions are driven by changes in chemical potential caused by infiltration or diffusion of components (Korzhinskii 1959), but can also be useful to illustrate changes in mineral assemblage and reaction sequence to be expected as a function of bulk composition (Marakushev & Kudryavtsev 1965, Lal *et al.* 1987, Schumacher & Robinson 1987, Hensen 1988). Compatibility relations in these diagrams can be shown by triangular FM–A–S plots (Lal *et al.* 1987).

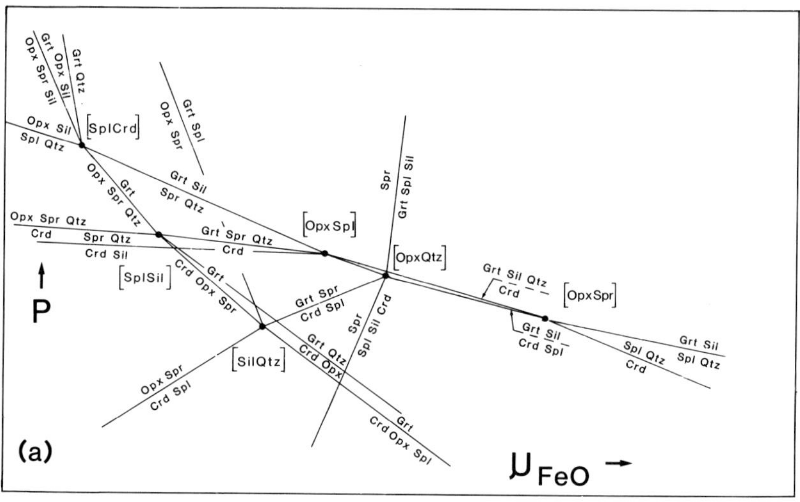

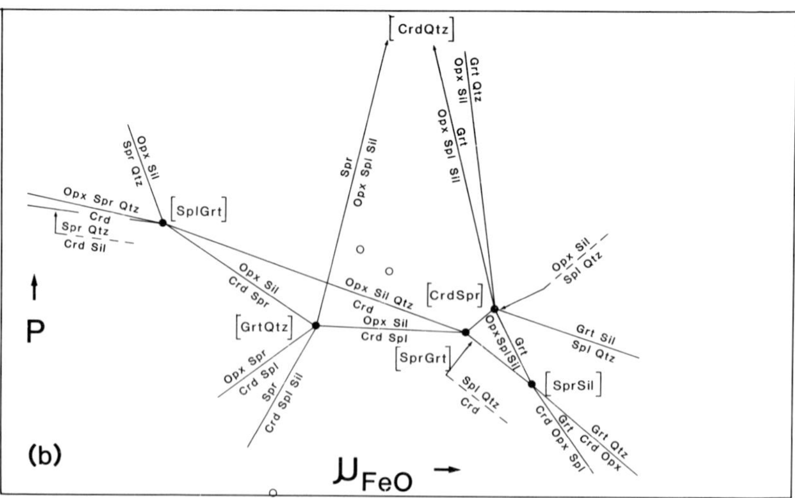

Figure 2.7 (a) P–μ_{FeO} section at a temperature between invariant points [Spl] and [Opx], (*re* Fig. 2.6b). (b) P–μ_{FeO} section at high f_{O_2} at a temperature below the invariant point [Grt] (*re* Fig. 2.5b).

$P-\mu_{FeO}$ diagrams can be regarded as a series of isothermal sections through the FMAS $P-T$ grid. The slopes of the isothermal univariant (= divariant) lines on the $P-\mu_{FeO}$ diagram are a function of the volume change (ΔV) and the ΔFeO, which is found by substituting the measured FeO content (in terms of atomic proportions) of coexisting minerals in the rocks into the balanced equations for the end-member minerals. The resulting slopes are qualitative only.

The reaction boundaries are assumed to be straight lines. In reality, because FeO (and Fe_2O_3) contents of the minerals will by definition vary across $P-\mu$ and $\mu-\mu$ diagrams, the reaction boundaries are most probably curved. However, since the diagrams show only relative relationships and the slopes are qualitative, as a first approximation these problems can be ignored. In $P-\mu_{FeO}$ sections at different temperatures, different $P-T$-univariant reactions are crossed, which appear as invariant points in the $P-\mu_{FeO}$ diagrams. For the low-f_{O_2} and high-f_{O_2} FMAS grids there are respectively five and four topologically distinct $P-\mu_{FeO}$ sections, depending on where the section is taken with respect to the invariant points in the $P-T$ grids. The diagram shown in Figure 2.6a constitutes an isothermal section between invariant points [Spl] and [Qtz] of Figure 2.6b. The relative positions of invariant points lacking a common absent phase (and thus not connected by a univariant line) cannot be directly determined, and have to be estimated by reference to the compositional vectors and the related geometry of skeletal diagrams in the $P-T$ grid (Fig. 2.6b; Hensen 1986, 1987, 1988).

$P-\mu_{FeO}$ sections for temperatures between [Spl] and [Opx] at low f_{O_2} and between [Grt] and [Spr] at high f_{O_2} are shown in Figures 2.7a & b. Note the correspondence between the μ_{FeO} co-ordinates of the invariant points and the X_{Mg} vectors on the $P-T$ diagrams (Fig. 2.6b). A $P-\mu_{FeO}$ section through a $P-T$ invariant point can have a total of nine non-degenerate reactions passing through a single point. Of the total of 15 divariant reactions associated with the six univariant reactions meeting in a $P-T$ invariant point, there are six reactions unique to the six sectors between adjacent univariants; these correspond to the lines connecting invariant points in the $P-\mu_{FeO}$ sections at temperatures above and below that of the $P-T$ invariant point. At the $P-T-\mu_{FeO}$ invariant point these six divariant reactions (isothermal univariant lines) are missing.

2.8 $\mu_{Fe_2O_3}-\mu_{FeO}$ diagrams

In $\mu_{Fe_2O_3}-\mu_{FeO}$ diagrams the effects of oxygen fugacity, expressed as $\mu_{Fe_2O_3}$, and Mg-number are shown at constant pressure and temperature. The slopes of the isothermal–isobaric univariant reactions are calculated by substituting the measured Fe_2O_3 and FeO contents of the minerals in the rocks into the end-

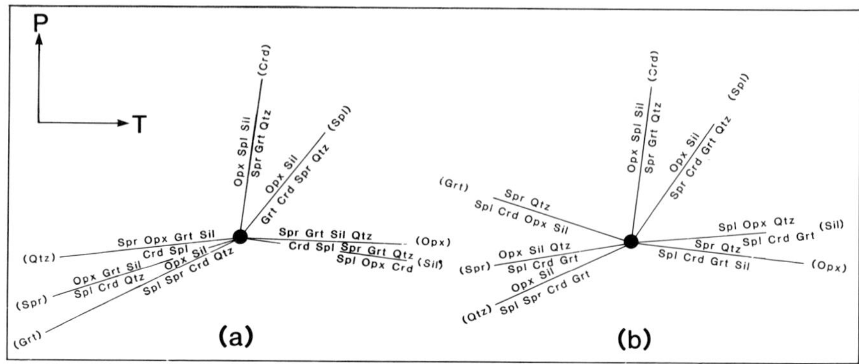

Figure 2.8 *P–T* bundles for O_2-absent univariant reactions in FMASO. (a) Topology based on mineral compositions in granulites (data from Lal *et al.* 1987, Table 7). This version has been adopted for the construction of the $\mu_{Fe_2O_3}$–μ_{FeO} diagrams (see text). (b) Alternative topology assuming a high $Fe_2O_3/(Fe_2O_3 + FeO)$ ratio in spinel. This topology is identical to Figure 7a of Powell & Sandiford (1988).

member equations. As with the P–μ_{FeO} diagrams, one set of FeO and Fe_2O_3 data are used to calculate all slopes for a range of $\mu_{Fe_2O_3}$–μ_{FeO} conditions.

Although, strictly speaking, a $\mu_{Fe_2O_3}$–μ_{FeO} diagram is only valid for a particular pressure and temperature, the topology of such a grid remains the same over a limited *P–T* interval bounded by univariant reactions in P–T–f_{O_2}. These univariant reactions, corresponding to the locus of the sliding invariant points with increasing f_{O_2}, intersect in an invariant inversion point in P–T–f_{O_2} space, at which the topology of the $n + 3$ grid is inverted (Fig. 2.5c; Hensen 1986, Powell & Sandiford 1988).

We have used the mineral compositions of Lal *et al.* (1987) to calculate the vapour(oxygen)-absent univariant reactions involving six phases in the system FMASO ($n + 3$). The resulting Schreinemakers bundle for these reactions in terms of pressure and temperature is shown in Figure 2.8. Each of the divariant sectors around the invariant point has its own characteristic $\mu_{Fe_2O_3}$–μ_{FeO} topology (a similar conclusion was reached by Watts 1974), with the invariant point (divariant reaction in P–T–f_{O_2}) marked by the absent phases of the bounding univariants being unique to that sector, e.g. [Crd, Qtz] only occurs as a stable invariant point in the (Crd–Qtz) sector of Figure 2.8. The μ–μ diagram of each *P–T* sector has a set of stable invariant points that can be derived by reference to the Schreinemakers bundle by noting the overlaps of the divariant fields, as shown in the insets of Figures 2.9–11 by overlapping circle segments.

The topology of the $\mu_{Fe_2O_3}$–μ_{FeO} diagram for the (Crd–Qtz) sector (Fig. 2.9) is almost identical to a diagram proposed by Lal *et al.* (1987, Fig. 9). Only the slopes of the Spr-out reactions have been changed to fit their data. The geometries of the diagrams for the (Spr–Qtz) (Fig. 2.10), and (Grt–Spr) (Fig. 2.11),

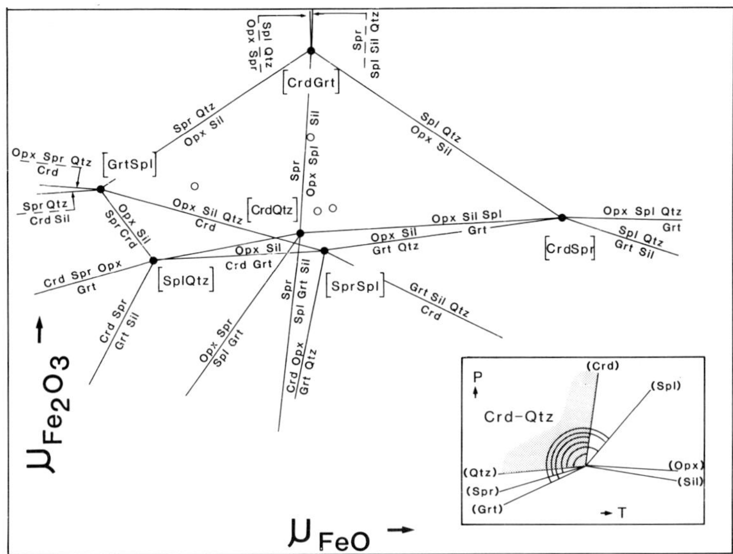

Figure 2.9 $\mu_{Fe_2O_3}$–μ_{FeO} diagram for *P–T* conditions within the Crd–Qtz sector (see inset and refer to Fig. 2.8a). The overlapping circle segments in the inset correspond to the stable isobaric–isothermal invariant points in the μ–μ grid. The diagram is slightly modified after Lal *et al.* (1987, Fig. 9). Open circles, metastable invariant points.

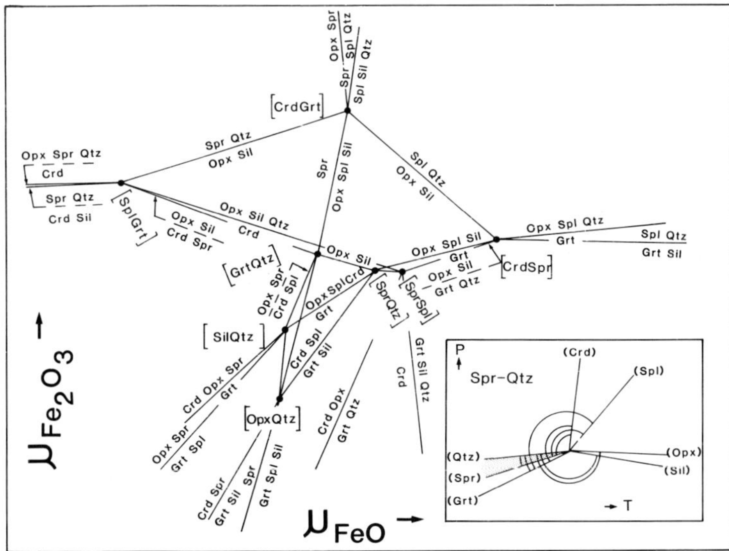

Figure 2.10 $\mu_{Fe_2O_3}$–μ_{FeO} diagram for *P–T* conditions within the Spr–Qtz sector (see inset and refer to Figs 2.8a & 9). The diagram contains four additional invariant points to the grid of Mohan *et al.* (1986, Fig. 8). Metastable invariant points have been omitted for clarity.

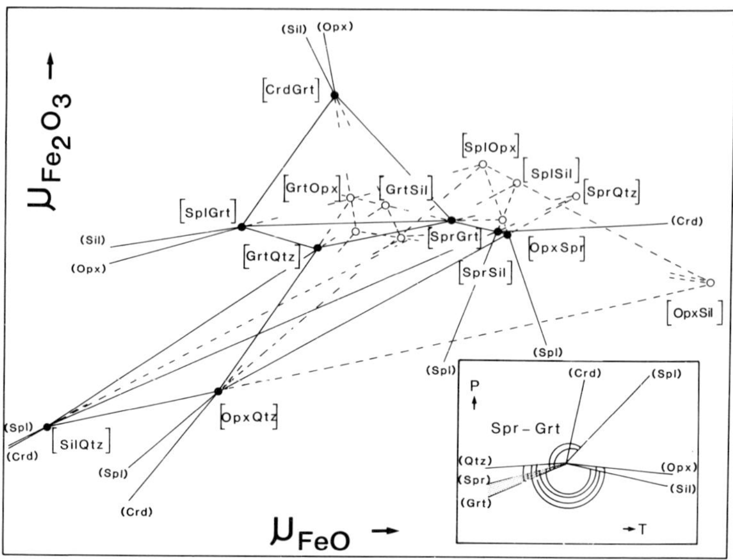

Figure 2.11 $\mu_{Fe_2O_3}$–μ_{FeO} grid for P–T conditions within the Spr–Grt sector (refer to inset and Fig. 2.8a), with eight isobaric–isothermal invariant points demanded by topological constraints (compare Ackermand *et al.* 1987, Fig. 7).

sectors show significant differences from those proposed by Mohan *et al.* (1986, Fig. 8) and Ackermand *et al.* (1987, Fig. 7) respectively. The (Spr–Qtz) diagram shows four additional invariant points [Crd, Grt], [Spl, Grt], [Spr, Crd], and [Opx, Qtz], whereas the (Grt–Spr) diagram includes two extra points: [Opx, Spr] and [Opx, Qtz] (Figs 2.10 & 11). The diagrams have been drawn taking topological constraints, imposed by metastable invariant points, into account. The μ–μ slopes for individual reactions have been varied from section to section in a few cases to achieve clarity of presentation. Four more unique topologies can be derived for the other four sectors of the bundle, with stable invariant points that can be readily derived by the sector rule. However, the topology of these $\mu_{Fe_2O_3}$–μ_{FeO} grids depends entirely on the validity of the P–T–f_{O_2} Schreinemakers bundle which, in turn, depends on the mineral compositions used to balance the univariant equations. As pointed out by Hensen (1986) and Powell & Sandiford (1988), the relatively low Fe_2O_3 contents for spinel determined by microprobe analysis may not represent the composition of that mineral at high temperature. Commonly observed magnetite lamellae in spinel suggest that it once contained more magnetite component in solution, and that reported Fe_2O_3 contents represent minimum values. If the $Fe^{3+}:Fe^{2+}$ ratio of spinel is greater than that of sapphirine at the pressure and temperature of interest, the P–T bundle takes on a different appearance (Fig. 2.8b), identical to that proposed by Powell & Sandiford (1988, Fig. 4a). Note that their alternative version (op. cit., Fig. 4b) for less oxidized spinel is different from ours because they did not take the Fe_2O_3 content of sillimanite into account.

Even though the $\mu_{Fe_2O_3}-\mu_{FeO}$ diagram of Lal *et al.* (1987, Fig. 9) contains most (but not all) of the divariant reactions that they propose on the basis of textural criteria and chemographic considerations, this does not signify that for their rocks the particular section is indeed valid, because the same reactions can all be the result of decompression *only* and thus may not, and we will argue *are* not, driven by changes in f_{O_2}. Until better data from experimental work or from spinel-bearing granulites become available, the correct set of $\mu_{Fe_2O_3}-\mu_{FeO}$ diagrams cannot be determined. However, the theoretical framework presented here provides the means to derive these diagrams once the problem of spinel composition has been solved.

Because it is obvious from the foregoing that variations in oxygen fugacity significantly affect the phase relationships in FMAS, the following question remains: Are the reaction textures observed in rocks caused by changes in f_{O_2}, or are changes in pressure (or temperature) mainly responsible? There is no *a priori* answer to this question, but if we consider the direction in which the various reactions in the rocks proceed, as deduced from textural relationships, for instance for the rocks from Paderu (Lal *et al.* 1987) or from Ganguvarpatti (Mohan *et al.* 1986), it appears that about half of the reactions would require an increase in f_{O_2}, whereas the other half need a decrease in f_{O_2}. On the other hand, all reactions proceed in the correct direction in response to a decrease in pressure. Similar arguments can be put forward with regard to changes in water activity (Hensen 1988). We conclude therefore that for the pressure–temperature range of interest, and with rocks behaving as essentially closed systems, variations in the activities of oxygen and water have at most a secondary effect for the rocks under discussion, and that reactions taking place in these rocks are primarily due to decompression and/or cooling. This also is borne out by the fact that neighbouring rocks in the instances cited show decompression reactions that are essentially independent of fluid activities (the breakdown of garnet and quartz to orthopyroxene and plagioclase).

2.9 Curved reaction boundaries and $P-T-X_{Fe-Mg-Al}$ relations in FMAS

If one or more of the participating phases in a univariant reaction is a solid solution with a considerable degree of pressure- or temperature-sensitive compositional variation, the reaction boundary is unlikely to be rectilinear. Even in the simple MAS system, two examples of curved reaction boundaries have been verified experimentally, namely enstatite + sillimanite = pyrope + quartz (Hensen & Essene 1971) and enstatite + sillimanite = sapphirine + quartz (Newton 1972, Chatterjee & Schreyer 1972, Hensen 1972). Thermochemical calculations suggest that the former reaction changes slope from positive at high temperature (>900°C) to negative at lower temperature (Aranovich & Podlesskii 1989). The curvature of the reaction boundaries results from the temperature(and pressure)-sensitive nature of Tschermaks solubility in enstatite. Because of such curvature the two reactions do not intersect to form an invariant point, and therefore the hypothetical *P–T* grid of Hensen & Essene (1971, reproduced in Fig. 2.2a) does not apply. However, there is experimental and field evidence for the existence of the reaction Opx + Sil = Grt + Spr + Qtz (Spl, Crd) in which Mg-numbers of coexisting ferromagnesian minerals increase with increasing pressure, and it would be expected to terminate in the MAS invariant point which we now know does not exist. The solution to this dilemma is shown in Figure 2.12 (after Hensen 1987, Fig. 8). The curved reaction boundaries in MAS are mimicked by the compositional isopleths of X_{Mg} for garnet (and also orthopyroxene and sapphirine). A maximum of X_{Mg} occurs for each of the three minerals. In this manner the Mg-numbers first increase towards a 'virtual' invariant point and then decrease after passing through a maximum value.

A second more geologically relevant case of a virtual invariant point in MAS is the intersection of the reactions enstatite + sillimanite = pyrope + quartz and cordierite = enstatite + sillimanite + quartz (also indicated schematically in Fig. 2.12).

There is now experimental evidence for the hypothetical topology suggested by Hensen (1987, Fig. 8) that involves a slope reversal and related compositional reversal for the reaction Crd + Grt = Opx + Sil + Qtz (Spl, Spr).

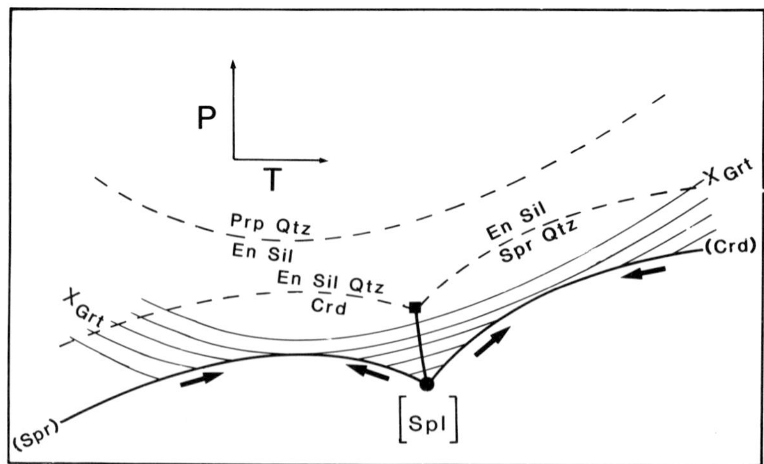

Figure 2.12 Schematic *P–T* diagram showing the effect of curvature of univariant reactions involving En in MAS (dashed lines) resulting in virtual (absent) invariant points in FMAS. One set of X_{Mg} isopleths for garnet is shown to illustrate the reversal in the X_{Mg} vectors along the curved univariant boundaries (after Hensen 1987).

Aranovich & Podlesskii (1989) and Bertrand *et al.* (1989) respectively have calculated and experimentally determined a slope reversal for this reaction.

In order to study the compositional variations in divariant assemblages on either side of a demonstrably curved univariant reaction, (Spl, Spr), two P–T grids with calculated isopleths have been constructed. The compositions of garnet, orthopyroxene, and cordierite as a function of pressure and temperature are determined by a set of net transfer and Fe–Mg exchange reactions. For the equilibria considered here the independent relations are:

$$\text{Grt} + \text{Qtz} = \text{Crd} + \text{Opx} \tag{1}$$

$$\text{Grt} + \text{Qtz} = \text{Opx} + \text{Sil} \tag{2}$$

$$\text{Grt} = \text{MgTs} + \text{Opx} \tag{3}$$

$$\text{Mg–Grt} + \text{Fe–Opx} = \text{Fe–Grt} + \text{Mg–Opx} \tag{4}$$

$$\text{Mg–Grt} + \text{Fe–Crd} = \text{Fe–Grt} + \text{Mg–Crd} \tag{5}$$

By assuming the position of X_{Mg}(Grt)-isopleths for equilibria (1) and (2) and by using the experimental data of Harley (1984a,b), Lee & Ganguly (1988), and Carswell & Harley (1989) for equilibria (3) and (4), the composition of orthopyroxene in terms of X_{Mg}(Opx) and X_{Al}(Opx) has been calculated at regular intervals along each Grt-isopleth. Two contoured P–T diagrams have been constructed: (a) a diagram assuming horizontal X_{Mg}(Grt)-isopleths for the low-pressure assemblage Grt–Crd–Opx–Qtz (Fig. 2.13); (b) a grid assuming sloping and curved X^{Grt} isopleths for both Grt–Crd–Opx–Qtz and the higher-pressure assemblage Grt–Opx–Sil–Qtz (Fig. 2.14).

The assumed position and horizontal orientation of X_{Mg}(Grt) isopleths for case (a) is an approximation combining the lower-P data of Hensen & Green (1972) at high temperature, with constraints imposed by the data of Aranovich & Podlesskii (1983) on the assemblage Grt–Crd–Sil–Qtz at 700°C and 750°C, and allowing for compositional uncertainties of X_{Mg}(Grt) = ±0.04 in both datasets. At high pressure and temperature the experimental data of Bertrand *et al.* (1989) indicate an X_{Mg}(Grt) value of 0.62 ± 0.02 near the invariant point [Spl].

For case (b), X_{Mg}(Grt)-isopleths have variable positive slopes and are convex towards the temperature axis. The position of these isopleths has been estimated using high-T (≥950°C) experimental constraints (Hensen & Green 1972, Bertrand *et al.* 1989), the calculated isopleth positions of Aranovich & Podlesskii (1989) at $T \leqslant 800$°C, and compositional data from metamorphic rocks, where the pressure and temperature have been determined by independent geothermobarometry.

The position of the univariant reaction is derived from the experimental

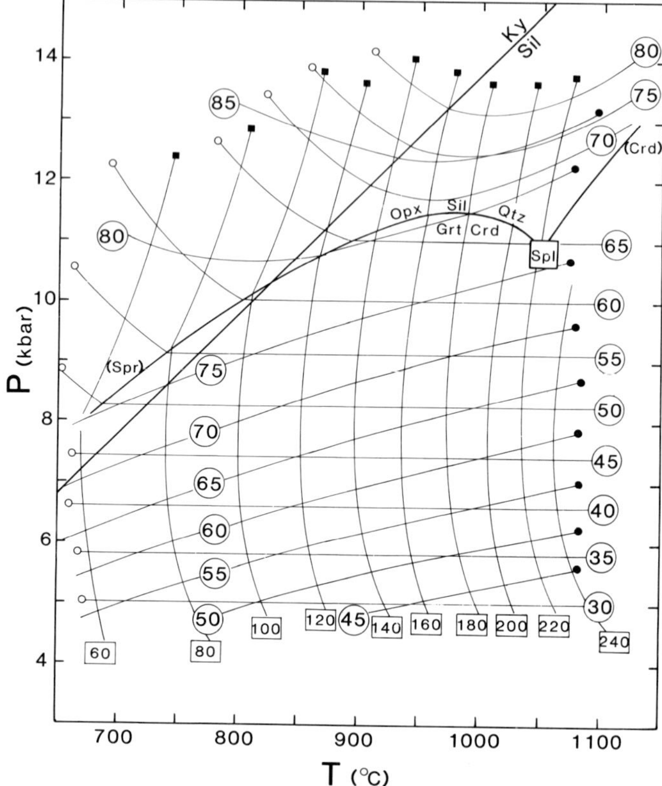

Figure 2.13 P–T grid with calculated X_{Mg} and X_{Al} isopleths for orthopyroxene assuming horizontal X_{Mg} isopleths for garnet. The garnet and orthopyroxene are in equilibrium with quartz and either cordierite (at pressures below those of the curve) or sillimanite (at higher pressures). The curved univariant reaction boundary combines the high-temperature data of Hensen & Green (1973) with new experimental data by Bertrand *et al.* (1989) and the calculated boundary of Aranovich & Podlesskii (1989). Note both garnet and orthopyroxene pass through a maximum X_{Mg} value along the curve. See text for further explanation. Isopleths are as follows. Thin lines with open circle symbols, X_{Mg} (Grt); X_{Mg} values in units × 100. Thin lines with solid circle symbols, X_{Mg} (Opx); X_{Mg} values in units × 100. Thin lines with solid square symbols, X_{Al} (Opx); X_{Al} values expressed as units of (Al cations/2, per 6 oxygen units) × 1000.

data of Bertrand *et al.* (1989) at high temperatures (>900°C) combined with the calculated boundary of Aranovich & Podlesskii (1989, Fig. 5) at lower T and for conditions of $P_{H_2O} = P_{total}$. The compositional variation of garnet along the univariant boundary at low temperature has been taken from the latter source, whereas the data of Bertrand *et al.* (1989) constrain the high-pressure, high-temperature end. Constraints from natural assemblages include: (a) the occurrence of garnet ($X_{Mg} = 0.30$) in Grt–Crd–Opx–Qtz rocks in terranes where independent P–T estimates yield $P = 5 \pm 1$ kbar, $T = 750$–900°C (Berg 1977, Grew 1981, Schenk 1984); (b) garnet

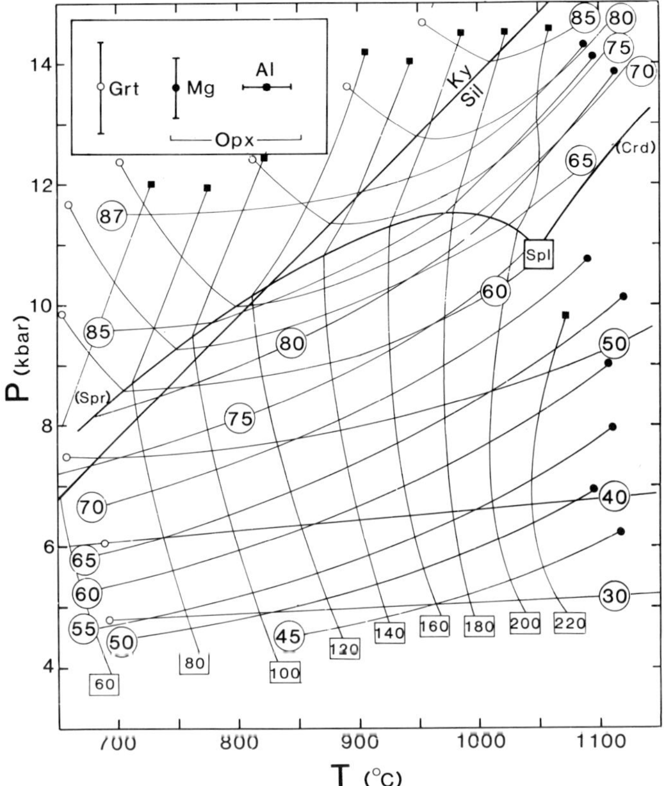

Figure 2.14 Best-fit P–T grid, calculated assuming curved X_{Mg} isopleths for garnet. Estimated uncertainties (errors) shown in inset. All symbols as in Figure 2.13.

($X_{Mg} = 0.45$–0.50) in areas constrained from independent thermobarometry to have equilibrated at 7 kbar and 750–850°C (Ellis 1983, Schenk 1984, Harley 1987); (c) maximum X_{Mg}(Grt) compositions at or near the univariant boundary of about 0.60–0.65 (Karsakov *et al.* 1975, Ellis *et al.* 1980, Harley 1985).

The garnet isopleths used in Figure 2.14 differ from those calculated by Aranovich & Podlesskii (1989) at the highest-T conditions (>900°C) because the above data are taken into account and, in general, have shallower dP/dT slopes. In the diagram with assumed horizontal X_{Mg}(Grt)-isopleths (Fig. 2.13) the X_{Mg}(Opx)-isopleths in the Grt–Crd–Opx–Qtz assemblages (Spl, Spr, Sil) have positive dP/dT slopes and are slightly concave to the T-axis. X_{Al}(Opx) isopleths are almost vertical and convex to the P-axis. The curvature of the X_{Mg}(Opx) isopleths is caused by the pressure dependence of the K_D(Mg–Fe) for Opx–Grt.

The best-fit P–T grid for assumed curved X_{Mg}(Grt) isopleths shows that the X_{Mg}(Opx) and X_{Al}(Opx) are relatively insensitive to the exact position of the

X_{Mg}(Grt)-isopleths (see uncertainty bars on Fig. 2.14). At pressures above the univariant boundary the strongly curved X_{Mg}(Grt) isopleths are after Aranovich & Podlesskii (1989). The negative slope of these isopleths towards lower temperature is accentuated by the inflection caused by the crossing of the sillimanite–kyanite transition. The main difference between the calculated diagram of Aranovich & Podlesskii (1989, Fig. 5) and our calculated grid is the slope and position of the X_{Mg}(Opx) isopleths and the fact that X_{Mg}(Opx) and X_{Mg}(Crd) pass through maximum values in the (Spl, Spr) reaction as well as X_{Mg}(Grt). These results are still valid even if the garnet isopleths of Aranovich & Podlesskii (1989) are used, as Grt–Opx experimental data (Harley 1984a,b; Lee & Ganguly, 1988) are used directly in our calculations, whereas such data are not incorporated in the data set and calculations of Aranovich & Podlesskii (1989). A schematic detail of the interval on the univariant curve where the compositional vectors for garnet, orthopyroxene, and cordierite compositions invert is shown in Figure 2.15. Note that the use of the skeletal diagrams to indicate the direction of the X_{Mg} vector discussed earlier (Fig. 2.6b) breaks down because in the inversion interval the vectors for garnet, orthopyroxene, and cordierite are opposite. The extent of the inversion interval depends on the relative slopes of the isopleths and the univariant boundary. The fact that no inversion of X_{Mg}(Opx) and X_{Mg}(Crd) occurs in the grid of Aranovich & Podlesskii (1989) is due to the steepness of their preferred isopleths for orthopyroxene and cordierite, dictated by the slopes of their garnet isopleths and magnitudes of Fe–Mg partitioning (K_D).

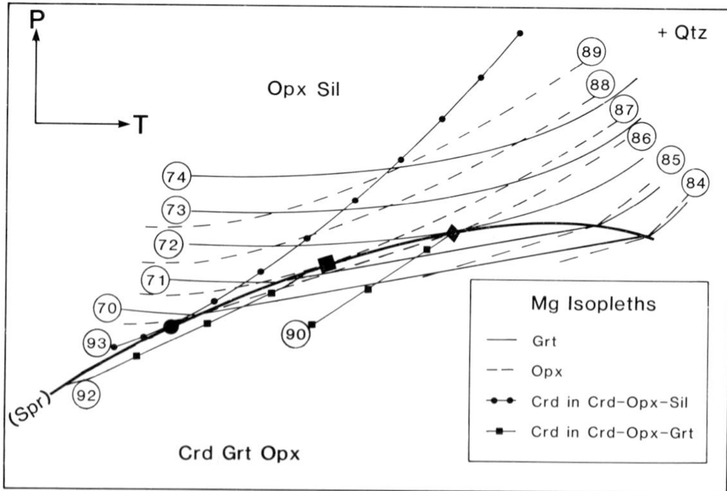

Figure 2.15 Schematic *P–T* diagram showing a detail of the inversion interval in which garnet, orthopyroxene, and cordierite (in that order with decreasing temperature) pass through a maximum X_{Mg} value along the curved univariant reaction boundary for Crd + Grt = Opx + Sil + Qtz (Spl, Spr). Compare with Figures 2.12–14. X_{Mg} isopleths are labelled with relative X_{Mg} values × 100.

The calculated *P–T–X* grid (Fig. 2.14) can be regarded as an approximation only, but it demonstrates the potential usefulness of this approach, particularly for rocks that have undergone reaction in response to a change in *P*, *T* conditions along a *P–T* path. By the use of the *P–T–X* diagram, zoning patterns in garnet and orthopyroxene can be employed to constrain possible *P–T* trajectories. The details of such zoning patterns for specific *P–T* paths can be deduced through inspection of Figures 2.13 or 14 and similar grids (Aranovich & Podlesskii 1989).

2.10 Effects of a_{H_2O}, a_{CO_2}, and vapour-absent conditions on the FMAS granulite grid

Many workers have considered the effect of volatiles, particularly H_2O, on stabilizing cordierite: for example, Newton (1972) concluded that $a_{H_2O} = 0$ conditions lead to a decrease in cordierite stability of 3–4 kbar. Aranovich & Podlesskii (1983, 1989) further considered the effects of H_2O, CO_2, and vapour-absent conditions and concluded:

(a) The stability of cordierite-bearing assemblages is maximized when $a_{H_2O} = 1$.
(b) CO_2-cordierite ($a_{CO_2} = 1$, i.e. pure CO_2 vapour phase) is stable to similar pressure or to a pressure of about 1 kbar less than water-bearing cordierite. Note that Bertrand *et al.* (1989) report no measurable difference between the H_2O- and CO_2-bearing systems.
(c) Under vapour-absent (dry) conditions, cordierite is stable only to 6–7 kbar and (Spl, Spr) occurs at pressures well below those considered here (Fig. 2.14).

If vapour-absent conditions applied throughout a metamorphic *P–T* path the maximum X_{Mg}(Grt) attainable in Grt–Crd–Opx–Qtz is only 0.45 (at $P \simeq 6$ kbar; Aranovich & Podlesskii 1989). Coexisting orthopyroxene would have $X_{Mg} = 0.72–0.69$ for $T = 750–850°C$ (our calculation). More importantly, in granulites with Grt–Opx–Sil–Qtz assemblages garnet with $X_{Mg} = 0.46$ should be observed. To our knowledge such Fe-rich garnet–orthopyroxene–sillimanite assemblages have not been reported. However, numerous examples of Grt–Opx–Crd–Qtz assemblages with garnet, $X_{Mg} \gg 0.46$ are known from Antarctica (Ellis *et al.* 1980, Harley 1985), the Aldan Shield (Karsakov *et al.* 1975), and elsewhere. Thus compositional data on Opx-bearing, Crd-absent granulites suggest that vapour-absent conditions do not determine cordierite stability in nature, and that cordierite typically persists to the high pressures of vapour-saturated breakdown curves, determined experimentally.

2.11 Phase relations in silica-undersaturated pelites

A self-consistent *P–T* grid involving corundum, constructed taking account of volume change constraints, experimental evidence from MAS, and information from granulite facies metapelites, is shown in Figure 2.16 (Hensen 1987). Note its relationships to the corundum-absent grid discussed above (Fig. 2.5a). The invariant point [Crn] in Figure 2.16 corresponds to [Qtz] in Figure 2.5a. Because of the incompatibility of corundum and quartz [Crn] and [Qtz] coincide, but the grids have no other invariant points in common. As shown in Figure 2.16, most of the silca-undersaturated grid occurs at a lower temperature than the quartz-bearing invariant points. This is consistent with the observation that Mg-rich biotite is a common constituent in silica-undersaturated pelites, with or without corundum. It is of interest therefore to consider phase relations involving biotite in the system KFMASH and their relationship to the anhydrous (excepting cordierite) parageneses considered so far.

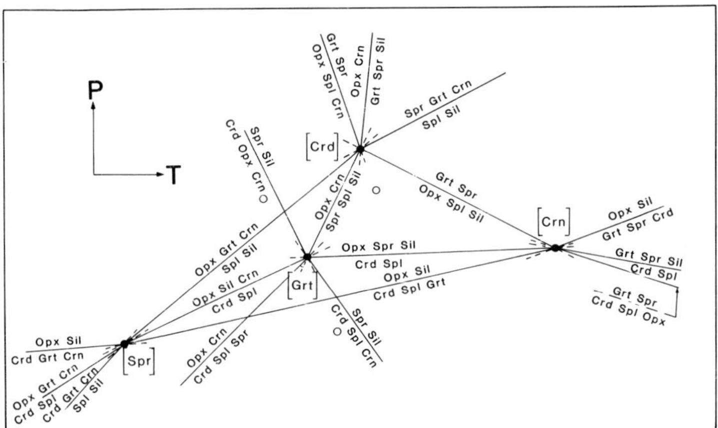

Figure 2.16 *P–T* grid for the silica-undersaturated portion of the system FMAS. The invariant point [Crn] corresponds to the point [Qtz] in Figure 2.5a.

2.12 Reactions involving biotite in KFMASH

Even though, in many granulites, biotite may be of retrograde origin, textural and compositional data suggest that often biotite is stable with, or in a reaction relationship with, the high-grade mineralogy of the rock. Most of these rocks may have been partially molten, with melt present as an additional phase at the peak of metamorphism (Powell 1983, Waters 1988, Hensen & Warren 1984). Under vapour-absent conditions or when a mixed gas phase is present, melt can occur instead of vapour in univariant dehydration reactions involving biotite (Grant 1985), so that the dehydration of biotite involves melting. The position in P and T of such a dehydration-melting reaction is independent of the activity of water, but a_{H_2O} is fixed at any point along it. It is analogous to the vapour-absent reactions in FMASO and corresponds to the polybaric univariant conditions of Greenwood (1975, p. 589). Because the positions of these reactions are not affected by a_{H_2O} they form an ideal framework for the phase relations in the amphibolite–granulite transition zone where water activities may be variable and are typically less than unity.

A qualitative KFMASH grid for biotite dehydration melting reactions, in the absence of corundum is shown in Figure 2.17. The melt composition used in the construction has the properties that $X_{Mg}(L) > X_{Mg}(Grt)$ (Ellis 1986, Vielzeuf & Holloway 1988) and $X_{H_2O}(L) > X_{H_2O}(Bt)$. The chosen melt composition lies close to the Grt–Crd tie line on an AFM diagram projected from Kfs and Qtz, and in the Qtz–Grt–Crd field on an $S(SiO_2)$–FM diagram projected from Kfs and Sil. In this diagram it has been assumed that Mg-rich biotite does not persist to temperatures as high as that of invariant point [Crn]

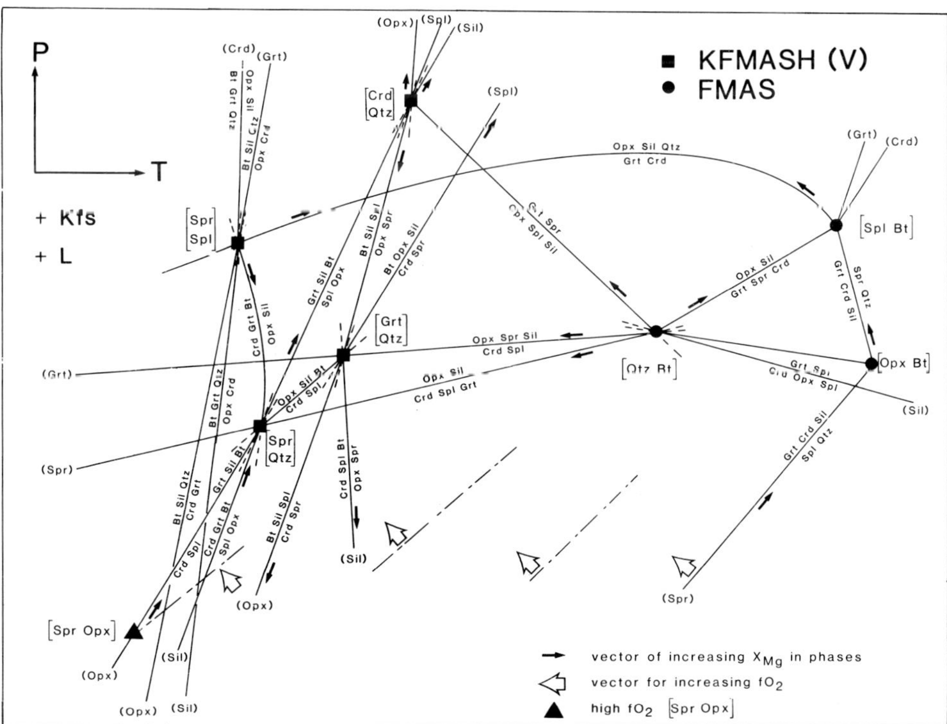

Figure 2.17 Theoretical P–T grid for vapour-absent dehydration melting reactions in the system KFMASH(O). Phase relations are projected from K feldspar and liquid (L). At high f_{O_2} an invariant point, [Spr, Opx], is created where Crd, Grt, Bt, Spl, Sil, Kfs, Qtz, and L coexist. The points [Spl, Bt], [Opx, Bt], and [Qtz, Bt] correspond to the low-f_{O_2} invariant points of Figure 2.5a.

of Figure 2.16, which is estimated to lie near 950°C at 8–9 kbar (Hensen 1987). As a result, three quartz-absent invariant points, [Crd, Qtz], [Grt, Qtz], and [Spr, Qtz], are generated. One quartz-present invariant point (marked [Spr, Spl] in Fig. 2.17) is found by taking the curved FMAS reaction (Spl, Spr) into account. This point has been previously proposed as intersection of dehydration reactions at $P_{H_2O} < P_{total}$ by Hess (1969), Hensen (1971) and Vielzeuf & Boivin (1984) and as a dehydration melting ('vapour-absent') invariant point by Grant (1985). The point [Spl, Spr] is estimated to occur at *P–T* conditions of 9–10 kbar and 850–900°C and involves a garnet with X_{Mg} greater than 0.40, the value reported by Vielzeuf & Holloway (1988) for the divariant dehydration melting reaction of biotite, producing garnet, at 10 kbar.

At low f_{O_2} spinel and quartz are not stable together in the biotite melting interval. However, at high f_{O_2} owing to the extension of the spinel stability field, a new invariant point [Spr, Opx] is created and spinel can occur with quartz (see Fig. 2.17) as a product of biotite dehydration melting, as has been reported from some low-pressure (3.5–5 kbar) granulite terranes (e.g. Stüwe & Powell 1989) and contact aureoles (Grant 1985), for which temperatures of *c.* 750°C have been inferred. The f_{O_2} effect invoked by Grant (1985) and in this chapter is similar to the effect of Zn in enhancing the Spl + Qtz stability field to lower temperature as proposed by Montel *et al.* (1986) to explain the occurrence of this pair in certain partially molten pelites.

The interaction of the corundum-present part of FMAS (Fig. 2.16) with the KFMASH grid of Figure 2.17 is considered in Figure 2.18. The main constraint used for the construction of the $n + 4$ multigrid, which includes corundum, is the stability of the reaction Spr + Kfs (+ L) = Opx + Bt + Sil + Crn, which is found in kornerupine-bearing granulites from the Western Harts Range, Arunta Block, Central Australia (Hensen & Warren 1983). Interaction of the two grids of Figures 2.16 & 17 is achieved by assuming the garnet-absent invariant point [Grt, Qtz] in the KFMASH grid to lie at a temperature slightly above that of [Grt, Crn] in the corundum-bearing grid. As a result the reactions (Grt, Crd), one from each grid, intersect to create a new ($n + 4$) invariant point [Crd, Grt]. This procedure causes the [Crd] invariant point in the KFMASH grid to become metastable, whereas the [Grt, Qtz] and [Spr, Qtz] points are maintained as [Grt, Crn] and [Spr, Crn]. A theoretical treatment of the combination of two $n + 3$ stable grids to create an $n + 4$ grid is given by Guo (1984). The other invariant points in the $n + 4$ system may be derived by application of Schreinemakers rules.

The proposed $n + 4$ grid (Fig. 2.18) must be regarded as a model diagram only, as insufficient critical data from granulite occurrences are presently available. The grid was constructed largely for application to the sapphirine granulites from the Arunta Block (Hensen 1983) and applies to the phase relationships observed there (Warren 1983). It is encouraging that the low-temperature–low-pressure portion of the diagram is entirely consistent with phase relationships reported independently from two high-grade aureoles

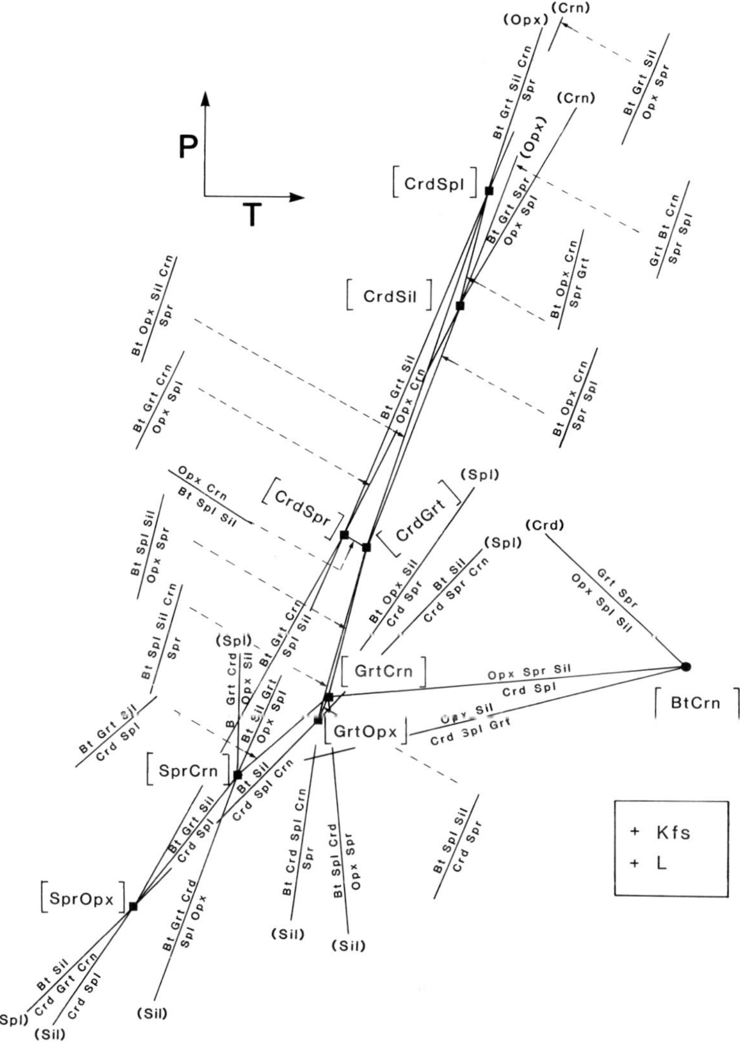

Figure 2.18 Theoretical P–T grid for KFMASH ($n + 4$), combining the grids of Figures 2.16 & 17. See text for further explanation.

displaying partial melting (Grant 1985, Pattison & Harte 1985, in press); in particular, the reaction sequence suggested by Grant (1985, Fig. 3.21) for the Laramie aureole, Wyoming is identical to that shown in Figure 2.19. Pattison & Harte (1985) considered mineral assemblages in several Scottish contact aureoles and proposed reactions with the numbers given in Figure 2.19, but in a different sequence. Pattison & Harte (in press) have now proposed a revised reaction sequence which is in accord with that given in Figures 2.18 & 19. The

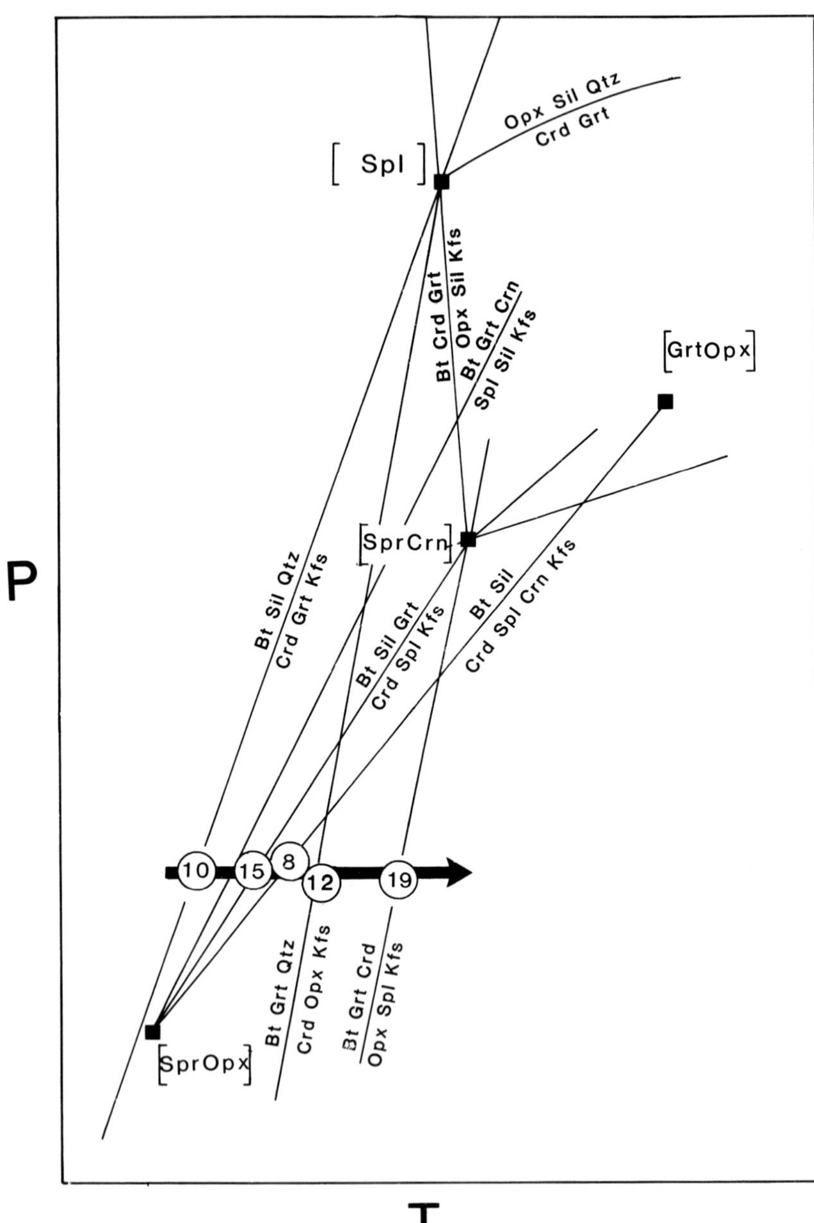

Figure 2.19 Section of KFMASH grid combining Figures 2.17 & 18. The order of reactions 10, 15, 8, 12, and 19 is that derived by Grant (1985) for the Laramie aureole, Wyoming, USA; the reaction numbers are those used by Pattison & Harte (1985 and in press).

indifferent crossings between the quartz-bearing, quartz-absent and corundum-present reactions appear to be demanded by all the studied aureoles. This excellent agreement between the theoretical grid and observed reaction sequences demonstrates the potential use of the grid in applications to high-grade rocks.

2.13 Concluding remarks

It must be emphasized that there are numerous pitfalls in establishing *P–T* grids with application to high-grade metamorphic rocks.

The single most effective approach is to integrate information on metamorphic mineral assemblages with the theoretical mass balanced grids in order to establish the stable configurations of such grids, always keeping in mind that even variations in minor components, often not considered, may sometimes change topological relationships. Used with the necessary care, the *P–T* (and *P–X*, $P–\mu$ etc.) graphical approach is an invaluable tool for the unravelling of the complex mineralogical and textural features of high-grade metamorphic rocks.

In this chapter we have developed the construction of a complex FMAS grid, accounting for mineral assemblages at both high and low f_{O_2}. The relationships between end-member grids and complex grids has been further clarified, and we have demonstrated how $P–\mu$ and $\mu–\mu$ grids are related to the more commonly used *P–T* and *P–X* sections.

Calculated contouring of divariant fields has been carried out for part of the FMAS system, and the significance of curved reaction boundaries, curved isopleths, and consequent X_{Mg} reversals along univariant curves has been explained in detail.

Dehydration melting reactions involving biotite in KFMASH have been explored and we suggest that vapour-absent melting reactions may constitute a convenient *P–T* framework for understanding phase relations in KFMASH within granulite facies conditions. There is a great need for further experimental work aimed at determining the vapour-absent melting relations of assemblages involving biotite. On the basis of our theoretical grids, and consideration of mineral compositional variations in solid–solid divariant reactions related to the KFMASH melting equilibria, it is apparent that existing experimental data (e.g. Vielzeuf & Holloway 1988) need to be complemented by experiments performed on more magnesian bulk-rock compositions involving biotite.

Acknowledgements

We would like to thank Drs J. R. Ashworth, L. M. Barron, B. Harte, and N. F. C. Hudson for their helpful critical comments on the manuscript. This work was done while B.J.H. was on study leave at the University of Edinburgh. The support of the British Council (B.J.H.) is gratefully acknowledged.

References

Ackermand, D., R. K. Herd, M. Reinhardt & B. F. Windley 1987. Sapphirine parageneses from the Caraiba complex, Bahia, Brazil: the influence of Fe^{2+}–Fe^{3+} distribution on the stability of sapphirine in natural assemblages. *Journal of Metamorphic Geology* **5**, 323–40.

Albee, A. L. 1965. A petrogenetic grid for the Fe–Mg silicates of pelitic schists. *American Journal of Science* **263**, 512–36.

Andrew, A. S. & J. Linde 1980. MRF, a FORTRAN IV computer program for the generation of univariant phase equilibria. *Computers & Geosciences* **6**, 227–36.

Annersten, H. & F. Seifert 1981. Stability of the assemblage orthopyroxene–sillimanite–quartz in the system MgO–FeO–Fe_2O_3–Al_2O_3–SiO_2–H_2O. *Contributions to Mineralogy and Petrology* **77**, 158–165.

Aranovich, L. Ya. & K. K. Podlesskii 1983. The cordierite–garnet–sillimanite–quartz equilibrium: experiments and applications. In *Kinetics and equilibrium in mineral reactions*, S. K. Saxena (ed.), 173–98. New York: Springer-Verlag.

Aranovich, L. Ya & K. K. Podlesskii 1989. Geothermobarometry of high-grade metapelites: simultaneously operating reactions. In *Evolution of metamorphic belts*, J. S. Daly, R. A. Cliff & B. W. D. Yardley (eds). Geological Society of London Special Publication, in press.

Ballevre, M., J. Pinardon, J. R. Kienast & J. P. Vuichard 1989. Reversal of Fe–Mg partitioning between garnet and staurolite: implications for high pressure metapelites. *Journal of Petrology* **30**, in press.

Berg, J. H. 1977. Regional geobarometry in the contact aureoles of the anorthositic Nain Complex, Labrador. *Journal of Petrology* **18**, 399–430.

Bertrand, P., D. J. Ellis & D. H. Green 1989. Stabilité des assemblages Sa–Qz et Hy–Sil–Qz dans le système FMAS, sous faible P_{H_2O} et f_{O_2}. *Comptes Rendus de l'Académie des Sciences, Paris, Série II* **308**, 1437–42.

Burt, D. M. 1971. Multisystems analysis of the relative stabilities of babingtonite and ilvaite. *Carnegie Institution of Washington Yearbook* **70**, 189–97.

Caporuscio, F. A. & S. A. Morse 1978. Occurrence of sapphirine plus quartz at Peekskill, New York. *American Journal of Science* **278**, 1334–42.

Carswell, D. A. & S. L. Harley 1989. Mineral thermometry–barometry. In *Eclogite facies rocks*, D. A. Carswell (ed.), Ch. 3. Glasgow: Blackie.

Chatterjee, N. D. & W. Schreyer 1972. The reaction enstatite$_{SS}$ + sillimanite $\rightleftharpoons$ sapphirine$_{SS}$ + quartz in the system MgO–Al_2O_3–SiO_2. *Contributions to Mineralogy and Petrology* **36**, 49–62.

Droop, G. T. R. & K. Bucher-Nurminen 1984. Reaction textures and metamorphic evolution of sapphirine bearing granulites from the Gruf complex, Italian Central Alps. *Journal of Petrology* **25**, 766–803.

Ellis, D. J. 1980. Osumilite–sapphirine–quartz granulites from Enderby Land, Antarctica: *P–T* conditions of metamorphism, implications for garnet–cordierite equilibria and the evolution of the deep crust. *Contributions to Mineralogy and Petrology* **74**, 201–10.

Ellis, D. J. 1983. The Napier and Rayner Complexes of Enderby Land Antarctica: contrasting styles of metamorphism and tectonism. In *Antarctic Earth Science*, R. L. Oliver, P. R. James & J. B. Jago (eds), 20–4. Canberra and Cambridge: Cambridge University Press.

Ellis, D. J. 1986. Garnet–liquid Fe^{2+}–Mg equilibria and implications for the beginning of melting in the crust and subduction zones. *American Journal of Science* **286**, 765–91.
Ellis, D. J., J. W. Sheraton, R. N. England & W. B. Dallwitz 1980. Osumilite–sapphirine–quartz granulites from Enderby Land, Antarctica – mineral assemblages and reactions. *Contributions to Mineralogy and Petrology* **72**, 123–43.

Finger, L. W. & D. M. Burt 1972. REACTION, a FORTRAN IV computer program to balance chemical reactions. *Carnegie Institution of Washington Yearbook* **71**, 616–20.

Grant, J. A. 1985. Phase equilibria in partial melting of pelitic rocks. In *Migmatites*, J. R. Ashworth (ed.), 86–144. Glasgow: Blackie.
Greenwood, H. J. 1975. Buffering of pore fluids by metamorphic reactions. *American Journal of Science* **275**, 573–93.
Grew, E. S. 1981. Granulite facies metamorphism at Molodezhnaya Station, East Antarctica. *Journal of Petrology* **22**, 297–336.
Guo, Q. T. 1984. Topological relations in multisystems of more than $n + 3$ phases. *Journal of Metamorphic Geology* **2**, 267–95.

Harley, S. L. 1984a. An experimental study of the partitioning of Fe and Mg between garnet and orthopyroxene. *Contributions to Mineralogy and Petrology* **86**, 359–73.
Harley, S. L. 1984b. The solubility of alumina in orthopyroxene coexisting with garnet in FeO–MgO–Al_2O_3–SiO_2 and CaO–FeO–MgO–Al_2O_3–SiO_2. *Journal of Petrology* **25**, 665–96.
Harley, S. L. 1985. Garnet–orthopyroxene bearing granulites from Enderby Land, Antarctica: metamorphic pressure–temperature–time evolution of the Archaean Napier Complex. *Journal of Petrology* **26**, 819–56.
Harley, S. L. 1986. A sapphirine–cordierite–garnet–sillimanite granulite from Enderby Land, Antarctica: implications for FMAS petrogenetic grids in the granulite facies. *Contributions to Mineralogy and Petrology* **94**, 452–60.
Harley, S. L. 1987. Metamorphic evolution of granulites from the Rauer Group, East Antarctica: decompression following Proterozoic collision. In *Abstracts 5th International Symposium on Antarctic Earth Science*, Cambridge 1987, p. 61.
Harte, B. & N. F. C. Hudson 1979. Pelite facies series and the temperatures and pressures of Dalradian metamorphism. In *The Caledonides of the British Isles – reviewed*, A. L. Harris, C. H. Holland & B. E. Leake (eds), 323–37. Geological Society of London Special Publication 8.
Hensen, B. J. 1971. Theoretical phase relations involving cordierite and garnet in the system MgO–FeO–Al_2O_3–SiO_2. *Contributions to Mineralogy and Petrology* **33**, 191–214.
Hensen, B. J. 1972. Phase relations involving pyrope, enstatite and sapphirine in the system MgO–Al_2O_3–SiO_2. *Carnegie Institution of Washington Yearbook* **71**, 421–427.
Hensen, B. J. 1983. Theoretical phase relations of ferromagnesian minerals in the silica-undersaturated part of the system FeO–MgO–Al_2O_3–SiO_2 (FMAS), KFMASH (additional H_2O) for aluminous high grade amphibolites and granulites. *Geological Society of Australia Abstract Series* **9**, 63–4.
Hensen, B. J. 1986. Theoretical phase relations involving cordierite and garnet revisited: the influence of oxygen fugacity on the stability of sapphirine and spinel in the system Mg–Fe–Al–Si–O. *Contributions to Mineralogy and Petrology* **92**, 362–7.
Hensen, B. J. 1987. *P–T* grids for silica-undersaturated granulites in the system MAS ($n + 4$) and FMAS ($n + 3$): tools for the derivation of *P–T* paths of metamorphism. *Journal of Metamorphic Geology* **5**, 255–71.
Hensen, B. J. 1988. Chemical potential diagrams and chemographic projections: application to sapphirine granulites from Kiranur and Ganguvarpatti. Evidence for rapid uplift in part of the South Indian Shield? *Neues Jahrbuch für Mineralogie, Abhandlungen* **158**, 193–210.
Hensen, B. J. & E. J. Essene 1971. Stability of pyrope–quartz in the system MgO–Al_2O_3–SiO_2. *Contributions to Mineralogy and Petrology* **30**, 72–83.

Hensen, B. J. & D. H. Green 1971. Experimental study of the stability of cordierite and garnet in pelitic compositions at high pressures and temperatures. I. Compositions with excess aluminosilicate. *Contributions to Mineralogy and Petrology* **33**, 309–30.

Hensen, B. J. & D. H. Green 1972. Experimental study of the stability of cordierite and garnet in pelitic compositions at high pressures and temperatures. II. Compositions without excess aluminosilicate. *Contributions to Mineralogy and Petrology* **35**, 331–54.

Hensen, B. J. & D. H. Green 1973. Experimental study of the stability of cordierite and garnet in pelitic compositions at high pressures and temperatures. III. Synthesis of experimental data and geological applications. *Contributions to Mineralogy and Petrology* **38**, 151–66.

Hensen, B. J. & R. G. Warren 1983. Kornerupine–sapphirine–granulites from the western Harts Ranges, Arunta Block. *Geological Society of Australia, Abstract Series* **9**, 70–2.

Hensen, B. J. & R. G. Warren 1984. Fluid evolution in granulites from the Arunta Block, Central Australia. *Geological Society of Australia, Abstract Series* **12**, 235–7.

Hess, P. C. 1969. The metamorphic paragenesis of cordierite in pelitic rocks. *Contributions to Mineralogy and Petrology* **24**, 191–207.

Holdaway, M. J. & S. M. Lee 1977. Fe–Mg cordierite stability in high-grade pelitic rocks based on experimental, theoretical and natural observations. *Contributions to Mineralogy and Petrology* **63**, 175–98.

Karsakov, L. P., V. I. Shuldiner & A. M. Lennikov 1975. Granulite complex of the eastern part of the Stanovoy Fold Province and the Chogar facies of depth. *Izvestiya Akademii Nauk SSSR, Seriya Geologicheskaya* **85**, 47–61 [in Russian].

Korzhinskii, D. S. 1959. *Physicochemical basis of analysis of the paragenesis of minerals.* New York: Consultants Bureau, Inc. (English translation).

Lal, R. K., D. Ackermand, P. Raith, P. Raase & F. Seifert 1984. Sapphirine-bearing assemblages from Kiranur, southern India: a study of chemographic relationships in the Na_2O–FeO–MgO–Al_2O_3–SiO_2–H_2O system. *Neues Jahrbuch für Mineralogie Abhandlungen* **150**, 121–52.

Lal, R. K., D. Ackermand & H. Upadhyay 1987. *P–T–X* relationships deduced from corona textures in sapphirine–spinel–quartz assemblages from Paderu, southern India. *Journal of Petrology* **28**, 1139–68.

Lee, H. Y. & J. Ganguly 1988. Equilibrium compositions of coexisting garnet and orthopyroxene: experimental determinations in the system FeO–MgO–Al_2O_3–SiO_2, and applications. *Journal of Petrology* **29**, 93–113.

Marakushev, A. A. & V. A. Kudryavtsev 1965. Hypersthene–sillimanite paragenesis and its petrological implication. *Doklady Akademii Nauk SSSR, Earth Science Sections* **164**, 179–82 (English translation 1966, pp. 145–8).

Mohan, A., D. Ackermand & R. K. Lal 1986. Reaction textures and *P–T–X* trajectory in the sapphirine–spinel-bearing granulites from Ganguvarpatti, Southern India. *Neues Jahrbuch für Mineralogie Abhandlungen* **154**, 1–19.

Montel, J. M., C. Weber & M. Pichavant 1986. Biotite–sillimanite–spinel assemblages in high-grade metamorphic rocks: occurrences, chemographic analysis and thermobarometric interest. *Bulletin de Minéralogie* **109**, 555–73.

Newton, R. C. 1972. An experimental determination of the high-pressure stability limits of magnesian cordierite under wet and dry conditions. *Journal of Geology* **80**, 398–420.

Niggli, P. 1954. *Rocks and mineral deposits.* New York: W. H. Freeman.

Pattison, D. & B. Harte 1985. A petrogenetic grid for pelites in the Ballachulish and other Scottish thermal aureoles. *Journal of the Geological Society of London* **142**, 7–28.

Pattison, D. R. M. & B. Harte, in press. Petrography and mineral chemistry of metapelites in the Ballachulish aureole. In *Equilibrium and kinetics in contact metamorphism: the Ballachulish igneous complex and its aureole*, G. Voll, J. Topel, D. R. M. Pattison & F. Seifert, (eds), Berlin: Springer-Verlag.

Powell, R. 1983. Fluids and melting under upper amphibolite facies conditions. *Journal of the Geological Society of London* **140**, 629–33.

Powell, R. & M. Sandiford 1988. Sapphirine and spinel phase relationships in the system $FeO–MgO–Al_2O_3–SiO_2–TiO_2–O_2$ in the presence of quartz and hypersthene. *Contributions to Mineralogy and Petrology* **98**, 64–71.

Ramberg, H. 1964. Chemical thermodynamics in mineral studies. In *Physics and chemistry of the Earth*, Vol. 5, L. H. Ahrens, F. Press & S. K. Runcorn (eds), 225–53. Oxford: Pergamon.

Richardson, S. W. 1968. Staurolite stability in a part of the system Fe–Al–Si–O–H. *Journal of Petrology* **9**, 467–88.

Robie, R. A., P. M. Bethke & K. M. Beardsley 1967. Selected X-ray crystallographic data, molar volumes and densities of minerals and related substances. *U.S. Geological Survey, Bulletin* 1248.

Sandiford, M., F. B. Neall & R. Powell 1987. Metamorphic evolution of aluminous granulites from Labwor Hills, Uganda. *Contributions to Mineralogy and Petrology* **95**, 217–25.

Schenk, V. 1984. Petrology of felsic granulites, metapelites, metabasics, ultramafics and metacarbonates from southern Calabria (Italy): prograde metamorphism uplift and cooling of a former lower crust. *Journal of Petrology* **25**, 255–98.

Schumacher, J. C. & P. Robinson 1987. Mineral chemistry and metasomatic growth of aluminous enclaves in gedrite cordierite–gneiss from southwestern New Hampshire, USA. *Journal of Petrology* **28**, 1033–73.

Stüwe, K. & R. Powell 1989. Low-pressure granulite facies metamorphism in the Larsemann Hills area, East Antarctica; petrology and tectonic implications for the evolution of the Prydz Bay area. *Journal of Metamorphic Geology* **7**, 465–83.

Thompson, A. B. 1976. Mineral reactions in pelitic rocks: I. Prediction of *P–T–X* (Fe–Mg) phase relations. II. Calculation of some *P–T–X* (Fe–Mg) phase relations. *American Journal of Science* **276**, 401–54.

Thompson, A. B. 1982. Dehydration melting of pelitic rocks and the generation of H_2O-undersaturated granitic liquids. *American Journal of Science* **282**, 1567–95.

Thompson, J. B. Jr 1955. The thermodynamic basis for the mineral facies concept. *American Journal of Science* **253**, 65–103.

Thompson, J. B. Jr 1957. The graphical analysis of mineral assemblages in pelitic schists. *American Mineralogist* **42**, 842–59.

Vielzeuf, D. 1983. The spinel and quartz associations in high grade xenoliths from Tallante (SE Spain) and their potential use in geothermometry and barometry. *Contributions to Mineralogy and Petrology* **82**, 301–11.

Vielzeuf, D. & P. Boivin 1984. An algorithm for the construction of petrogenetic grids – application to some equilibria in granulitic paragneisses. *American Journal of Science* **284**, 760–91.

Vielzeuf, D. & J. R. Holloway 1988. Experimental determination of the fluid-absent melting relations in the pelitic system: consequences for crustal differentiation. *Contributions to Mineralogy and Petrology* **98**, 257–76.

Warren, R. G. 1983. Prograde and retrograde sapphirine in metamorphic rocks of the Central Arunta Block, Central Australia. *BMR Journal of Australian Geology and Geophysics* **8**, 139–45.

Waters, D. J. 1988. Partial melting and the formation of granulite facies assemblages in Namaqualand, South Africa. *Journal of Metamorphic Geology* **6**, 387–404.

Watts, B. J. 1974. Fluid-bearing and fluid-absent invariant points in the system CaO–MgO–Al_2O_3–CO_2–H_2O for a greenschist facies assemblage: a correction and some further implications. *Contributions to Mineralogy and Petrology* **47**, 153–64.

Zen, E-An 1966. Construction of pressure-temperature diagrams for multicomponent systems after the method of Schreinemakers – a geometric approach. *U.S. Geological Survey*, *Bulletin* 1225, 56 pp.

CHAPTER THREE

A model for rates of disequilibrium melting during metamorphism

David C. Rubie & Adrian J. Brearley

3.1 Introduction

In numerous studies of partial melt[illegible] rimarily on topics such as equilibrium phase [illegible] ng, and the textural relations between liquid [illegible] molten aggregates. In contrast, the *rate* of m[illegible] eceived little or no attention. An evaluation of [illegible] umber of reasons, such as the following.

(a) In geochemical studies, it is generally assumed that chemical equilibrium is maintained between the liquid and crystalline phases at the site of partial melting. Such an assumption then allows conclusions to be made about the nature of the source materials on the basis of the geochemistry of crystallized magmas together with major and trace element partitioning data obtained from experiments. Equilibrium at regions of partial melting is most likely to be achieved when melting occurs slowly. However, if melting occurs rapidly, and if the melt is subsequently removed rapidly from the site of origin (e.g. Spence & Turcotte 1985, Turcotte *et al.* 1987), equilibration between the liquid and residual phases is not likely to be achieved. A case for disequilibrium during the partial melting of peridotites, for example, has been argued by Prinzhofer & Allègre (1985).

(b) The rate of melting may affect mechanisms of melt migration. In general, this process has been discussed in terms of models of textural equilibration in which the distribution of melt is controlled by the dihedral angle between the liquid and two adjacent solid grains (e.g. Waff & Bulau 1979, Jurewicz & Watson 1985, McKenzie 1985, Wickham 1987). If melting occurs rapidly, however, the liquid can migrate along *fractures* induced by the volume increase produced by the melting reaction (Mawer *et al.* 1988). In terms of rates of migration, this is likely to be a much more effective mechanism than the textural equilibration model, at least for moving melts over relatively short distances, as in the formation of migmatites.

The purpose of this chapter is to discuss mechanisms by which very rapid partial melting could occur. A simple numerical model is formulated in order to make a preliminary and semi-quantitative evaluation of possible melting rates. Although the discussion is focused primarily on crustal melting, the general conclusions may also be applicable to partial melting in the mantle, below mid-ocean ridges for example.

3.2 Factors controlling melting rates

The rate of melting, in the crust or mantle, can be controlled by one of the following three factors:

(a) The rate of supply of the heat which is required for endothermic melting reactions. When melting is heat-flow-controlled, the rate of heat supply is balanced by the rate of consumption of the latent heat of melting.
(b) The kinetics of the melting reaction.
(c) The rate of supply of externally derived fluid (in the case of fluid-present melting).

The relative importance of each of these three factors differs according to whether melting occurs under equilibrium or disequilibrium conditions, as discussed below (see also Lasaga 1986).

3.2.1 Equilibrium melting

During equilibrium melting, the volume fraction of melt (melt fraction) and the compositions of coexisting phases present at any given time are a function of thermodynamic variables such as P, T, and a_{H_2O}. Estimates of the melt fraction as a function of temperature have been made, on the basis of experimental data and model mass balance calculations, for granitic, pelitic, and mafic systems under both H_2O-saturated and vapour-absent conditions (e.g. Wyllie 1977, Clemens & Vielzeuf 1987, Vielzeuf & Holloway 1988). Such estimates indicate that the variation of melt fraction with temperature can be highly non-linear between the solidus and liquidus, especially for vapour-absent melting. For example, in metapelites under vapour-absent conditions, the melt fraction increases from ~ 0.1 to ~ 0.5 over a very narrow temperature interval as a result of the breakdown of biotite, according to the results of Vielzeuf & Holloway (1988, Fig. 3).

The rate of equilibrium melting during metamorphism is, in general, controlled by the rate of heat supply and therefore depends on the rate at which temperature increases with time. The rate of melting in various tectonic settings can therefore be estimated from thermal modelling combined with the estimates of equilibrium melt fraction as a function of temperature. Thermal evolution in continental collision zones, involving crustal thickening with subsequent partial melting, has been modelled by England & Thompson (1986), Zen (1988), and De Yoreo *et al.* (1989). In these models, the evolution of temperature with time depends significantly on thermal parameters such as the rate of heat flow from the mantle, heat capacity, and thermal diffusivity of crustal rocks, and the latent heat of melting (see England & Thompson 1984). Various simplifying assumptions have generally been made. For example, England & Thompson (1986) considered melting to occur by a series of discrete reactions, such as the fluid-absent melting of biotite, and they did not consider the effect of the latent heat of melting. In contrast, De Yoreo *et al.* (1989) assumed that melting occurs continuously over the temperature interval between the solidus and liquidus, and they also studied the effects of a range of possible values of latent heat of melting on thermal evolution and melt fraction as a function of time. The results of the modelling of De Yoreo *et al.* (1989) suggest that several million years are required for the melt fraction to reach a significant value at a given depth in the crust.

3.2.2 Disequilibrium melting

Disequilibrium melting can occur if the temperature significantly exceeds that of the solidus before melting begins. Then, when melting starts, excess heat is available to contribute to the latent heat of melting. The amount of excess heat available (per unit mass of rock) is given by

$$\Delta q = (T - T_s)C_p \qquad (3.1)$$

where T_s is the solidus temperature and C_p is the heat capacity of the rock undergoing melting. Under such circumstances, as shown below, the rate of melting is not controlled by the rate of heat supply but either by reaction kinetics or by the rate of supply of fluid.

Overstepping of the solidus temperature before melting begins may occur by one or more of the following mechanisms:

(a) Infiltration of H_2O into ‘dry’ rocks at a temperature in excess of the H_2O-saturated solidus.
(b) The lack of mutual contact between the reactant mineral grains.
(c) For an incongruent melting reaction, substantial overstepping of the solidus temperature may be required for the new crystalline phases to nucleate (Rubie & Brearley 1988, 1990).

FLUID INFILTRATION

It is likely that fluid-absent conditions persist for long periods of time in the lower and middle crust and that fluid-present conditions exist for periods of relatively limited duration during devolatilization or fluid infiltration (e.g. Thompson 1983, Rubie 1986, Clemens & Vielzeuf 1987). Both hydration and solid–solid reactions can occur in response to fluid infiltration, at conditions which deviate significantly from equilibrium (Rubie 1986, 1989; Wayte *et al.* 1989). Such reactions fail to occur where fluid is absent as a catalyst (in the case of a solid–solid reaction) or reactant (in the case of a hydration reaction). Therefore, models have been proposed in which the timing of some reactions is controlled not by the crossing of equilibrium boundaries in P–T space but by the timing of fluid infiltration (Yardley & Baltatzis 1985; Rubie 1986, 1989). Retrograde reactions frequently occur in shear zones as a consequence of fluid

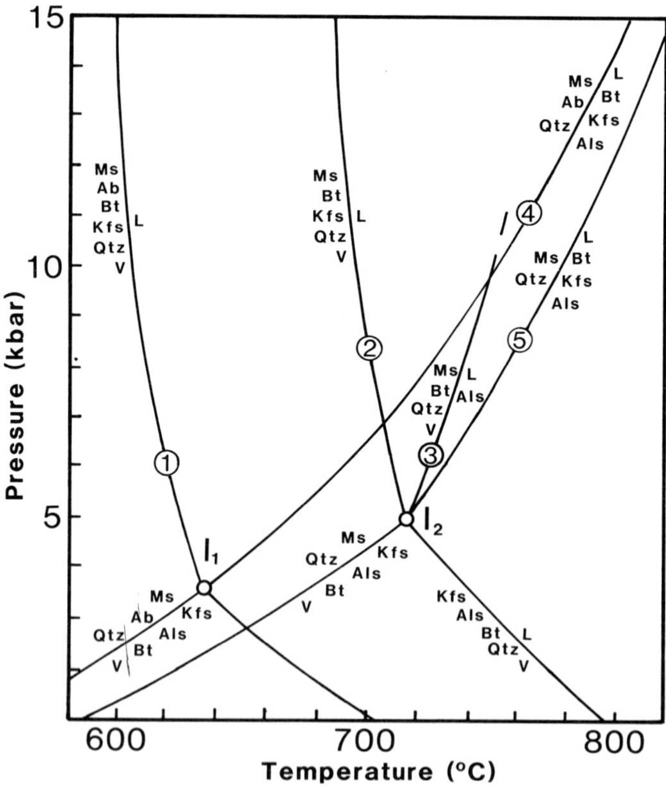

Figure 3.1 Pressure–temperature diagram, showing a selection of melting curves relevant to the assemblage Ms + Ab + Bt + Kfs + Qtz + H_2O(V) (after Thompson 1982, Figs 5–7). Reactions emanating from the invariant point I_1 are in the system K_2O–Na_2O–FeO–Al_2O_3–SiO_2–H_2O (KNFASH) and those emanating from I_2 are in the K_2O–FeO–Al_2O_3–SiO_2–H_2O (KFASH) system. The numbered reaction curves correspond to Reactions R–1 to R–5 in the text.

infiltration, and isotopic evidence and mass balance considerations indicate that the volumes of infiltrating fluid can be very large (Beach 1980, Rumble *et al.* 1982, Kerrich *et al.* 1984).

We propose that melting reactions can also result from the infiltration of fluid into 'dry' rocks, at temperatures significantly in excess of the H_2O-saturated solidus. The involvement of large volumes of externally derived fluid in some examples of crustal melting is indicated by isotopic evidence (Wickham & Taylor 1985). Because of the low porosity of metamorphic rocks, large volumes of fluid cannot be present at the initiation of melting, but must be introduced by infiltration during melting. During prograde metamorphism, there is no reason why fluid should necessarily be available at the time when the temperature reaches the H_2O-saturated solidus. At ~10 kbar, the vapour-absent solidus for pelites is >100°C above the H_2O-saturated solidus (Fig. 3.1; Thompson 1982, Fig. 8). Therefore, under vapour-absent conditions, a very large overstepping of the H_2O-saturated solidus is possible (e.g. $T - T_s = 100°C$) with melting initially occurring only when fluid infiltration occurs or when a vapour-absent melting reaction is reached. Fluid infiltration could occur in response to initial vapour-absent melting if the latter initiates deformation, in shear zones for example.

LACK OF MUTUAL CONTACT BETWEEN REACTANT MINERAL GRAINS

Coarse-grained rocks, which have been metamorphosed without being strongly deformed, often show evidence of pronounced disequilibrium on a thin-section scale. In eclogite facies metagabbros, for example, rates of intra-crystalline diffusion are often slow, even when small traces of H_2O are present. Consequently, adjacent reactant mineral grains (e.g. plagioclase and augite) can react as almost closed systems, with little or no chemical exchange between them (e.g. Rubie 1989). It is possible that this type of behaviour also occurs when the *P*, *T* conditions required for equilibrium melting are reached, particularly if several reactant phases are involved. For example, the H_2O-saturated solidus in a model pelitic system is often defined by a reaction such as

$$\mathrm{Ms + Ab + Bt + Kfs + Qtz + H_2O \rightarrow L} \qquad \text{(R-1)}$$

(Fig. 3.1), which involves five or more different crystalline reactant phases (see also Grant 1985). If there are no points in the rock at which all these reactant phases are in physical contact, and if there is poor chemical communication between these phases (i.e. slow rates of grain-boundary diffusion), then melting will fail to occur at the solidus. The temperature may then increase during metamorphism until a reaction boundary is reached which involves a smaller number of phases which are in contact, such as (for example)

$$\mathrm{Ms + Bt + Kfs + Qtz + H_2O \rightarrow L} \qquad \text{(R-2)}$$

If this reaction also fails to occur due to a lack of physical contact between Ms, Bt, Kfs, and Qtz, the temperature can continue to increase until a reaction involving only two or three phases is reached, such as

$$\mathrm{Ms + Bt + Qtz + H_2O \rightarrow L + Als} \quad \text{(R-3)}$$

$$\mathrm{Ms + Ab + Qtz \rightarrow L + Bt + Kfs + Als} \quad \text{(R-4)}$$

or

$$\mathrm{Ms + Qtz \rightarrow L + Bt + Kfs + Als} \quad \text{(R-5)}$$

(Fig. 3.1). If melt produced by one of these latter reactions migrates through the rock, pathways of rapid diffusion will develop between all the phases, thus allowing Reaction R-1 to occur under conditions of significant overstepping ($T - T_s = 80–180°C$ at 10 kbar, according to Fig. 3.1). Such a process may also be important for fluid-absent melting in mantle peridotites, depending on the textural relations between the constituent mineral grains.

INCONGRUENT MELTING REACTIONS

The nucleation of new crystalline phases (such as aluminosilicates) in a melting reaction can be very sluggish and can require an overstepping of tens of degrees on a laboratory timescale (see below; Rubie & Brearley 1990, Fig. 2). In the absence of the new crystalline phases, only small amounts of liquid can form, and melting will be inhibited until the temperature required for nucleation is reached. This effect may be particularly significant for vapour-absent melting reactions, such as Reaction R-4 (Fig. 3.1), which require the nucleation of several new crystalline phases. However, it is very difficult to estimate the overstepping required for nucleation during metamorphism because of the much longer timescale than in laboratory experiments.

3.3 Experimental kinetic studies

In order to evaluate rates of disequilibrium melting during metamorphism (i.e. initially with a significant temperature overstep of the solidus), experimental data for the kinetics of melting are required. Unfortunately, there have been few experimental studies of the kinetics of melting reactions relevant to crustal rocks. Some of the most detailed studies have been carried out at atmospheric pressure on anhydrous systems such as plagioclase and plagioclase + diopside (Tsuchiyama & Takahashi 1983; Tsuchiyama 1985a,b, 1986). However, the melting of rock samples containing hydrous phases, at elevated pressure, is more relevant to crustal melting during metamorphism. The early stages of melting in a variety of quartzofeldspathic rocks, in the presence of excess H_2O, have been studied up to 4 kbar by Mehnert *et al.* (1973) and Büsch *et al.* (1974), but it is difficult to extract quantitative kinetic data from their results. Rubie & Brearley (1988, 1990) have studied the kinetics of partial melting in samples of quartz–muscovite schist under H_2O-saturated conditions at 1 kbar. The results of this latter study are used below as the basis of a model melting reaction which is developed to evaluate possible rates of disequilibrium melting in the crust.

3.3.1 Kinetics of melting of muscovite + quartz

The kinetics of the melting reaction

$$\text{muscovite} + \text{quartz} + H_2O \rightarrow \text{liquid} + \text{mullite} + \text{biotite} \qquad \text{(R-6)}$$

have been studied in the temperature range 680–757°C at 1 kbar (Rubie & Brearley, 1988, 1990). Experiments were performed for times ranging from 12 h to 4 months on cylindrical samples of a natural quartz–muscovite schist in the presence of excess H_2O (1 wt%). The experimental conditions actually lie in the subsolidus stability field of K-feldspar + aluminosilicate + biotite, and the melting reaction R-6 occurs metastably, probably because metastable liquid can nucleate more readily than the stable crystalline phases Kfs and Als (see Rubie & Brearley 1987). In addition, mullite is also metastable with respect to either andalusite or sillimanite. Other reactions also occur locally at intracrystalline sites in muscovite (Rubie & Brearley 1987), but compared with Reaction R-6 make a relatively minor contribution to the consumption of this phase.

The experimental run products were examined by light-optical microscopy, transmission electron microscopy (TEM) and backscattered electron imaging (BSEI). Analytical data were obtained by electron microprobe analysis and by analytical electron microscopy (AEM) (see Brearley & Rubie 1990).

Initial reaction involves the formation of a layer of liquid, up to a few microns wide, along quartz–muscovite grain boundaries. Biotite and mullite nucleate sequentially only after a significant temperature-dependent incubation period, which varies from ~30 h at 757°C to >2 months at 680°C (Fig. 3.2). Prior to the nucleation of mullite, melting is very limited because the liquid becomes supersaturated in the mullite component and muscovite dissolution is consequently inhibited. In regions where mullite has nucleated locally, embayments develop in the adjacent muscovite (Fig. 3.3), indicating that the reaction rate increases following mullite nucleation. This increase in reaction rate results from an enhanced rate of muscovite dissolution due to a reduction in the supersaturation of the liquid in the mullite component (Brearley & Rubie 1990).

After all the product phases have nucleated, the reaction zone between muscovite and quartz progressively widens until the muscovite eventually

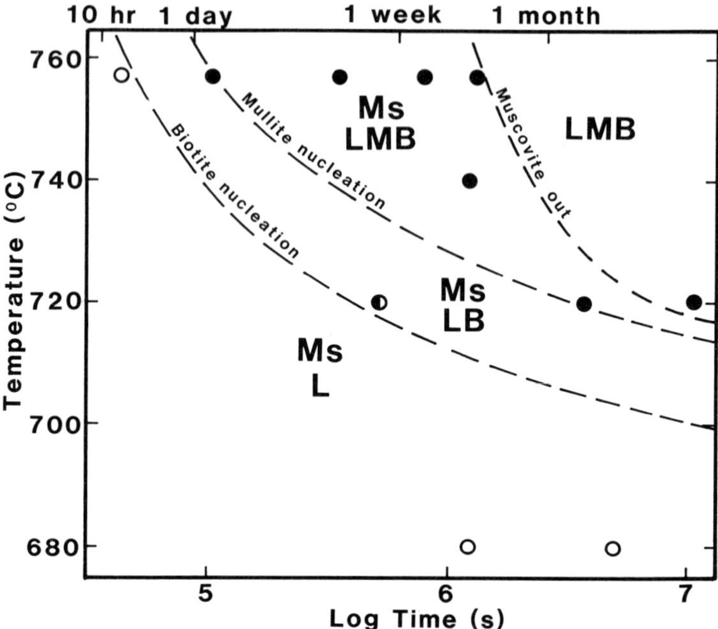

Figure 3.2 Time–temperature-transformation diagram, showing the delayed nucleation of biotite and mullite during the H_2O-saturated melting of muscovite + quartz. The presence of product phases is indicated by: open circles, liquid (L); half-closed circle, liquid + biotite (LB); solid circles, liquid + biotite + mullite (LMB). The curves indicating the nucleation of the crystalline phases and the completion of reaction are semi-schematic because their positions are poorly constrained by the experimental data, and also because nucleation occurs over a finite time period. Of these three curves, only the position of the muscovite-out curve depends on the grain size of the rock.

becomes completely pseudomorphed. The reaction zone develops consistently with a layered structure in which a layer of liquid + mullite + biotite (LMB) is situated adjacent to muscovite, and a layer of liquid (L) is situated adjacent to quartz.

The liquid is peraluminous and has a composition range which lies close to the eutectic on the $KAlSi_3O_8$–SiO_2 join. However, its exact composition varies with temperature, time, and position relative to the reactant phases (Brearley & Rubie 1990). Distinct gradients in Si, Al, and K content are present in the melt, across the reaction zone between muscovite and quartz, and the analytical data suggest that these composition gradients change at the boundary between the LMB and L layers (Rubie & Brearley 1990).

Three factors indicate that the rate of the melting reaction (R-6) is diffusion-controlled after mullite has nucleated (as opposed to being interface-controlled):

(a) The development of a reaction zone with a mineralogically layered struc-

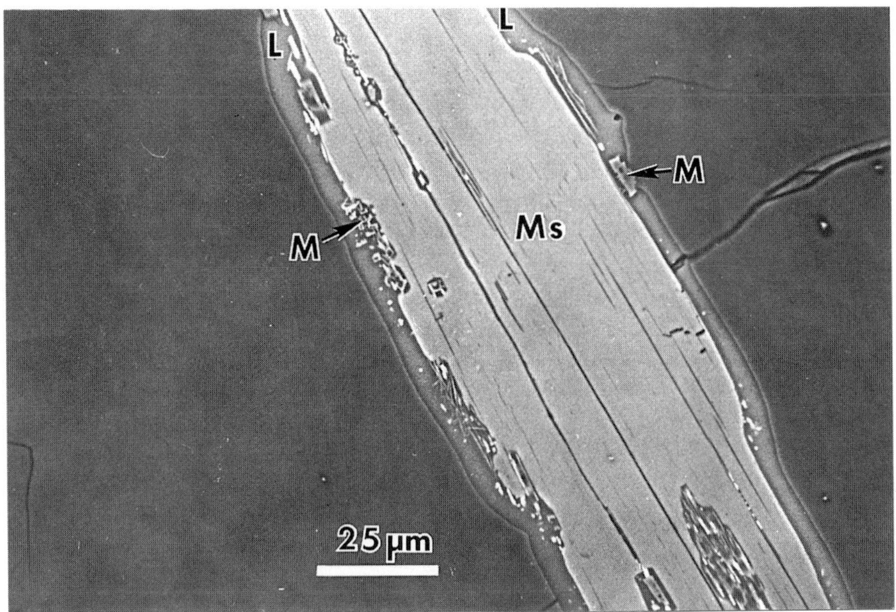

Figure 3.3 Backscattered electron image showing H_2O-saturated partial melting of muscovite (Ms) + quartz. An early stage of melting is illustrated in which a layer of liquid (L) has developed along the muscovite–quartz grain boundary with only the localized nucleation of mullite (M). Embayments have developed in the muscovite where mullite has nucleated, as a result of an enhanced reaction rate. Small grains of biotite (high contrast) are also present. (757°C, 1 kbar, 29 h.)

ture, between incompatible mineral assemblages, is characteristic of diffusion-controlled reactions (Joesten 1977, Fisher 1978).

(b) The presence of composition gradients in the liquid indicates that the reaction rate is controlled by diffusion of Si, Al, and K through the melt.

(c) The width of the reaction zone (x) as a function of time (t) can be modelled by the equation

$$x^2 = D(t - t') \tag{3.2}$$

where D approximates a diffusion coefficient and t' approximates the time of mullite nucleation (Fig. 3.4; see also Rubie & Brearley 1990). This equation is a slight modification of the 'parabolic' law, $x^2 = Dt$, which is indicative of a diffusion-controlled reaction rate.

The growth of the layered reaction zone has been analysed by Rubie & Brearley (1988, 1990) using the theory of non-equilibrium thermodynamics following, in part, the approach of Joesten (1977). It was assumed that reactions between phases and diffusing components occur only at the contacts

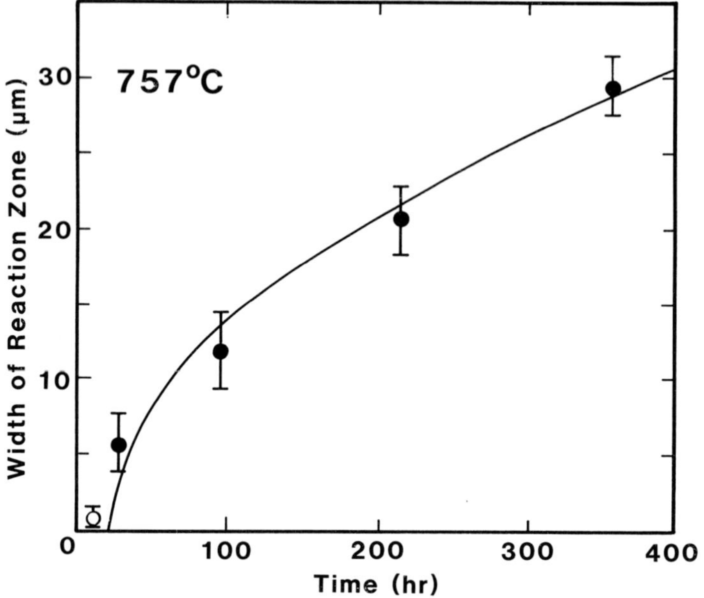

Figure 3.4 Width of the reaction zone, between muscovite and quartz, as a function of time at 757°C. Open circle, liquid only; solid circles, liquid + mullite + biotite. The curve is a least-squares fit of equation (3.2) to the data after mullite has nucleated, with $D = 7 \times 10^{-16}\,m^2\,s^{-1}$ and $t' = 8.5 \times 10^4\,s$.

between mineralogically distinct layers (cf. Fig. 3.5). Preliminary results, based on the assumption that the diffusion of each component is independent of the chemical potential and concentration gradients of the other components, gave the following apparent coefficients of chemical diffusion for Si and Al in the H_2O-saturated liquid at 757°C and 1 kbar (Rubie & Brearley 1988):

$$D_{Si} = 1.5(\pm 0.8) \times 10^{-15}\,m^2\,s^{-1}, \qquad D_{Al} = 1.0(\pm 0.2) \times 10^{-16}\,m^2\,s^{-1}$$

3.4 Description and results of a model

3.4.1 A model melting reaction

In order to model rates of disequilibrium melting in the crust, we use the simple model univariant melting reaction

$$\text{muscovite} + \text{quartz} + H_2O \rightarrow \text{liquid} + \text{aluminosilicate} \qquad \text{(R-7)}$$

in the system K_2O–Al_2O_3–SiO_2–H_2O (KASH), together with the experimental kinetic data of Rubie & Brearley (1988) as summarized in the previous section.

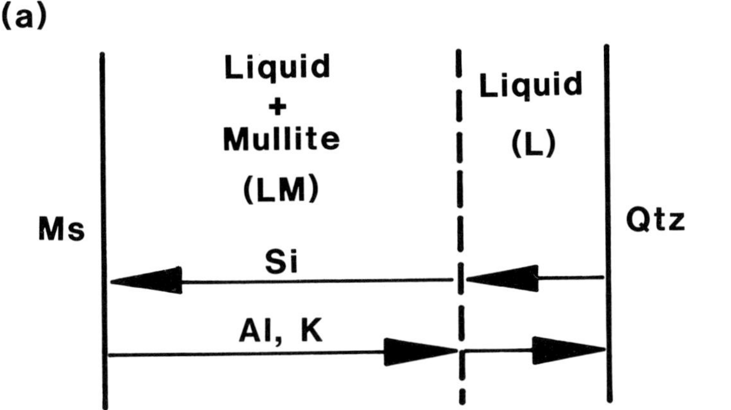

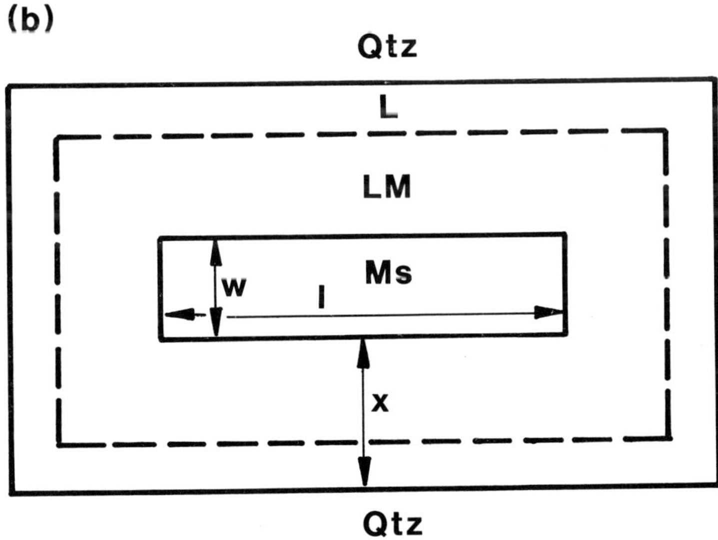

Figure 3.5 Details of the model melting reaction between muscovite and quartz. (a) Schematic illustration showing the layered reaction zone between Ms and Qtz (cf. Fig. 3.3). The directions of diffusion of the components Si, Al, and K are shown. The layers in the reaction zone grow as the result of reactions, at the three layer boundaries, between the phases and diffusing components in a system which is closed to all components except H_2O. (b) A section through a three-dimensional reaction zone around a grain of relict muscovite which has the dimensions $l \times l \times w$. The total width of the layered reaction zone is x. Note that the original quartz–muscovite grain boundary is unlikely to coincide with the boundary between the L and LM layers (Joesten 1977).

Although many melting reactions in crustal rocks will be multivariant, considering a univariant reaction can be justified on several grounds. First, some important melting reactions, such as that involving the vapour-absent breakdown of biotite (Vielzeuf & Holloway 1988), may occur over such a narrow temperature interval that they can be regarded as being effectively univariant. In addition, with an overstep of a multivariant field in *P–T* space, a reaction also becomes effectively univariant. For example, if plagioclase of intermediate composition starts to melt at a temperature above the liquidus, it will react directly to liquid and the composition changes that are required for equilibrium melting do not occur. Finally, kinetic data for more complex reactions at high pressure are currently not available.

We consider Reaction R-7 in the simplified KASH system because of the lack of data for rates of diffusion of minor components such as Na_2O, FeO, and MgO in the liquid. Excluding the components FeO and MgO, including all alkalis as K_2O, and estimating the H_2O-content from Clemens (1984, Fig. 1) and Johannes (1985, Fig. 2.3), the average melt composition in Rubie & Brearley's (1990) experiments at 757°C is 73.81 wt% SiO_2, 14.40 wt% Al_2O_3, 7.79 wt% K_2O, and 4 wt% H_2O. Balancing Reaction R-7 using this composition gives

$$\mathrm{KAl_3Si_3O_{10}(OH)_2\,(Ms) + 4.86\,SiO_2\,(Qtz) + 0.34\,H_2O\,(V)}$$

$$\rightarrow \mathrm{KAl_{1.71}Si_{7.43}O_{16.59}(OH)_{2.68}\,(L) + 0.43\,Al_3SiO_{6.5}\,(Mul)} \quad \text{(R-8)}$$

Although a more realistic model metamorphic reaction would include an aluminosilicate phase with the composition Al_2SiO_5, Reaction R-8 is written with mullite as a product phase in order to develop a model based as closely as possible on the experimental results of Rubie & Brearley (1988, 1990).

On the basis of the experimental results, Reaction R-8 is modelled as forming a layered reaction zone, between the reactant phases, consisting of a layer of liquid + mullite (LM) adjacent to muscovite and a liquid layer (L) adjacent to quartz (Fig. 3.5). The growth of these layers, which is controlled by diffusion of Si, Al, and K through the liquid, can be modelled by a series of mass balance, conservation, and diffusion equations. These equations are most conveniently written in terms of the oxide components SiO_2, $AlO_{3/2}$, and $KO_{1/2}$ without, however, making any inferences about the nature of the complexing of diffusing components (see Fisher 1973, p. 901). The layers grow as the result of reactions between phases and diffusing components, and it is assumed that the reactions occur at the layer contacts and not within the layers. The validity of this assumption has been discussed by Joesten (1977).

For reaction at a contact between two layers, consisting of the phases X and Y + Z respectively, stoichiometric coefficients (ν) are constrained by the mass balance equations

$$\sum_{\phi=1}^{p} \nu_{\phi}^{X-YZ} n_i^{\phi} + \nu_i^{X-YZ} = 0 \tag{3.3}$$

written for each component i in the system, where ν_{ϕ}^{X-YZ} is the number of moles of phase ϕ consumed or released by the reaction, ν_i^{X-YZ} is the number of moles of component i consumed or released by the reaction, n_i^{ϕ} is the number of moles of component i in one formula unit of phase ϕ, and p is the number of phases present.

The number of moles of component i involved in the reaction at a layer contact is equal to the difference between the flux of i into the boundary ($J_i^{X/in}$) and the flux away from the boundary ($J_i^{YZ/out}$), and is given by the conservation equation:

$$\nu_i^{X-YZ} = J_i^{YZ/out} - J_i^{X/in} \tag{3.4}$$

Finally, diffusion through a layer consisting of the phases Y + Z can be described by Fick's first law (assuming steady state conditions):

$$J_i^{YZ} = - D_i^{YZ} (dc_i/dx)^{YZ} \tag{3.5}$$

where J_i^{YZ} is the flux of i, D_i^{YZ} is the diffusion coefficient for component i through the layer YZ, and $(dc_i/dx)^{YZ}$ is the concentration gradient of i in layer YZ. Although it is more rigorous to write diffusion equations in terms of chemical potential gradients ($d\mu_i/dx$) and phenomenological (Onsager) diffusion coefficients (L_{ij}) (e.g. Joesten 1977), (3.5) is used here because concentration gradient data are available, whereas chemical potential gradients cannot be estimated readily. Also, diffusion equations written in terms of chemical potential gradients are only valid if local equilibrium is maintained, whereas such a condition is not required by equation (3.5). When diffusion occurs entirely through *one* or two or more phases in a layer (e.g. through liquid in a layer consisting of liquid + mullite), (3.5) must be modified as follows:

$$J_i^{YZ} = - D_i^{Y} \Phi_Y^{YZ} \tau (dc_i^Y/dx)^{YZ} \tag{3.6}$$

In (3.6), D_i^Y is the diffusion coefficient for component i in phase Y (the phase through which diffusion occurs), Φ_Y^{YZ} is the volume fraction of phase Y in the layer YZ, c_i^Y is the concentration of i in phase Y, and τ is tortuosity – a dimensionless parameter (< 1) which is necessary because components cannot diffuse through the layer in a straight line due to the presence of phases other than Y (cf. Walther & Wood 1984, Eqn 11). With steady state conditions, the concentration gradient is given by

$$(\mathrm{d}c_i^{\mathrm{Y}}/\mathrm{d}x)^{\mathrm{YZ}} = \Delta c_i^{\mathrm{Y/YZ}}/x^{\mathrm{YZ}} \tag{3.7}$$

where $\Delta c_i^{\mathrm{Y/YZ}}$ is the change in concentration of i in the phase Y across the YZ layer, and x^{YZ} is the width of that layer.

The temperature-dependence of the diffusion coefficient D_i^{Y} in (3.6) is given by

$$D_i^{\mathrm{Y}} = D_{0(i)}^{\mathrm{Y}} \exp(-Q_i/RT) \tag{3.8}$$

where $D_{0(i)}^{\mathrm{Y}}$ is a constant, Q_i is the activation energy for the diffusion of component i, R is the gas constant, and T is absolute temperature. The dependence of the concentration difference $\Delta c_i^{\mathrm{Y/YZ}}$ on temperature is more difficult to evaluate. In the absence of thermodynamic data for silicate melts, the effects of two possible *empirical* models are investigated. At the equilibrium temperature for the melting reaction (T_{e}) the concentration gradients for all components must be zero. In the first model, it is assumed that the concentration difference for component i has a linear relationship to the deviation from the equilibrium temperature:

$$\Delta c_i = k_i(T - T_{\mathrm{e}}) \tag{3.9}$$

where k_i is a constant. In the second model, the concentration difference varies exponentially with the deviation from T_{e}:

$$\Delta c_i = \alpha_i[\,1 - \exp(-\beta_i(T - T_{\mathrm{e}}))] \tag{3.10}$$

where α_i and β_i are constants.

The mass balance, conservation, and diffusion equations (3.3, 4 & 6) for the Ms | LM | L | Q layer sequence are listed in full in Appendix 1: 16 equations result for a three-component system, containing as unknowns the 16 stoichiometric coefficients (ν) for phases and components in reactions at layer boundaries, diffusion coefficients for Si and Al in the liquid, concentration gradients for Si and Al in the liquid across the LM and L layers respectively, and tortuosity (τ) for the LM layer. The equations can therefore be solved for the stoichiometric coefficients if the diffusion coefficients, concentration gradients and tortuosity are specified.

The kinetics of melting are modelled in a rock consisting of muscovite + quartz. Parameters which must be specified are the modal proportion of muscovite in the rock prior to melting, $(X_{\mathrm{Ms}})_0$, and the muscovite grain size. Prior to melting, each muscovite grain initially has the dimensions $l_0 \times l_0 \times w_0$.

These parameters define the number of muscovite grains per unit volume of rock as

$$N_{Ms} = (X_{Ms})_0 / l_0^2 w_0 \tag{3.11}$$

taking the grain size of muscovite to be uniform.

The geometric basis for the reaction model is shown in Figure 3.5. Because the coefficients (ν_ϕ) have the units of moles $m^{-2} s^{-1}$ (see Appendix 1, Eqns 3.6.5–6.9), their values can be used to determine the rate of melting. The values of dc_i^L/dx decrease with time, as the width (x) of the reaction zone increases, so that only instantaneous rates can be calculated from the above equations (at specified instantaneous values of dc_i^L/dx). Thus, the rate of migration of the boundary Ms | LM is given by

$$(dx/dt)^{Ms-ML} = \nu_{Ms}^{Ms-LM} \bar{V}_{Ms} \tag{3.12}$$

where $\bar{V}_{Ms}$ is the molar volume of muscovite. Because the rate given by Equation 3.12 is instantaneous, the displacement (Δx) of the Ms | LM boundary which occurs over a time interval Δt is given by the relationship

$$\Delta x = \lim_{\Delta t \to 0} (dx/dt)^{Ms-ML} \Delta t \tag{3.13}$$

and can therefore be approximated by

$$\Delta x \simeq (dx/dt)^{Ms-ML} \Delta t \tag{3.14}$$

provided that Δt is very small. The change in volume of a muscovite grain in time Δt can then be calculated from the grain size. Consider a grain of relict muscovite with the dimensions, at time t, of $l \times l \times w$ (Fig. 3.5b). The change in volume of the muscovite grain which occurs when the Ms | LM boundary is further displaced by a distance Δx is given by

$$\Delta V_{Ms} = (l - 2\Delta x)^2 (w - 2\Delta x) - l^2 w \tag{3.15}$$

so that the total change in volume of muscovite per unit volume of rock is given by

$$\Delta X_{Ms} = N_{Ms} [(l - 2\Delta x)^2 (w - 2\Delta x) - l^2 w] \tag{3.16}$$

where N_{Ms} (the number of muscovite grains per unit volume of rock) is defined by (3.11). The corresponding changes in the volume fractions of quartz (ΔX_{Qtz}), liquid (ΔX_L) and mullite (ΔX_{Mul}) can then be calculated from the stoichiometric coefficients of Reaction R-8 together with the molar volumes

($\bar{V}$) of these phases, as follows:

$$\Delta X_{L} = \Delta X_{Ms} \bar{V}_{L} / \bar{V}_{Ms}, \qquad \Delta X_{Qtz} = 4.86\, \Delta X_{Ms} \bar{V}_{Qtz} / \bar{V}_{Ms},$$
$$\Delta X_{Mul} = 0.43\, \Delta X_{Ms} \bar{V}_{Mul} / \bar{V}_{Ms} \quad (3.17a\text{–}c)$$

The molar volumes of muscovite (140.7 cm^3), quartz (22.7 cm^3), and mullite (67.3 cm^3) have been taken from Robie *et al.* (1978), for the compositions listed in Reaction R-8. The molar volume of the liquid (280 cm^3) has been calculated by the method of Bottinga *et al.* (1982), using a value for the partial molar volume of H_2O estimated from data for the $NaAlSi_3O_8$–H_2O system (Burnham & Davis 1971).

3.4.2 *Thermal model*

If the rate of a melting reaction is fast during prograde metamorphism, the consumption of latent heat of melting may cause the temperature to drop so that the reaction rate becomes slower. At a certain reaction rate, the rate of consumption of latent heat will exactly balance the rate of supply of heat which is causing metamorphism. The temperature will then remain constant (i.e. buffered by the reaction) until melting is complete. This interrelationship between reaction rate and rate of external heat supply has been modelled for subsolidus metamorphic reactions by Ridley (1985) and Lasaga (1986). In order to study the interrelationship during melting, we use a simple zero-dimensional thermal model, similar to that of Ridley (1985), in which the conservation of energy equation is solved by an explicit difference approximation (see Smith 1985). We consider melting in a volume of rock which is sufficiently small that temperature gradients within it are negligible. An external heat source causes the temperature to increase with time, provided that the rock is not melting, at a rate characteristic of prograde metamorphism. At a chosen temperature, in excess of the equilibrium melting temperature, melting is allowed to start. Then, for successive short time steps, of duration Δt, the extent of reaction, the heat consumed by melting, and a resulting new temperature are calculated. For each time step, the diffusion coefficients (D_i), and the concentration gradients (dc_i/dx) are calculated as a function of temperature from (3.7), (3.8) and either (3.9) or (3.10). The mass balance, conservation, and diffusion equations (3.3, 4 & 6), as listed in Appendix 1, are then solved to obtain the ν_ϕ terms. The changes in the volume fractions of the phases are calculated from (3.12) and (3.14–17), from which a new melt fraction is obtained. The heat consumed during each time step Δt, by each increment of the melting reaction, is given by

$$\Delta q = L(\Delta X_{Ms}\rho_{Ms} + \Delta X_{Qtz}\rho_{Qtz})/\rho_{T} \quad (3.18)$$

where L is the latent heat of melting per mass of reactants consumed by melting, and ρ_{Ms}, ρ_{Qtz}, and ρ_T are the densities of muscovite, quartz, and the whole rock respectively. The consumption of heat results in the temperature change

$$\Delta T = \Delta q / C_p + \Delta t (dT/dt)^* \tag{3.19}$$

from which a new temperature can be calculated. In (3.19), C_p is the heat capacity of the rock and $(dT/dt)^*$ is the rate of temperature increase which occurs, in the absence of melting, due to the external heat source.

3.4.3 *Model parameters*

Values of parameters used in the modelling are listed in Table 3.1. Because some of these parameters are poorly known quantitatively, the effect of varying them over a range of possible values is investigated.

The muscovite grain size is varied over a range characteristic of fine-, medium- and coarse-grained metamorphic rocks (Table 3.1). The modal proportion of muscovite, prior to melting, is varied between 0.1 and 0.56. With the latter value, quartz and muscovite are present in the proportions defined by the coefficients of Reaction R-8, so that at the completion of melting both muscovite and quartz are completely consumed.

As far as possible, kinetic parameters are based on the experimental results of Rubie & Brearley (1988, 1990), obtained at 1 kbar. The lowest temperature at which melting occurred in these experiments was 680°C. At these conditions, melting was extremely limited and the crystalline phases failed to nucleate (Fig. 3.2). The equilibrium temperature (T_e) for the model melting reaction is therefore taken as 677°C. Diffusion coefficients for Si and Al in the melt have only been estimated experimentally at 757°C (see above). The values listed in Table 3.1 are based on these results, but have been modified slightly to give a similar time-dependence for layer growth in the KASH system, as was observed in the natural samples (Fig. 3.4). Activation energies for the diffusion of Si and Al in silicate liquids are required to estimate diffusion coefficients at other temperatures. Rubie & Brearley's (1990) data suggest an activation energy $\leq$ 383 kJ mol^{-1} but, otherwise, data for hydrous melts of granitic composition, in the appropriate temperature range, are scarce. Baker (1988) obtained activation energies in the range 220–325 kJ mol^{-1} for diffusion of Si in rhyolite melts, both anhydrous and containing 0.5 wt% F, between 1 bar and 10 kbar, at temperatures $\geq$1200°C. For Mg–Al interdiffusion in an anhydrous melt between plagioclase and diopside at 1 bar, 1250–1307°C, Tsuchiyama (1985a) obtained $Q \simeq 460$ kJ mol^{-1}. Data for the diffusion of various trivalent cations in anhydrous felsic melts at 1 bar give activation energies in the range 175–420 kJ mol^{-1} (Henderson *et al.* 1986). Considering that diffusion in hydrous melts should have a lower activation energy than for anhydrous melts, we have used values in the range 150–300 kJ mol^{-1}. Such a wide range is probably realistic when the possible effects of pressure and H_2O

Table 3.1 Model parameter values.

Muscovite grain size:
20 x 100 x 100 μm - 400 x 2000 x 2000 μm

Proportion of muscovite in the rock prior to melting:
$(X_{Ms})_0$ = 0.1 - 0.56

Equilibrium melting temperature:
T_e = 677°C

Initital overstepping of the equilibrium melting temperature:
$\Delta T = T - T_e$ = 5° - 100°C

Diffusion coefficients for Si and Al in the liquid at 757°C: (Rubie & Brearley 1988)
$D^L{}_{Si}$ = 3×10^{-15} $m^2\ s^{-1}$

$D^L{}_{Al}$ = 1×10^{-16} $m^2\ s^{-1}$

Activation energy for diffusion:

Q = 150 - 300 kJ mol^{-1}

Concentration changes for Si and Al in the melt across the LM and L layers at 757°C: (Rubie & Brearley 1990)

$\Delta c_{Si}{}^{L/LM}$ = 1000 mol m^{-3} $\Delta c_{Si}{}^{L/L}$ = 850 mol m^{-3}

$\Delta c_{Al}{}^{L/LM}$ = -1000 mol m^{-3} $\Delta c_{Al}{}^{L/L}$ = -600 mol m^{-3}

Heating rate from an external heat source:
$(dT/dt)^*$ = 20°C - 200°C/Ma

Latent heat of melting:
L = 100-500 J/g

Heat Capacity:
C_p = 1.0 kJ kg^{-1} K^{-1}

Tortuosity: (Rubie & Brearley 1990)
τ = 0.6

---

content are considered. It is assumed that the activation energies for diffusion of Si and Al are the same. D_0 in (3.8) is calculated from values of the activation energy and the diffusion coefficients at 757°C.

The change in concentration (Δc_i^L) across the L and LM layers is estimated from either (3.9) or (3.10). The parameter β_i in (3.10) is taken as 0.1 in order to give a strongly non-linear trend, and the parameters k_i and α_i, in (3.9) and (3.10) respectively, are then determined from the experimental data obtained at 757°C. As an example, the variations of Δc_{Si}^L with temperature across the LM layer, given by the two models (3.9 & 10), are shown in Figure 3.6.

The rate of heating during prograde regional metamorphism, in the absence of melting, is likely to be of the order of 20°C Ma^{-1} (England & Thompson

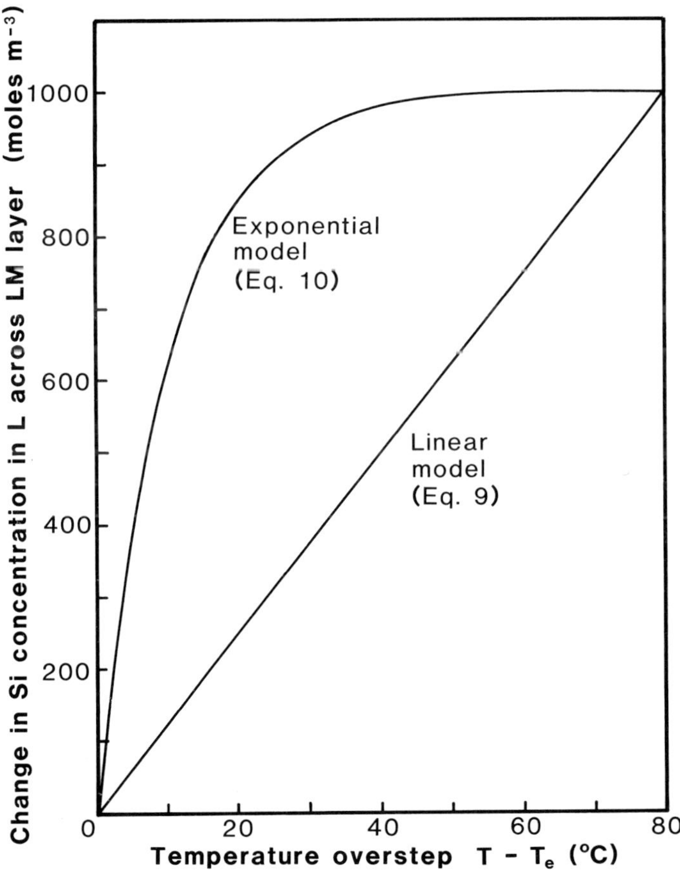

Figure 3.6 Two empirical models for the change of concentration of Si in the liquid across the LM layer ($\Delta c_{Si}^{L/LM}$), as a function of the deviation of temperature from the equilibrium melting temperature. In the exponential model (Eqn 3.10), a value of $\beta_{Si} = 0.1$ has been used in order to give a trend substantially different from the linear model (Eqn 3.9). In both models, ($\Delta_{Si}^{L/LM}$) is constrained by the experimental data at $T - T_e = 80$°C (Rubie & Brearley 1990).

1984, De Yoreo *et al.* 1989). During contact metamorphism, the heating rate will be considerably faster; e.g. $200^\circ C\,Ma^{-1}$. Values used for the latent heat of melting (L) and heat capacity (C_P) are similar to those employed in other studies (e.g. De Yoreo *et al.* 1989).

The time step for each calculation (Δt) must be sufficiently small for the approximation given by (3.14) to be accurate. Values used have been small enough so that doubling the time step changes the calculated values of temperature and melt fraction by $<1\%$. Because there is generally a very large reduction in the reaction rate as the temperature decreases during melting (see below), it was found necessary occasionally to increase the value of Δt during a sequence of calculations. For an initial temperature overstep of 100°C, Δt was typically increased from 1 s at the start of melting to >1 year as the temperature approached T_e.

3.4.4 Results

The same type of reaction behaviour occurs for the whole range of model parameters listed in Table 3.1. Typical results, for a rock initially consisting of 40% Ms + 60% Qtz, with a muscovite grain size of $50 \times 500 \times 500\ \mu m$, and with Δc_i^L given by (3.9), are shown in Figure 3.7. When melting starts at a temperature in excess of T_e, the initial rate of melting is extremely rapid. For example, with an initial temperature overstep of 100°C, the melt fraction can increase from 0 to 0.5 in ~ 1 year (Figs 3.7a & b). During this period of rapid melting, the temperature falls rapidly (Figs 3.7c & d), due to the consumption of the latent heat of melting, thus causing a progressive decrease in the reaction rate. The rate of melting during this stage is dependent on the kinetic parameters and to a lesser extent on the latent heat of melting, but is independent of the rate of heat flow from the external source. The reaction rate is therefore *kinetic controlled.* Provided that the melting reaction does not reach completion first, the temperature drops to within a fraction of a degree of the equilibrium temperature (T_e) for the melting reaction, at which point the heat consumed by melting becomes balanced by the heat introduced from the external source. At this stage the reaction becomes *heat flow controlled*, and the temperature remains buffered within a fraction of a degree of T_e until melting is complete (Figs 3.7a & c). The time to reach the completion of melting then depends on the rate of heat supply from the external source (which determines $(dT/dt)^*$) and the latent heat of melting, but is independent of the reaction kinetics. Typically, this time is of the order of 1–10 Ma. After the completion of melting, the temperature starts to increase once more, at a rate given by $(dT/dt)^*$ (Fig. 3.7c).

The effects on reaction behaviour, during the kinetic control stage, of varying the various model parameters within the ranges given in Table 3.1, are shown in Figure 3.8 as a series of plots of melt fraction against time. Corresponding temperature–time curves have similar but inverted forms (cf.

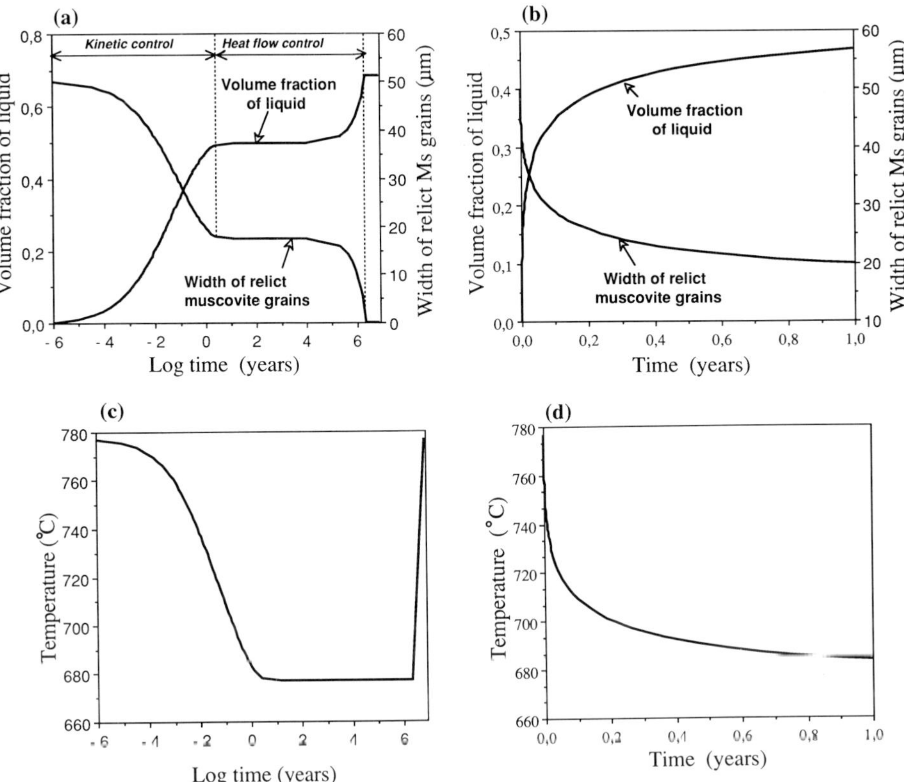

Figure 3.7 Typical results of the modelling plotted as melt fraction, width of relict muscovite grains, and temperature as a function of time. (a) and (c) show the complete reaction history, whereas (b) and (d) show the early kinetic control stage. Values of model parameters used to obtain these results are as follows: initial Ms grain size = 50 × 500 × 500 μm; initial Ms content $(X_{Ms})_0 = 0.4$; initial $(T - T_e) = 100°C$; activation energy = 235 kJ mol^{-1}; latent heat of melting = 200 kJ kg^{-1}; $(dT/dt)^* = 20°C/Ma$. The linear concentration difference model is used (Eqn 3.9, Fig. 3.6). The initial volume fraction of muscovite of 0.4 gives a melt fraction at the completion of melting of 0.68.

Figs 3.7b & d). Changing the model parameters can significantly affect the melt fraction at which melting becomes heat flow controlled, and can also determine whether the melting reaction reaches completion before the heat flow control stage is reached. Although the rate of melting is also affected, this always remains extremely rapid in terms of a geological timescale, provided that the reaction is kinetic controlled. The time, after the initiation of melting, at which heat flow control starts is likewise always short.

The initial *overstep of the equilibrium melting temperature* and the *latent heat of fusion* are the two important parameters which determine the melt fraction when the reaction becomes heat flow controlled, and therefore determine the quantity of liquid that forms during rapid melting (Figs 3.8a & b).

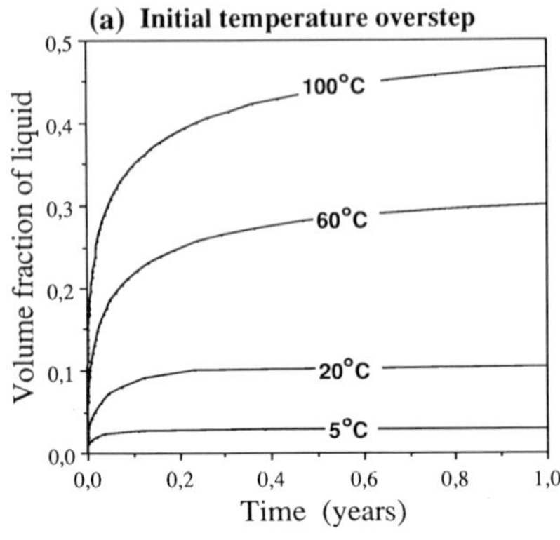
(a) Initial temperature overstep
Volume fraction of liquid
Time (years)
100°C
60°C
20°C
5°C

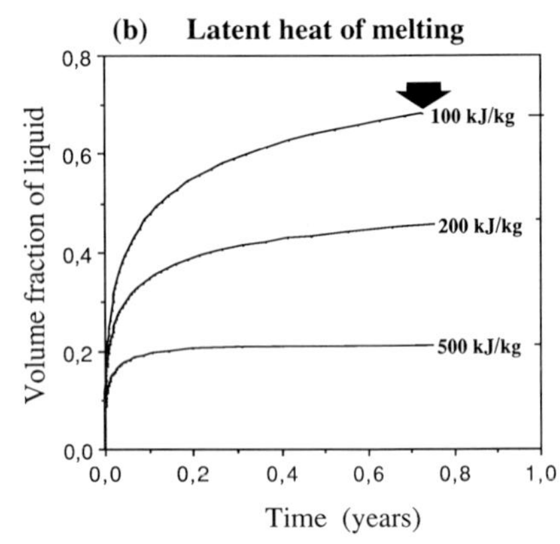
(b) Latent heat of melting
Volume fraction of liquid
Time (years)
100 kJ/kg
200 kJ/kg
500 kJ/kg

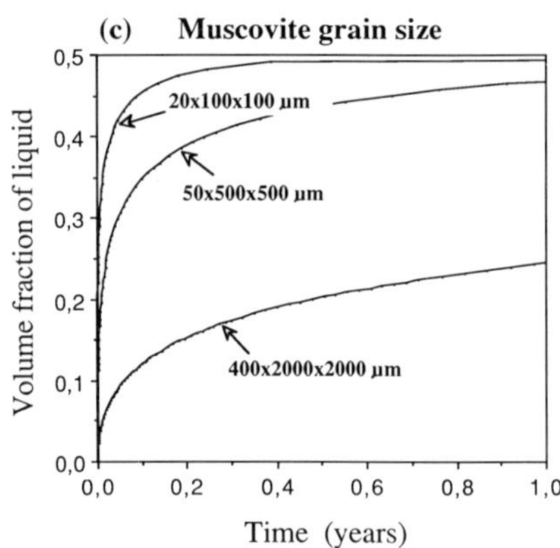
(c) Muscovite grain size
Volume fraction of liquid
Time (years)
20x100x100 µm
50x500x500 µm
400x2000x2000 µm

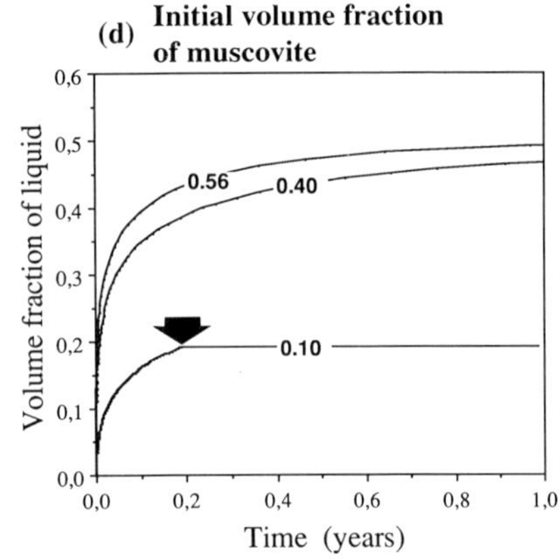
(d) Initial volume fraction of muscovite
Volume fraction of liquid
Time (years)
0.56
0.40
0.10

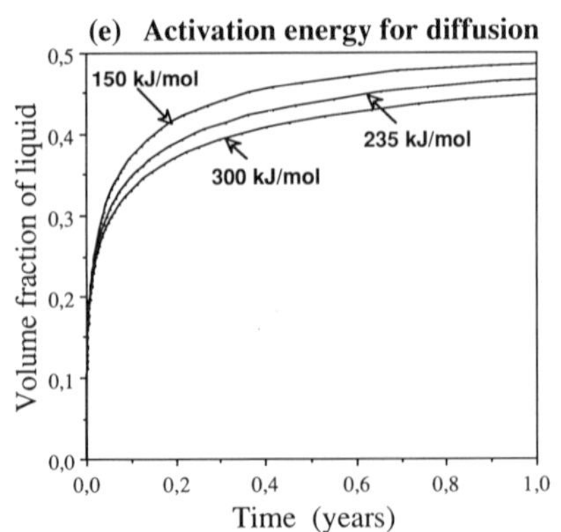
(e) Activation energy for diffusion
Volume fraction of liquid
Time (years)
150 kJ/mol
235 kJ/mol
300 kJ/mol

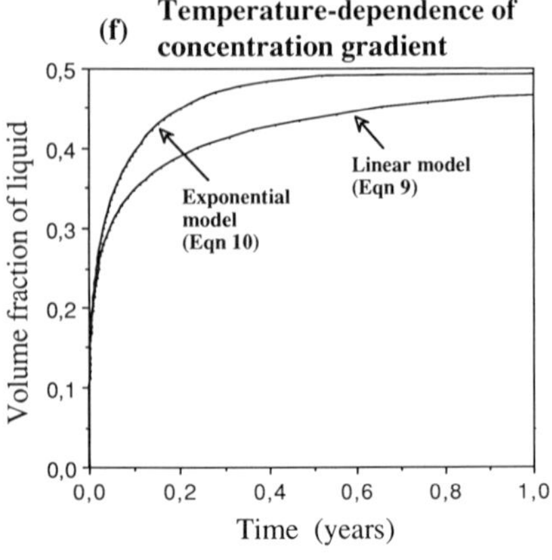
(f) Temperature-dependence of concentration gradient
Volume fraction of liquid
Time (years)
Linear model
(Eqn 9)
Exponential
model
(Eqn 10)

With a large temperature overstep and a small latent heat of melting, large quantities of liquid form rapidly, and the reaction may reach completion before the heat-flow-control stage is reached (Fig. 3.8b).

Increasing the *grain size*, while keeping the initial muscovite content constant, reduces the melting rate, when kinetic controlled, because: (a) necessary diffusion distances become larger for a given extent of reaction; and (b) the number of muscovite grains per unit volume of rock decreases (3.11) so that the area of muscovite–quartz grain boundary per unit volume also decreases (Fig. 3.8c).

Varying the initial *muscovite content* of the rock, $(X_{Ms})_0$, has two effects:

(a) An increase in $(X_{Ms})_0$ with constant grain size increases the area of muscovite–quartz grain boundary per unit volume of rock, and produces a faster initial reaction rate (Fig. 3.8d). However, as a consequence, the temperature initially drops more quickly, thus tending to counteract this effect to some extent.
(b) Below a critical value of $(X_{Ms})_0$, the melting reaction reaches completion before the temperature drops to T_e, so the reaction rate never becomes heat flow controlled (Fig. 3.8d).

Varying the *activation energy* for diffusion has only a minor effect (Fig. 3.8e), as does varying the *temperature-dependence of the concentration gradients* (Δc_i^L) (Fig. 3.8f). Also, varying the *rate of heat input* from the external source, within the limits given in Table 3.1, only affects the rate of melting during the heat flow control stage and has no effect during the kinetic control stage.

The effects of changing the variables pressure and H_2O content of the liquid have not been investigated due to a lack of relevant data. The H_2O content of water-saturated granitic liquids increases with pressure so that at 10 kbar, for example, diffusion rates may be considerably faster than in the experiments of Rubie & Brearley (1988, 1990). Conversely, in melts which are H_2O-undersaturated, diffusion rates may be slower than in the experiments. However, even if kinetic-controlled melting rates are changed by one or two orders of magnitude at these different conditions, melting will still remain very rapid on a geological timescale.

Figure 3.8 The effects of varying the values of the model parameters on calculated rates of kinetic-controlled melting (cf. Table 3.1). Unless otherwise stated, the values of the parameters are the same as those listed for Figure 3.7. The maximum melt fraction (at completion of reaction) is usually 0.68 (with $(X_{Ms})_0 = 0.4$); in (d) the maximum melt fractions are 0.19 with $(X_{Ms})_0 = 0.1$ and 0.91 with $(X_{Ms})_0 = 0.56$. With a small latent heat of melting and/or a small muscovite content, the melting reaction can reach completion before the heat-flow control stage is reached, at the points indicated by the bold arrows in (b) and (d); otherwise, the curves level off when the reaction becomes heat-flow controlled.

3.5 Discussion

The results of the modelling show that if melting initiates at a temperature which significantly oversteps the equilibrium melting temperature, the rate of melting can be extremely rapid. For example, with an initial overstep of 100°C, 20–50% of a rock could melt within 1 year. The uncertainties in the parameters in the numerical model (Table 3.1) do not significantly affect this conclusion. Such rapid rates of melting in the crust and mantle are likely to have major effects on the geochemistry of magmas, mechanisms of melt segregation and migration, and the rheology of the region of partial melting.

During rapid melting, equilibrium major and trace element partitioning (e.g. Fe–Mg) between the liquid and solid residue cannot occur because intracrystalline diffusion distances will be negligible on the timescale under discussion (see Prinzhofer & Allègre 1985). Thus the geochemistry of rapidly formed magmas will not give realistic information on the nature of the source region if equilibrium melting is assumed. This will be particularly the case if the melts are extracted from the source region very soon after their formation.

In general, melting reactions involve a positive volume increase at the site of melting. This must even be the case for H_2O-saturated melting, provided that the necessary fluid is introduced from an external source during melting. Because of the volume increase, rapid melting will result in the development of a high melt pressure which should favour hydraulic fracturing and cataclastic deformation. Such processes may be important for the formation of layered migmatites (Mawer *et al.* 1988), and could lead to the rapid extraction of melts along propagating fractures, especially in the case of relatively low-viscosity magmas (Spence & Turcotte 1985, Turcotte *et al.* 1987).

The model discussed above does not take into account rates of supply of hydrous fluid during vapour-present melting. When disequilibrium melting initiates in response to the infiltration of H_2O into dry rocks, the rate of supply of fluid, rather than reaction kinetics, could be the rate-controlling step. In this case, melting rates may be significantly slower than those calculated above (Figs 3.7 & 8), but are very difficult to estimate quantitatively. Fluid infiltration could be a rapid process if melting is accompanied by cataclastic deformation involving dilatancy. The operation of such a process is suggested by experimental deformation studies of partially melted rocks as well as field studies of migmatites (van der Molen & Paterson 1979, Dell'Angelo & Tullis 1988, Mawer *et al.* 1988, McLellan 1988).

The uncertainty in disequilibrium melting rates is less for vapour-absent melting reactions which do not depend on the supply of fluid. In mantle peridotites for example, disequilibrium melting may occur as the result of a lack of mutual contact between reactant mineral grains, as discussed above. Although the kinetics of the melting process will differ in detail from the model discussed above, rates of disequilibrium melting are again likely to be very rapid, and probably of the same order of magnitude as the rates shown in Figures 3.7 & 8.

Acknowledgements

This work was initiated while D.C.R. held a research fellowship at the Department of Geology, University of Manchester, with financial support from the Natural Environment Research Council (Grant GR3/6409). A.J.B. was funded by NASA grant NAG 9–30 to Klaus Keil (Principal Investigator). We thank G. E. Lloyd for assistance with backscattered electron imaging (Fig. 3.3). Analytical electron microscopy was carried out in the Electron Microbeam Analysis Facility, Department of Geology and Institute of Meteoritics, University of New Mexico.

References

Baker, D. R. 1988. Chemical diffusion in intermediate to silicic melts: effects of pressure and fluoridation. *EOS, Transactions, American Geophysical Union* **69**, 511.

Beach, A. 1980. Retrogressive metamorphic processes in shear zones with special reference to the Lewisian complex. *Journal of Structural Geology* **2**, 257–63.

Bottinga, Y., D. Weill & P. Richet 1982. Density calculations for silicate liquids. I. Revised method for aluminosilicate compositions. *Geochimica et Cosmochimica Acta* **46**, 909–19.

Brearley, A. J. & D. C. Rubie 1990. Effects of H_2O on the disequilibrium breakdown of muscovite + quartz. Submitted to *Journal of Petrology*.

Burnham, C. W. & N. F. Davis 1971. The role of H_2O in silicate melts I. *P–V–T* relations in the system $NaAlSi_3O_8$–H_2O to 10 kilobars and 1000°C. *American Journal of Science* **270**, 54–79.

Büsch, W., G. Schneider & K. R. Mehnert 1974. Initial melting at grain boundaries. Part II: Melting in rocks of granodioritic, quartzdioritic and tonalitic composition. *Neues Jahrbuch für Mineralogie, Monatshefte*, 345–70.

Clemens, J. D. 1984. Water contents of silicic to intermediate magmas. *Lithos* **17**, 273–87.

Clemens, J. D. & D. Vielzeuf 1987. Constraints on melting and magma production in the crust. *Earth and Planetary Science Letters* **86**, 287–306.

Dell'Angello, L. N. & J. Tullis 1988. Experimental deformation of partially melted granitic aggregates. *Journal of Metamorphic Geology* **6**, 495–515.

De Yoreo, J. J., D. R. Lux & C. V. Guidotti 1989. The role of crustal anatexis and magma migration in the thermal evolution of regions of thickened continental crust. In *Evolution of metamorphic belts*, J. S. Daly, R. A. Cliff & B. W. D. Yardley (eds). Geological Society of London Special Publication, in press.

England, P. C. & A. B. Thompson 1984. Pressure–temperature–time paths of regional metamorphism. I. Heat transfer during the evolution of regions of thickened continental crust. *Journal of Petrology* **25**, 894–928.

England, P. C. & A. B. Thompson 1986. Some thermal and tectonic models for crustal melting in continental collision zones. In *Collision tectonics*, M. P. Coward & A. C. Ries (eds), 83–94. Geological Society of London Special Publication 19.

Fisher, G. W. 1973. Nonequilibrium thermodynamics as a model for diffusion-controlled metamorphic processes. *American Journal of Science* **273**, 897–924.
Fisher, G. W. 1978. Rate laws in metamorphism. *Geochimica et Cosmochimica Acta* **42**, 1035–50.

Grant, J. A. 1985. Phase equilibria in partial melting of pelitic rocks. In *Migmatites*, J. R. Ashworth (ed.), 86–144. Glasgow: Blackie.

Henderson, P., J. Nolan & R. K. Lowry 1986. Cation diffusion in natural silicate melts. *Materials Science Forum* **7**, 257–66.

Joesten, R. 1977. Evolution of mineral assemblage zoning in diffusion metasomatism. *Geochimica et Cosmochimica Acta* **41**, 649–70.
Johannes, W. 1985. The significance of experimental studies for the formation of migmatites. In *Migmatites*, J. R. Ashworth (ed.), 36–85. Glasgow: Blackie.
Jurewicz, S. R. & E. B. Watson 1985. The distribution of partial melt in a granitic system: the application of liquid phase sintering theory. *Geochimica et Cosmochimica Acta* **49**, 1109–21.

Kerrich, R., T. E. La Tour & L. Willmore 1984. Fluid participation in deep fault zones: evidence from geological, geochemical and $^{18}O/^{16}O$ relations. *Journal of Geophysical Research* **89**, 4331–43.

Lasaga, A. C. 1986. Metamorphic reaction rate laws and development of isograds. *Mineralogical Magazine* **50**, 359–73.

Mawer, C. K., D. C. Rubie & A. J. Brearley 1988. A model for rapid melting in crustal shear zones: Implications for mechanisms of melt migration. *EOS, Transactions, American Geophysical Union* **69**, 1411.
McKenzie, D. 1985. The extraction of magma from the crust and mantle. *Earth and Planetary Science Letters* **74**, 81–91.
McLellan, E. L. 1988. Migmatite structures in the Central Gneiss Complex, Boca de Quadra, Alaska. *Journal of Metamorphic Geology* **6**, 517–42.
Mehnert, K. R., W. Büsch & G. Schneider 1973. Initial melting at grain boundaries of quartz and feldspar in gneisses and granulites. *Neues Jahrbuch für Mineralogie, Monatshefte* 165–83.
van der Molen, I. & M. S. Paterson 1979. Experimental deformation of partially-melted granite. *Contributions to Mineralogy and Petrology* **70**, 299–318.

Prinzhofer, A. & C. J. Allègre 1985. Residual peridotites and the mechanisms of partial melting. *Earth and Planetary Science Letters* **74**, 251–65.

Ridley, J. 1985. The effect of reaction enthalpy on the progress of a metamorphic reaction. In *Metamorphic reactions: kinetics, textures and deformation. Advances in physical geochemistry*, Vol. 4, A. B. Thompson and D. C. Rubie (eds), 80–97. New York: Springer-Verlag.
Robie, R. A., B. S. Hemingway & J. R. Fisher 1978. Thermodynamic properties of minerals and related substances at 298.15 K and 1 bar (10^5 pascals) pressure and at higher temperatures. *U.S. Geological Survey, Bulletin* 1452.
Rubie, D. C. 1986. The catalysis of mineral reactions by water and restrictions on the presence of aqueous fluid during metamorphism. *Mineralogical Magazine* **50**, 399–415.
Rubie, D. C. 1989. Role of kinetics in the formation and preservation of eclogites. In *Eclogite facies rocks*, D. A. Carswell (ed.). Glasgow: Blackie (in press).
Rubie, D. C. & A. J. Brearley 1987. Metastable melting during the breakdown of muscovite + quartz at 1 kbar. *Bulletin de Minéralogie* **110**, 533–49.
Rubie, D. C. & A. J. Brearley 1988. The application of non-equilibrium thermodynamics to

partial melting and rates of multicomponent diffusion in granitic melts. *EOS, Transactions, American Geophysical Union* **69**, 1475–6.
Rubie, D. C. & A. J. Brearley 1990. Kinetics of partial melting of muscovite + quartz and rates of multicomponent diffusion in H_2O-saturated granitic liquid at 1 kbar. Submitted to *Contributions to Mineralogy and Petrology*.
Rumble, D., III, J. M. Ferry, T. C. Hoering & A. J. Boucot 1982. Fluid flow during metamorphism at the Beaver Brook fossil locality, New Hampshire. *American Journal of Science* **282**, 886–919.

Smith, G. D. 1985. *Numerical solution of partial differential equations: finite difference methods*, 3rd edn. Oxford: Oxford University Press.
Spence, D. A. & D. L. Turcotte 1985. Magma driven propagation of cracks. *Journal of Geophysical Research* **90**, 575–80.

Thompson, A. B. 1982. Dehydration melting of pelitic rocks and the generation of H_2O-undersaturated granitic liquids. *American Journal of Science* **282**, 1567–95.
Thompson, A. B. 1983. Fluid-absent metamorphism. *Journal of the Geological Society of London* **140**, 533–47.
Tsuchiyama, A. 1985a. Partial melting kinetics of plagioclase–diopside pairs. *Contributions to Mineralogy and Petrology* **91**, 12–23.
Tsuchiyama, A. 1985b. Dissolution kinetics of plagioclase in the melt of the system diopside–albite–anorthite, and the origin of dusty plagioclase in andesites. *Contributions to Mineralogy and Petrology* **89**, 1–16.
Tsuchiyama, A. 1986. Melting and dissolution kinetics: application to partial melting and dissolution of xenoliths. *Journal of Geophysical Research* **91**, 9395–406.
Tsuchiyama, A. & E. Takahashi 1983. Melting kinetics of a plagioclase feldspar. *Contributions to Mineralogy and Petrology* **84**, 345–54.
Turcotte, D. L., S. H. Emerman & D. A. Spence 1987. Mechanics of dyke injection. In *Mafic dyke swarms*, H. C. Halls & W. F. Fahrig (eds), 25–9. Geological Association of Canada Special Paper 34.

Vielzeuf, D. & J. R. Holloway 1988. Experimental determination of the fluid-absent melting relations in the pelitic system. *Contributions to Mineralogy and Petrology* **98**, 257–76.

Waff, H. S. & J. R. Bulau 1979. Equilibrium fluid distribution in an ultramafic partial melt under hydrostatic stress conditions. *Journal of Geophysical Research* **84**, 6109–14.
Walther, J. V. & B. J. Wood 1984. Rate and mechanism in prograde metamorphism. *Contributions to Mineralogy and Petrology* **88**, 246–59.
Wayte, G. J., R. H. Worden, D. C. Rubie & G. T. R. Droop 1989. A TEM study of disequilibrium plagioclase breakdown at high pressure: the role of infiltrating fluid. *Contributions to Mineralogy and Petrology* **101**, 426–37.
Wickham, S. M. 1987. The segregation and enplacement of granitic magmas. *Journal of the Geological Society of London* **144**, 281–97.
Wickham, S. M. & H. P. Taylor 1985. Stable isotopic evidence for large scale sea water infiltration in a regional metamorphic terrane; the Trois Seigneurs Massif, Pyrenees, France. *Contributions to Mineralogy and Petrology* **91**, 122–37.
Wyllie, P. J. 1977. Crustal anatexis: an experimental review. *Tectonophysics* **43**, 41–71.

Yardley, B. W. D. & E. Baltatzis 1985. Retrogression of staurolite schists and the sources of infiltrating fluids during metamorphism. *Contributions to Mineralogy and Petrology* **89**, 59–68.

Zen, E-An 1988. Thermal modelling of stepwise anatexis in a thrust thickened sialic crust. *Transactions of the Royal Society of Edinburgh: Earth Sciences* **79**, 223–36.

Appendix 1

Using phase compositions given in Reaction R-8, the mass balance, conservation and diffusion equations (3.3, 4 & 6) are written for the layer sequence Ms | LM | L | Q as follows (cf. Joesten 1977):

Mass balance equations (3.3)

At the Ms | LM layer contact:

$$3.0\nu_{Ms}^{Ms-LM} + 7.43\nu_{L}^{Ms-LM} + 1.0\nu_{M}^{Ms-LM} + \nu_{SiO_2}^{Ms-LM} = 0 \quad (3.3.1)$$

$$3.0\nu_{Ms}^{Ms-LM} + 1.71\nu_{L}^{Ms-LM} + 3.0\nu_{M}^{Ms-LM} + \nu_{AlO_{3/2}}^{Ms-LM} = 0 \quad (3.3.2)$$

$$1.0\nu_{Ms}^{Ms-LM} + 1.0\nu_{L}^{Ms-LM} + \nu_{KO_{1/2}}^{Ms-LM} = 0 \quad (3.3.3)$$

At the LM | L layer contact:

$$7.43\nu_{L}^{LM-L} + 1.0\nu_{M}^{LM-L} + \nu_{SiO_2}^{LM-L} = 0 \quad (3.3.4)$$

$$1.71\nu_{L}^{LM-L} + 3.0\nu_{M}^{LM-L} + \nu_{AlO_{3/2}}^{LM-L} = 0 \quad (3.3.5)$$

$$1.0\nu_{L}^{LM-L} + \nu_{KO_{1/2}}^{LM-L} = 0 \quad (3.3.6)$$

At the L | Q layer contact:

$$7.43\nu_{L}^{L-Q} + \nu_{Q}^{L-Q} + \nu_{SiO_2}^{L-Q} = 0 \quad (3.3.7)$$

$$1.71\nu_{L}^{L-Q} + \nu_{AlO_{3/2}}^{L-Q} = 0 \quad (3.3.8)$$

$$1.0\nu_{L}^{L-Q} + \nu_{KO_{1/2}}^{L-Q} = 0 \quad (3.3.9)$$

Conservation equations (3.4)

At the Ms | LM layer contact (assuming that diffusion in Ms is negligible):

$$J_{SiO_2}^{LM} - \nu_{SiO_2}^{Ms-LM} = 0 \quad (3.4.1)$$

$$J_{AlO_{3/2}}^{LM} - \nu_{AlO_{3/2}}^{Ms-LM} = 0 \quad (3.4.2)$$

$$J_{KO_{1/2}}^{LM} - \nu_{KO_{1/2}}^{Ms-LM} = 0 \quad (3.4.3)$$

At the LM | L layer contact:

$$J_{SiO_2}^{LM} - J_{SiO_2}^{L} + \nu_{SiO_2}^{LM-L} = 0 \quad (3.4.4)$$

$$J_{AlO_{3/2}}^{LM} - J_{AlO_{3/2}}^{L} + \nu_{AlO_{3/2}}^{LM-L} = 0 \quad (3.4.5)$$

$$J_{KO_{1/2}}^{LM} - J_{KO_{1/2}}^{L} + \nu_{KO_{1/2}}^{LM-L} = 0 \quad (3.4.6)$$

At the L | Q layer contact (assuming negligible diffusion in quartz):

$$J^{L}_{SiO_2} + \nu^{L-Q}_{SiO_2} = 0 \tag{3.4.7}$$

$$J^{L}_{AlO_{3/2}} + \nu^{L-Q}_{AlO_{3/2}} = 0 \tag{3.4.8}$$

$$J^{L}_{KO_{1/2}} + \nu^{L-Q}_{KO_{1/2}} = 0 \tag{3.4.9}$$

The flux terms (J) can then be eliminated by substituting (3.4.1–3) and (3.4.7–9) into (3.4.4–6), thereby reducing the number of conservation equations to three:

$$\nu^{Ms-LM}_{SiO_2} + \nu^{LM-L}_{SiO_2} + \nu^{L-Q}_{SiO_2} = 0 \tag{3.4.10}$$

$$\nu^{Ms-LM}_{AlO_{3/2}} + \nu^{LM-L}_{AlO_{3/2}} + \nu^{L-Q}_{AlO_{3/2}} = 0 \tag{3.4.11}$$

$$\nu^{Ms-LM}_{KO_{1/2}} + \nu^{LM-L}_{KO_{1/2}} + \nu^{L-Q}_{KO_{1/2}} = 0 \tag{3.4.12}$$

Diffusion equations (3.6)

The three components considered here do not diffuse independently because of mass balance constraints and therefore diffusion equations are only required for two components. Si and Al are chosen because the concentration gradient data for these components are better defined than for K (Rubie & Brearley 1990).

In the LM layer:

$$J^{LM}_{SiO_2} = -D^{L}_{Si}\Phi^{LM}_{L}\tau(dc^{L}_{SiO_2}/dx)^{LM} \tag{3.6.1}$$

$$J^{LM}_{AlO_{3/2}} = -D^{L}_{Al}\Phi^{LM}_{L}\tau(dc^{L}_{AlO_{3/2}}/dx)^{LM} \tag{3.6.2}$$

In the L layer (in which $\Phi = 1$ and $\tau = 1$):

$$J^{L}_{SiO_2} = -D^{L}_{Si}(dc^{L}_{SiO_2}/dx)^{L} \tag{3.6.3}$$

$$J^{L}_{AlO_{3/2}} = -D^{L}_{Al}(dc^{L}_{AlO_{3/2}}/dx)^{L} \tag{3.6.4}$$

Eliminating the flux terms in (3.6.1–4), using (3.4.1–3) and (3.4.7–9) gives:

$$\nu^{Ms-LM}_{SiO_2} = -D^{L}_{Si}\Phi^{LM}_{L}\tau(dc^{L}_{SiO_2}/dx)^{LM} \tag{3.6.5}$$

$$\nu^{Ms-LM}_{AlO_{3/2}} = -D^{L}_{Al}\Phi^{LM}_{L}\tau(dc^{L}_{AlO_{3/2}}/dx)^{LM} \tag{3.6.6}$$

$$\nu^{L-Q}_{SiO_2} = D^{L}_{Si}(dc^{L}_{SiO_2}/dx)^{L} \tag{3.6.7}$$

$$\nu^{L-Q}_{AlO_{3/2}} = D^{L}_{Al}(dc^{L}_{AlO_{3/2}}/dx)^{L} \tag{3.6.8}$$

The resulting 16 equations (3.3.1–9, 4.10–12 & 6.5–8) can be solved for the 16 unknown stoichiometric reaction coefficients (ν) for specified values of the diffusion coefficients, concentration gradients and tortuosity (τ). The term Φ_{L}^{LM} is given by the stoichiometric coefficients of liquid and mullite in the reaction at the Ms | LM boundary, together with molar volume data, and can be calculated by solving the equations iteratively.

CHAPTER FOUR

Formation and composition of H_2O-undersaturated granitic melts

W. Johannes & F. Holtz

4.1 Introduction

Granites are a very common component of the continental crust, and granitic melts are the most important medium for the transport of silicon, alkali metals and some associated, less abundant, elements from lower to higher crustal levels. Granites have been the subject of numerous petrographic and experimental investigations. The first systematic experimental research was performed in the synthetic haplogranitic system Qz–Ab–Or–H_2O (Tuttle & Bowen 1958, Luth, Jahns & Tuttle 1964). Other investigations with natural rocks as the starting material were performed by Winkler and co-workers (summarized in Winkler 1979). The results of these H_2O-saturated experiments emphasized that granitic magmas can exist at relatively low temperatures within the continental crust, and that many crystalline rocks can be the source of granitic melts.

Comparison of results obtained from field studies and experimental investigations clearly shows that granitic melts are usually not H_2O-saturated but are undersaturated (e.g. Conrad *et al.* 1988). The water content of a natural granitic melt may change considerably during the course of its crystallization. The amount of water present in the melt may be very low at great depth and high temperature, and may increase continuously during ascent and crystallization of the magma. The very last melts should be H_2O-saturated or almost saturated if there are no other components (e.g. CO_2 or N_2) reducing the H_2O activity in the system.

Phase relationships occurring in natural rocks and mixtures of natural and synthetic minerals at hypersolidus temperatures, different pressures and for variable water contents have been investigated (e.g. Huang & Wyllie 1975, 1986; Stern & Wyllie 1981 – for high pressures up to 35 kbar; Whitney 1975, Green 1976, Clemens & Wall 1981, Naney 1983 – for lower pressures). Combined with field and geochemical observations (e.g. Presnall & Bateman 1973, Bateman & Chappell 1979, Chappell, White & Wyborn 1987, White & Chappell 1988) these studies provide helpful information concerning the genesis, source-rock composition, ascent, crystallization and geochemical evolution of granitic magmas.

In addition, dehydration melting (also called fluid-absent or vapour-absent melting) has been the subject of several experimental investigations and theoretical considerations (Thompson 1982, Grant 1985, Clemens & Vielzeuf 1987, Vielzeuf & Holloway 1988, Le Breton & Thompson 1988, Whitney 1988). At the present time, published experimental, petrographic and geochemical investigations are sufficient to constrain ideas on the petrogenesis of autochthonous or anatectic granitoids. The mineral assemblages of the country rocks associated with *in situ* migmatites and granites are useful in defining the limits of the *P*, *T* conditions of their formation. Knowledge about intrusive granites is much less complete. In this case the pressure–temperature régime in which melting occurs, the degree of partial melting, and the mechanisms and extent of segregation of partial melts are very often not clearly defined. Estimates of the amount and composition of residual minerals and the amount of water present in the magmas show a wide range of uncertainty. In the same way, the effects of fractional crystallization and of changing H_2O activity levels in granitic magmas during their ascent towards higher crustal levels are not well constrained.

In recent years an increasingly popular idea has been that many intrusive granites may come from the lower crust, and that there is a genetic relationship between granites and granulites (Fyfe 1973; Lamb & Valley 1984, 1988). In order to elucidate their relationship, and to contribute to the constraining of magma derivation and ascent, we will discuss the following subjects:

(a) the minimum levels of H_2O activity at which granitic melts can be generated in the lower crust;
(b) the composition of H_2O-undersaturated partial melts in the system Qz–Ab–Or–H_2O–CO_2;
(c) the An content of partial melts;
(d) the amounts of mafic components (Mg, Fe, Al) dissolved in granitic melts;
(e) the liquidus temperatures of H_2O-undersaturated granitic melts;
(f) the amount of partial melt generated at given *P*, *T* and water content conditions.

Finally, we discuss the development of these magmas and the conditions for magma ascent.

4.2 The initiation of melting at lower-crustal *P* and *T*, and at a reduced level of H_2O activity

Geobarometric and geothermometric evaluations in granulitic rocks are reported in numerous papers (e.g. Johnson *et al.* 1983, Schenk 1984, Barbey *et al.* 1986, Bhattacharya & Sen 1986, Schreurs & Westra 1986, Lamb & Valley 1988). The reconstructed petrogenetic temperatures and pressures range up to 1100°C and 20 kbar. Any combination of medium to high temperatures with deep-crustal pressures (or pressures observed in subduction zones) seems to be possible. In Figure 4.1 we disregard extreme possibilities and indicate the *P–T* field estimated for most of the granulites. The appearance of orthopyroxene and other anhydrous minerals due to the breakdown of biotite or hornblende within this *P–T* box requires H_2O-undersaturation (H_2O activity $a_{H_2O} < 1$; Bohlen *et al.* 1983, Lamb & Valley 1988). Since a genetic relationship between granulites and granites is now often postulated, the main question is whether the reduced activity of water in granulitic rocks is not too low to produce partial melting and generate granitic magmas. In other words: Can granitic melts be formed at the same P–T–a_{H_2O} conditions as granulites? A positive answer to this question leads to a further one: Are granulites are by-products of partial melting and do they represent a residual assemblage, with many granitoids corresponding to the extracted or partially extracted melts? In order to answer

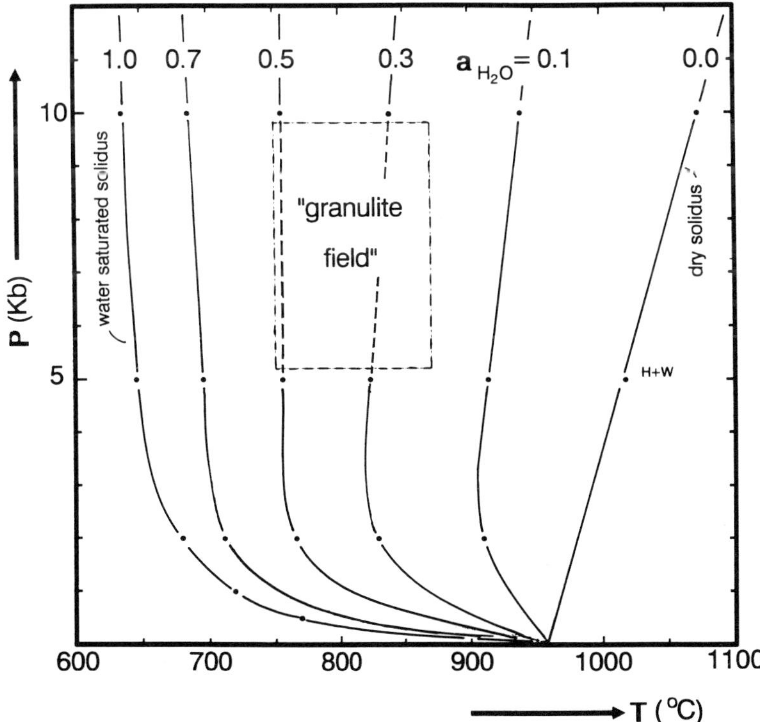

Figure 4.1 Solidus curves of the system Qz–Ab–Or–H_2O–CO_2. Each solidus curve is for one specified a_{H_2O}. Data are from Johannes (1984) for $a_{H_2O} = 1$, Huang & Wyllie (1975) for $a_{H_2O} = 0$, and Ebadi & Johannes (unpublished) for a_{H_2O} between 0 and 1.

these questions we require initiation of melting data determined at reduced levels of H_2O activity.

Initiation of melting curves for H_2O-undersaturated conditions are shown in Figure 4.1, from Ebadi & Johannes (unpublished data). They are determined experimentally at NNO buffer conditions and are valid for eutectic or minimum compositions. Reduced levels of H_2O activity were obtained by using a mixture of H_2O and CO_2. The amount of gas in the experimental run was much higher than the amount of melt generated.

The initiation of melting curves of Figure 4.1 are presented for different levels of H_2O activity. The fluid phase was present in excess during the runs, and the activities were calculated using the modified Redlich–Kwong equation of Kerrick & Jacobs (1981). It should be noticed that the solidus *P*- and *T*-values are not solely constrained by the water dissolved in the melt but also by some dissolved CO_2. The amount of dissolved CO_2 is small according to results obtained in feldspar melts (Boettcher *et al.* 1987, Stolper *et al.* 1987). In a quenched melt obtained at $P = 5$ kbar, $T = 800°C$ for a_{H_2O} around 0.4, we observed approximately 0.12 wt% CO_2. A simple estimation based on CO_2 solubilities in albitic melts obtained by Boettcher *et al.* (1987) shows that the amount of dissolved CO_2 in our experiments could reduce the solidus temperature by 5–10°C. More information is needed to define the effect of CO_2 on the solidus temperatures exactly, but this is not yet available. The influence of CO_2 on the solidus temperature, in addition to that of water, increases with increasing pressure and the CO_2 content of the fluid, X_{CO_2}.

In order to form granitic melts within the *P–T* field of granulites represented on Figure 4.1, a_{H_2O} can be approximately as low as 0.25. For $a_{H_2O} = 0.25$ the solidus is below 850°C. Temperatures of 850°C and above are sufficient to melt a rock with a granitic minimum melt composition completely if there is enough vapour with such a value of a_{H_2O}. The minimum level of H_2O activity necessary to melt a rock with a minimum melt composition completely increases with decreasing temperature. This minimum level of H_2O activity is slightly greater than 0.5 at 750°C.

4.3 Composition of H_2O-undersaturated melts in the system Qz–Ab–Or–H_2O–CO_2

The compositions of H_2O-saturated melts of eutectic or minimum composition in the haplogranite system have been well known since Tuttle & Bowen (1958) and Luth *et al.* (1964) published their experimental results. The agreement between experimental data and petrographic observations convinced many workers of the relevance of these experimental investigations to natural processes. However, as a result of more recent work it has become accepted that H_2O-saturated melting is more the exception than the rule, not only in the Earth's mantle but also within the continental crust. Therefore more information about the composition of H_2O-undersaturated melts is needed in order to interpret the genesis of crustal granitoids.

To date, only scarce experimental data on the Qz : Ab : Or ratios in H_2O-undersaturated melts are available. Nekvasil & Burnham (1987) and Nekvasil (1988) calculated cotectic lines and liquidus surfaces of the system Qz–Ab–Or–H_2O for 2 and 10 kbar total pressure and various levels of H_2O activity. These results show a decreasing Qz content and an increasing Ab content of the minimum composition with a decreasing level of H_2O activity at constant pressure. The Or content remains virtualiy constant. According to the calculated results, the effect of a decreasing level of H_2O activity is similar to the effect of increasing pressure under H_2O-saturated conditions.

Our experimental results differ from the calculations of Nekvasil (1988). The

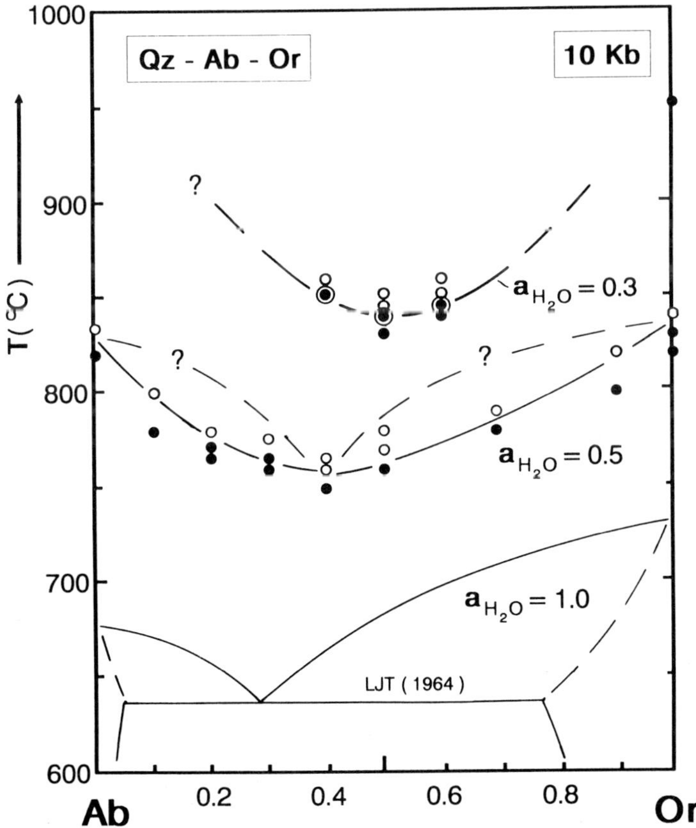

Figure 4.2 Projection of the solidus surfaces and cotectic curves determined in the system Qz–Ab–Or–H_2O–CO_2 at 10 kbar. H_2O-saturated data (LJT on the figure) are from Luth, Jahns & Tuttle (1964). Data for H_2O activities of 0.5 and 0.3 are from Ebadi & Johannes (unpublished). Dots indicate no melt and circles the presence of melt. Dots surrounded by circles mean that under these conditions the reaction product may or may not contain melt.

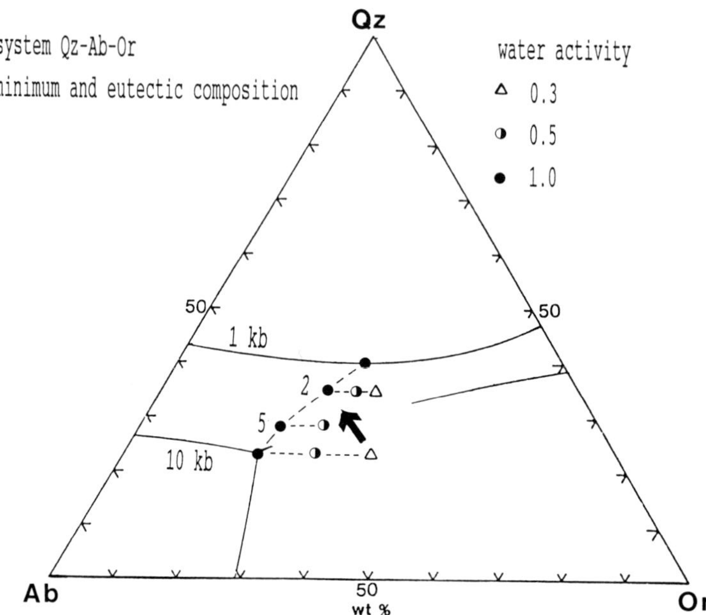

Figure 4.3 Effect of reduced H_2O activities on minimum (or eutectic) composition in the system Qz–Ab–Or–H_2O–CO_2 for 2, 5, and 10 kbar. Data are from Tuttle & Bowen (1958) and Luth *et al.* (1964) for H_2O-saturated conditions (black dots). H_2O-undersaturated data for 2 kbar are from Holtz *et al.* (1988 and unpublished data). Preliminary results for 5 and 10 kbar are from Ebadi & Johannes (unpublished). The arrow indicates the hypothetical evolution of residual melt compositions developing during the ascent of a magma formed at 10 kbar and $a_{H_2O} = 0.3$.

experimental data obtained at 2, 5 and 10 kbar (Ebadi *et al.* 1987, Holtz *et al.* 1988, Ebadi & Johannes unpublished) show almost constant Qz, decreasing Ab, and increasing Or contents, with a decreasing level of H_2O activity. The most recent data obtained at 10 kbar are presented in Figure 4.2.

Some of the experimentally determined eutectic and minimum compositions are included in Figure 4.3. These data can be used to reconstruct the compositional evolution of residual melts. According to data on the minimum or eutectic composition of H_2O-saturated melts, during ascent and crystallization such melts will move from Ab-rich to more Qz- and Or-rich compositions (see heavy dots in Fig. 4.3). The new information about the composition of H_2O-undersaturated melts, however, suggests a very different evolution. If H_2O-undersaturated magma crystallizes at constant pressure, the H_2O activity of the residual melt will increase. Its composition moves from right to left in Figure 4.3: the Ab content of the melt increases at the expense of Or, whereas the amount of Qz remains constant. If a melt crystallizes while the magma is ascending, pressure decreases and the level of H_2O activity increases. If we assume that a melt was formed at $P = 10$ kbar and $a_{H_2O} = 0.3$, that it crystallizes during its ascent and is completely solid at a depth of 7 km ($P \approx 2$ kbar), the Ab content of this melt remains approximately constant, its Or content decreases, and its Qz content increases (see arrow in Fig. 4.3). This trend is almost perpendicular to the evolutionary trend indicated for H_2O-saturated melts.

4.4 Amounts of the An component in partial granitic melts

The distribution of plagioclase components between partial melts and coexisting plagioclases is one of the most important factors controlling the differentiation of crustal rocks. The melting behaviour of plagioclase at 1 atm was first described by Bowen (1913). Yoder *et al.* (1957) investigated melting of plagioclase in the system Ab–An–H_2O at 5 kbar, and Yoder (1968) also determined liquidus and cotectic temperatures in the system Qz–Ab–An–H_2O at 5 kbar water pressure. Johannes (1978, 1984) established the solidus surfaces of the systems Qz–Ab–An–H_2O and Qz–Ab–An–Or–H_2O for water pressures up to 10 kbar. He also determined the initiation of melting temperatures in these systems for various plagioclase compositions between An_0 and An_{100}. Equilibrium distribution of Ab and An between melt and crystals could not be demonstrated in quartz–plagioclase and quartz–alkali feldspar–plagioclase assemblages. A more detailed study of the kinetics of melting reactions involving plagioclase feldspars showed that equilibrium compositions may only be

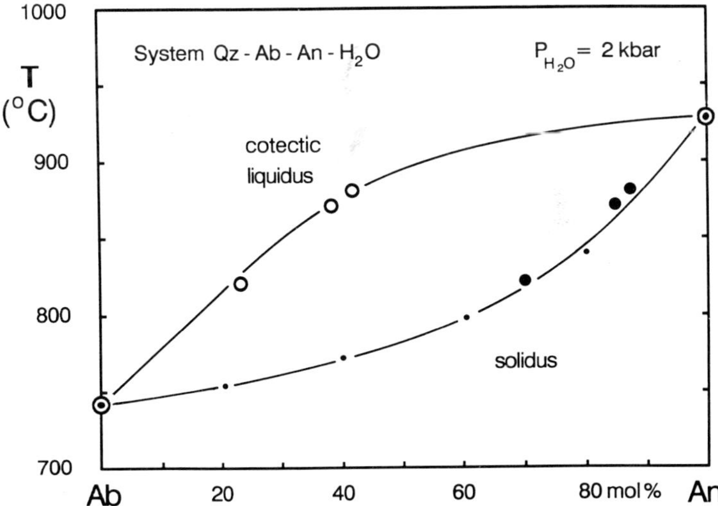

Figure 4.4 Projection of the solidus surface and cotectic curve determined in the system Qz–Ab–An–H_2O for $P_{H_2O} = 2$ kbar (based on Figure 7 of Johannes 1989, by permission of Springer-Verlag). Dots and circles, experimental results. Results indicated by large and small dots have been obtained by two different experimental methods (for more details see Johannes 1989).

obtained at temperatures above 800°C (Johannes 1989). The distribution of feldspar components determined in the system Qz–Ab–An–H_2O at $P_{H_2O} = 2$ kbar is shown in Figure 4.4.

Examination of Figure 4.4 shows that a significant amount of the Ab component is incorporated into the melt (along with Or), whereas the An component mainly remains in the coexisting plagioclases. The difference between the Ab : An ratios of the melt and the crystals is considerably higher than in the simple Ab–An system (Bowen 1913). The results mean that partial melts formed in rocks with plagioclases of intermediate composition are rich in Ab and poor in An. Partial melting and segregation of partial melts can be a very effective process, leading to the formation of alkali-rich rocks from basic protoliths. Even in basic rocks containing a plagioclase of composition An_{60}, the very first melts are rich in Ab and poor in An component (see Fig. 4.4). The same consideration is true for residual melts of crystallizing magmas. It is assumed that the partition of plagioclase components in granitic magmas is as pronounced as is indicated for quartz–plagioclase assemblages (tonalites) in Figure 4.4. Only a very high degree of partial melting can produce an Ab : An ratio in the melt similar to that of the source rock.

4.5 Amounts of mafic components (Fe, Mg, Al) dissolved in granitic melts

In addition to quartz and feldspars, granitoids contain variable amounts of mafic minerals. The presence of these minerals may result from assimilation processes or magma mixing; they can be unmelted restite materials or the crystalline products of components dissolved in the melt. In order to clarify the origin of the mafic minerals we have to know the solubility of mafic components in granitic melts. Data on solubilities of such components are almost absent (Clemens & Wall 1981; Naney 1983; Puziewicz & Johannes 1988, 1989). This is particularly true for H_2O-undersaturated conditions.

Phase equilibria and compositions of Fe–Mg–Al minerals and melts in H_2O-saturated peraluminous systems were investigated by Puziewicz & Johannes (1988) at water pressures of 2 and 5 kbar. According to these authors, the solubility of MgO and FeO in the melt is low. At 5 kbar and temperatures up to 850°C the maximum solubility of MgO in melts saturated with Al_2O_3 is 0.5 wt%, and that of FeO is 2.0 wt%. These results are in good agreement with those of Miller *et al.* (1985). Clemens & Wall (1981) observed higher solubilities for MgO (around 0.9 wt%) and FeO (around 3.1 wt%) at 5 kbar and 850°C. Although these values are different, they remain relatively low. The maximum solubility of Al_2O_3 corresponds to 3.5% normative corundum at 2 kbar and 5–5.5% corundum at 5 kbar.

Solubility of FeO and MgO in H_2O-undersaturated haplogranitic melts was investigated at 2 kbar (Puziewicz & Johannes 1989). The runs were performed between 750 and 850°C, and the level of H_2O activity ranged between 0.5 and 1.00. No change in the solubility of MgO could be observed with the changing level of H_2O activity. For FeO, a very small increase of solubility with an increasing level of H_2O activity was noted.

If all the MgO and most of the FeO contained in granitic melts is taken to form Fe-rich biotite, the total amount of biotite is 5 wt%. In other words: the maximum amount of biotite which can be dissolved in H_2O-saturated and -undersaturated haplogranitic melts at pressures up to 5 kbar and temperatures up to 850°C is 5%. Many granitoids have more than a 5% content of mafic minerals. This probably means that a lot of mafic (and felsic) minerals can be carried along as unmelted restite material. The chemical analyses of synthetic melts may support the restite model of Chappell & White (1974) and Chappell *et al.* (1987) for peraluminous granitoids containing important amounts of mafic minerals.

4.6 Liquidus temperatures and minimum water contents of H_2O-undersaturated haplogranitic melts

In order to discuss the formation of partial melts, not only do solidus temperatures and compositions of the first melts have to be constrained, but one must also know liquidus temperatures and the minimum amount of water necessary to form the melts. In Figure 4.5 preliminary liquidus curves are shown for minimum and eutectic melt compositions and a given amount of water.

The curves are constructed in a very simple way. All curves are parallel to the dry melting curve of Huang & Wyllie (1975). They are drawn through the data points indicating the solubility of water in the melts on the H_2O-saturated solidus. The solubility data used for the construction are from Tuttle & Bowen (1958), Luth *et al.* (1964), and Steiner (1970). The shape of the curves is similar to those calculated by J. R. Holloway (see Johannes 1985). We decided to present the simply constructed curves of Figure 4.5 because they are in good agreement with our preliminary experimental results (Holtz *et al.*, unpublished data) also reported on this figure (open circles).

Each curve in Figure 4.5 represents the minimum water content of the melts as a function of *P* and *T*. Within the 'granulite field' the minimum amount of water necessary to generate haplogranitic melts is between 2 and 8 wt%: it is 2 wt% at 5 kbar and 870°C and 8 wt% at 10 kbar and 750°C. The minimum content (at 10 kbar and 750°C) is 8 wt% water, whereas the maximum solubility of water at the same *P* and *T* conditions is approximately 16 wt% (Luth *et al.* 1964). Granitic melts formed at relatively low temperature and high pressure must be rich in water.

If quartz and alkali feldspar are present in a haplogranitic magma and in equilibrium with the melt, this means that the magma is a buffer system: if

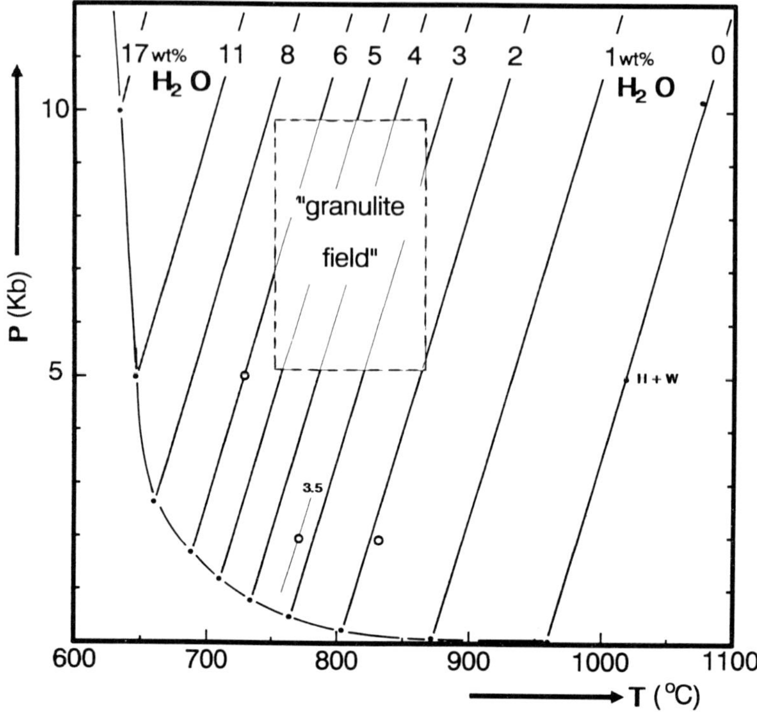

Figure 4.5 Liquidus curves for minimum melt compositions and specified water contents in the system Qz–Ab–Or–H_2O. Each curve shows the equilibrium *P* and *T* for one specified minimum water content. Further information on the construction of the diagram is given in the text. H + W refers to Huang & Wyllie (1975).

external water is added, new melt will form in order to consume this water; if the H_2O activity is reduced (e.g. by adding CO_2 or N_2 to the system), part of the melt will crystallize. The liquidus curves in Figure 4.5 are the *P–T* paths of ascending magmas with a constant water content and a constant crystal : melt ratio.

4.7 Amount of partial melts generated in fluid-absent conditions at given *P*, *T*, and water content

The importance of volatile constituents in the genesis of igneous rocks is acknowledged, but neither their total amounts nor their relative proportions are precisely known (Burnham 1979). There is a special need for information about the relation between the amount of water and the amount of melt in H_2O-undersaturated magmas (Clemens & Vielzeuf 1987). There are two major ways in which H_2O-undersaturated melting occurs in nature; either through lack of water or through dilution of water with other volatile components (e.g. CO_2). The first case means that no fluid phase is present (all the water is dissolved in the melt). In the second case a fluid phase is present but the level of H_2O activity is reduced. The terms 'dehydration melting' and 'fluid-absent melting' often encountered in the literature (e.g. Thompson 1982, Le Breton & Thompson 1988, Clemens & Vielzeuf 1987) correspond to the first case.

In this section we estimate the amount of melt generated in the system Qz–Ab–Or–H_2O at $P = 5$ kbar and given temperatures and amounts of water. For our calculations we consider that an initial amount of H_2O is present in the system (free H_2O) and that melting rapidly leads to H_2O-undersaturated conditions. The results are presented in Figure 4.6. In this diagram the water activity is 1 at 645°C (the H_2O-saturated solidus temperature of the system Qz–Ab–Or–H_2O at $P_{H_2O} = 5$ kbar). The level of H_2O activity prevailing at higher temperatures is <1 and can be read from Figure 4.1.

The curves of Figure 4.6 have been constructed with the help of Figure 4.5. In Figure 4.5 a horizontal line drawn at 5 kbar gives us the combination of T and wt% H_2O at which eutectic or minimum compositions are completely melted. The amount of water necessary to produce X% of melt at a given temperature is X% of the amount of water necessary for 100% melting at this

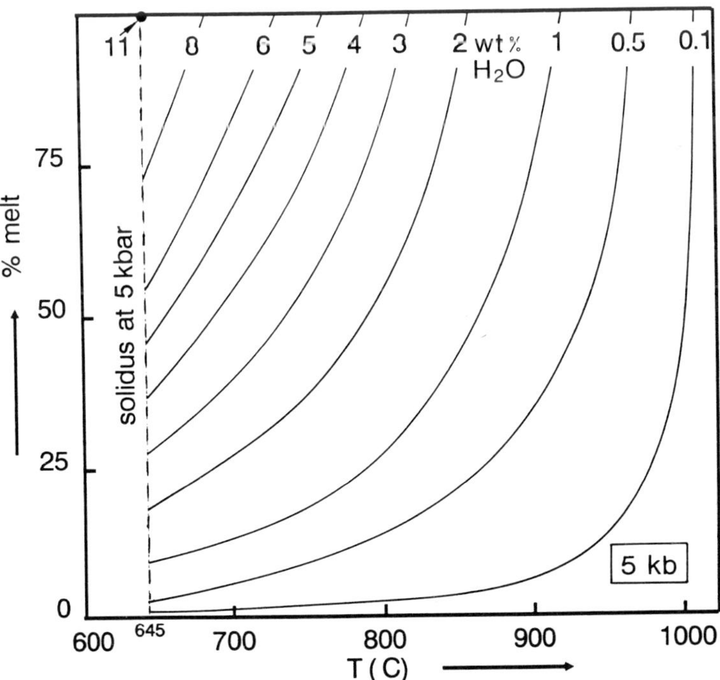

Figure 4.6 The amount of melt (given in wt%) formed by fluid-absent melting reactions in a hypothetical quartzofeldspathic rock as a function of T and water content of the source rock (reported on the curves) at 5 kbar (for further information see text).

temperature. Similar constructions have been made by Clemens & Vielzeuf (1987) for pelitic, basic, and intermediate compositions.

From Figure 4.6 we can read that only a small amount of water (1 wt% or less) is needed to produce relatively large amounts of melt at high temperature (e.g. at 900°C only 0.5 wt% H_2O is needed for 35 wt% melt). The melt fraction increases considerably for small temperature increases at high temperatures (see curves for 1, 0.5, and 0.1 wt% water above 800°C in Fig. 4.6). At low temperatures much greater amounts of water are required to form substantial amounts of melt.

The amount of melt produced by dehydration melting, as discussed by Thompson (1982), Clemens & Vielzeuf (1987), and Le Breton & Thompson (1988), can also be defined with the help of Figure 4.6. If we assume that a rock contains 20% biotite, the maximum amount of H_2O which can be released by dehydration is 0.8%. According to Figure 4.6 this 0.8 wt% water results in the formation of approximately 10 wt% (20%, 60%) of melt at 700°C (800°C, 900°C).

4.8 Summary and conclusions

With the help of the information given in the previous sections, the amount and composition of partial melts generated at given *P* and *T* in quartzo-feldspathic rocks can be estimated. The diagrams show that initiation of melting, liquidus temperatures, and the amount and composition of melt strongly depend on the prevailing *level of H_2O activity* and *amount of water available* for melting.

The *minimum level of water activity* necessary to produce melts may be as low as 0.3 around 850°C (Fig. 4.1). Such H_2O activity is low enough to cause dehydration reactions (Bohlen *et al.* 1983, Vielzeuf & Holloway 1988). This means that granitic melts may be formed under the same conditions as granulites. Granitic melts may absorb any H_2O produced by dehydration. Quartzo-feldspathic assemblages may even cause dehydration in the residual or surrounding rocks through melting. Obviously, granitic melts may dry the lower crust and transport water upwards.

The presence of CO_2, reducing the level of H_2O activity, may play an important role in the formation and crystallization of partial melts. In quartzo-feldspathic rocks, even if the a_{H_2O} of the coexisting vapour phase is as low as 0.25, partial melts can still be produced at around 850°C (Fig. 4.1). CO_2 may help to transport H_2O between rock volumes of different compositions. In a homogeneous vapour phase composed of CO_2 and H_2O, H_2O can be transported from rock volumes with higher levels of H_2O activity to those with a lower level of activity. Water may thus move from dehydrating to melting areas (or vice versa).

Within the granulite field (Fig. 4.1), the *amount of water* necessary to obtain a reasonable degree of melting (25–40 wt%) at low temperature and high pressure is high. Water has to be added to the system in order to obtain such a degree of melting. This is especially true for normal quartzofeldspathic rocks (e.g. metagreywackes) that are not very rich in hydrous mafic minerals. Assuming a complete dehydration of a rock with 20% biotite, only 0.8 wt% H_2O can be incorporated into the melt. With help of this amount of water 10 wt% melt can be produced at 10 kbar and 750°C, and 40 wt% at 5 kbar and 870°C (see upper left and lower right corners of the granulite field in Fig. 4.5). Melt of 25–40 wt% is probably sufficient for a magma to be able to ascend (Arzi 1978, van der Molen & Paterson 1979). It is shown in Figure 4.5 that temperatures must be high to produce 25–40% of melt if the amount of water available for partial melting is as low as 0.8 wt%.

The experimental results and the calculated data presented in the previous sections enable us to trace the evolution of a haplogranitic magma during its ascent to higher levels of the crust. Sykes & Holloway (1987) calculated the *P–T* trajectories of an albitic magma during ascent, considering different crystallization rates and heat exchanges with the surrounding rocks. In the following discussion we consider the *P–T* trajectories of an ascending magma composed of a haplogranitic melt with quartz and alkali feldspar as restitic phases.

We may distinguish three different cases (Fig. 4.7):

(1) *path A* for a magma which crystallizes during ascent and loses energy to its surroundings;
(2) *path B* for a magma in which the crystal : melt ratio is constant during its ascent and which cools only slightly;
(3) *path C* for a magma which moves upwards as a closed system, and is also closed for heat exchange.

Path A: intensive crystallization may occur in magma which moves very slowly within a cooler environment. In such a magma the crystal : melt ratio increases, and the level of H_2O activity and amount of water in the residual melt also increase (Figs 4.1 & 7). Within the melt, the Qz component will increase at the expense of Or (see arrow in Fig. 4.3). The final melt crystallizing is on the H_2O-saturated solidus (in Fig. 4.7 at 3 kbar).

Path B: this curve marks the trajectory of an ascending magma in which the crystal : melt ratio remains constant. The temperature decrease is mainly due to heat loss to the country rocks. The minimum water content remains constant, although the level of H_2O activity increases and becomes 1 on the solidus at relatively low pressure (Fig. 4.7).

Path C: in this case we consider no heat exchange with the surrounding rocks. An adiabatic ascent of a magma composed of 100% melt would mean a cooling of the melt of approximately 1°C per kbar due to its expansion with decreasing pressure (Jaeger 1968, Maaløe 1985). On such a near-vertical

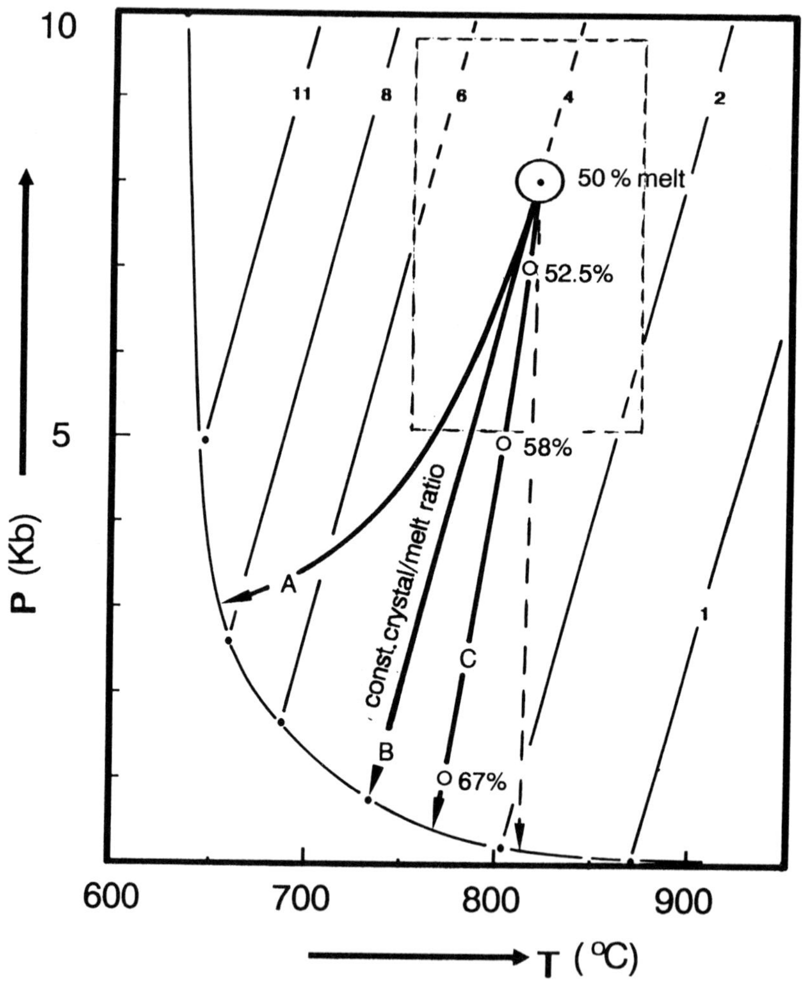

Figure 4.7 Comparison of hypothetical ascent trajectories. The starting conditions are 8 kbar and 820°C. The magma is composed of 50 wt% melt and 50 wt% quartzofeldspathic crystals and contains 2 wt% total amount of water. In path A crystallization occurs with decreasing *P* and *T*. Path B represents a trajectory with constant crystal : melt ratio. Path C is the calculated adiabatic trajectory. Melting in the magma occurs along this path; the amounts of melt in the magma are reported for 7, 5, and 1 kbar in the diagram. The dashed line is the adiabatic *P–T* path of a pure melt given for comparison.

path (see dashed line in Fig. 4.7), the minimum amount of water which must be present in the melts decreases. This means that melting of restitic quartz–feldspar assemblages may occur. In this way melt is formed and temperature decreases.

We have calculated the *P–T* trajectory and the melt : crystal ratio of a magma formed at 8 kbar and 820°C and containing 50 wt% melt at these conditions. The amount of water in the melt is 4 wt% and the level of H_2O activity is between 0.3 and 0.4 (Fig. 4.1). The composition of the melt is approximately 25% Qz, 35% Or, and 40% Ab (see Fig. 4.3 and discussion above). The *P–T* trajectory was calculated incrementally, taking account of the following parameters: specific heat of melt and crystals; latent heat of melting for quartz and alkali feldspars; density of melt and of minerals; and coefficient of thermal expansion. After a pressure decrease of 1 kbar, the amount of melt has increased from 50 to 52.5 wt%. At 5 kbar (1 kbar), 58 wt% (67 wt%) of melt is present in the magma (Fig. 4.7).

The magma crystallizes on the solidus at approximately 0.5 kbar and 770°C. At this temperature the solubility of H_2O is around 3 wt%. The level of H_2O activity is almost constant in the magma between 8 and 5 kbar, but increases at lower pressures and is 1 on the solidus. Here it contains approximately 68.5 wt% H_2O-saturated melt. Its composition is rich in Qz with approximately equal amounts of Ab and Or (see Tuttle & Bowen 1958).

To summarize, partial melts formed in quartzofeldspathic rocks in the lower crust can be formed at low levels of H_2O activity: they are H_2O-undersaturated, and poor in An and in mafic components. Compared with H_2O-saturated melts, they are poorer in Ab and richer in Or. They are mostly of leucogranitic composition. The level of H_2O activity in H_2O-undersaturated melts increases during ascent and crystallization of the magma; the Qz and Ab contents of the residual melts increase at the expense of Or and An. If there is no heat exchange between the ascending magma and the surrounding rocks, the amount of partial melt increases while the temperature decreases only slightly (see path C in Fig. 4.7).

References

Arzi, A. A. 1978. Critical phenomena in the rheology of partially melted rocks. *Tectonophysics* **44**, 173–84.

Barbey, P., J. Bernard-Griffiths & J. Convert 1986. The Lapland charnockitic complex: REE geochemistry and petrogenesis. *Lithos* **19**, 95–111.

Bateman, P. C. & B. W. Chappell 1979. Crystallization, fractionation and solidification of the Tuolumne Intrusive Series, Yosemite National Park, California. *Geological Society of America Bulletin* **90**, 465–82.

Bhattacharya, A. & S. K. Sen 1986. Granulite metamorphism, fluid buffering, and dehydration melting in the Madras charnockites and metapelites. *Journal of Petrology* **27**, 1119–41.

Boettcher, A. L., R. W. Luth & B. S. White 1987. Carbon in silicate liquids: the systems $NaAlSi_3O_8$–CO_2, $CaAl_2Si_2O_8$–CO_2 and $KAlSi_3O_8$–CO_2. *Contributions to Mineralogy and Petrology* **97**, 297–304.

Bohlen, S. R., A. L. Boettcher, V.J. Wall & J. D. Clemens 1983. Stability of phlogopite–quartz and sandine–quartz: a model for melting in the lower crust. *Contributions to Mineralogy and Petrology* **83**, 270–7.

Bowen, N. L. 1913. The melting phenomena of the plagioclase feldspars. *American Journal of Science* **35**, 577–99.

Burnham, C. W. 1979. The importance of volatile constituents. In *The evolution of the igneous rocks*, H. S. Yoder, (ed.), 439–82. Princeton, NJ: Princeton University Press.

Chappell, B. W. & A. J. R. White 1974. Two contrasting granite types. *Pacific Geology* **8**, 173–4.

Chappell, B. W., A. J. R. White & D. Wyborn 1987. The importance of residual source material (restite) in granite petrogenesis. *Journal of Petrology* **28**, 1111–38.

Clemens, J. D. & V. J. Wall 1981. Origin and crystallization of some peraluminous (S-type) granitic magmas. *Canadian Mineralogist* **19**, 111–31.

Clemens, J. D. & D. Vielzeuf 1987. Constraints on melting and magma production in the crust. *Earth and Planetary Science Letters* **86**, 287–306.

Conrad, W. K., I. A. Nicholls & V. J. Wall 1988. Water-saturated and -undersaturated melting of metaluminous and peraluminous crustal compositions at 10 kb: evidence for the origin of silicic magmas in the Taupo volcanic zone, New Zealand, and other occurrences. *Journal of Petrology* **29**, 765–803.

Ebadi, A., D. Ziegenbein & W. Johannes 1987. Schmelzbildung im Granitsystem Qz–Ab–Or bei Wasseraktivitäten kleiner 1. *Fortschritte der Mineralogie* **65**, 41.

Fyfe, W. S. 1973. The granulite facies, partial melting and the Archaean crust. *Philosophical Transactions of the Royal Society of London* **A273**, 457–61.

Grant, J. A. 1985. Phase equilibria in partial melting of pelitic rocks. In J. R. Ashworth (ed.), *Migmatites*, 86–144. Glasgow: Blackie.

Green, T. H. 1976. Experimental generation of cordierite- or garnet-bearing granitic liquids from a pelitic composition. *Geology* **4**, 85–8.

Holtz, F., W. Johannes, P. Barbey & M. Pichavant 1988. Liquidus phase relations in the system Qz–Ab–Or at 2 kbar: the effect of a_{H_2O}. *EOS, Transactions, American Geophysical Union*, **69**, 513.

Huang, W. L. & P. J. Wyllie 1975. Melting reactions in the system $NaAlSi_3O_8$–$KAlSi_3O_8$–SiO_2 to 35 kilobars, dry and with excess water. *Journal of Geology* **83**, 737–48.

Huang, W. L. & P. J. Wyllie 1986. Phase relationships of gabbro–tonalite–granite–water at 15 kbar with applications to differentiation and anatexis. *American Mineralogist* **71**, 301–16.

Jaeger, J. C. 1968. Cooling and solidification of igneous rocks. In *Basalts*, H. H. Hess & A. Poldervaart (eds), 503–36. New York: Interscience.

Johannes, W. 1978. Melting of plagioclase in the system Ab–An–H_2O and Qz–Ab–An–H_2O at $P_{H_2O} = 5$ kbar, an equilibrium problem. *Contributions to Mineralogy and Petrology* **66**, 295–303.

Johannes, W. 1984. Beginning of melting in the granite system Qz–Or–Ab–An–H_2O. *Contributions to Mineralogy and Petrology* **86**, 264–73.

Johannes, W. 1985. The significance of experimental studies for the formation of migmatites. In *Migmatites*, J. R. Ashworth (ed.), 36–85. Glasgow: Blackie.

Johannes, W. 1989. Melting of plagioclase–quartz assemblages at 2 kbar water pressure. *Contributions to Mineralogy and Petrology*, **103**, 270–6.
Johnson, C. A., S. R. Bohlen & E. J. Essene 1983. An evaluation of garnet–clinopyroxene geothermometry in granulites. *Contributions to Mineralogy and Petrology* **84**, 191–8.

Kerrick D. M. & G. K. Jacobs 1981. A modified Redlich–Kwong equation for H_2O, CO_2, and H_2O–CO_2 mixtures at elevated pressures and temperatures. *American Journal of Science* **281**, 735–67.

Lamb, W. M. & J. W. Valley 1984. Metamorphism of reduced granulites in low-CO_2 vapour-free environment. *Nature* **312**, 56–8.
Lamb, W. M. & J. W. Valley 1988. Granulite facies amphibole and biotite equilibria, and calculated peak-metamorphic water activities. *Contributions to Mineralogy and Petrology* **100**, 349–60.
Le Breton, N. & A. B. Thompson 1988. Fluid-absent (dehydration) melting of biotite in metapelites in the early stages of crustal anatexis. *Contributions to Mineralogy and Petrology* **99**, 226–37.
Luth, W. C., R. H. Jahns & O. F. Tuttle 1964. The granite system at pressures of 4 to 10 kilobars. *Journal of Geophysical Research* **69**, 759–73.

Maaløe, S. 1985. *Principles of igneous petrology*. Berlin: Springer.
Miller, C. F., E. B. Watson & R. P. Rapp 1985. Experimental investigation of mafic mineral–felsic liquid equilibria: preliminary results and petrogenetic implications. *EOS, Transactions, American Geophysical Union* **66**, 1130.
van der Molen, I. & M. S. Paterson 1979. Experimental deformation of partially-melted granite. *Contributions to Mineralogy and Petrology* **70**, 299–318.

Naney, M. T. 1983. Phase equilibria of rock-forming ferromagnesian silicates in granitic systems. *American Journal of Science* **283**, 993–1033.
Nekvasil, H. 1988. Calculated effect of anorthite component on the crystallization paths of H_2O-undersaturated haplogranitic melts. *American Mineralogist* **73**, 966–81.
Nekvasil, H. & C. W. Burnham 1987. The calculated individual effects of pressure and water content on phase equilibria in the granite system. In *Magmatic processes: physicochemical principles*, B. O. Mysen (ed.), 433–45. Geochemical Society Special Publication 1.

Presnall, D. C. & P.C. Bateman 1973. Fusion relations in the system $NaAlSi_3O_8$–$CaAl_2Si_2O_8$–$KAlSi_3O_8$–SiO_2–H_2O and generation of granitic magmas in the Sierra Nevada Batholith. *Geological Society of America Bulletin* **84**, 3181–202.
Puziewicz, J. & W. Johannes 1988. Phase equilibria and compositions of Fe–Mg–Al minerals and melts in water-saturated peraluminous granitic systems. *Contributions to Mineralogy and Petrology* **100**, 156–68.
Puziewicz, J. & W. Johannes 1989. Experimental study of biotite–orthopyroxene–magnetite–granitic melt assemblages: Results obtained at water-saturated and -undersaturated conditions and controlled oxygen fugacity. *Contributions to Mineralogy and Petrology*, in press.

Schenk, V. 1984. Petrology of felsic granulites, metapelites, metabasics, ultramafics, and metacarbonates from Southern Calabria (Italy): prograde metamorphism, uplift and cooling of a former lower crust. *Journal of Petrology* **25**, 255–98.
Schreurs, J. & L. Westra 1986. The thermotectonic evolution of a Proterozoic, low pressure, granulite dome, West Uusimaa, SW Finland. *Contributions to Mineralogy and Petrology* **93**, 236–50.
Steiner, J. C. 1970. *An experimental study of the assemblage alkali feldspar + liquid + quartz in the system* $NaAlSi_3O_8$–$KAlSi_3O_8$–SiO_2–H_2O *at 4000 bars*. PhD Thesis, Stanford University.

Stern, C. R. & P. J. Wyllie 1981. Phase relationships of I-type granite with H_2O to 35 kilobars: the Dinkey Lakes Biotite–Granite from the Sierra Nevada Batholith. *Journal of Geophysical Research* **86**, 10 412–22.

Stolper, E., G. Fine, T. Johnson & S. Newman 1987. Solubility of carbon dioxide in albitic melt. *American Mineralogist* **72**, 1071–85.

Sykes, M. L. & J. R. Holloway 1987. Evolution of granitic magmas during ascent: a phase equilibrium model. In *Magmatic processes: physicochemical principles*, B. O. Mysen (ed.), 447–61. Geochemical Society Special Publication 1.

Thompson, A. B. 1982. Dehydration melting of pelitic rocks and the generation of H_2O-undersaturated granitic liquids. *American Journal of Science* **282**, 1567–95.

Tuttle, O. F. & N. L. Bowen 1958. Origin of granite in the light of experimental studies in the system $NaAlSi_3O_8$–$KAlSi_3O_8$–SiO_2–H_2O. *Geological Society of America, Memoir* 74.

Vielzeuf, D. & J. R. Holloway 1988. Experimental determination of the fluid-absent melting relations in the pelitic system: consequences for crustal differentiation. *Contributions to Mineralogy and Petrology* **98**, 257–76.

White, A. J. R. & B. W. Chappell 1988. Some supracrustal (S-type) granites of the Lachlan Fold Belt. *Transactions of the Royal Society of Edinburgh: Earth Sciences* **79**, 169–81.

Whitney, J. A. 1975. The effects of pressure, temperature and X_{H_2O} on phase assemblages in four synthetic rock compositions. *Journal of Geology* **83**, 1–31.

Whitney, J. A. 1988. The origin of granite: the role and source of water in the evolution of granitic magmas. *Geological Society of America Bulletin* **100**, 1886–97.

Winkler, H. G. F. 1979. *Petrogenesis of metamorphic rocks*, 5th edn. New York: Springer.

Yoder, H. S. 1968. Albite–anorthite–quartz–water at 5 kbar. *Carnegie Institution of Washington Yearbook* **66**, 477–8.

Yoder, H. S., D. B. Stewart & J. R. Smith 1957. Feldspars. *Carnegie Institution of Washington Yearbook* **56**, 206–14.

CHAPTER FIVE

Garnet porphyroblast-bearing leucosomes in metapelites: mechanisms, phase diagrams, and an example from Broken Hill, Australia

Roger Powell & Jon Downes

5.1 Introduction

Mafic porphyroblasts are commonplace in leucosomes in metamorphic rocks at uppermost amphibolite facies and in the granulite facies. Large, 5 cm or more in diameter, garnet porphyroblasts are common in leucosomes in low- to medium-pressure granulite facies metapelites such as those in the Broken Hill area, Australia. Although there is significant consensus regarding the origin of those porphyroblasts – that they form as a consequence of incongruent melting reactions in which hydrous minerals are consumed – there are various interesting questions which need to be answered. For example: if the porphyroblasts are melting products, why then are they preserved during cooling when the melt crystallizes and the hydrous mineral becomes stable again? Intuitively, one would expect complete, or at least major retrogression of the porphyroblasts. This is not usually observed, and is generally absent in the garnet porphyroblasts at Broken Hill.

A description of the garnet porphyroblast-bearing leucosomes at Round Hill, Broken Hill, is followed by a discussion of the problems associated with the formation and preservation of their textures, with particular reference to phase diagrams in the system K_2O–FeO–MgO–Al_2O_3–SiO_2–H_2O (KFMASH).

5.2 Round Hill metapelites

The aluminous metapelites at Round Hill have a host-rock mineral assemblage of quartz, alkali feldspar, plagioclase, sillimanite, biotite, and cordierite, and contain leucosomes which are dominantly alkali feldspar, quartz, garnet and plagioclase, with minor sillimanite, cordierite, and biotite. The leucosomes are coarser-grained than their hosts, and their textures are allotriomorphic-granular to hypidiomorphic-granular, whereas their hosts are typically granoblastic. The leucosomes, which are up to 10–15 cm thick and several metres long, are parallel to a composite S_0–S_1 foliation defined by sillimanite, and folded by a strong D_2, which has an axial plane foliation defined by sillimanite–biotite in host *and* leucosome. Large garnet porphyroblasts, up to 10 cm across, but commonly about 4 cm across, occur only in the leucosomes. The occasional large garnet porphyroblast without leucosome is interpreted to have been originally within leucosome, on the basis of the close similarity and the close association with garnets which are still within leucosomes. These large garnets record S_0 compositional layering and S_1 sillimanite fabrics, which may occur at high angles to S_2, and rarely occur at high angles with each other in S_1 fold noses. Small garnets associated with S_2 occur in some rocks.

The chemistry of the minerals is summarized in Figure 5.1; note that the compositions of the minerals are the same in the leucosomes and host. This observation provides important constraints on the leucosome-forming and -crystallization processes (see later). The minerals are all generally unzoned. Thermometry and barometry indicate equilibration at 750–800°C at about 5 kbar for these mineral compositions (J. Downes, unpublished data). Leucosome modes are presented in Figure 5.2. There is a decrease in plagioclase and an increase in alkali feldspar as the thickness of leucosome decreases. If the leucosome is considered to be partial melt, most involve less to much less plagioclase, and many involve more alkali feldspar, than expected. If the garnets are considered to have crystallized *from* the melt (rather than just crystallized *in* the melt), the leucosome compositions would be even more inappropriate as partial melts.

As discussed further below, the reaction which can be considered to be responsible for the garnet-bearing leucosomes in the *model* system, KFMASH, is:

biotite + sillimanite + quartz = garnet + cordierite + orthoclase + melt (A)

In metapelites, considered in terms of KFMASH, starting with quartz + alkali feldspar + sillimanite + cordierite + biotite, progress of this reaction to the right involves consumption of biotite with a concomitant melting step and the production of garnet. With this model, several questions need to be addressed:

(a) If segregation of leucosome is by a physical process in which melt, gener-

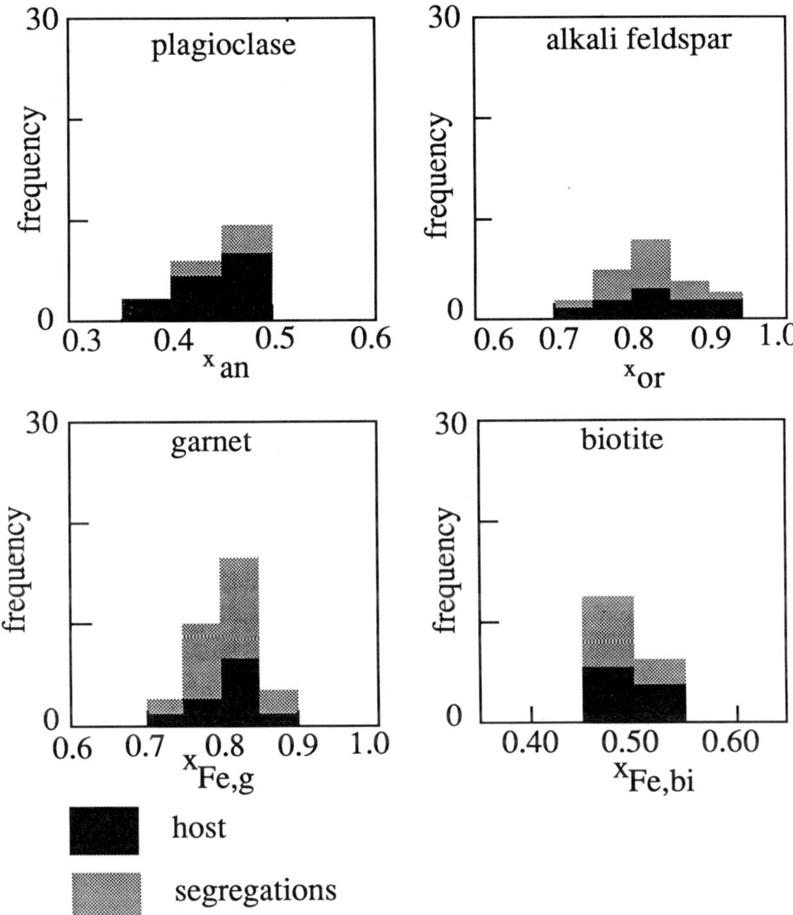

Figure 5.1 Histograms of mineral compositions for leucosomes and hosts. Note the similarity of compositions of the minerals in leucosomes and hosts. In the case of garnet, the comparison is between leucosome garnet and the later small garnets.

ated at appropriate grain contacts, moves to its current position, why then are the garnets within leucosomes?

(b) If the above reaction is responsible for the formation of both the garnet and the melt, then why is the reaction not reversed, and the garnet completely retrogressed, during cooling?

(c) A pronounced fractionation of mineral compositions between host and leucosome might be expected as a consequence of partial melting. Why are the compositions of the minerals in the leucosomes and their hosts the same?

(d) Why do the leucosome compositions not correspond to melt compositions?

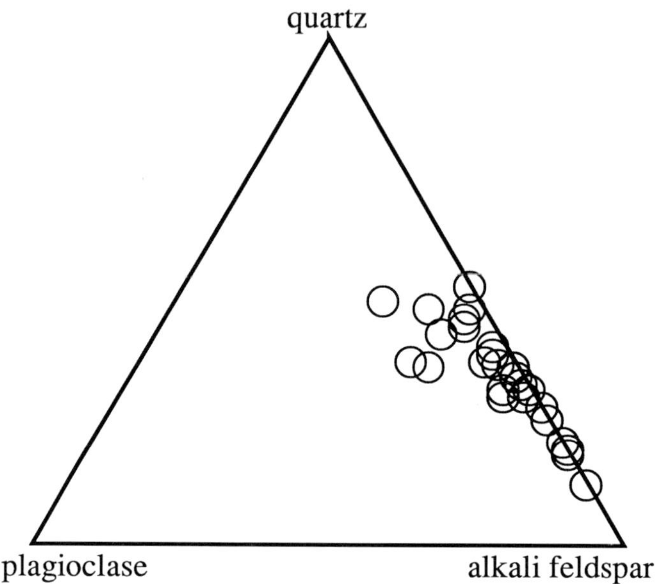

Figure 5.2 Modal proportions of quartz, plagioclase, and alkali feldspar in leucosomes in metapelites. The modes towards the alkali feldspar apex are for thinner leucosomes. Note the non-melt nature of most of the modes.

These questions may be answered partly by appraising the processes involved in leucosome segregation and crystallization, and partly by looking at the phase equilibria in relation to these processes, and to the pressure–temperature path followed during metamorphism.

5.3 Processes

Three observations on the Round Hill rocks force a reappraisal of processes involved in the formation of leucosomes: (a) the leucosomes do not have melt compositions; (b) the minerals in and adjacent to leucosomes have the same compositions; and (c) the garnet porphyroblasts occur only in the leucosomes.

The sceptical view, from the first observation, is that partial melting was not involved and that the leucosomes are a solid-state phenomenon related to garnet porphyroblast growth, with the leucosome being a solid-state diffusion-controlled 'corona'. However, closely associated metabasic rocks are orthopyroxene + clinopyroxene-bearing and, from the thermometry results, it is clear that metamorphism proceeded well above the H_2O-saturated solidus. Although it is conceivable that metamorphism at low a_{H_2O} might give rise to such textures, it is unlikely that such aluminous metapelites could have low a_{H_2O} unless they had been previously metamorphosed, and there is no evidence of this at Broken Hill. Moreover, as discussed by Powell (1983) and Stüwe & Powell (1989c), with increasing temperature, even a fluid-absent but biotite-bearing rock will be internally buffered to higher a_{H_2O} along a dehydration reaction until partial melting starts. Essentially the same equilibria are involved as in rocks with initially high a_{H_2O}, at least for intermediate Fe–Mg pelites; it is very difficult to have any significant reaction which does *not* involve partial melting for this assemblage at these temperatures.

If it is accepted that partial melting *was* involved, the textures observed may be considered with the help of the concept of equilibration volume. Within a particular mineral assemblage, chemical equilibration involves the removal of chemical potential gradients by diffusion. For each element, the scale of such equilibration depends on the factors which control diffusion; temperature, the duration of that temperature, the presence of a grain-boundary fluid phase, and so on. The equilibration volume corresponds to the scale of this equilibration; it is different for different elements and, for example, changes with temperature. Each mineral in a particular assemblage will tend to have a constant composition across the equilibration volume or, more strictly, a constant exchange vector (e.g. $MgFe_{-1}$) across the equilibration volume, which corresponds to the more slowly diffusing element in the exchange. It does *not* preclude bulk compositional variations within that scale, *as long as* variations in bulk composition can be accomplished with different proportions of the same set of minerals with the same compositions.

The second observation can be understood in the context of equilibration volumes. The similarity in mineral compositions between host and leucosome may be accounted for if the equilibration volumes when the melt in the leucosome crystallized were rather larger than the scale of the leucosomes for most elements, with consequent equilibration between the minerals in the host and those crystallizing in the leucosome.

The occurrence of relatively large equilibration volumes at the time of melt crystallization is consistent not only with the minerals having the same composition in leucosomes and host, but also with the non-melt compositions of the leucosomes. For example, plagioclase, to grow from the melt as it crystallizes, may grow on pre-existing plagioclase grains in the host, rather than nucleating in the melt. If this happens, then the volume of leucosome obviously decreases during crystallization. Although this is a plausible mechanism for compositional changes in the leucosome following peak metamorphism, additional possibilities are considered below, following discussion of the location of garnet porphyroblasts.

The temperature of major melt generation and the temperature of melt crystallization are likely to be comparable, and it is logical to contend that similar equilibration volumes were involved at these stages in the prograde and retrograde histories. Implicit in this contention is that heating rates were comparable to cooling rates. If equilibration volumes were large during the higher-temperature part of the prograde history, an alternative view of the leucosome-forming process is appealing: that it is primarily chemical, and related to garnet nucleation and growth, rather than physical. Once the garnet-forming reaction, for example, reaction A, is crossed on the prograde path, garnet nucleates in a few places; these garnets grow instead of more garnet nuclei forming. Given that the growing garnets are the loci of the creation of products, melt would *form* around the garnet, with garnet *and* melt components diffusing from the host to these sites.

Leucosome formation controlled by garnet nucleation and growth obviously accounts for the association of garnet porphyroblasts with leucosomes. Given that reactions such as A all involve alkali feldspar on the high-temperature side, alkali feldspar should also form within the melt, in association with the garnet, even though alkali feldspar is abundant in the host, contributing to the abundance of alkali feldspar in the leucosome modes.

The difficulty of assessing the original melt composition by examination of the leucosomes is exacerbated by the possibility of melt loss from the leucosomes. As discussed, for example, by Le Breton & Thompson (1988), the proportion of melt in a rock leading to melt segregation is not well established, particularly if the rock is undergoing strain. In the context of the model presented above, this proportion of melt would correspond to the proportion of melt *in the leucosome* which could lead to melt loss. In this model, the proportion of melt originally present in a leucosome is controlled primarily by the amount of biotite which reacts, and the amount of reactants remaining and products generated in the volume which becomes the leucosome. Melt loss could occur if this proportion of melt exceeded that which is stable in a rock undergoing strain; at a minimum the leucosome would presumably be grain supported. In the absence of knowledge of this stable proportion of melt, it is interesting to speculate on what the proportion of melt remaining after melt loss from the leucosome could be, while leaving the leucosome recognizable as such. It is tempting to suggest that the proportion of melt remaining could be small.

5.4 Phase diagram considerations

In this section, equilibration along pressure–temperature (P–T) paths for a pelitic model system is considered in the light of the ideas presented in the last section. The system K_2O–FeO–MgO–Al_2O_3–SiO_2–H_2O (KFMASH) is adopted because it can be used to account for many of the features observed in pelites. In projection from quartz, sillimanite, and alkali feldspar, KFMASH reduces to the effectively ternary system FeO–MgO–H_2O (FMH), which is of low enough dimension to be simply portrayed, yet is sufficiently complex to be used to account for metamorphism and partial melting in many pelites (Thompson 1982). Clearly, omission of Fe_2O_3 and TiO_2 from the model system precludes consideration of the Fe–Ti oxides and omission of CaO and

Na_2O precludes consideration of plagioclase; the qualitative effects of adding additional components to KFMASH are alluded to later.

The starting point is the pressure–temperature diagram for KFMASH in projection from quartz, sillimanite, and alkali feldspar, Figure 5.3, based on Grant (1985, Figs 3.11 & 14), combined with the results of calculations. These calculations use the new expanded internally consistent thermodynamic dataset of Holland & Powell (1990) with the software THERMOCALC (Powell & Holland 1988). The grid, Figure 5.3, is topologically identical to Grant's grid, but there are minor differences. From the calculations, cd + g = opx + l should have a positive dP/dT, and therefore the slope of cd + g = opx + bi

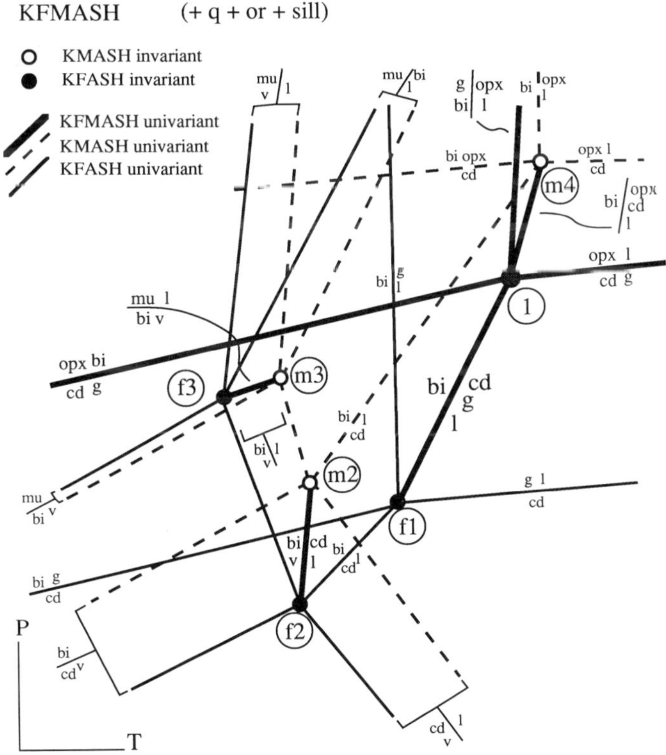

Figure 5.3 The *P–T* grid for KFMASH in projection from quartz, orthoclase, and sillimanite, after Grant (1985) as discussed in the text. The abbreviations used are: q, quartz; or, orthoclase; sill, sillimanite; v, H_2O; l, melt; mu, muscovite; bi, biotite; g, garnet; cd, cordierite; opx, orthopyroxene. Labels for invariant points are circled; the KFASH invariant points are prefixed by 'f'; the KMASH invariant points by 'm'. KFMASH univariant lines are thicker lines; KFASH and KMASH univariant lines are thinner lines, with the KFASH lines solid and the KMASH lines dashed. Note that the compositions of the minerals change along the univariant lines, with the Tschermak's substitutions in the phases changing along the KFASH and KMASH lines, and $X_{Fe} = Fe/(Fe + Mg)$ and Tschermak's substitutions in the phases changing along KFMASH lines, with melt compositions changing in additional ways.

should also be positive. These conclusions are more or less inescapable, even though silicate melt cannot yet be included in the calculations; the conclusions are based on calculations undertaken at a series of a_{H_2O}.

This grid can be used to consider the generation of the garnet porphyroblast-bearing leucosomes. To acclimatize, note that the relatively steep reactions through f2 and f3 and through m2 and m3 form the solidus for bulk compositions starting with a hydrous mineral and a fluid phase. Also, the shallower

Table 5.1 *P–T–X* relationships calculated by THERMOCALC in KFMASH for the univariant reaction bi + sill = g + cd, with q, or, H_2O in excess. *X* is Fe/(Fe + Mg); *y*(bi) is Tschermak's substitution in biotite, with *y*(bi) = 0 in phlogopite and annite, and *y*(bi) = 1 in eastonite.

X_{H_2O}	P(kbar)	T(°C)	X(cd)	X(g)	X(bi)	y(bi)
1.0	3.0	615	0.915	0.987	0.941	0.557
	4.0	660	0.723	0.946	0.796	0.537
	5.0	704	0.572	0.892	0.668	0.517
	6.0	748	0.452	0.827	0.555	0.499
	7.0	792	0.354	0.749	0.455	0.483
	8.0	835	0.273	0.659	0.364	0.467
0.75	3.0	599	0.814	0.971	0.866	0.545
	4.0	644	0.635	0.923	0.721	0.523
	5.0	689	0.495	0.862	0.595	0.503
	6.0	734	0.385	0.788	0.485	0.484
	7.0	778	0.295	0.699	0.388	0.466
	8.0	823	0.220	0.597	0.300	0.450
0.5	2.0	529	0.916	0.990	0.941	0.554
	3.0	576	0.697	0.948	0.772	0.528
	4.0	622	0.533	0.892	0.629	0.504
	5.0	668	0.407	0.821	0.507	0.482
	6.0	714	0.308	0.733	0.401	0.462
	7.0	760	0.227	0.628	0.308	0.444
	8.0	805	0.160	0.507	0.224	0.427
0.25	2.0	491	0.743	0.966	0.807	0.530
	3.0	538	0.548	0.915	0.639	0.500
	4.0	585	0.406	0.844	0.502	0.474
	5.0	632	0.298	0.754	0.388	0.449
	6.0	679	0.214	0.643	0.289	0.427
	7.0	726	0.145	0.510	0.204	0.408
	8.0	772	0.0876	0.354	0.127	0.390

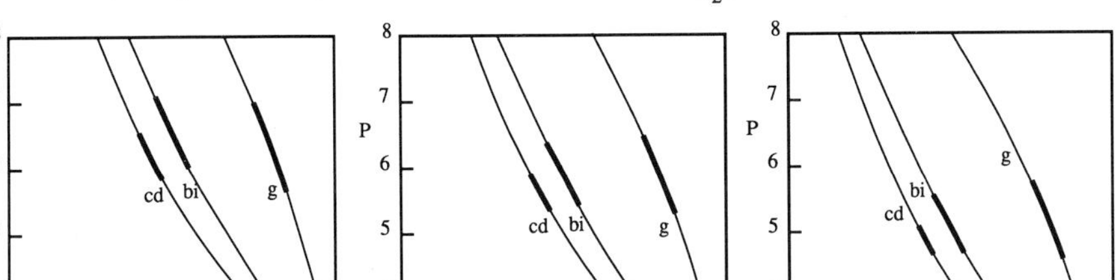

Figure 5.4 Calculated $P–X_{Fe}$ diagrams along the KFMASH univariant line involving cd + g + bi + sill + q + or at fixed a_{H_2O}, showing how the compositions of the minerals change with pressure along the reaction line. The highlighted parts of the lines correspond to the mineral compositions observed at Round Hill.

positive-slope reactions through f3 and m3 mark the breakdown of mu + q with increasing temperature in KFMASH. Further, invariant point 1, and the reactions emanating from it, are fundamental in subdividing metamorphic conditions. In terms of the model system, the reaction bi + sill = cd + g + l can be considered to mark the lower-pressure transition from the amphibolite facies to the granulite facies. At higher pressure, the assemblage opx + sill is stabilized via the reactions bi + sill = cd + opx + l and cd + g = opx + sill + l. Moreover, as suggested earlier, the reaction which connects f1 and 1 (reaction A) may well be responsible for the garnet-bearing leucosomes.

Results for reaction A generated using the software THERMOCALC (Powell & Holland 1988) are summarized in Table 5.1 and Figure 5.4. It is important to realize the relationship between these results and the corresponding reaction on Figure 5.3. Whereas reaction A is fluid-absent and a_{H_2O} varies continuously with pressure along it, the calculations were performed at a series of values of a_{H_2O}. Thus, in terms of a $P–a_{H_2O}–X_{Fe}$ block diagram, the diagrams in Figure 5.4 are constant a_{H_2O} sections, and the 'real' reaction corresponds to an oblique section. Note, on Figure 5.4, how quickly the compositions of the coexisting minerals change with pressure along the reaction. This rapid change is mainly controlled by the strong pressure dependence of the divariant FMASH equilibrium, cd + g + sill + q. This change allows an estimation of pressure for the Round Hill rocks; for a_{H_2O} of 0.5–0.75 (Phillips 1980; J. Downes & V. Wall, unpublished data), pressures of the order of 5–6 kbar are implied for the observed mineral compositions, the same as the conditions estimated by conventional barometry. In the light of the above equilibration volume argument, this pressure would correspond to the early retrograde history, when the melt was crystallizing.

5.4.1 Prograde history

Interpretation of diagrams such as Figure 5.3, particularly where mineral compositions change rapidly along reaction lines, requires care. The first step is to devise an appropriate compatibility diagram; following Thompson (1982), the most informative is one with H_2O at an apex. In Figures 5.5a–c is shown a series of such compatibility diagrams for isobaric increase of temperature between f2–f3 and m2–m3 on Figure 5.3, in projection from sill, or, and q, in the effectively ternary system FMH. The series depicts the l + bi + H_2O tie triangle swinging across from FeO-rich compositions to MgO-rich compositions. By the time the temperature is above the line m2–m3, assemblages which started as bi + H_2O have become bi + l. Whereas in the prograde history, fluids evolved by reactions are presumed to be lost from rocks, the same is not necessarily true when the reactions generate melt rather than fluid; in fact, the assumption is often made that no melt loss is involved.

For rocks which started out as fluid-present, there is a narrow range of normally accessible bulk compositions in the compatibility diagram, distal to H_2O within tie lines and tie triangles involving H_2O, because fluid-present rocks are suggested to have rather a small proportion of fluid on grain boundaries. In Figure 5.5, the accessible range is close to the 'H_2O-rich' edge (actually the aluminous edge) of the band of biotite compositions (dashed line in Fig. 5.5c). *If* melt is not lost in the same way as fluid is supposed to be, the range of bulk compositions of originally fluid-present rocks will be frozen-in from when the minerals are cut off from the H_2O apex in the compatibility diagram as melt is stabilized (at the dashed line in Fig. 5.5c for the case of Figs 5.5a–c). If the prograde *P–T* path passes between f2 and m2, then more Mg-rich compositions will have been dehydrated by the divariant bi = cd + H_2O, and, assuming that the fluid generated by this biotite breakdown is lost, the range of normally accessible bulk compositions will have a dog-leg, as in Figure 5.5d. At still lower pressures, the range of accessible compositions will follow the top of the cordierite field. Such H_2O-poor compositions can produce little melt. For the great majority of regional metamorphic terrains, *P–T* paths pass well above f2, and the range of accessible compositions will be as portrayed in Figure 5.5c. In the discussion below, we will focus on the range of normally accessible bulk compositions.

As temperature increases from that of Figure 5.5c, two tie triangles are encountered, both of which restrict the stability of biotite: one stabilizes garnet + melt, and the other cordierite + melt (Fig. 5.5e). With further increase in temperature, these equilibria progressively reduce the compositional range of stability of biotite, until, at reaction A, the KFMASH univariant reaction between f1 and l (Fig. 5.3), biotite is finally consumed. For our 'accessible' bulk compositions, what happens on crossing A depends on the X_{Fe} of the bulk composition compared to the X_{Fe} of the last biotite. For bulk compositions with X_{Fe} less than the biotite, garnet will appear in

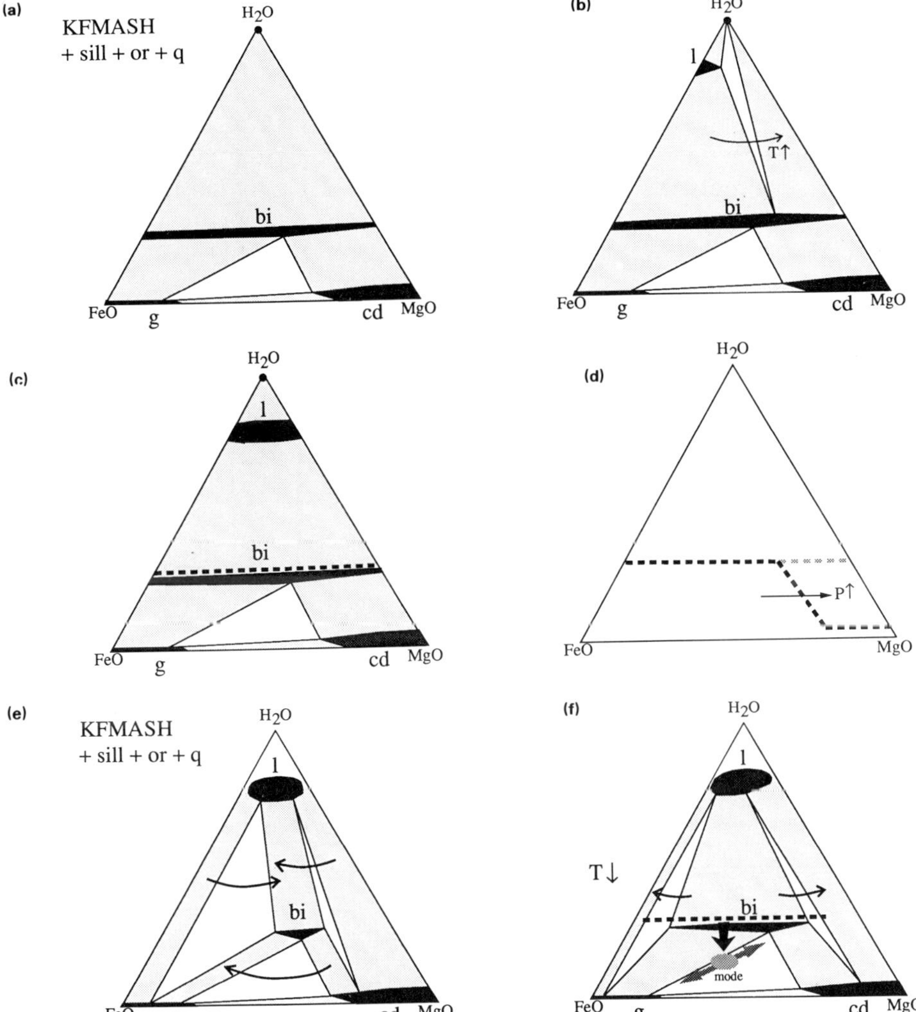

Figure 5.5 The FeO–MgO–H_2O compatibility diagram for KFMASH in projection from quartz, orthoclase, and sillimanite. (a–c) Crossing of the solidus for biotite + H_2O assemblages; the biotite + melt + H_2O tie triangle rotates rapidly from the FeO–H_2O to the MgO–H_2O subsystems with increasing temperature. On (c) the heavy dashed line is the range of normally observable bulk compositions for originally H_2O-present bulk compositions. On (d) the normally observable bulk compositions for somewhat lower pressures, more magnesian biotite having already broken down to cordierite + water before melting. (e) The tie triangle rotations as the reaction bi + sill = cd + g is approached with increasing temperature. (f) A synoptic diagram showing the observed mode in the rocks; how the bulk composition of the rock must have moved by melt loss (heavy black arrow); the rotations of tie triangles with decreasing temperature re-establishing a complete range of biotite compositions across the triangle; and the stippled arrow across the mode indicating the fractionation of bulk compositions between leucosome and host with cooling.

cordierite–biotite-bearing rocks as the reaction is crossed, whereas for bulk compositions with X_{Fe} greater than the biotite, cordierite will appear in garnet–biotite-bearing rocks.

A powerful way of portraying phase relationships in systems with variable composition phases is with *pseudosections* (Hensen 1971). For example, a *P–T* pseudosection is a *P–T* diagram for just one bulk composition in a model system; only those relationships in the *P–T projection* (e.g. Fig. 5.3) which are 'seen' by the bulk composition are portrayed. Figure 5.6 is a *P–T*

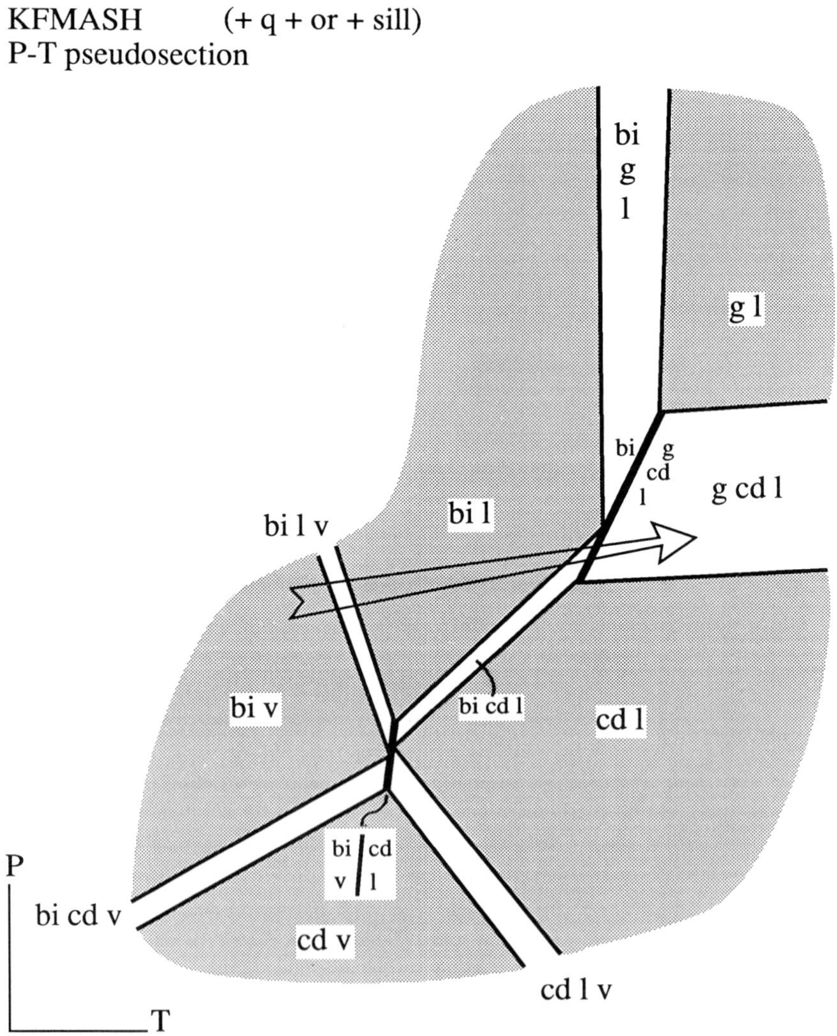

Figure 5.6 A *P–T* pseudosection based on Figure 5.3 for an intermediate X_{Fe} on the dashed line in Figure 5.5c. The arrow is referred to in the text.

pseudosection for an intermediate X_{Fe} composition along the dashed line in Figure 5.5c. Comparing Figure 5.6 with Figure 5.3, KMASH and KFASH reactions do not appear at all; instead they give rise to three phase fields which represent KFMASH continuous reactions. Thus, the bi + l + v field occurs between, and more or less parallel to, the KMASH and KFASH bi + v = l reactions running between m3 and m2, and f3 and f2, respectively. KFMASH discontinuous reactions and intersections are shown only if they can be 'seen' by the bulk composition. Thus intersection 1, affecting more Mg-rich bulk compositions than the pseudosection bulk composition, is not shown; instead the g + l two-phase field occurs in this part of the *P–T* pseudosection. Similarly, only a part of reaction A, bi = g + cd + l, is shown; at lower *P*, instead, the cd + l field occurs and at higher *P*, the g + l field.

For progressively more Fe-rich bulk composition than the bulk composition used in Figure 5.6, along the dashed line in Figure 5.5c, all the phase relationships on Figure 5.6 slide to lower pressures; for more Mg-rich bulk compositions they slide to higher pressures. The relevance of this can be seen by observing the prograde evolution along the arrow in Figure 5.6 for different bulk compositions. For the bulk composition of the pseudosection in Figure 5.6, cordierite appears shortly before reaction A is crossed, and garnet forms. For more Mg-rich compositions, reaction A will not be seen and garnet not developed. For more Fe-rich compositions, garnet will appear before reaction A is crossed, when cordierite is developed. For much more Fe-rich compositions, reaction A is not crossed and cordierite not developed.

Another way of seeing these effects is with T–a_{H_2O} diagrams (e.g. Powell 1983); Figure 5.7a is drawn for the conditions along the arrow on Figure 5.6. The intersection on Figure 5.7a corresponds to reaction A on Figure 5.3; the reactions to the three-phase fields. (The labelling of these reactions does *not* correspond to the labelling of the end-member system reactions in Figure 5.3. With respect to Figure 5.5c, the former involve the phases in projection from H_2O onto the FeO–MgO line; the latter involve the end-members of the phases, for example along the FeO–H_2O side of the compatibility diagram.) Figures 5.7b–d are T–a_{H_2O} pseudosections for bulk compositions along the dashed line in Figure 5.5c. On these diagrams, fixed bulk composition, i.e. internally buffered, paths (e.g. Powell 1978), involve negligible reaction within two-phase fields; substantial reaction occurs only along the reactions and through the intersection. Clearly, the phase relationships along the arrowed paths on Figures 5.7b–d correspond to the phase relationships discussed in relation to the *P–T* pseudosection above.

Application of Figure 5.6 to the Round Hill metapelites to account for the observed relationships is complicated by the dependence on the *P–T* path; previous workers in the Broken Hill and Olary Blocks have concluded that an anticlockwise *P–T* path was followed. On this basis, isobaric and increasing pressure prograde paths will be considered. Accepting the argument presented earlier that garnet nucleation and growth control the formation of garnet porphyroblast-bearing leucosomes, the *P–T* path is required to involve garnet growth that is more or less coincident with an increase in the proportion of melt. This appears to preclude an increasing pressure path below the low-pressure end of reaction A into the bi + g + l field, because the major melting step would take place across the bi + cd + l field. Such a path might be expected to produce cordierite porphyroblast-bearing leucosomes. Thus, the required path must cross reaction A, with either garnet or cordierite appearing immediately prior to the crossing. Our prejudice is that the latter case occurs, given that the lower-grade metapelitic rocks to the north of Round Hill are cordierite–biotite-bearing, and do not involve garnet. In the absence of cordierite, pelites following a *P–T* path crossing the bi + g + l field into the g + l field above reaction A will also develop garnet porphyroblast-bearing leucosomes.

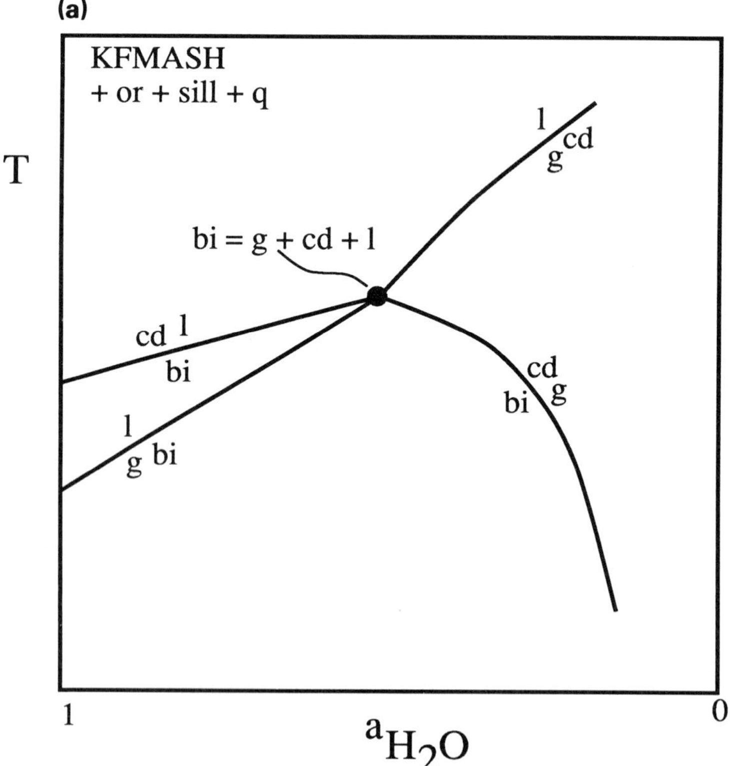

Figure 5.7 T–a_{H_2O} diagrams for the conditions of the arrow on Figure 5.6. (a) T–a_{H_2O} projection; (b–d) T–a_{H_2O} pseudosections for a range of bulk compositions along the dashed line in Figure 5.5c. The arrowed lines are internal buffering paths for an initial assemblage of bi + v.

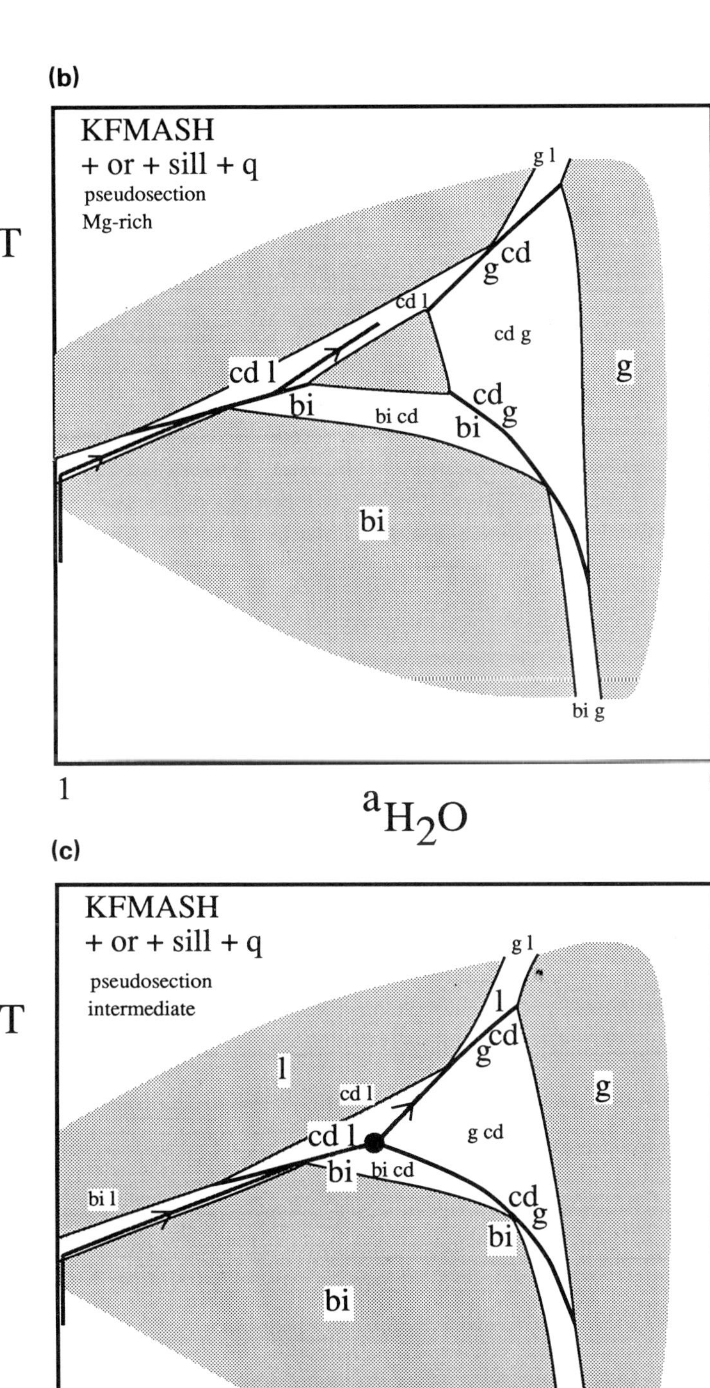
(b)
KFMASH
+ or + sill + q
pseudosection
Mg-rich
T
g l
g cd
cd l
cd g
cd l
g
bi
bi cd
cd g
bi
bi
bi g
1
0
a_{H_2O}
(c)
KFMASH
+ or + sill + q
pseudosection
intermediate
T
g l
l
g cd
l
cd l
g
cd l
g cd
bi
bi cd
bi l
cd g
bi
bi
bi g
1
0
a_{H_2O}

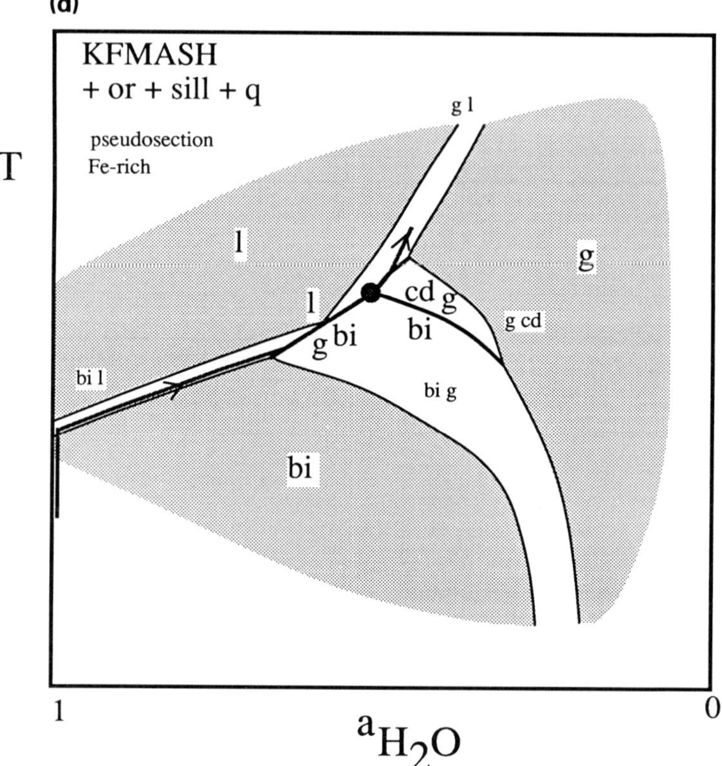

Figure 5.7 (*Continued*)

5.4.2 Retrograde history

In the retrograde history, as reaction A is crossed to lower temperature, biotite stability is re-established and, with further cooling, the divariant equilibria, bi = g + l and bi = cd + l, rapidly convert the compatibility diagram to the appearance of Figure 5.5c. This is also very clear on Figure 5.6. For rocks along the 'accessible' range, Figure 5.5c, garnet produced by reaction A should all be converted back to biotite! It is clear that, to preserve the garnet, Figure 5.6 must be inappropriate. Some of the possible explanations have been discussed by Waters (1988): most involve some form of disequilibrium. As discussed earlier, equilibration volumes appear to have been relatively large early in the retrograde history; this is inconsistent with explanations of garnet preservation based on *dis*equilibrium. The simplest explanation is to assume that some melt is lost, presumably at temperatures at or above reaction A. Such melt loss moves a bulk composition away from melt and so, on the compatibility diagram being used, the dominant effect is to move it away from the H_2O apex, across the biotite range of compositions. As long as the melt loss occurs before reaction A is crossed during the retrograde history, garnet survives. On Figure 5.5f, the black arrow represents melt loss, moving a bulk composition from the normal range of accessible compositions (dashed line) to the observed mode of the rocks (stippled ellipse). Melt loss can also be portrayed on the T–a_{H_2O} pseudosections of Figure 5.7. For example, on Figure 5.7d, following melt loss the retrograde path would involve reacting out of melt (rather than garnet) at the intersection, and the resulting path would descend the bi = cd + g reaction to lower temperatures and a_{H_2O}. Departure from this reaction into the bi + g field would eventually occur, accounting for the observed, relatively late, fine-grained alteration of cordierite to biotite and sillimanite.

Given the rapid change of garnet and cordierite compositions with pressure, textures involving these two minerals may be illuminating. Certainly, with even a small amount of decompression, cordierite reaction haloes on garnet seem to be able to form easily (e.g. Stüwe & Powell 1989a), although the converse of garnet overgrowing cordierite occurs less readily (but see Stüwe & Powell 1989b). No such textures are observed at Round Hill, suggesting that cooling was essentially isobaric. The isobaric rotation of the bi = cd + g tie triangle with decreasing *T and* a_{H_2O} involves a small decrease in the X_{Fe} of the minerals, but these changes are unimportant compared with the rotation caused by pressure. The only indication that garnet is being stabilized during the retrograde history, and thus that a small pressure increase occurs with cooling, is the occurrence of the small late garnets in the host associated with D_2, even though these have compositions that are similar to those of the leucosome garnets.

These considerations concern the post-equilibration P–T path, i.e. the P–T path after the retrograde crossing of reaction A. In addition, there is an important implication for the earlier P–T path from the observed textures. If, as seems likely, garnet first forms across A, then there is an important implication for the pressure at which reaction A is crossed in the prograde history. With reference to Figure 5.5f, if the black arrow must emanate from a point on the dashed line which was originally in the cd + bi + l tie triangle, then the last biotite consumed in the prograde crossing of A must be rather more Fe-rich than the first biotite formed in the retrograde crossing of A. From Table 5.1, this implies that the P–T path through the metamorphic peak involved an *increase* of pressure. This is consistent with the conclusions of previous workers in the Broken Hill and adjacent Olary Blocks.

5.4.3 *Expanding the model system*

The KFMASH system does not model metapelites in a number of important regards. In KFMASH, biotite will react out on crossing reaction A, whereas in the 'real' system, biotite survives becoming more ferric and Ti-rich, for example. In this case the KFMASH univariant reaction A is smeared out to become a higher-variance field. Similarly, on addition of Na_2O and CaO, and the inclusion of plagioclase, KFMASH relationships are smeared out. Although the resulting relationships are more complex, the essence of the above discussion still holds. For example, reaction A, on addition of Na_2O and CaO, and the inclusion of plagioclase, becomes the divariant field, bi + cd + g + plag + l.

The experimental studies of metapelite melting by Vielzeuf & Holloway (1988) and Le Breton & Thompson (1988) were undertaken in 'real' systems. The results of Vielzeuf & Holloway are not relevant here because they do not include alkali feldspar; however, the results of Le Breton & Thompson may be related to Figure 5.3, and particularly to Figure 5.6. Their results may be considered in terms of KFMASH, the experimentally determined melting 'reaction' corresponding to the bi + g + l divariant field on Figure 5.6. At 10 kbar, this field extends from 800°C to 850°C, and so, if it has the positive slope suggested by Le Breton & Thompson, it should extend from 750°C to 800°C in the vicinity of reaction A on Figure 5.6. These conditions may be related to the thermometry results on the Round Hill rocks, given that the X_{Fe} of the biotites in the rocks and the experiments are approximately the same. There is agreement in the temperature ranges, although a strict comparison is unwarranted because the experiments were undertaken in a 'real' system; thus the experimentally determined bi + g + plag + l melting reaction has a much higher variance.

5.5 Conclusions

The garnet porphyroblast-bearing leucosomes in (nearly) granulite facies metapelites at Round Hill, Broken Hill, can be accounted for as the consequence of crossing an incongruent melting reaction, with the garnet as a reaction product. The melt occurs around the garnet as a diffusion-controlled texture due to the difficulty of nucleating garnet, with a consequence that melt segregation is a chemical rather than a physical process. An implication of several features of these rocks, particularly the similarity of mineral compositions in the leucosomes and their hosts and the non-melt composition of the leucosomes, is that equilibration volumes were large under near-metamorphic-peak conditions. From simple phase-diagram arguments, the preservation of the garnet porphyroblasts suggests some melt loss at near-metamorphic-peak conditions.

Many of these conclusions are relevant to the formation of mafic porphyroblast-bearing leucosomes in general. Preservation of the porphyroblasts usually requires some melt loss. Difficulties with mineral composition relationships can usually be understood in terms of equilibration volume arguments. Of course, detailed arguments must be developed for each particular example, including consideration of P–T paths.

Acknowledgements

We would like to thank Dave Waters and John Clemens for careful reviews; R.P. would like to thank Michel Guiraud, Mike Sandiford, Thomas Will, Neil Phillips, and Tim Holland for valuable discussions. Of course, the remaining infelicities are ours!

References

Grant, J. A. 1985. Phase equilibria in partial melting of pelitic rocks. In *Migmatites*, J. R. Ashworth (ed.), 86–144. Glasgow: Blackie.

Hensen, B. J. 1971. Theoretical phase relations involving cordierite and garnet in the system $MgO–FeO–Al_2O_3–SiO_2$. *Contributions to Mineralogy and Petrology* **33**, 191–214.

Holland, T. J. B. & R. Powell 1990. An enlarged and updated internally consistent thermodynamic dataset with uncertainties and correlations: the system $K_2O–Na_2O–CaO–MgO–MnO–FeO–Fe_2O_3–Al_2O_3–TiO_2–SiO_2–C–H_2–O_2$. *Journal of Metamorphic Geology* **8**, in press.

Le Breton, N. & A. B. Thompson 1988. Fluid absent (dehydration) melting of biotite in metapelites in the early stages of crustal anatexis. *Contributions to Mineralogy and Petrology* **99**, 226–37.

Phillips, G. N. 1980. Water activity changes across an amphibolite–granulite facies transition, Broken Hill, Australia. *Contributions to Mineralogy and Petrology* **75**, 377–86.

Powell, R. 1978. *Equilibrium thermodynamics in petrology*. London: Harper & Row.

Powell, R. 1983. Fluids and melting under upper amphibolite facies conditions. *Journal of the Geological Society of London* **140**, 629–33.

Powell, R. & T. J. B. Holland 1988. An internally consistent dataset with uncertainties and correlations: 3. Applications to geobarometry, worked examples and a computer program. *Journal of Metamorphic Geology* **6**, 173–204.

Stüwe, K. & R. Powell 1989a. Low-pressure granulite facies metamorphism in the Larsemann Hills area, East Antarctica; petrology and tectonic implications for the evolution of the Prydz Bay area. *Journal of Metamorphic Geology* **7**, 465–83.

Stüwe, K. & R. Powell 1989b. Metamorphic evolution of the Bunger Hills, East Antarctica: evidence for substantial post-metamorphic peak compression with minimal cooling in a Proterozoic orogenic event. *Journal of Metamorphic Geology* **7**, 449–64.

Stüwe, K. & R. Powell 1989c. Metamorphic segregations associated with garnet and orthopyroxene porphyroblast growth: two examples from the Larsemann Hills, East Antarctica. *Contributions to Mineralogy and Petrology*, in press.

Thompson, A. B. 1982. Dehydration melting of pelitic rocks and the generation of H_2O-undersaturated granitic liquids. *American Journal of Science* **282**, 1567–95.

Vielzeuf, D. & J. R. Holloway 1988. Experimental determination of the fluid-absent melting relations in the pelitic system. *Contributions to Mineralogy and Petrology* **98**, 257–76.

Waters, D. J. 1988. Partial melting and the formation of granulite facies assemblages in Namaqualand, South Africa. *Journal of Metamorphic Geology* **6**, 387–404.

CHAPTER SIX

Isotopic modification of the continental crust: implications for the use of isotope tracers in granite petrogenesis

Stephen M. Wickham

6.1 Introduction

The use of isotope systematics in the study of igneous and metamorphic processes has revolutionized our understanding of the geochemical evolution of the Earth. This technique has been particularly important in studies of crustal magmatism, and has permitted the identification of granitoid source regions and the evaluation of magmatic mixing process (e.g. O'Neil & Chappell 1977, Taylor & Silver 1978, Allègre & Ben Othman 1980, Farmer & DePaolo 1983, Fleck & Criss 1985, Taylor 1980, DePaolo 1981, Vitrac-Michard *et al.* 1980). More recently, these studies have been expanded to include metamorphic processes and fluid (in addition to magma) transport at all levels in the crust.

Isotope geochemistry is uniquely powerful in tackling these problems, because isotopic compositions are in general not significantly modified by igneous processes involving chemical fractionation, and are therefore able to 'see through' these complicating effects back to the early stages of magma formation (Taylor & Sheppard 1986). Furthermore, different major lithological reservoirs often have distinctive isotopic compositions, or in the case of radiogenic isotopes have compositions that evolve through time at different rates, such that they can be used to calculate the relative contribution made by these different reservoirs to a given magma body (e.g. Vitrac-Michard *et al.* 1980). In well constrained examples they may also identify the origin and amount of hydrothermal fluid that has interacted with a given body of rock (e.g. Norton & Taylor 1979).

This chapter will be primarily concerned with the evolution of crustal $^{87}Sr/^{86}Sr$ and $^{18}O/^{16}O$ ratios during metamorphism, melting, and pluton emplacement. In modelling granitoid petrogenesis, workers have often used characteristic isotopic compositions of the various possible source materials (mantle, crustal igneous, crustal metasedimentary, etc.) in order to estimate their relative proportions within a particular granitoid. Here it will be argued that this approach can be a gross oversimplification, that may easily lead to major errors of interpretation. This is because modelling of granitoid petrogenesis must take into account a wide range of isotopic mixing processes that occur not only at the magmatic stage (e.g. assimilation-fractional crystallization processes – see Section 6.2 below), but also at the prograde subsolidus stage as a result of interactions with hydrothermal fluids. It has recently become apparent that very significant (and hitherto poorly recognized) changes in the isotopic composition of large volumes of the crust may occur at this time (Bickle *et al.* 1988, Wickham & Peters 1988). These phenomena, which appear to occur during the prograde (heating) stage in certain types of metamorphic terrane, may result in profound isotopic modification of source regions, and must also be taken into account in any petrogenetic modelling that attempts to identify specific magma sources.

Thus, in evaluating likely granite source materials, it is not valid to assume that particular crustal lithologies (e.g. shales or greywackes) evolve as chemically closed systems throughout their subsolidus (metamorphic) histories, and it can be misleading to assign to these materials characteristic, fixed isotopic compositions (in the case of stable isotopes) or fixed parent/daughter isotope ratios (in the case of radiogenic isotope systems). Isotopic mixing and exchange can occur at the prograde metamorphic *and* the magmatic stage, and both must be assessed if serious errors of interpretation are to be avoided.

6.2 Geochemical evolution of silicic magmas

Most studies of the petrogenesis of granitoids have tended to focus on the early stages of melting (e.g. the many studies of migmatites from high-grade metamorphic sequences; Mehnert 1968, Gupta & Johannes 1982, Ashworth 1985) or the final stages of pluton emplacement (e.g. White & Chappell 1988, Fourcade & Allègre 1981). It has only rarely been possible to examine a km-sized silicic magma body adjacent to its source rock and hence directly link this magma to its source (e.g. Joplin 1942, Wickham 1987a). This is because most bodies, once formed, rise through several kilometres to much higher levels in the crust. Although isotopic studies have improved the situation to the extent that granitoids have been specifically exploited as a means by which to sample and define the major lithological units within the lower crust (Farmer & DePaolo 1983), the overall process leading from source rock to pluton is nevertheless exceedingly complex. Only in the very simplest of cases will a granitoid directly image its source region. The final isotopic composition of a given pluton reflects a number of different processes which we will first review.

The geochemical evolution of crustally derived silicic magmas and their source regions may be conveniently classified in the following way (see Fig. 6.1).

(a) *Magma directly images source.* This is the simplest scenario and involves the direct inheritance of the isotopic composition of the source region. The source undergoes closed-system melting in response to heating by any appropriate mechanism. Melting proceeds to the point where a magma is formed, either by small-scale segregation of liquid at low melt fraction

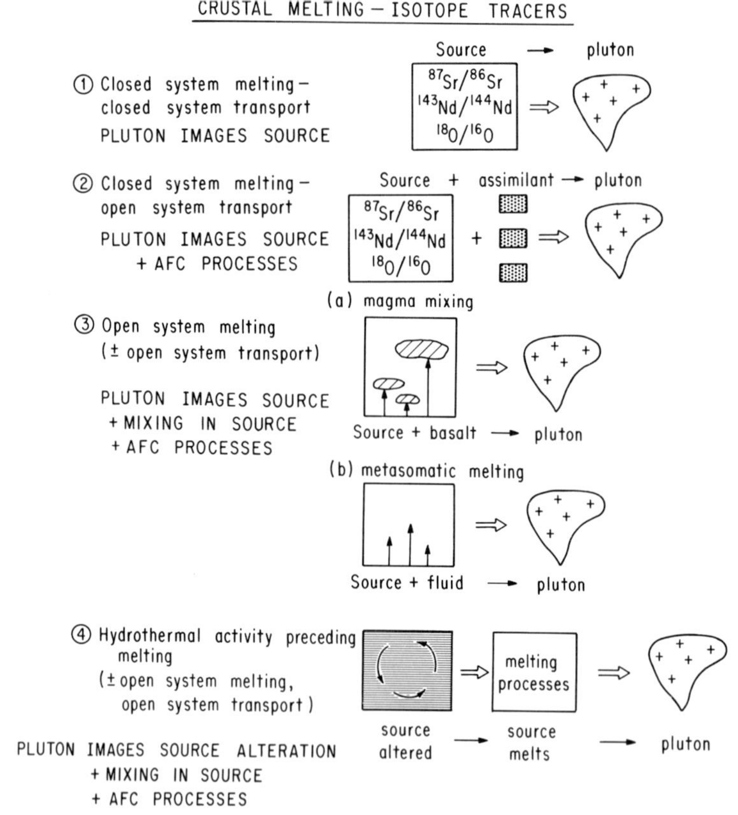

Figure 6.1 Illustration of the main factors influencing the isotopic composition of a silicic magma. In the simplest scenario (1) the source rock undergoes closed-system melting and generates a magma which ascends *without interaction with other rocks* to its emplacement position, thereby inheriting exactly the same composition as the source. In (2) melting is again closed-system, but transport is accompanied by assimilation of country rock and the isotopic composition of the pluton is controlled by the composition of both source *and* assimilant. In (3) a further degree of complexity is added, with the source undergoing open-system melting as a result of (a) magma mixing or (b) fluid infiltration. The most complex situation (4) involves pre-anatectic subsolidus modification of the source as a result of hydrothermal activity accompanying prograde metamorphism. Any or all of the other sequences ((1), (2), or (3)) may potentially be added to this last scenario.

(<30% melt) or by large-scale melting to in excess of 50% (Wickham 1987b). The magma ascends to its ultimate level of emplacement in the crust, and cools and crystallizes with no chemical interaction with adjacent country-rock lithologies at any level. The pluton thus inherits exactly the same isotopic composition as its source region (see Fig 6.1).

[To be absolutely correct, a low-melt fraction segregation from a quartzofeldspathic rock such as a pelite may be slightly more ^{18}O-rich than its source, because it will be derived mostly from the melting of quartz and feldspar, and these minerals are the most ^{18}O-rich in a typical crustal source rock. However, as with fractional crystallization, the magnitude of this effect is unlikely to be more than a few tenths of a per mil (Taylor & Sheppard 1986) inasmuch as fractionation factors between minerals are small at magmatic temperatures. Small changes in Rb/Sr ratios may also occur in similar ways at this time.]

(b) *Magma assimilates material during ascent.* This process involves closed-system melting within the source, followed by contamination of the magma after it has left the source and while it is in transit to its emplacement level. The physical and chemical laws constraining this process were originally recognized by Bowen (1928) and, in recent times, many workers have attempted to model the influence of assimilation-fractional crystallisation (AFC) processes on isotope systematics (e.g. Taylor 1980, DePaolo 1981). At its simplest, such a process involves three end-members (magma, country rock, and cumulates), all of which may have different concentrations of the element of interest, and two of which (magma and country rock) may have widely different isotopic compositions. Inasmuch as magmas are rarely superheated, assimilation does not lead to the formation of any more liquid than was originally present. It simply speeds up crystallization, releasing a latent heat which promotes fusion of the assimilant. Thus, the most contamination occurs in the latest liquid to crystallize. It is quite likely that natural systems will be affected by these processes in far more complex ways than the simple models that have been devised to date but, despite these difficulties, this type of process convincingly explains the commonly observed positive correlation between initial $^{87}Sr/^{86}Sr$ and $^{18}O/^{16}O$ ratios in silicic magma series (e.g. Vitrac-Michard *et al.* 1980, Fleck & Criss 1985). Such correlations reflect mixing between isotopically primitive sources (mantle-derived rocks or rocks with a large proportion of mantle-derived material within them) and isotopically evolved crustal rocks (typically, but not necessarily, sediments). Of course, these (virtually ubiquitous) correlations may arise in other ways that do not involve AFC (magma mixing in the source for example, see below) and, furthermore, it is conceivable that AFC processes could occur cryptically, without imparting an isotopic signature, if the contaminant had a very similar geochemical signature to the original magma. However, notwithstanding these difficulties, isotopic variation in a suite of igneous rocks can often place tight constraints on the nature and extent of assimilation processes, allowing us to 'strip off' the effects of secondary contamination and still accurately identify silicic magma source regions, and study source melting processes.

(c) *Open-system source melting* (± *AFC processes*). A further degree of complexity is introduced if we consider open-system melting processes in the source region. Such effects could involve a number of scenarios which, for the sake of brevity, are subdivided into two categories (Fig. 6.1); those involving (i) magma mixing in the source region and (ii) source interaction with volatile-rich fluid. In case (i), mixing is typified by the interaction of a magmatic heat source (e.g. basalt) with the region within the crust that is being melted (the basalt could also be the principal source of heat as well as being a contaminant; e.g. Huppert & Sparks 1988). Such mixing between mafic and silicic magma has been convincingly documented in a number of cases (e.g. Sparks & Marshall 1986, Frost & Mahood 1987), although we have very little detailed understanding of the kinematics of this process. This type of open-system process is (relatively) easy to recognize because the contaminant is often basaltic (or at least relatively primitive) and has a fairly consistent geochemical signature that is often distinct from that of the rocks that are being melted. In case (ii), the effects are likely to be much more unpredictable, cryptic, and difficult to detect. This is because of the wide range of possible geochemical signatures that can result from metasomatic effects. Metasomatic melting is important because this process can theoretically flux the melting of even very refractory crustal lithologies (e.g. quartzite) via the introduction of alkalis and alumina. Many data exist to support the concept of geochemical changes during granitoid petrogenesis, largely collected by geologists seeking to document subsolidus 'granitization' processes (e.g. Reynolds 1946). Some of these data relate to processes occurring at relatively shallow crustal levels during the later stages of pluton emplacement, but there are also plenty of examples of metasomatic changes accompanying or even promoting migmatization (e.g. Olsen 1984). Since metasomatism invariably involves interaction with an H_2O- or CO_2-rich fluid (often also rich in alkalis), this type of process has enormous potential to disturb Rb/Sr ratios and O and Sr isotopic compositions. This type of process overlaps in part with (d) below but is here taken to refer to metasomatism *during melting* of the source. However, it may be impossible to make this distinction on the basis of isotopic parameters alone, and in most cases it will be extremely difficult to prove that a metasomatic fluid and a melt *coexisted* within the source region.

(d) *Source modified before melting* (± *open-system melting, AFC processes*). Obviously, a virtually limitless variety of geochemical processes may affect a crustal source region during its (potentially very long) history prior to a melting event. This discussion will be concerned mainly with chemical changes occurring prior to melting, but during the same thermal (metamorphic) event responsible for generating the silicic magma body in question. Although sedimentary rocks are undoubtedly a major source component in silicic magmas, it is not the sediments themselves that melt but their metamorphosed equivalents. Before a sedimentary unit will melt, it (and the crust around it) must be heated to temperatures in excess of 700–800°C. If heat transport is conductive, this process must take at least tens of thousands (and possibly millions) of years, and it is likely to be accompanied by various types and degrees of chemical and isotopic change. An example that has long been recognized is devolatilization, but it is becoming increasingly apparent that there may be loss or gain of other elements, with or without associated isotopic changes (Fleck & Criss 1985, Wickham *et al.* 1987, Bickle *et al.* 1988). Recent studies have suggested major shifts in oxygen isotopic composition during metamorphism (Garlick & Epstein 1967, Wickham & Taylor 1985), and it follows that if the most abundant element in the rock has been exchanged over wide regions, other elements may have been similarly affected during heating to solidus temperatures. It is therefore dangerous to use the chemical and isotopic composition of low-grade sedimentary rocks as an end-member in modelling the source materials of silicic magma bodies, even if it is certain that these same sediments were a major source material for the magma in question. For example, it is commonly assumed that shales have $\delta^{18}O$ values of about +12 to +18 per mil, and that their metamorphosed equivalents and silicic magmas derived therefrom will have a similar isotopic composition. However, some high-grade terranes contain pelitic rocks and migmatite leucosomes (inferred to be derived by melting of the same material) with $\delta^{18}O$ values as low as +7 (Wickham *et al.* 1987, and unpublished data). This implies that crustal metasediments can in some circumstances take on any $\delta^{18}O$ value within the range of normal crustal values, and anywhere within the range of values observed in normal igneous rocks.

In seeking to understand the origin of granitoids, it is clearly very important to evaluate the nature and extent of these subsolidus, pre-anatectic geochemical changes. In particular, it is very pertinent to examine those terranes in which granitoids and the low-grade and high-grade equivalents of their source materials can be unambiguously identified and are accessible, and for which isotope data are available. In this way we are best able to recognize compositional and isotopic changes that occur at the prograde metamorphic (subsolidus) stage preceding melting. The remainder of this chapter will review data from several terranes in which significant subsolidus, pre-anatectic changes in the Sr and O isotopic composition of the metasedimentary source rocks of granitoid plutons have demonstrably occurred.

6.3 Anomalously low $^{87}Sr/^{86}Sr$ ratios in some crustally derived granitoids

6.3.1 The Trois Seigneurs Massif, Pyrenees, France

Hercynian basement terranes exposed in the Pyrenees comprise Palaeozoic (and possibly Late Proterozoic) metasedimentary rocks, metamorphosed in high-thermal gradient environments during the Late Carboniferous and intruded by various kinds of granitoid (see various papers in Banda & Wickham 1986). These granitoids can be broadly classified into two separate groups: a volumetrically extensive group of post-metamorphic plutons, dominantly composed of biotite and hornblende granodiorite; and a volumetrically minor group of synmetamorphic peraluminous plutons that are intimately associated with the high-grade rocks.

In the Trois Seigneurs Massif, a typical prograde Hercynian metamorphic sequence is particularly well exposed. The metasediments, which are dominantly pelitic but also contain a few layers of metacarbonate and quartzite, range continuously over a distance of only a few kilometres from fossiliferous chlorite-grade shales and phyllites through andalusite- and sillimanite-bearing mica-schists to anatectic migmatites and a strongly peraluminous biotite- and cordierite-bearing granite complex (the 'biotite granite'). This sequence has been the subject of a number of detailed field, petrological, and geochemical studies (Wickham 1987a, Wickham & Taylor 1985, Bickle *et al.* 1988). These indicate that the biotite granite complex is derived from at least three sources with distinct Sr and O isotopic compositions. The dominant component (particularly important within the upper portion of the pluton) was derived by melting and homogenization of the adjacent pelitic metasedimentary sequence under water-rich conditions (Wickham 1987a).

Detailed oxygen, hydrogen, and strontium isotopic studies of the entire massif (including both the low- and high-grade metasediments and the granitic rocks) are discussed by Wickham & Taylor (1985, 1987) and Bickle *et al.* (1988). The data show that the schists at grades higher that the 'andalusite in' isograd have lower $\delta^{18}O$ values (+ 11 to + 13) than their low-grade protoliths, the shales and phyllites which have $\delta^{18}O$ of + 14 to + 16. Carbonate horizons (10–30 m thick) within the higher-grade rocks were even more profoundly depleted in ^{18}O, producing homogeneous whole-rock values almost identical to those in the surrounding mica-schists. The values measured in the synmetamorphic granitoids were also similar to those in the high-grade pelites, except at deep levels within the biotite granite complex (Bickle *et al.* 1988).

The $^{18}O/^{16}O$ systematics at Trois Seigneurs were interpreted by Wickham & Taylor (1985, 1987) to result from the pervasive interaction of the high-grade metasediments with a large quantity of a relatively ^{18}O-poor aqueous hydrothermal fluid during Hercynian metamorphism and anatexis. Fluid infiltration is unlikely to have occurred during retrograde cooling because disequilibrium isotopic effects are not observed in these rocks. The high peak-metamorphic thermal gradient (Zwart 1962, Wickham 1987a) precludes major fluid circulation at this time, and it is most likely that fluid circulation took place during prograde metamorphism, and had largely ceased by the time that peak-metamorphic temperatures were reached (see Wickham & Taylor 1985, 1989 for a full discussion of this point). Prograde fluid circulation was responsible for homogenizing oxygen isotopic compositions and for the shift of about 3 per mil in the isotopic composition of the high-grade rocks relative to their low-grade precursors.

Strontium isotope studies (Bickle *et al.* 1988) confirm the interaction of the terrane with hydrothermal fluids during the Hercynian metamorphism. These data show that the low-grade shales and phyllites, the high-grade pelites, and the structurally higher parts of the biotite granite complex all had very similar $^{87}Sr/^{86}Sr$ values of about 0.715 at 310 Ma during the Hercynian metamorphism. Good Hercynian Rb/Sr isochrons were obtained for sets of small slices cut from larger samples of metasediment from all grades within the terrane, and for samples from the upper and intermediate parts of the biotite granite complex (see Fig. 6.2). Similar systematics indicating strontium isotope homogenization in metasediments during metamorphism have been observed elsewhere (e.g. Pidgeon & Compston 1965), but this is by far the most detailed study of a single prograde metamorphic sequence that has yet been made. A suite of unsliced metasediment samples collected from throughout the metamorphic sequence also gave a Hercynian isochron, although much more crudely defined. The combined data indicate Sr isotope homogenization at least on an outcrop scale, and possibly on a regional scale. The homogenization of $^{87}Sr/^{86}Sr$ to 0.713–0.717 at 310 Ma is probably connected with the Hercynian metamorphism because of the coincidence in ages. Moreover, the oxygen isotope shifts are definitely Hercynian because they are restricted to the higher-grade Hercynian-age metamorphic rocks. If, as seems plausible, the Sr isotope effects occurred at the same time, this further endorses an Hercynian age for them as well. The strontium isotope effects (and possibly other geochemical effects) must have occurred (or at least commenced) at the very earliest stages of the heating process, because they are observed within the lowest-grade rocks at Trois Seigneurs.

The Sr and O isotope data from Trois Seigneurs are plotted in Figure 6.3 (taken from Bickle *et al.* 1988), where the relative homogeneity of the metasediments is clearly seen. Also plotted in this diagram are the model $^{87}Sr/^{86}Sr$ values that these rocks would have been expected to attain at 310 Ma, had they started out as Lower Palaeozoic sediments with initial $^{87}Sr/^{86}Sr$ ratios of 0.707 and retained their measured Rb/Sr ratios throughout. The model values (0.709–0.736) of the higher-grade samples are in all but one sample significantly higher than are actually observed. This implies not only that Sr isotope ratios were homogenized during metamorphism, but also that they were

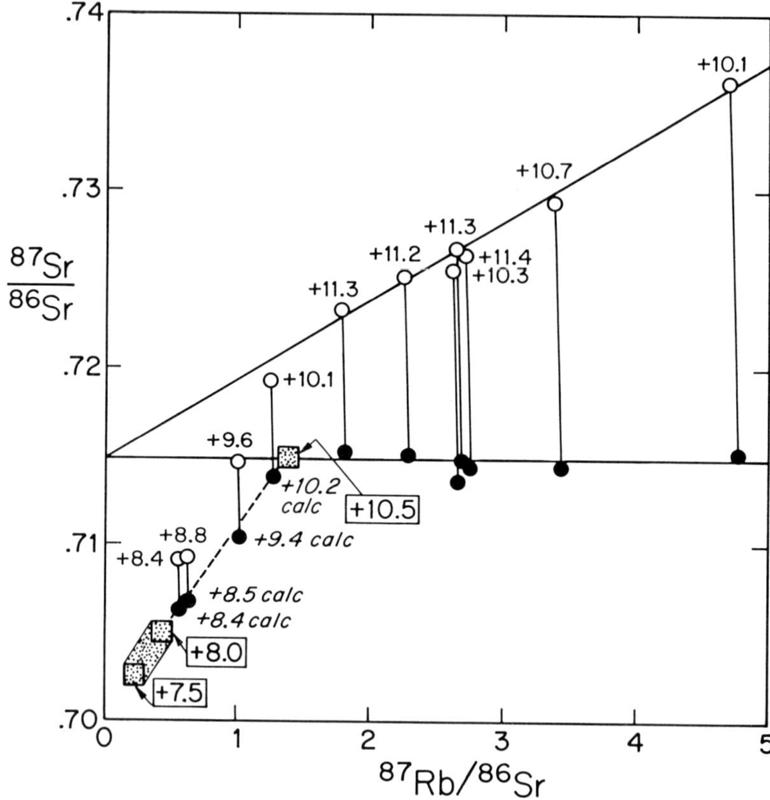

Figure 6.2 Rb–Sr isochron diagram reproduced from Bickle *et al.* (1988) by permission of Springer-Verlag, showing measured data for the Trois Seigneurs biotite granite (open circles) and also plotting the calculated values that these samples would have had at 310 Ma (solid circles). $\delta^{18}O$ values are shown above each of the data points. The 310 Ma values of seven of the samples plot close to a horizontal line at $^{87}Sr/^{86}Sr = 0.7149$ (these samples define a good Hercynian isochron). The other four samples define a mixing line apparently involving a low-^{87}Sr, low-$\delta^{18}O$ end-member and a high-^{87}Sr, high-$\delta^{18}O$ end-member. The $\delta^{18}O$ values of mixtures also vary approximately linearly along this line. Using end-member $\delta^{18}O$ values of +10.5 and either +7.5 or +8.0, the calculated $\delta^{18}O$ of each of the four samples (shown in italics) can be seen to match the measured value fairly closely.

substantially lowered as a result of mixing with relatively unradiogenic strontium ($^{87}Sr/^{86}Sr < 0.715$). As yet, we are unable to define the overall scale of this process, but it appears to affect pelitic metasediments throughout the Trois Seigneurs sequence. Interestingly, the same homogenization is observed within the lower-grade samples (shales and phyllites), although there is more scatter in the data. If anything, the low-grade samples have even lower $^{87}Sr/^{86}Sr$ ratios than the high-grade mica-schists, indicating that Sr isotopic alteration must have commenced at very low temperatures (<300°C?).

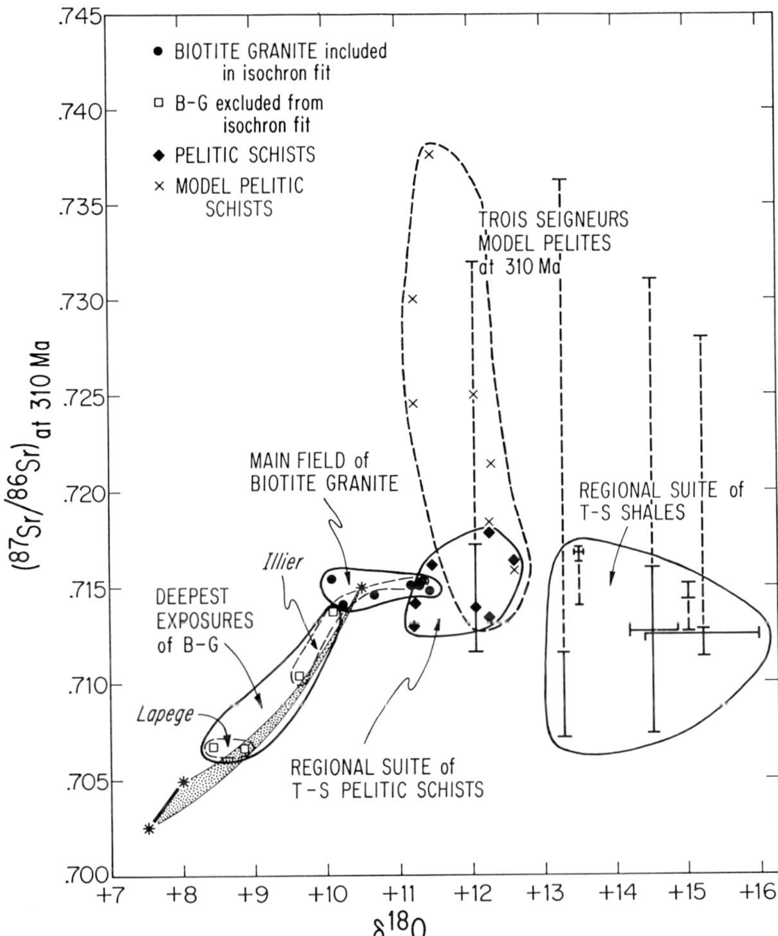

Figure 6.3 Plot of $^{87}Sr/^{86}Sr$ *vs.* $\delta^{18}O$, showing the biotite granite (B–G) data from Figure 6.2 together with data for a regional suite of Trois Seigneurs (T–S) pelitic schists, shales, and phyllites (from Bickle *et al.* 1988 by permission of Springer-Verlag). The vertical and horizontal brackets within the shale field indicate the range of isotopic compositions at the five outcrop localities that were studied, with the centres of the brackets indicating the average value at each locality. Also shown are the following. (a) A calculated field (crosses enclosed by the dashed envelope) of model pelitic schists representing the $^{87}Sr/^{86}Sr$ values that these samples would have attained at 310 Ma if they had evolved from sediments deposited at 560 Ma with an initial $^{87}Sr/^{86}Sr$ ratio of 0.707, assuming that they behaved as closed systems with respect to Rb and Sr (note that if they had also behaved as closed systems with respect to oxygen, the $\delta^{18}O$ values of these model schists would also have been much higher prior to metamorphism than is shown in the figure, namely about +15). (b) The calculated range (vertical dashed lines) of model $^{87}Sr/^{86}Sr$ ratios that the regional suite of shales and phyllites would have attained at 310 Ma if they had evolved from sediments deposited at 450 Ma with an initial $^{87}Sr/^{86}Sr = 0.707$, again assuming closed-system behaviour. The measured T–S pelitic schists define a fairly tight cluster of points, with much more homogeneous $^{87}Sr/^{86}Sr$ values at 310 Ma than the range of values shown by the two groups of model pelites. This suggests that, like the $\delta^{18}O$ values, the $^{87}Sr/^{86}Sr$ ratios were changed and partially homogenized in the metasediments during metamorphism. The mixing line for the biotite granite suite from Figure 6.2 is also shown (it transforms to the curved triangular stippled region on this diagram, with the asterisks representing model end-members). The homogenized Trois Seigneurs pelites clearly constitute a plausible high-^{18}O, high-^{87}Sr end-member source material for the main body of the biotite granite.

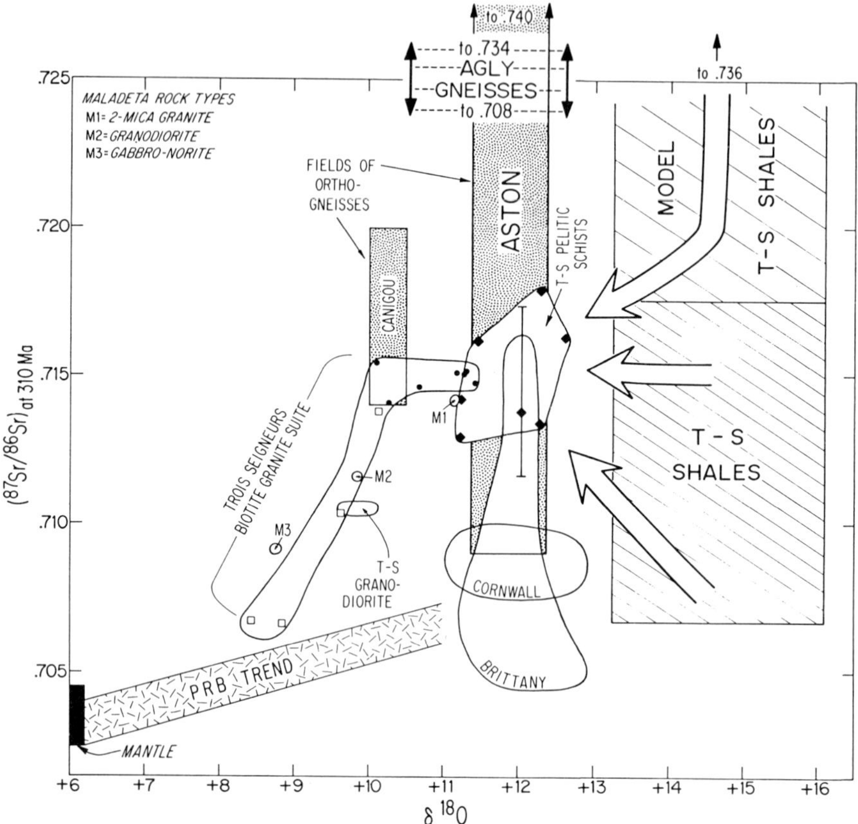

Figure 6.4 Plot of $^{87}Sr/^{86}Sr$ at 310 Ma *vs.* $\delta^{18}O$ (from Bickle *et al.* 1988, by permission of Springer-Verlag), comparing the Trois Seigneurs (T–S) granite and pelite data with data from the Maladeta plutonic complex in the central Pyrenees (M1, M2, and M3; Vitrac-Michard *et al.* 1980). Also shown are fields for the measured values of Trois Seigneurs shales and phyllites at 310 Ma, together with the calculated range of model values at 310 Ma for these same rocks (see Bickle *et al.* 1988). The Trois Seigneurs pelitic schists are an obvious high-^{18}O, high-^{87}Sr end-member for the biotite granite suite, and they have much more homogeneous $^{87}Sr/^{86}Sr$ ratios than either the model shales or the basal gneisses and orthogneisses from other parts of the Pyrenees (Jäger & Zwart 1968; Vitrac-Michard & Allègre 1975; Ben Othman *et al.* 1984; Majoor & Priem 1987; Wickham & Taylor 1985, 1987, unpublished data). The large arrows indicate schematically how the $\delta^{18}O$ and $^{87}Sr/^{86}Sr$ values of the shales and phyllites have been changed during prograde Hercynian metamorphism. Fields for the upper mantle, for the main part of the Peninsular Ranges Batholith (PRB) from Taylor & Silver (1978), and for the Hercynian granites from Brittany and Cornwall, south-west England (Sheppard 1986) are also shown. Despite their having major metasedimentary components in their sources, these other Hercynian granitic rocks also have relatively low $^{87}Sr/^{86}Sr$ ratios, suggesting that the source rocks for these magmas may also have been affected by the same sort of $^{87}Sr/^{86}Sr$ homogenization and modification processes that appear to have affected the Trois Seigneurs pelites.

Unfortunately, we have no way of knowing what the original sedimentary $\delta^{18}O$ values of the metasediments were, and therefore we cannot say if oxygen isotope alteration also commenced at this time.

One of the most important aspects of this data set is that it allows us to observe the succession of geochemical changes that accompanies the generation of a crustally derived magma body. Although most of the biotite granite is derived from relatively high Rb/Sr pelites, its final Sr and O isotopic composition is very different to the values that it would have inherited had its source material evolved as a closed system during metamorphism and anatexis. This composition would have been about $\delta^{18}O = +15$, with an initial $^{87}Sr/^{86}Sr$ of anywhere between 0.720 and 0.740, quite different to the measured values in the granite and in the higher-grade mica-schists of about 0.715 (at 310 Ma) and +11.5. If the biotite granite had evolved to the point where it had been able to leave its source region and migrate to higher crustal levels, where it could be sampled in isolation from its source materials, its measured initial $^{87}Sr/^{86}Sr$ and $\delta^{18}O$ values would be anomalously low when compared to the age and Rb/Sr ratio of the low-grade pelitic sediments that form a major component of its source. These features are illustrated in Figure 6.4 (Bickle *et al.* 1988), in another Sr–O plot which shows the inferred evolution of the Palaeozoic shales during their metamorphic and anatectic history, involving reduction of both $^{87}Sr/^{86}Sr$ and $\delta^{18}O$ at the prograde metamorphic stage, followed by a further tightening of $^{87}Sr/^{86}Sr$ (presumably reflecting magmatic mixing) going from the high-grade pelites to the biotite granite.

6.3.2 *The Hercynian of Brittany*

The fields for other sets of data from Hercynian granites in Brittany and south-west England (from the compilation of Sheppard 1986) are also shown in Figure 6.4. The Brittany data were recently presented and discussed in detail by Peucat *et al.* (1988). All of these granites are peraluminous and have $\delta^{18}O$ values in excess of +11, requiring them to be dominantly derived from a sedimentary source; yet their initial $^{87}Sr/^{86}Sr$ ratios are relatively low (mostly between 0.704 and 0.712). A wide variety of potential source lithologies are exposed in Brittany, including a variety of different Palaeozoic and Proterozoic rocks. The most obvious source materials for these granites are the Palaeozoic and Precambrian metasediments that are exposed throughout the Brittany area, and which have also been studied by various workers. Peucat *et al.* (1988) carefully evaluated these data and used them to estimate the $^{87}Sr/^{86}Sr$ ratios in these metasediments at 340–300 Ma (the age range of the granitoids). They found that virtually all of the local metasediments would have had much higher $^{87}Sr/^{86}Sr$ ratios at 340 Ma than those of the granites which, at first sight, appears to rule these lithologies out as potential source components. Indeed, Peucat *et al.* (1988) describe this striking mismatch, whereby none of the local metasediments apparently match the isotopic composition of the granites, as the 'strontium paradox'. They conclude that the real source for the granitoids was 'young unradiogenic rocks which have now disappeared.' A more plausible alternative is that the Palaeozoic and Proterozoic metasediments in Brittany (and by analogy Cornwall too) evolved in a similar way to those in the Pyrenees, and had their $^{87}Sr/^{86}Sr$ ratios homogenized and lowered by interaction with aqueous fluids, immediately prior to melting. In this way the actual source materials of the granites had a similar major-element geochemistry, but a quite different isotopic composition *at the time that they melted* than they did a short while earlier before they were heated to melting temperatures.

It is important to recognize a major difference between the two situations in Brittany and the Pyrenees. In the latter example, metasedimentary source rocks have identical $^{87}Sr/^{86}Sr$ ratios to the granites but do not match the closed-system values that they should have attained; in the former example, the metasediments and granites do not match, and we have to infer that they were altered during the prograde heating event that preceded melting. Part of the problem in Brittany may be one of sampling, because Rb/Sr ratios may vary considerably within the metasediments, and a large data set is required to place firm constraints on bulk Sr isotope evolution (e.g. Bickle *et al.* 1988). Furthermore, even the lowest-grade metasediments in the Trois Seigneurs Massif are chlorite-grade shales and slates, and are therefore not necessarily as 'low-grade' as some of those sampled in Brittany.

As a footnote, it is worth mentioning that other large-scale Hercynian chemical and isotopic alteration has also been noted, in the Pyrenees and elsewhere. These additional examples include the Aston orthogneiss complex within the Pyrenean axial zone (Majoor & Priem 1987), where Sr isotope ratios were homogenized at 0.719 over wide areas at 315 Ma. Furthermore, Vitrac-Michard *et al.* (1981) identified regional homogenization and systematic regional variation (on a scale of hundreds of km) in Pb isotope and U/Pb ratios in feldspars from Hercynian granites from a large part of Western Europe, which they attributed to regional fluid-flow effects. Future studies in these areas will help to identify the extent and the timing of metamorphic isotope alteration events, which clearly play a crucial rule in the geochemical evolution of the crust.

6.3.3 Eastern Nevada

The geology of the Basin and Range Province is much too complex to review in anything but a cursory way here, and further information about this terrane may be found in the many articles and books (e.g. Crittenden *et al.* 1980, Ernst 1988). The significance of the area for this discussion is that it exposes miogeoclinal Palaeozoic and Precambrian sedimentary rocks both as fossiliferous, very-low-grade sediments, *and* as high-grade metamorphic rocks, and that in some areas *the same stratigraphic units* can be identified at high, low, and intermediate metamorphic grade. The high-grade terranes (which have been termed 'metamorphic core complexes') are intruded by a wide variety of different types of granitoid of Jurassic, Cretaceous, and Tertiary ages, reflecting a complex thermal history associated with several distinct orogenic events throughout this period. Several field and geochemical studies have been made of these terranes in Eastern Nevada (e.g. Kistler *et al.* 1981, Farmer & DePaolo 1983, Miller *et al.* 1988, Snoke & Miller 1988), and this discussion will focus on one particular area (the Ruby Mountains, and its continuation to the north-east, the East Humboldt Range) which, in terms of geochemistry, is one of the best documented (e.g. Kistler *et al.* 1981).

The great advantage of this terrane over many others is that is permits unambiguous identification of individual stratigraphic units at a variety of different metamorphic grades. Thus, the variation of the geochemistry of the metasediments through time (and the geochemical response to the various Mesozoic and Tertiary orogenic events) can be studied. The drawback is that the terrane has a complex polymetamorphic history involving several Mesozoic and Tertiary thermal events of poorly constrained magnitude and tectonic significance. However, as far as the evolution of granitoid source rock is concerned, this does not present significant difficulties in the analysis of Sr isotope evolution.

The Northern Ruby – East Humboldt Range exposes high-grade metasedimentary rocks of the miogeoclinal succession, typically intruded by large numbers of metre-scale granitic pods. These pods have variable composition, and are so abundant that in places they make up over 50% of the terrane. A few large plutons (hundred metre to km scale) also occur, some of them intruding at higher structural levels (i.e. within lower-grade metasediments), and these have been assigned Mesozoic or Tertiary ages (Kistler *et al.* 1981, Wright & Snoke 1986). The rocks have undoubtedly experienced more than one high-temperature event, the most recent probably associated with Tertiary Basin and Range extension (as evidenced by many 35–25 Ma K–Ar and Ar–Ar ages throughout the terrane; Dallmeyer *et al.* 1986), and the intrusion ages for most of the granitoids probably lie between 160 and 25 Ma (Kistler *et al.* 1981, Wright & Snoke 1986, Lush *et al.* 1988). However, there is still considerable uncertainty over some of these ages (particularly those determined by the Rb–Sr whole-rock technique), and over what proportion of the (literally thousands) of distinct smaller bodies is Tertiary.

On the basis of a large body of oxygen isotope data, Solomon & Taylor (1989) have argued that the Mesozoic and Tertiary granitoids of the entire eastern half of Nevada are dominantly derived from miogeoclinal sedimentary material rather than cratonal basement (as preferred by Farmer & DePaolo 1983). These data are plotted in Figure 6.5 and, in fact, cover a broad range of $\delta^{18}O$ values, from +6.5 to +13.0. In the Ruby Mountains, the granitoids analysed by Kistler *et al.* (1981) and Wickham *et al.* (1987) fall within the

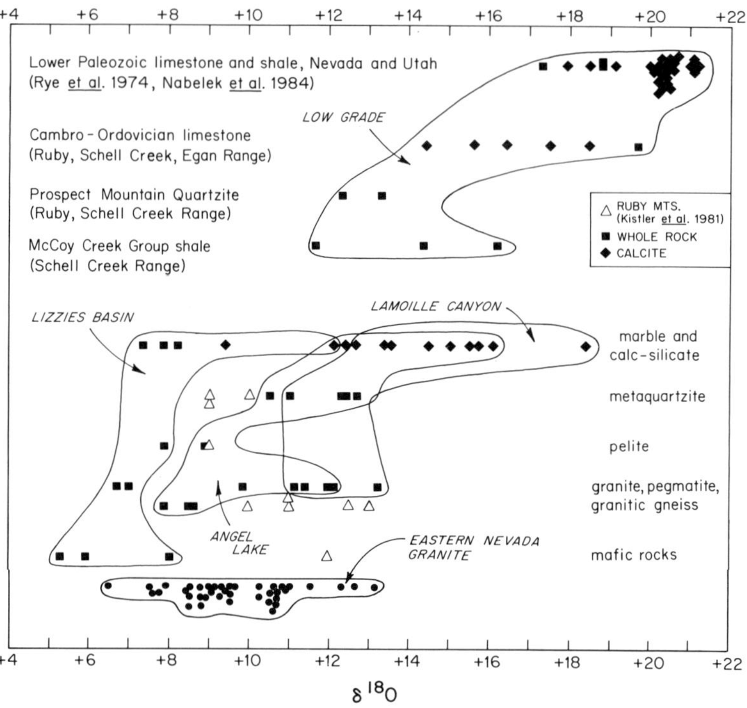

Figure 6.5 A compilation of $^{18}O/^{16}O$ data from high-grade rocks of the Ruby – East Humboldt Ranges, and from low-grade sediments from elsewhere in the Northern Great Basin, which include some possible protoliths to the high-grade metasediments and granites (where not otherwise indicated, data from Wickham *et al.* 1987 and unpublished). The high-grade lithologies are fairly homogeneous at a given locality, but vary considerably between localities, with Lizzies Basin being isotopically lightest and Lamoille canyon being heaviest. Most of the high-grade lithologies have $\delta^{18}O$ values that are much lower than their present-day low-grade equivalents, and almost all the carbonates are much lower than typical sedimentary carbonate values of +20 to +30. These data imply that some of the metasediments have undergone shifts of 5–10 per mil to lighter values during their metamorphic history. Mesozoic and Tertiary granitic rocks span a range of values from +6.5 to +14.0, although the granites at Lizzies Basin (+7) are mineralogically and compositionally similar to those from Lamoille Canyon (+11 to +14), implying that they are both derived from similar source material. The range is similar to that observed by Solomon & Taylor (1989) for many granite plutons from throughout eastern Nevada (solid circles at bottom of diagram), and suggests that the metasedimentary source material for these bodies may have been variably depleted in ^{18}O during its metamorphic history, as an alternative to their being derived from a mixture of lithologically distinct sources.

higher part of this range (+11 to +13) and must have been derived almost exclusively from material that had been through a sedimentary weathering cycle. Most of these bodies are peraluminous two-mica granites that could in principle (on the basis of their chemistry) have been derived from a sedimentary or an igneous source, or a mixture of the two (White *et al.* 1986). The high $\delta^{18}O$ values of two-mica leucogranites from the Ruby Mountains make it clear that here, at least, the granites have mostly come from a sedimentary source, the most likely one being the pelitic and psammitic sediments of the Late Precambrian McCoy Creek Group, the lowest sedimentary unit of the miogeoclinal sequence that is observed in eastern Nevada. This 5 km thick section of shales and quartzites is the only major miogeoclinal unit that has not been clearly identified in the Ruby – East Humboldt Range, and it has probably been almost totally obscured by a combination of high-grade metamorphism, substantial melting, extensional attenuation, and other deformation processes.

Very similar types of granitoid and metamorphic rock occur in the East Humboldt Range, the northeasterly continuation of the Ruby Mountains, but here the $\delta^{18}O$ values that have been measured span a range of values from +6.7 to +12.0 (Wickham *et al.* 1987). Interestingly, this is almost exactly the same range of values that is observed for all granitoids in the eastern half of Nevada by Solomon & Taylor (1989) (see Fig. 6.5). However, the similarities of mineralogy, morphology, and geochemistry of the East Humboldt Range leucogranites to those in the Ruby Mountains make it likely that they have both been derived from similar source material, despite the fact that the $\delta^{18}O$ values span a wide range. The most obvious major source material is again McCoy Creek Group metasediments, implying that the $\delta^{18}O$ values of some of these rocks have been drastically lowered at some stage of their metamorphic history (Wickham *et al.* 1987). These are all fresh, high-grade rocks which show equilibrium high-temperature fractionations, and there is no possibility that these values have been lowered during subsolidus retrograde cooling after the magmatic stage. Indeed, the lack of retrogression, coupled with regional evidence of a widespread Tertiary heating event (as indicated by Ar–Ar and U–Pb ages; Dallmeyer *et al.* 1986, Wright & Snoke 1986) suggest that this latest high-temperature equilibration event was also Tertiary.

The low-^{18}O metamorphic rocks and granites imply major shifts in oxygen isotope composition during metamorphism, resulting from equilibration with a large ^{18}O-depleted oxygen reservoir. Theoretically, the igneous rocks could have been derived from a primitive low-^{18}O source material (e.g. mafic igneous rocks), but their mineralogical and chemical similarities to the very-high-^{18}O granites in the immediately adjacent northern Ruby Range make it more likely that these granites came from a similar metasedimentary source. This automatically implies that a bulk shift in oxygen isotopic composition occurred, either in the metasediments prior to melting, or in the granites derived therefrom.

Strontium isotope data are available for some of the Ruby metasediments and granites (Kistler *et al.* 1981), and for a few of the other eastern Nevada granitoids. Almost all of these granites have anomalously low initial $^{87}Sr/^{86}Sr$ ratios compared to the values that typical McCoy Creek Group shales would have evolved to at the same time. This is illustrated in Figure 6.6, which also

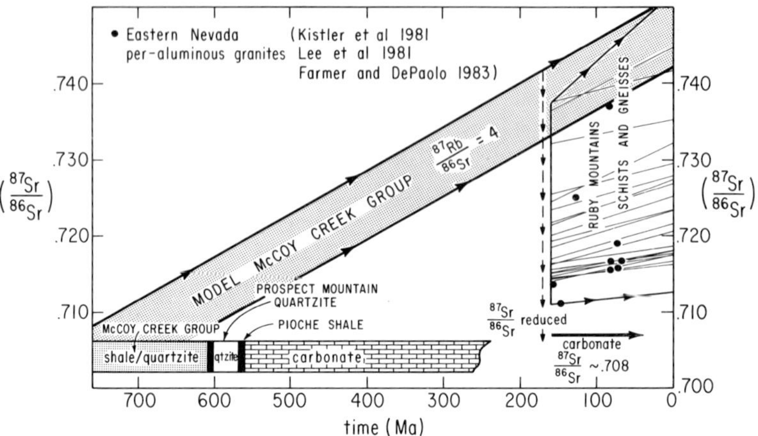

Figure 6.6 The evolution of $^{87}Sr/^{86}Sr$ ratios with time from the Late Proterozoic to the present day for eastern Nevada granites and metamorphic rocks. The trapezium on the right of the diagram outlines the field of schists and gneisses from the Ruby Mountains analysed by Kistler *et al.* (1981). Inasmuch as the age of many of these rocks is in question, evolution lines back to 160 Ma have been calculated for these samples (fine unbroken lines). The black circles within this field represent the initial ratio and age of peraluminous granites from eastern Nevada (Kistler *et al.* 1981, Lee *et al.* 1981, Farmer & DePaolo 1983). The horizontal lithostratigraphic column depicts the general distribution of lithologies within the miogeoclinal sedimentary sequence. The oldest unit (the McCoy Creek Group) is the youngest major lithology that could form a significant source component for some of the granites in this area. The predicted evolution of the $^{87}Sr/^{86}Sr$ ratio of this unit is shown (diagonal stippled band) assuming an initial depositional value of 0.708 and a bulk Rb/Sr ratio of 1.38 ($^{87}Rb/^{86}Sr = 4.0$). This value is probably considerably lower than the actual average Rb/Sr ratio of this unit but, even so, the expected Mesozoic and Tertiary values are considerably more radiogenic than most of the measured values of the metamorphic rocks or granites. This suggests that the $^{87}Sr/^{86}Sr$ ratios of the high-grade rocks have been lowered (dashed vertical line), perhaps through mixing or exchange with miogeoclinal carbonate strontium, which would have maintained a fairly constant $^{87}Sr/^{86}Sr$ ratio of about 0.708 since deposition (black arrow).

shows the evolution lines for Ruby Mountains metasediments (Kistler *et al.* 1981) and initial ratios for several eastern Nevada granitoids (Lee *et al.* 1981, Farmer & DePaolo 1983).

Although the ages of several of these plutons are not particularly well constrained, this does not affect the conclusions drawn here because they are certainly all Jurassic or younger, and are all constrained to lie within the box on the right-hand size of the diagram. This box also contains the Sr isotope evolution lines of the suite of metamorphic rocks analysed by Kistler *et al.* (1981), and thus represents almost all the published Sr isotope data from eastern Nevada. The *youngest* sedimentary material that could have formed the protolith to any of the granitoids is material from the McCoy Creek Group, which is also depicted in Figure 6.6, together with the rest of the miogeoclinal stratigraphic column. Even assuming what is probably an unreasonably low, conservative estimate of the $^{87}Rb/^{86}Sr$ ratio in the McCoy Creek Group rocks of 4.0 (equivalent to an Rb/Sr of about 1.4), it is clear that these rocks should have evolved to much more radiogenic $^{87}Sr/^{86}Sr$ by Mesozoic and Tertiary times than is observed within most of the metamorphic rocks and granites.

One interpretation of this disparity is that McCoy Creek Group material was *not* a major source for these rocks, and that they are mostly derived from the Archaean basement underlying the miogeoclinal sequence, but this is belied by the very high $\delta^{18}O$ values of some rocks (e.g. the Ruby Mountains two-mica granites), and by the simple observation that the McCoy Creek material must have melted substantially when subjected to the very high temperatures implied by mineral assemblages and textures in the other miogeoclinal high-grade rocks.

If we accept that McCoy Creek Group material formed a major protolith to the high-^{18}O granites, then there must have been a major change in the Sr isotopic composition of these metasediments at some stage within the Phanerozoic, involving shifts to lower $^{87}Sr/^{86}Sr$ ratios. Unradiogenic strontium (i.e. with $^{87}Sr/^{86}Sr < 0.715$) must have been added to the system, or widespread isotopic exchange must have occurred, in order to lower the values from predicted $^{87}Sr/^{86}Sr$ ratios well in excess of 0.740 to values mostly less than 0.720 at the time of Mesozoic and Tertiary metamorphism. An obvious source for this low-^{87}Sr strontium is the approximately 8 km thick section of miogeoclinal Palaeozoic carbonate rocks present throughout the Basin and Range Province. These rocks are very Sr-rich and would have had a depositional $^{87}Sr/^{86}Sr$ close to 0.708, which would then have remained virtually constant throughout the Phanerozoic, owing to the extremely low Rb/Sr ratios. Their high Sr content of several thousand ppm is about one order of magnitude higher than that of most of the other miogeoclinal lithologies and, consequently, only a small fraction of carbonate strontium need be mobilized in order to lower $^{87}Sr/^{86}Sr$ in the pelites, and granites derived therefrom, to the Mesozoic and Tertiary values that are observed.

It is interesting to note that these carbonate lithologies, even where they are very low-grade, often have ^{18}O-depleted isotopic compositions when compared with normal carbonate sediments (e.g. +14 to +22 – see Fig. 6.5), suggesting that they have undergone some oxygen isotope exchange with a less ^{18}O-rich reservoir. This is also consistent with open-system behaviour during metamorphism. Alternatively, low ^{87}Sr may have been derived from interstitial formation waters, or from trace amounts of carbonate that were present within the original McCoy Creek sediments themselves. Inasmuch as most lithologies, particularly those from the Lizzies Basin region of the East Humboldt Range (see Fig. 6.5), have undergone massive oxygen isotope exchange in order to bring $\delta^{18}O$ values down by 5–10 per mil, it is likely that significant transport and exchange of Sr could have occurred at the same time. The intercalated metre-scale layers of amphibolite facies marble, pelite, and granite provide the ideal situation for advective transport of strontium out of the carbonates and into the adjacent lithologies. In this way, Sr isotopic compositions within the miogeoclinal metasediments could have been profoundly reset, resulting in the anomalously low $^{87}Sr/^{86}Sr$ ratios observed in the Mesozoic and Tertiary granites (Fig. 6.6).

Owing to extreme uncertainty in the precise age of most of these plutons, and poor understanding of the tectonic setting of Mesozoic and Tertiary thermal events in this area, it is hard to constrain the exact timing of isotopic alteration. Inasmuch as the rate of isotopic exchange will be enhanced by a heating event (due, for example, to higher diffusion rates and exchange rates, or to enhanced fluid fluxes), any of these thermal events may have promoted mixing, isotopic exchange, and lowering of $^{87}Sr/^{86}Sr$ and $\delta^{18}O$. However, the best clues to the exact timing of this process come from the high-grade metasediments in the East Humboldt Range. Extensive high-temperature fluid-rock interaction is convincingly demonstrated in these rocks, with widespread biotitization of amphibolites, spectacular development of coarse calc-silicate minerals such as grossular and clinozoisite in metacarbonate layers, cross-cutting calcite–dolomite–tremolite veins, and widespread pegmatitic textures. As noted above, all of these assemblages are very fresh and show no sign of retrograde alteration such as chloritization of biotite. Inasmuch as a Tertiary age has been obtained for a widespread group of biotite monzogranite pods in the East Humboldt Range, by U–Pb dating of zircon (Wright & Snoke 1986), and given widespread Tertiary K–Ar and Ar–Ar ages in this area, and in the Ruby Mountains, this suggests that the most recent high-grade event experienced by these rocks was also Tertiary. If it was older, the assemblages (in particular, biotite) would be expected to show Tertiary retrograde alteration in response to the Tertiary heating event. The same argument can be made on the basis of high-temperature oxygen isotope fractionations that are still preserved in the high-grade rocks (Wickham *et al.* 1987, and unpublished data).

The Tertiary event has been linked to regional extension throughout the Basin and Range Province (e.g. Snoke & Miller 1988), and this presents more circumstantial evidence that isotopic homogenization occurred at this time. During a rifting event, hydrothermal activity in the upper part of the crust is ubiquitous and may extend to depths of 10–12 km, at least during the prograde stage of regional metamorphism (Wickham & Taylor 1985, 1987). A rift-zone model was used to explain Sr and O isotope systematics in the Pyrenees (see above) and could equally well apply in the case of the Basin and Range. At the very least, a rifting event presents the most favourable tectonic setting for deep hydrothermal circulation that will promote pervasive isotopic exchange. Further study of this area should provide answers to these questions by documenting the extent of exchange in areas that have not been strongly affected by Tertiary thermal events, and by further detailed geochronology.

6.4 Discussion and conclusions

In this chapter we have examined the relationship between the isotopic composition of granites and their source rocks, and have attempted to highlight data conflicting with the usually tacit assumption that granites image their sources. In the previous section we have presented descriptions of three areas in which the Sr isotopic composition of metasedimentary source rocks appears to have been modified, probably resulting from fluid-mediated exchange with low-$^{87}Sr/^{86}Sr$ rocks (e.g. carbonates) or pore fluids. There are many other examples in the literature (although, in general, these are not so clear-cut). In the New England Batholith of Australia, the Bundara Suite plutons have $\delta^{18}O$ values of +11.0 to +12.4 (Flood & Shaw 1977), and also have mineralogical and chemical characteristics consistent with their derivation from a pelitic metasedimentary source; yet these rocks have extraordinarily low $^{87}Sr/^{86}Sr$ initial ratios of 0.703–0.706 (Hensel *et al.* 1985). Although it is difficult to constrain the age of the source material very accurately, once again, this appears to have been depleted in ^{87}Sr prior to melting.

Another Australian example is the Cooma granodiorite (Joplin 1942, White & Chappell 1988), a classic 400 Ma regional aureole S-type which has an initial $^{87}Sr/^{86}Sr$ ratio of 0.718 (Pidgeon & Compston 1965), similar to the homogenized value observed in the adjacent Lower Palaeozoic metasediments from which it is believed to be derived. Similar systematics are also reported in a more recent Sr and O isotopic study by Munksgaard (1988). Wyborn & Chappell (1983) and Munksgaard (1988) report Rb/Sr ratios in the low-grade Cambro-Ordovician equivalents of these metasediments which imply $^{87}Sr/^{86}Sr$ values that are more radiogenic at 400 Ma than those measured by Pidgeon & Compston (1965) or Munksgaard (1988) in the high-grade mica-schists, suggesting lowering of the ratio during prograde heating. It is also worth noting that $^{87}Sr/^{86}Sr$ is relatively uniform in the Cooma metasediments at 400 Ma despite the highly variable Rb/Sr ratios in the various metasedimentary layers, implying that there was extensive isotopic homogenization in these rocks at the subsolidus stage, as noted in the Trois Seigneurs prograde sequence in the Pyrenees.

A detailed discussion of all regions showing subsolidus modification of silicic magma source material is well beyond the scope of this chapter, but the preceding descriptions of isotopic changes can be summarized as follows:

(a) High-grade metamorphic terranes associated with anatectic rocks commonly show homogenization and lowering of $\delta^{18}O$ values over km-sized areas (in both the anatectic *and* the subsolidus rocks), suggesting pervasive interaction with metamorphic fluids.
(b) $^{87}Sr/^{86}Sr$ is also often homogenized in these terranes, again in subsolidus metasedimentary units such as shales, as well as in the anatectic rocks,

although the database is smaller than for oxygen. Sr-rich metacarbonates in these terranes do not seem to be so strongly affected.

(c) Crustally derived granitoids associated with or derived from these terranes commonly have similar isotopic compositions, or have compositions implying that a significant fraction of the magma was derived from the local high-grade rocks.

(d) $^{87}Sr/^{86}Sr$ ratios in high-grade metasedimentary source rocks (and in granitoids derived therefrom) are often lower than would be expected given the age (since deposition) and the Rb/Sr ratio of these rocks, although this is often difficult to quantify owing to uncertainties in depositional age, initial depositional ratio, Rb/Sr ratios and the variation of Sr abundance in sedimentary sequences.

Taken together, points (b) and (d) imply that $^{87}Sr/^{86}Sr$ ratios in crustal metasediments are often homogenized and lowered in regional metamorphic rocks during a prograde heating event leading to anatexis. It is logical to assume that at least in part, this chemical transformation accompanies the major shifts and homogenization of oxygen isotopic composition that have also occurred within these terranes. The data from the Pyrenees imply that the changes in strontium isotopic composition commenced at low temperatures, at least as low as chlorite-grade, but the data from Brittany suggest that some sort of low-temperature limit may exist, because some of the sediments here were not affected.

In most cases, $^{87}Sr/^{86}Sr$ ratios in pelitic and psammitic metasediments appear to have been lowered, suggesting mixing with reservoirs containing abundant low-^{87}Sr strontium. The most likely source of this strontium is either from metasedimentary carbonates (particularly in regions such as the Basin and Range where carbonates are very common) or from sedimentary pore fluids. These fluids may have been trapped in metasedimentary layers at the time of deposition, or transported from the surface in deeply penetrating hydrothermal solutions (Wickham & Taylor 1985, 1989). In either case, they could contain several hundreds or even thousands of ppm of relatively unradiogenic strontium.

Such mobility of strontium in the crust is entirely consistent with the global strontium cycle and the observed evolution of the strontium isotopic composition of sea and river water (Holland 1984). The continents are the prime terrestrial source of radiogenic strontium which is transported to the oceans via fluvial systems, and which balances inputs of unradiogenic strontium from exchange of sea water with MORB. This balance has maintained $^{87}Sr/^{86}Sr$ in the oceans at about 0.707–0.709 throughout much of the Phanerozoic (Burke *et al.* 1982). It is notable that almost all river input of continental radiogenic strontium has $^{87}Sr/^{86}Sr$ between 0.705 and 0.720 (see Table 6.8 in Holland 1984), which is low compared to the age and Rb/Sr ratio of some of the bedrock lithologies in continental interiors. Of course, this partly reflects mixing of Sr (derived from many lithologies) at the fluvial stage, and may also be influenced by preferential weathering of unradiogenic carbonate, but it also implies that few large continental areas of regionally high $^{87}Sr/^{86}Sr$ metasediment or granite exist. This provides further evidence suggesting efficient homogenization of $^{87}Sr/^{86}Sr$ ratios in the continental crust prior to exposure at the surface, this ratio being buffered to low values by mixing and exchange at the metamorphic and magmatic stage.

The most important conclusions to be drawn from this discussion of strontium and oxygen isotope systematics in metamorphic rocks and granites is that major changes in isotopic composition may occur at the metamorphic stage, prior to anatexis. In using these isotopic parameters as tracers of granite source material, it is extremely important that these potential early stage changes are taken into account. Thus, it is not longer valid to claim on the basis of an initial $^{87}Sr/^{86}Sr$ ratio of 0.710 and a $\delta^{18}O$ of +8 that a granite has not been derived from sedimentary material. Pelitic metasediments with this isotopic composition exist (in eastern Nevada, for example) and may have formed a significant source component to some of the granites of that region. In seeking to identify granite source materials, the isotopic compositions of the various crustal lithologies must not be so conservatively restricted as has often been the case in the past.

Acknowledgement

This work was supported in part by NSF grant EAR8720097.

References

Allègre, C. J. & D. Ben Othman 1980. Nd–Sr isotopic relationship in granitoid rocks and continental crust development: a chemical approach to orogenesis. *Nature* **286**, 335–41.

Ashworth, J. R. (ed.) 1985. *Migmatites*. Glasgow: Blackie.

Banda, E. & S. M. Wickham (eds) 1986. The Geological Evolution of the Pyrenees. *Tectonophysics* **129**.

Ben Othman, D., S. Fourcade & C. J. Allègre 1984. Recycling processes in granite–granodiorite complex genesis: the Querigut case studied by Nd–Sr isotope systematics. *Earth and Planetary Science Letters* **69**, 290–300.

Bickle, M. J., S. M. Wickham, H. J. Chapman & H. P. Taylor Jr 1988. A strontium, neodymium and oxygen isotope study of hydrothermal metamorphism and crustal anatexis in the Trois Seigneurs Massif, Pyrenees, France. *Contributions to Mineralogy and Petrology* **100**, 399–417.

Bowen, N. L. 1928. *The evolution of the igneous rocks*. Reprinted 1956, New York: Dover.

Burke, W. H., R. E. Dennison, E. A. Hetherington, R. B. Koepnick, H. F. Nelson & J. B. Otto 1982. Variation of seawater $^{87}Sr/^{86}Sr$ throughout Phanerozoic time. *Geology* **10**, 516–19.

Crittenden, M. D. Jr, P. J. Coney & G. H. Davis 1980. Cordilleran metamorphic core complexes. *Geological Society of America, Memoir* 153, 490 pp.

Dallmeyer, R. D., A. W. Snoke & E. H. McKee 1986. The Mesozoic–Cenozoic tectonothermal evolution of the Ruby Mountains–East Humboldt Range, Nevada: a Cordilleran metamorphic core complex. *Tectonics* **5**, 931–54.

DePaolo, D. J. 1981. Trace element and isotopic effects of combined wallrock assimilation and fractional crystallization. *Earth and Planetary Science Letters* **53**, 189–202.

Ernst, G. (ed.) 1988. *Metamorphism and crustal evolution of the Western United States*. Rubey Volume VII. Englewood Cliffs, NJ: Prentice-Hall.

Farmer, G. L. & D. J. DePaolo 1983. Origin of Mesozoic and Tertiary granite in the Western United States and implications for pre-Mesozoic crustal structure I. Nd and Sr isotopic studies in the Geocline of the Northern Great Basin. *Journal of Geophysical Research* **88**, 3379–401.

Fleck, R. J. & R. E. Criss 1985. Strontium and oxygen isotopic variations in Mesozoic and Tertiary plutons of central Idaho. *Contributions to Mineralogy and Petrology* **90**, 291–308.

Flood, R. H. and S. E. Shaw 1977. Two 'S-type' granite suites with low initial $^{87}Sr/^{86}Sr$ ratios from the New England batholith, Australia. *Contributions to Mineralogy and Petrology* **61**, 163–73.

Fourcade, S. & C. J. Allègre 1981. Trace element behaviour in granite genesis: a case study. The calc-alkaline plutonic association from the Querigut complex (Pyrenees, France). *Contributions to Mineralogy and Petrology* **76**, 177–95.

Frost T. P. & G. A. Mahood 1988. Field, chemical and physical constraints on mafic–felsic magma interaction in the Lamarck Granodiorite, Sierra Nevada, California. *Geological Society of America Bulletin* **99**, 272–91.

Garlick, G. D. & S. Epstein 1966. Oxygen isotope ratios in coexisting minerals of regionally metamorphosed rocks. *Geochimica et Cosmochimica Acta* **31**, 181–214.

Gupta, L. N. & W. Johannes 1982. Petrogenesis of a stromatic migmatite (Nelaug, South Norway). *Journal of Petrology* **23**, 548–67.

Hensel, H. D., M. T. McCulloch & B. W. Chappell 1985. The New England Batholith: constraints on its derivation from Nd and Sr isotopic studies of granitoids and country rocks. *Geochimica et Cosmochimica Acta* **49**, 369–84.

Holland, H. D. 1984. *The chemical evolution of the atmosphere and oceans*. Princeton, NJ: Princeton University Press.

Huppert, H. E. & R. S. J. Sparks 1988. The generation of granitic magmas by intrusion of basalt into continental crust. *Journal of Petrology* **29**, 599–624.

Jäger E. & H. J. Zwart 1968. Rb–Sr age determinations of some gneisses and granites of the Aston–Hospitalet massif (Pyrenees). *Geologie en Mijnbouw* **47**, 349–57.

Joplin, G. A. 1942. Petrological studies in the Ordovician of New South Wales. I. The Cooma complex. *Proceedings of the Linnean Society of New South Wales* **67**, 159–96.

Kistler, R. W., E. D. Ghent & J. R. O'Neil 1981. Petrogenesis of garnet two-mica granites in the Ruby Mountains, Nevada. *Journal of Geophysical Research* **86**, 10 591–606.

Lee, D. E., R. W. Kistler, I. Friedman & R. E. van Loenen 1981. Two-mica granites of northeastern Nevada. *Journal of Geophysical Research* **86**, 10 607–16.

Lush, A. P., A. J. McGrew, A. W. Snoke & J. E. Wright 1988. Allochthonous Archean basement in the northern East Humboldt Range, Nevada. *Geology* **16**, 349–53.

Majoor, F. J. M. & H. N. A. Priem 1987. Rb–Sr whole-rock investigations in the Aston massif, central Pyrenees. *Geologische Rundschau* **76**, 787–94.

Mehnert, K. R. 1968. *Migmatites and the origin of granitic rocks*. Amsterdam: Elsevier.

Miller, E. L., P. B. Gans, J. E. Wright & J. F. Sutter 1988. Metamorphic history of the east-

central Basin and Range Province: tectonic setting and relationship to magmatism. In *Metamorphism and crustal evolution of the Western United States*, W. G. Ernst (ed.), 649–82. Rubey Volume VII. Englewood Cliffs, NJ: Prentice-Hall.
Munksgaard, N. C. 1988. Source of the Cooma Granodiorite, New South Wales; a possible role of fluid–rock interactions. *Australian Journal of Earth Sciences* **35**, 363–77.

Nabelek, P. I., T. C. Labotka, J. R. O'Neil & J. J. Papike 1984. Contrasting fluid/rock interaction between the Notch Peak granitic intrusion and argillites and limestones in western Utah: evidence from stable isotopes and phase assemblages. *Contributions to Mineralogy and Petrology* **86**, 25–34.
Norton, D. & H. P. Taylor Jr 1979. Quantitative simulation of the hydrothermal systems of crystallizing magmas on the basis of transport theory and oxygen isotope data: an analysis of the Skaergaard intrusion. *Journal of Petrology* **20**, 421–86.

Olsen S. N. 1984. Mass-balance and mass-transfer in migmatites from the Colorado Front Range. *Contributions to Mineralogy and Petrology* **85**, 30–44.
O'Neil, J. R. & B. W. Chappell 1977. Oxygen and hydrogen isotope relations in the Berridale batholith. *Journal of the Geological Society of London* **133**, 559–71.

Peucat, J. J., P. Jegouzo, P. Vidal & J. Bernard-Griffiths 1988. Continental crust formation seen through the Sr and Nd isotope systematics of S-type granites in the Hercynian belt of western France. *Earth and Planetary Science Letters* **88**, 60–8.
Pidgeon, R. T. & W. Compston 1965. The age and origin of the Cooma granite and its associated metamorphic zones, New South Wales. *Journal of Petrology* **6**, 193–222.

Reynolds, D. L., 1946. The sequence of geochemical changes leading to granitization. *Quarterly Journal of the Geological Society of London* **102**, 389–446.
Rye, R. O., B. R. Doe & J. D. Wells 1974. Stable isotope and lead isotope studies of the Cortez, Nevada, gold deposit and surrounding area. *Journal of Research of the United States Geological Survey* **2**, 13–23.

Sheppard, S. M. F. 1986. Igneous rocks: III. Isotopic case studies of magmatism in Africa, Eurasia and Oceanic Islands. In *Stable Isotopes in High Temperature Geological Processes*, J. W. Valley, H. P. Taylor, Jr & J. R. O'Neil (eds), 319–72. *Reviews in Mineralogy*, vol. 16. Washington, D.C.: Mineralogical Society of America.
Snoke, A. W. & D. M. Miller 1988. Metamorphic and tectonic history of the northeastern Great Basin. In *Metamorphism and crustal evolution of the Western United States*, W. G. Ernst (ed.), 606–48. Rubey Volume VII. Englewood Cliffs, NJ: Prentice-Hall.
Solomon, G. C. & H. P. Taylor Jr 1989. Isotopic evidence for the origin of Mesozoic and Cenozoic granitic plutons in the northern Great Basin. *Geology* **17**, 591–4.
Sparks, R. S. J. & L. A. Marshall 1986. Thermal and mechanical constraints on mixing between mafic and silicic magmas. *Journal of Volcanology and Geothermal Research* **29**, 99–124.

Taylor, H. P. Jr 1980. The effects of assimilation of country rocks by magmas on $^{18}O/^{16}O$ and $^{87}Sr/^{86}Sr$ systematics in igneous rocks. *Earth and Planetary Science Letters* **47**, 243–54.
Taylor, H. P. Jr & S. M. F. Sheppard 1986. Igneous rocks: I. Processes of isotopic fractionation and isotope systematics. In *Stable Isotopes in High Temperature Geological Processes*, J. W. Valley, H. P. Taylor, Jr & J. R. O'Neil (eds) 227–72. *Reviews in Mineralogy*, vol. 16. Washington, D.C.: Mineralogical Society of America.
Taylor, H. P. Jr & L. T. Silver 1978. Oxygen isotope relationships in plutonic igneous rocks of the Peninsular Ranges batholith, southern and Baja California. In *Short papers of the fourth International Conference on Geochronology, Cosmochronology, and Isotope Geology*, 423–6. U.S. Geological Survey Open-file report 78–701.

Vitrac-Michard, A. & C. J. Allègre 1975. A study of the formation and history of a piece of continental crust by $^{87}Rb-^{87}Sr$ method: the case of the French Oriental Pyrenees. *Contributions to Mineralogy and Petrology* **50**, 257–85.

Vitrac-Michard, A., F. Albarède, C. Dupuis & H. P. Taylor, Jr 1980. The genesis of Variscan (Hercynian) plutonic rocks: inferences from Sr, Pb and O studies of the Maladeta igneous complex, Central Pyrenees, Spain. *Contributions to Mineralogy and Petrology* **72**, 57–72.

Vitrac-Michard, A., F. Albarède & C. J. Allègre 1981. Lead isotopic composition of Hercynian granitic K-feldspars constrains continental genesis. *Nature* **291**, 460–4.

White A. J. R. & B. W. Chappell 1988. Some supracrustal (S-type) granites of the Lachlan Fold Belt. *Transactions of the Royal Society of Edinburgh: Earth Sciences* **79**, 169–81.

White A. J. R., J. D. Clemens, J. R. Holloway, L. T. Silver, B. W. Chappell & V. J. Wall 1986. S-type granites and their probable absence in southwestern North America. *Geology* **14**, 115–18.

Wickham, S. M. 1987a. Crustal anatexis and granite petrogenesis during low pressure regional metamorphism: the Trois Seigneurs Massif, Pyrenees, France. *Journal of Petrology* **28**, 127–69.

Wickham, S. M. 1987b. The segregation and emplacement of granitic magmas. *Journal of the Geological Society of London* **144**, 281–97.

Wickham, S. M. & M. T. Peters 1988. Fluid and melt transport in anatectic environments. GSA abstract, Denver, Colorado. *Geological Society of America, Abstracts with Programs* **20**, A304.

Wickham, S. M. & H. P. Taylor Jr 1985. Stable isotopic evidence for large-scale seawater infiltration in a regional metamorphic terrane: the Trois Seigneurs Massif, Pyrenees, France. *Contributions to Mineralogy and Petrology* **91**, 122–37.

Wickham, S. M. & H. P. Taylor Jr 1987. Stable isotope constraints on the origin and depth of penetration of hydrothermal fluids associated with Hercynian regional metamorphism and crustal anatexis in the Pyrenees. *Contributions to Mineralogy and Petrology* **95**, 255–68.

Wickham, S. M. & H. P. Taylor Jr 1989. Hydrothermal systems associated with regional metamorphism and crustal anatexis: examples from the Pyrenees, France. In *Special volume on fluids and crustal processes*, National Research Council (in press).

Wickham, S. M., H. P. Taylor Jr & A. W. Snoke 1987. Fluid–rock–melt interaction in metamorphic core complexes – a stable isotope study of the Ruby Mountains – East Humboldt Range, Nevada. GSA Abstract, Hilo, Hawaii. *Geological Society of America, Abstracts with Programs* **19**(6), 463.

Wright, J. E. & A. W. Snoke 1986. Mid-Tertiary mylonitization in the Ruby Mountain – East Humboldt Range metamorphic core complex, Nevada. *Geological Society of America, Abstracts with Programs* **18**, 795.

Wyborn, L. A. I. & B. W. Chappell 1983. Chemistry of the Ordovician and Silurian greywackes of the Snowy Mountains, southeastern Australia: an example of chemical evolution of sediments with time. *Chemical Geology* **39**, 81–92.

Zwart, H. J. 1962. On the determination of polymetamorphic mineral associations, and its application to the Bosost area (Central Pyrenees). *Geologische Rundschau* **52**, 38–65.

CHAPTER SEVEN

Fluids and melting in the Archaean deep crust of southern India*

Robert C. Newton

7.1 Introduction

7.1.1 Granites and granulites

A close relationship between granulite facies metamorphism and crustal melting has often been inferred in studies of Precambrian high-grade terrains. Quensel (1952) described granite–granulite associations in four continents, and Rama Rao (1940) called attention to the particularly close association of migmatites, granites, and granulites near the orthopyroxene isograd in southern India. The concept of dehydration melting in the generation of granulites has been the subject of much discussion (Phillips 1980, Waters 1988). The temperature–pressure range of granulite facies metamorphism (700–900 + °C, 4–10 kbar) embraces the region where many rocks would begin to melt if any water is present (Bohlen *et al.* 1983).

The simplest and most generalized explanation of the granite–granulite relationships is that the large granulite facies terrains are dried and incompatible-element-depleted portions of the deep crust, from which variable amounts of granite magmas have been removed during profound metamorphic episodes. This hypothesis was developed as a general mechanism by Fyfe (1973) and applied to various granulite terrains by Schmid (1978), Nesbitt (1980), and Pride & Muecke (1980). A major problem with this mechanism is the difficulty of removal of small amounts of viscous granitic melts from their source rocks (Wickham 1987); thinner and more penetrating fluids may be necessary to remove incompatible elements and to produce some of the metasomatic effects attributable to some granulite facies metamorphism, such as changes in the oxygen and carbon isotopes in the Lofoten–Vesteralen area of Norway (Baker & Fallick 1988). The absence of systematic isotopic changes in

* XXth Hallimond Lecture, British Mineralogical Society, 15 December 1988.

some other granulite facies terrains, as in the Adirondack Mountains of New York (Valley & O'Neil 1984), is evidence against the involvement of pervasive metasomatizing fluids. The extreme dryness of the Adirondack terrain may require yet another desiccation mechanism, such as widespread, shallow, contact metamorphism prior to high-pressure regional metamorphism (McLelland & Husain 1986).

Dehydration melting as a mechanism for the origin of granulites has been argued persuasively from phase equilibrium principles by Powell (1983) and from field documentation by Waters (1988). According to this hypothesis, small amounts of H_2O initially present in pore spaces and hydrous minerals are absorbed into developing granitic melts, with continual reduction of the H_2O activity, leading to the stability of anhydrous mafic minerals, especially pyroxenes. Extraction of melt is not likely in small amounts: the system may remain largely closed to material transport. The ultimate means of removal of the H_2O while preserving anhydrous assemblages has not been clarified. Strong channelization of fluids evolved from freezing migmatites may inhibit retrograde hydration (Waters 1988).

The possible metasomatic effects of vapour-undersaturated partial melting without melt extraction are obviously limited. Large-scale metasomatic alteration, along with other features such as the presence of abundant fluid inclusions, especially in the most altered rocks, may require the presence of an active vapour phase in the genesis of some high-grade terrains (Touret 1985). The terrain to be described, that of southern India, features relationships of granite and granulite which suggest fluid movement as a common mechanism in their formation.

7.1.2 The Dharwar Craton

The Dharwar Craton of southern India is an Archaean continental fragment composed principally of low-K tonalitic to trondhjemitic gneisses (Peninsular Gneiss) with minor infolded older supracrustals (Sargur Group) and capped by a younger series of volcanic and sedimentary 'greenstone' basins (Dharwar Supergroup) (Pichamuthu & Srinivasan 1984). The pronounced north–south elongation of every element in the craton is one of the most conspicuous map features (Fig. 7.1). According to Drury & Holt's (1980) study of Landsat images and ground geology, this structural grain results from large-scale transcurrent faulting. The shearing is more pronounced in zones of 'high finite strain'. These are thinner in the northern craton and widen southwards. Intensity of deformation increases in an apparently continuous way from north to south across the craton. The northern greenstone basins are more nearly equant in map aspect; basins become attenuated southwards, with long, hook-like tails. South of Mysore, only isolated relics of the volcanic–sedimentary sequences remain. The grade of metamorphism advances steadily, from low greenschist in the north to granulite facies in the south. Simultane-

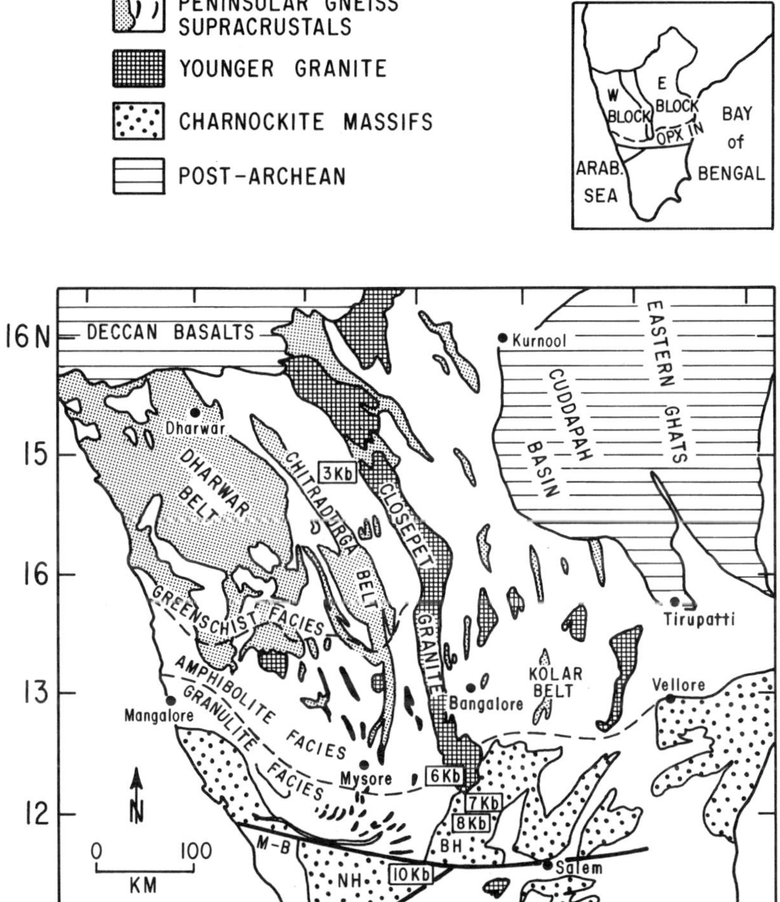

Figure 7.1 Generalized geology of Dharwar Craton, southern India. Dharwar Supergroup metasediments and metavolcanics are stippled. Sargur Group high-grade supracrustals are shown in black. Metamorphic facies boundaries are taken from Pichamuthu (1965) and Subramaniam (1967). Palaeopressures in rectangles are from Hansen *et al.* (1984), Harris & Jayaram (1982), and Raith *et al.* (1988). BH, Biligirirangan (BR) Hills granulite massif; NH, Nilgiri Hills granulite massif; M–B, Moyar–Bhavani Proterozoic shear zone (Drury & Holt 1980).

ously, palaeopressures of gneissic and mafic rocks increase from about 3 kbar in central Karnataka to near 8 kbar in northern Tamil Nadu (Hansen *et al.* 1984, Raase *et al.* 1986). However, it has not yet been demonstrated conclusively that the southward increases in grade and palaeopressure relate to a single metamorphic episode. The southernmost portion of the craton is cut by large east–west Proterozoic shear zones. South of these shear zones, exposed rocks are dominantly granulites and retrogressed granulites, and it is not certain if they should be considered part of the Dharwar Craton. The progression to granulite facies north of the shear zones apparently represents a tilted and bevelled profile of Late Archaean crust.

7.1.3 Closepet Granite

The Closepet Granite is a long narrow body of granitic rocks which traverses the Dharwar Craton from north to south and follows the general grain of the craton: the north–south elongational strain presumably accounts in some way for the peculiar ribbon-like aspect of the Closepet outcrop. The Closepet terrain ranges from nebulously migmatized Peninsular Gneiss to small discrete domical plutons of coarse porphyritic granite. In the northern craton, the granite is in the form of high-level intrusions cutting gneiss and greenstone basins, with contact metamorphism, as in the Sandur Belt. At the southern end where it abuts on the granulite facies terrain, it becomes segmented and breached by migmatitic Peninsular Gneiss (Fig. 7.2). The Closepet Granite is thus a feature of the middle and upper crust which grades downwards into migmatitic roots in the lower crust, similar in aspect to Hutchison's (1970)

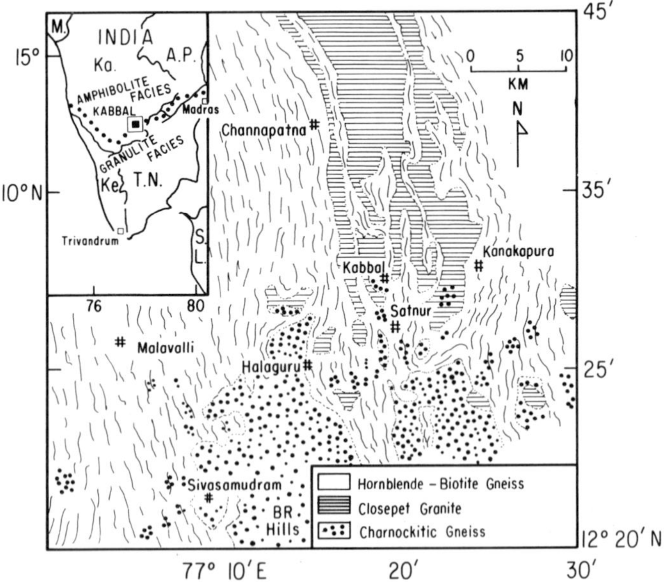

Figure 7.2 Relations of Closepet Granite and charnockitic rocks in the Kabbal – Satnur – BR Hills area of southern Karnataka, compiled from the maps of Suryanarayana (1960), Devaraju & Sadashivaiah (1969), Mahabaleswar & Naganna (1981), and Mahabaleswar *et al.* (1986). The abrupt termination of the Closepet Granite at the onset of charnockite is evident. Abbreviations on inset map: Ka., Karnataka State; A.P., Andhra Pradesh State; Ke., Kerala State; T.N., Tamil Nadu State; M., Maharashtra State; S.L., Sri Lanka.

Table 7.1 Chemical analyses of Closepet Granite and related rocks and of Western Arabian Shield granites.

	S CLOSEPET			S CLOSEPET-GRANULITE TRANSITION ZONE								W DHARWAR			W ARABIA
	Granitic Rocks			Granite		Amphibolite f. Gneiss			Granulite f. gneiss			Peninsular Gneiss			Monzo-granite
wt%	1	2	3	K116	K134	K143	K136	M20	K76	K150	K159	11	9	18	
SiO_2	73.20	72.80	65.20	62.40	69.30	60.44	67.20	70.50	62.30	66.40	72.93	68.90	73.20	74.90	72.64
TiO_2	0.28	0.20	0.64	0.31	0.26	0.26	0.32	0.31	0.62	0.39	0.27	0.80	0.15	0.63	0.34
Al_2O_3	13.90	14.10	15.40	17.64	11.10	14.34	16.66	12.10	15.64	14.34	10.80	14.70	14.70	13.00	14.14
Fe_2O_3				4.82	3.69	5.02	3.14	5.28	2.62	2.21	1.76	1.60			1.44
	2.14	1.58	5.17										2.20	3.00	
FeO				3.12	2.21	4.18	2.26	3.42	4.64	4.34	3.80	2.50			0.79
MnO	0.02	0.01	0.08	0.18	0.20	0.19	0.18	0.22	0.15	0.12	0.17	----	----	----	0.05
MgO	0.44	0.37	1.48	1.36	1.85	3.20	1.85	1.40	2.12	1.80	1.62	1.30	0.35	0.47	0.45
CaO	1.63	1.43	3.40	2.69	2.91	5.80	4.60	4.16	7.10	5.25	3.13	2.80	2.00	1.70	1.63
Na_2O	3.49	3.41	4.10	3.30	3.58	4.38	3.10	1.35	3.10	4.36	5.00	4.40	5.80	4.40	3.99
K_2O	4.44	5.38	3.67	3.78	3.98	1.12	0.56	0.36	0.76	0.68	0.48	2.20	1.30	1.00	3.90
P_2O_5	0.08	0.07	0.32	----	----	----	----	----	----	----	----	----	----	----	0.96
LOI	0.49	0.62	0.48	1.10	0.92	0.81	0.90	0.96	0.80	0.11	0.03	----	----	----	0.53
ppm															
Rb	120	108	103	132	116	201	51	153	38	44	51	47	34	27	114
Sr	260	545	669	749	428	532	180	273	284	1102	666	267	577	72	239
Y	13	6	35	12	2	8	91	63	41	34	58	13	26	122	24
Zr	191	141	208	252	159	269	648	299	881	219	1281	150	105	557	190
Ba	667	1428	827	----	----	----	----	----	245	1615	1325	290	336	110	648
La	68	32	76	35	24	44	69	33	34	95	110	----	----	----	28
Ce	124	52	154	67	41	140	160	60	78	201	239	----	----	----	50

S CLOSEPET: 1: Tumkur - KD granite (avg. of 10) 2: Chana- KD granite (avg. of 3) 3: Tumkur - KD granodiorite (avg. of 7) (Allen *et al.* 1986)

GRANULITE TRANSITION ZONE: Granites and gneisses from the southernmost Closepet region, near Kanakapura, Karnataka (Mahabaleswar *et al.* 1986)

W DHARWAR CRATON (Compilation of Rogers *et al.* 1986): 11: Peninsular Gneiss, S Kanara, Karnataka (Balasubrahmanyam, 1978) 9: Peninsular Gneiss, Chickmagalur, Karnataka (Taylor *et al.* 1984, avg. of 14) 18: Gneiss, Holenarsipur, Karnataka (Monrad 1983, avg. of 7)

W ARABIAN SHIELD: Monzogranite association, younger granitoids, Central Hijaz region, W Arabian Shield (Jackson *et al.* 1984, avg. of 17) picture of the Coast Range Batholith of the Prince Rupert area in British Columbia.

Rocks which have been called Closepet Granite grade from granite to granodiorite and are alkaline to calc-alkaline, according to the criteria of Rogers & Greenberg (1981). Some analyses of Closepet Granite from the literature are presented in Table 7.1. Near the town of Ramnagaram (formerly Closepet) the granite consists of several phases, including migmatitic gneiss, grey clinopyroxene-bearing granite with abundant pink megacrysts, and pink leucogranite in discrete vein networks. In some quarry exposures near the southern terminus, as at Yelachi Palayam and Kabbal, Karnataka, evidence of metasomatism exists in the form of nebulous migmatites, K-metasomatized metabasic lozenges, thin veins consisting of nearly pure pink feldspar showing ductile deformation, and thicker veins made up of K-feldspar or albitized K-feldspar, epidote, and carbonate (Radhakrishna 1958). Some analyses of Closepet Granite, particularly the grey varieties, show very high concentrations of Ba, Sr, and light rare earth elements (Table 7.1), suggestive of metasomatic enrichment. Rare earth element patterns in some of Allen *et al.*'s (1986) Closepet classes are highly depleted in heavy rare earths, with a large positive Eu anomaly. Weaver (1980) interpreted similar patterns in Madras granitic charnockites as the result of K-metasomatism of trondhjemitic gneisses; however, rare earth patterns in granites are often quite variable and not generally diagnostic. Petrographic evidence of K-metasomatism exists in the form of marginal K-feldspar replacement of plagioclase and biotite replacement of amphibole (Divakara Rao *et al.* 1972).

The low-K country-rock gneisses of the southern Closepet zone are quite typical of grey gneisses of the western Dharwar Craton in their major element chemistries (Table 7.1), and little distinction can be made between amphibolite-facies gneisses and granulite-facies (pyroxene-bearing) gneisses. However, some of the Closepet-zone granulite-facies gneisses are considerably enriched in Sr, Ba, Zr, and Ce relative to typical cratonal gneisses, which is suggestive of metasomatic alteration.

7.1.4 Granite–charnockite relations

The relationship between the southern Closepet Granite and the charnockitic terrain is a major problem of the Dharwar Craton. The relationships in southern Karnataka at the orthopyroxene isograd, compiled from several published maps are shown in Figure 7.2. The remarkable mutual avoidance of granite and charnockitic gneiss is evident. Numerous descriptions of the quarry at Kabbaldurga, Karnataka, have emphasized the intimate relations between granite and charnockite in the zone of overlap near the orthopyroxene isograd (Friend 1983, Hansen *et al.* 1987, Stähle *et al.* 1987). Dark charnockite occupies short shear veins and diffuse patches. Many of the charnockite-filled shears run nearly north–south and show warping of foliation and offsets in the right-lateral sense, but some of them are broad and diffuse, with an overall trend nearly at right angles to the straighter discrete shears. Other charnockitic patches occupy the margins of metabasic lenses and granite dykes. The overall impression is that granite is generally somewhat earlier than charnockite, but examples can be found of diffuse charnockitic patches grading continuously into coarse pink leucogranite, indicating the near-contemporaneity of the phases (Friend 1983). Allanite from granite at Kabbal yields a concordant U–Pb age of 2.52 Ga (Grew & Manton 1984), and this age coincides with radiometric ages of high-grade rocks over a large part of the southern part of the craton (Peucat *et al.* 1987). Localization of charnockite in shear zones suggests that deformation features were effective fluid pathways for charnockitizing and granitizing fluids (Friend 1985).

Fluid inclusions further reveal fluid activity in the formation of granite and charnockite. Dense CO_2-rich inclusions are abundant in quartz and feldspar of the charnockite veins, nebulous granite, and granite dykes, but are absent in unaltered grey gneiss host rocks. Densitites of the majority of CO_2 inclusions from localities near the orthopyroxene isograd are consistent with entrapment at high pressures and temperatures. Hansen *et al.* (1984) found a steady increase of CO_2 densities in charnockitic veins and quartz–feldspar segregations from Kabbal southward into the charnockitic terrain. The densities would be consistent with entrapment of the CO_2 at the peak metamorphic conditions of about 750°C and 5–8 kbar inferred from mineralogical geothermometry and geobarometry, based on the CO_2 equation of state of Touret & Bottinga (1979).

Orthopyroxene appears at Kabbal and similar localities by breakdown of calcic amphibole, possibly by reaction with biotite (Hansen *et al.* 1987). Rock and mineral analyses and modal counts of charnockitic veins and adjacent host gneisses lead to the following generalized open-system reaction:

$$\begin{aligned}&\text{hornblende} + \text{biotite} + \text{quartz} + \text{magnetite} + (Na_2O + SiO_2 \text{ from vapour})\\&= \text{orthopyroxene} + \text{K-feldspar} + \text{plagioclase} + \text{ilmenite}\\&+ (MgO + CaO + FeO \text{ removed in vapour}) + (H_2O + F + O_2) \quad (7.1)\end{aligned}$$

It is significant that an orthopyroxene-forming reaction which satisfies the analyses and modal data cannot be balanced in a closed system. The necessity of an open-system reaction emphasizes the rôle of external fluids in the origin of the charnockite at Kabbal. Mafic components were removed in the metasomatism and replaced by SiO_2, and perhaps Na_2O, at nearly constant volume. Hansen *et al.* (1987) found no clear-cut evidence for introduction of K_2O in charnockite formation, although Stähle *et al.* (1987) inferred this from their analyses.

Biotite and quartz are stable together at the orthopyroxene isograd in southern India. The CO_2-rich inclusions are consistent with the assemblage biotite–quartz–orthopyroxene–K-feldspar, which requires a low H_2O activity of less than 0.3 (Hansen *et al.* 1984): charnockite appears to have formed as a subsolidus assemblage at Kabbal and similar localities. Twelve km south of Kabbal, in the banded charnockites of the Halaguru region (Fig. 7.2), some orthopyroxene appears in pegmatoidal segregations and dykes (Devaraju & Sadashivaiah 1969), and may have been produced by a melting reaction. Still further south, orthopyroxene is ubiquitous. Some may have formed from subsolidus reactions, some from melting reactions, and some may have formed in earlier Archaean (pre-2.5 Ga) cycles of deep-crustal metamorphism (Pichamuthu 1965, Mahabaleswar *et al.* 1986).

7.1.5 Proposed origins of Closepet Granite

Theories of the origin of Closepet Granite include purely igneous intrusion and differentiation (Allen *et al.* 1986), *in situ* anatexis (Friend 1984), and combined metasomatic infiltration and partial melting (Radhakrishna 1956, Divakara Rao *et al.* 1972). Various uncertainties attend all of these proposals.

Allen *et al.* (1986) proposed a complex scheme of fractionation of clinopyroxene, plagioclase, and alkali feldspar which relates several types of granite and granodiorite collected in the region between Kabbal and Tumkur. A good fit of the major and most trace elements could be made for most of their granites, assuming batch melting of a tonalitic Peninsular Gneiss-like precursor, followed by variable amounts of fractionation. This scheme has the problem that the proposed differentiation sequences have little supporting evidence from chronology based on field observations, and ignores the evidence for fluid involvement almost completely. Furthermore, the heat source for extensive crustal melting is not apparent, since coeval intermediate and mafic plutonics have not been recognized. Finally, the field and petrographic evidence that Kabbal-type charnockites formed from subsolidus breakdown of amphibolite (Reaction 7.1) suggests that some of the chemical features of the gneisses and granites are also metasomatic in origin.

The palaeopressure, and hence depth-zone, arrangement of the Closepet and charnockite terrains seems consistent with the classical theory that deep-crustal granulite facies terrains result from removal upward of granitic anatectic melts (Fyfe 1973). This complementary theory of granites and depleted granulites has found some support in comparing the major element chemistries of granulite facies gneiss and non-granulite facies gneiss composites (Nesbitt 1980, Pride & Muecke 1980). However, some authors have pointed out that many rocks from granulite facies terrains have major element compositions more representative of liquid abstracts than of out-melted residues (Weaver & Tarney 1983). Moreover, certain large-ion lithophile (LIL) elements cannot be modelled in their distributions by such granitic out-melting. Rb, which is severely depleted in some very high-grade granulites, as in the BR Hills (Fig. 7.2), is not effectively removed in granitic melts, but may require a transporting vapour phase to be mobilized (Lamb *et al.* 1986). The solubilities of zircon and phosphate minerals, which are the usual hosts for U and Th, are so small in granitic liquids that out-melting could well enrich the deep crust in these elements, rather than deplete them (Harrison *et al.* 1986). Finally, the high viscosity of granitic melts presents an almost insurmountable physical barrier to the extraction of relatively small percentages of such liquid from the lower crust (Wickham 1987).

In situ partial melting is an attractive alternative, and infiltration of H_2O to effect melting is implied in Friend's (1983) analysis. A sequence of fluids, at first H_2O-rich, and subsequently CO_2-rich, explains the general alteration sequence. Gopalakrishna *et al.* (1986) emphasized the rôle of K-metasomatism in the origin of the Closepet Granite, and relatively early K-mobility would be consistent with the greater solubility of silicate materials in H_2O-richer solutions. Input of only H_2O would not explain the localization of granite in the middle and upper crust, since hydrous melting should have occurred even more extensively in the hotter, higher-pressure lower crust through which the H_2O passed. Establishment of a vertical K enrichment gradient would help to explain concentration of melting in mid-crustal levels, however, since K_2O is a hyperfusible constituent in rock systems.

7.1.6 Scope of the present work

The present work investigates the relationship between vapour-saturated partial melting of felsic and intermediate rocks and orthopyroxene production, making use of recent experimental petrology of K_2O-bearing silicate systems, with concentration on the possible rôle of a fluid phase. The fluids are postulated to be H_2O–CO_2 solutions of variable proportions in equilibrium with rock systems (that is, containing dissolved mineral components). In addition to the $CO_2 : H_2O$ ratio, other pertinent parameters are K activity, expressed as the K : Na ratio of fluids, the temperature, and the pressure. These are varied to construct phase equilibrium diagrams showing the interplay of partial melting and subsolidus dehydration of biotite. Emphasis is placed on defining the conditions under which the middle regions of the crust may be melted, and subsolidus alteration of the upper and lower crust occurs, in the presence of a CO_2–H_2O fluid. From this information one may explore the possibility that the Closepet Granite, the incipient charnockite at its southern end, the undepleted banded charnockites with orthopyroxene pegmatoids, and the deep-crustal depleted charnockites of the high-grade massif to the south can all be explained by the passage upward of metasomatizing solutions, perhaps of subcrustal origin, along a crustal shear zone.

7.2 Melting relations of a model crust

7.2.1 KMASH system

The phase equilibrium relations among phlogopite, sanidine, orthopyroxene, liquid, and aqueous fluid for the SiO_2-saturated (quartz-present) portion of the system $K_2O–MgO–Al_2O_3–SiO_2–H_2O$ (KMASH) were deduced by Luth (1963). The schematic arrangement of univariant *P–T* curves is shown in Figure 7.3. Slight degeneracy (an effective four-component system) results from use of the good approximation that the K : Al ratio is unity in all participating phases. Luth located some of the *P–T* curves approximately but did not demonstrate reversible equilibrium. Six phases coexist in a stable manner at the low-pressure invariant point: phlogopite, sanidine, quartz, enstatite, liquid, and vapour. The univariant curves are labelled by the missing phases in parentheses. Thus, the subsolidus equilibrium labelled (Liq) lacks a liquid

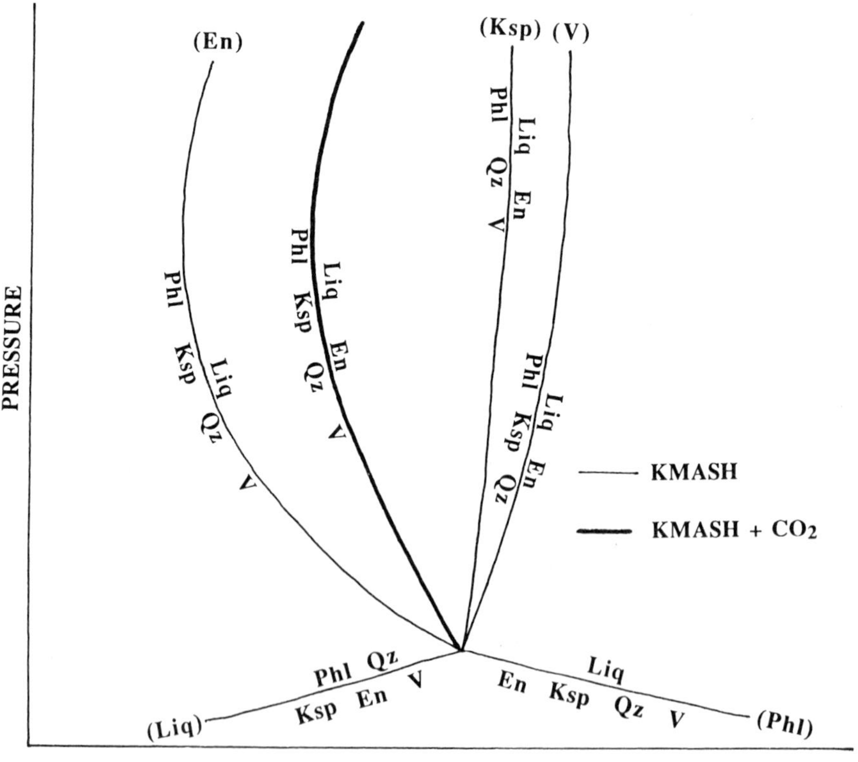

Figure 7.3 Schematic univariant melting and subsolidus univariant equilibria involving phlogopite (Phl), K-feldspar (Ksp), enstatite (En), quartz (Qz), liquid (Liq), and vapour (V). The arrangement of curves follows Wendlandt (1981) and Peterson & Newton (1987, 1988a).

phase:

(Liq) $KAlMg_3Si_3O_{10}(OH)_2 + 3SiO_2 = 3MgSiO_3 + KAlSi_3O_8 + H_2O$
Phlogopite quartz enstatite sanidine vapour

(En) phlogopite + sanidine + quartz + H_2O = liquid
(Ksp) phlogopite + quartz + H_2O = enstatite + liquid
(Phl) enstatite + sanidine + quartz + H_2O = liquid
(V) phlogopite + quartz = enstatite (+ sanidine?) + liquid (7.2)

The last equilibrium is vapour-absent melting. According to the experiments of Peterson & Newton (1987), the liquid composition is nearly on the phlogopite–H_2O join, so that the presence of sanidine among the product assemblage, as stated by Luth (1963), is doubtful. This equilibrium provides an important constraint on deep-crustal melting under volatile-deficient conditions, and could be an important mechanism for generating granulite facies terrains such as the Namaqualand terrain of southern Africa (Waters 1988).

The location of the vapour-present melting equilibria as defined by the reversed experiments of Peterson & Newton (1987, 1988a,b) is shown in Figure 7.4. The sanidine-absent equilibrium (Ksp) is displaced to considerably higher temperatures from the previous determination of Bohlen *et al.* (1983), who placed it almost coincident with the enstatite-absent (En) curve of Peterson & Newton (1987). The revised positions of these melting curves are important in the phase equilibrium relations to be developed, in defining limits at high levels of K_2O and H_2O activity.

One of the most important determinants of the phase diagram geometry is the location of the subsolidus equilibrium (Liq). The positions of contours in P–T–X_{H_2O} (or a_{H_2O}) space depend critically on the pressure of (Liq) at given temperatures in the KMASH system. The definitive reversal data which define the positions of the (Liq) curve and of the invariant point at the intersection of curves are shown in Figure 7.5. The curve drawn through the reversals of Peterson & Newton (1988b) and of Wood (1976) leads to the equilibrium free energy (ΔG°) expression:

$$\Delta G^\circ_{(Liq)} = -118.7T(K) + 73\,491 \text{ joules} \quad (7.3)$$

This expression makes use of the thermodynamic properties of H_2O of Burnham *et al.* (1969) and the thermal expansions of the participating phases tabulated by Holland & Powell (1985). The temperature coefficient of the free energy implies a maximally (Si, Al)-disordered phlogopite, as advocated by Clemens *et al.* (1987) on the basis of their solution calorimetry. The internally consistent thermodynamic data set of Berman (1988), which accepts maximal disorder in estimating the entropy of phlogopite, predicts pressures of equilibrium (Liq) that are only slightly higher than shown in Figure 7.5, but the Holland & Powell (1985) data set, which includes an ordered phlogopite with-

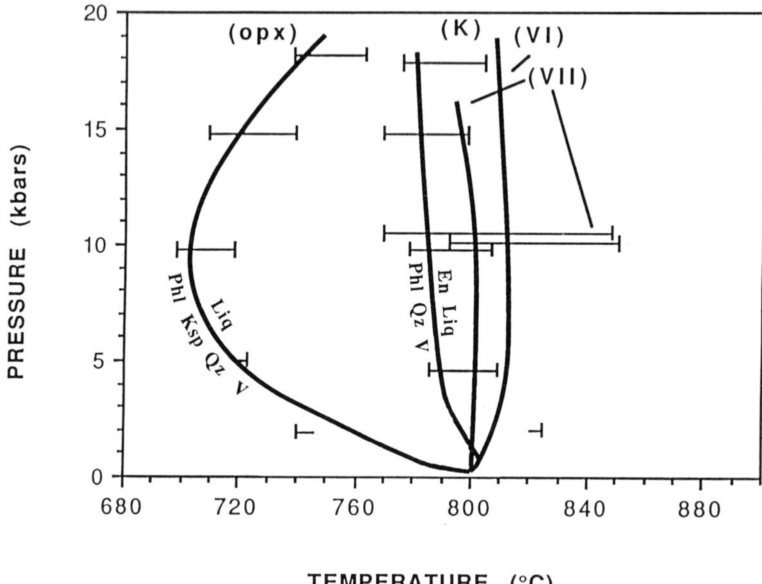

Figure 7.4 Experimental reversal brackets of melting equilibria of Peterson & Newton (1987, 1988a). Abbreviations are as in Figure 7.3, except K = K-feldspar. The designations (VI) and (VII) refer, respectively, to the vapour-absent melting of phlogopite + quartz and of phlogopite + K-feldspar + quartz. The large reversal intervals associated with them attest to the experimental difficulty of reversing vapour-absent melting in siliceous systems. The two half-reversals at 2 kbar are associated with the equilibria nearest to them (after Peterson & Newton 1989, Fig. 2, by permission of the University of Chicago Press. Copyright © 1989 by the University of Chicago).

out configurational entropy, predicts pressures much higher (about 1200 bar at 750°C). The important consequence of the low-pressure Peterson & Newton (1988b) version, described by expression (7.3) is a quite low level of H_2O activity for the coexistence of enstatite, K-feldspar, and vapour at elevated pressures.

7.2.2 KMASH–CO_2 system

The KMASH phase relations of Figure 7.4 can be extended with the addition of CO_2, with modification of the vapour and liquid phases. Each of the vapour-saturated univariant lines in the KMASH system becomes a divariant surface in the KMASH–CO_2 system, as shown in Figure 7.6. Theoretically, all of the surfaces must intersect in a single curve in multi-dimensional space:

$$\text{phlogopite} + \text{sanidine} + \text{quartz} + \text{vapour} = \text{enstatite} + \text{liquid} \quad (7.4)$$

The *P–T* projection of this curve is shown schematically in Figure 7.3.

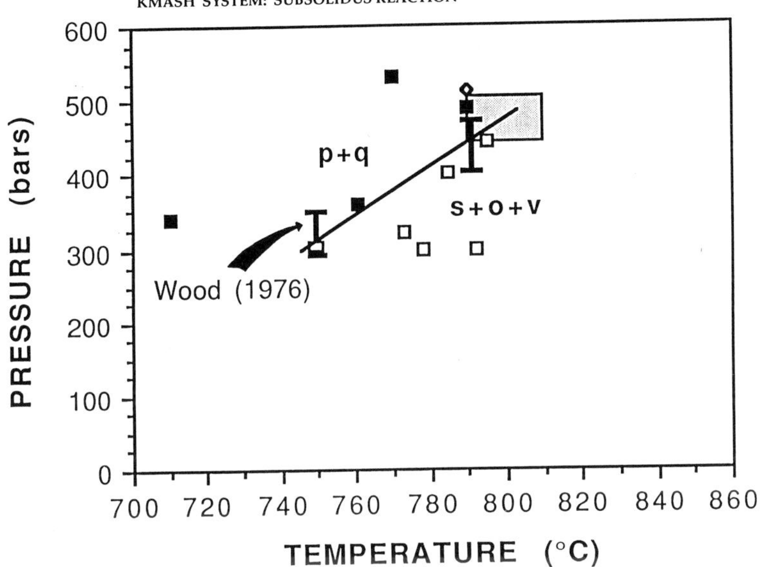

Figure 7.5 Reversed experimental data for the subsolidus equilibrium phlogopite + quartz = enstatite + K-feldspar + H_2O. Squares are individual runs of Peterson & Newton (1988b); brackets are intervals confining the equilibrium curve of Wood (1976). The shaded rectangle constrains the location of the KMASH invariant point of Figure 7.3 based on reversed melting experiments. P, phlogopite; Q, quartz; S, sanidine (K-feldspar); O, orthopyroxene (enstatite); V, vapour (H_2O) (after Peterson & Newton 1989, Fig. 5, by permission of the University of Chicago Press. Copyright © 1989 by the University of Chicago).

Wendlandt (1981) located this curve by experiments at 3, 8 and 15 kbar with crystalline starting materials and hydrous oxalic acid. His determination of the univariant curve, shown in Figure 7.6, is surprising in that the phlogopite-bearing assemblages melt under CO_2-rich vapours at temperatures nearly as low as, or lower than, under pure H_2O vapour. In particular, the (Ksp) melting curve is considerably depressed by the addition of CO_2 to the KMASH system. Wendlandt (1981) did not reverse the univariant reaction: his results rigorously indicate only upper temperature limits on melting.

Addition of CO_2 to the haplogranitic system sanidine–albite–quartz–H_2O considerably raises the solidus, rather than lowering it (Holtz *et al.* 1988). Moreover, the presence of phlogopite has almost no effect on melting in the haplogranitic system, and liquids near the solidus take less than 1 wt% MgO (Puziewicz & Johannes 1988). From these results it is clear that neither CO_2 nor MgO alone could explain Wendlandt's results: the two components must co-operate in some fundamental way to lower the solidus, as by a solute $MgCO_3$ component. An apparently analogous effect is seen in the simple system Na_2O–MgO–Al_2O_3–SiO_2–H_2O, where the hydrous melting point of

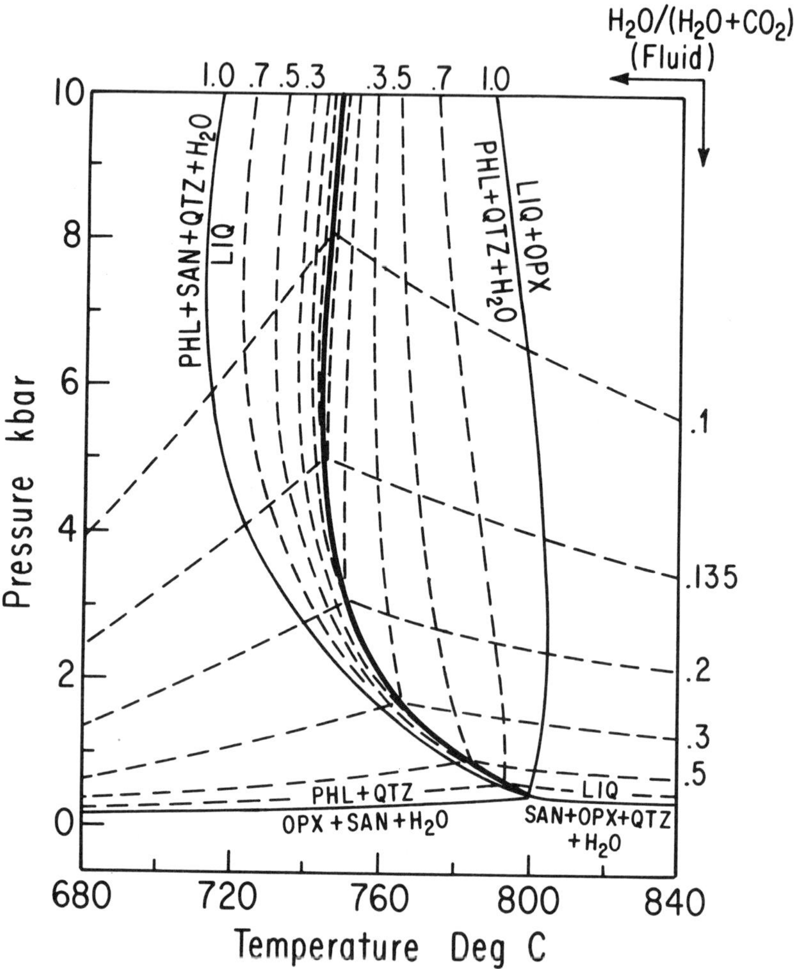

Figure 7.6 Univariant vapour-saturated equilibria in the KMASH system (lighter solid lines) and the KMASH–CO_2 system (heavy solid line). Contours of constant mole fraction of H_2O on divariant surfaces in the latter system are identifiable by near-parallelism to the solid KMASH curves. Abbreviations as in Figure 7.3, except SAN = sanidine; OPX = orthopyroxene. See text for details of construction.

pargasite is depressed by the addition of CO_2 in the range 1–7 kbar (Holloway 1973). The univariant KMASH–CO_2 liquids at pressures above about 4 kbar are not likely to be truly granitic, as noted by Wendlandt (1981), but become increasingly more mafic and lower in SiO_2 as the CO_2 content increases.

The univariant space curve in the KMASH–CO_2 system emanates from the invariant point at 450 bar and 800°C in the KMASH system, and is stable to at least 15 kbar. At higher pressures the reaction may change with the appearance of magnesite on the solidus (Wendlandt 1981). Divariant surfaces meeting in the univariant curve are shown in Figure 7.6, in the form of contours of constant H_2O content of the vapour phase. They were calculated in the following way. First, contours of constant mole fraction of H_2O in the vapour, $X^V_{H_2O}$, on the subsolidus surface (Liq) were calculated from the expression:

$$\Delta G^0_{(Liq)} + RT\ln f_{H_2O} + P\Delta V_{S^*} = 0 \quad (7.5)$$

where f_{H_2O} denotes the fugacity of H_2O in CO_2–H_2O mixtures at T and P, the equilibrium pressure of (Liq), and ΔV_{S^*} denotes the solid volume change of the reaction. The fugacity of H_2O was taken from Burnham *et al.* (1969) and Kerrick & Jacobs (1981). ΔV_{S^*} was evaluated using the thermal expansions and compressibilities of Holland & Powell (1985). Because of its occurrence in the expressions for f_{H_2O} and ΔV_{S^*}, P had to be found by iteration. Next, contours on the surfaces (En) and (Ksp) were constructed entirely from the geometrical consideration that they be roughly congruent with their KMASH univariant bounding curves. This supposes that they are 'well behaved'; that is, that they do not have eccentric features such as maxima or minima. This is reasonable, since the bounding curves are close together in P–T space. Contours on the (Phl) surface are less well constrained by these considerations, especially at low H_2O contents and high pressures, but this lack of constraint is not important in the construction of the topologies which follow.

7.2.3 *Effect of plagioclase and fluid K : Na*

Addition of the Na_2O component allows the construction of phase equilibrium diagrams to study the effect of K-metasomatism, including metasomatic melting and charnockite formation, on a model low-K tonalitic or trondhjemitic crust. A key additional equilibrium is that of alkali exchange between coexisting alkali feldspar and plagioclase:

$$\underset{\substack{\text{plagioclase}\\ \text{solid solution}}}{NaAlSi_3O_8} + K = \underset{\substack{\text{sanidine}\\ \text{solid solution}}}{KAlSi_3O_8} + Na \quad (7.6)$$

The nature of the alkali-bearing solute species is not indicated in the above equation. This is not known for H_2O–CO_2 mixtures at elevated temperatures and pressures. The solution of alkalis could be as ions, alkali carbonate neutral complexes, or dissolved silicate constituents. The latter is an important possibility, since comparative chemical studies of charnockitization in South India by Hansen *et al.* (1987) and Stähle *et al.* (1987) showed mobility of virtually all major and minor rock components in charnockite-forming solutions.

It is expected that alkali solubilities in carbonate solutions will be somewhat different from those in chloride solutions. Rubie & Gunter (1983) emphasized that the fluid K : Na of carbonate solutions in equilibrium with two feldspars at 600°C and 1 kbar, as shown in the exchange experiments of Iiyama (1965), is somewhat smaller than in chloride solutions at the same conditions. However, these experimental data are quite limited. An additional complication is the possibility of dissolved alkali silicate species, for which there are no experimental free energy data. For these reasons, the K : Na ratios in exchange equilibria are calculated here as if the solutions were 2M chlorides. This assumption will not change the phase diagram geometries to be derived in any fundamental way, as the absolute value of the K : Na ratio is not a critical consideration in the present analysis. The following is a free energy expression for Reaction (7.6), which was derived by interpolation between the experimental exchange data of Orville (1963) and the ΔG° values below 300°C given by Helgeson (1969):

$$\Delta G_6^\circ = 1.2T - 9337 \text{ joules} \tag{7.7}$$

The slight effect of a moderate amount of $CaAl_2Si_2O_8$ (anorthite) component in the plagioclase is calculated using the plagioclase free energy expressions of Newton *et al.* (1980).

One additional subsolidus equilibrium is necessary to construct the topologies:

$$\text{plagioclase} + \text{enstatite} + H_2O + K = \text{phlogopite} + \text{quartz} + Na \tag{7.8}$$

This reaction states that, at a sufficiently low activity of K, phlogopite must break down and be replaced by enstatite, even for elevated levels of H_2O activity. The reaction has a standard free energy given by $\Delta G_6^\circ - \Delta G_{(\mathrm{Liq})}^\circ$:

$$\Delta G_8^\circ = +\,119.9T - 82\,828 \text{ joules} \tag{7.9}$$

Four alkali-exchange melting reactions are important in the topologies:

$$\text{enstatite} + \text{plagioclase} + \text{quartz} + H_2O + K = \text{liquid} + Na \tag{7.10}$$

$$\text{liquid} + \text{enstatite} + K = \text{phlogopite} + \text{quartz} + H_2O + Na \tag{7.11}$$

$$\text{phlogopite} + \text{plagioclase} + \text{quartz} + H_2O + K = \text{liquid} + Na \tag{7.12}$$

$$\text{liquid} + K = \text{phlogopite} + \text{sanidine} + \text{quartz} + H_2O + Na \tag{7.13}$$

These reactions are univariant lines in the plane of temperature versus K : Na (fluid) for constant pressure, $H_2O : (H_2O + CO_2)$ in the fluid, and mole fraction of $CaAl_2Si_2O_8$ in the plagioclase. Experimental petrology of melting reactions helps to place constraints on Reactions (7.10–13). The topology of

Figure 7.6 provides asymptotes at infinite K : Na (fluid) for Reactions (7.11) and (7.13). These upper-temperature limits will be virtually realized at high ratios (of the order of ten). An additional assumption is made that plagioclase saturation lowers the melting temperatures of phlogopite–sanidine–quartz assemblages by the same amounts as it does in the system albite–sanidine–quartz–H_2O–CO_2 (Holtz *et al.* 1988) compared to the system sanidine–quartz–H_2O–CO_2 (Bohlen *et al.* 1983). This assumption places the invariant points involving plagioclase about 60° lower than the asymptotic values of Reactions (7.11) and (7.13). A final constraint is that the melting curves of albite for variable $H_2O : (H_2O + CO_2)$ (vapour) of Eggler & Kadik (1979) provide upper temperature limits at zero K : Na (fluid) for Reaction (7.10). The

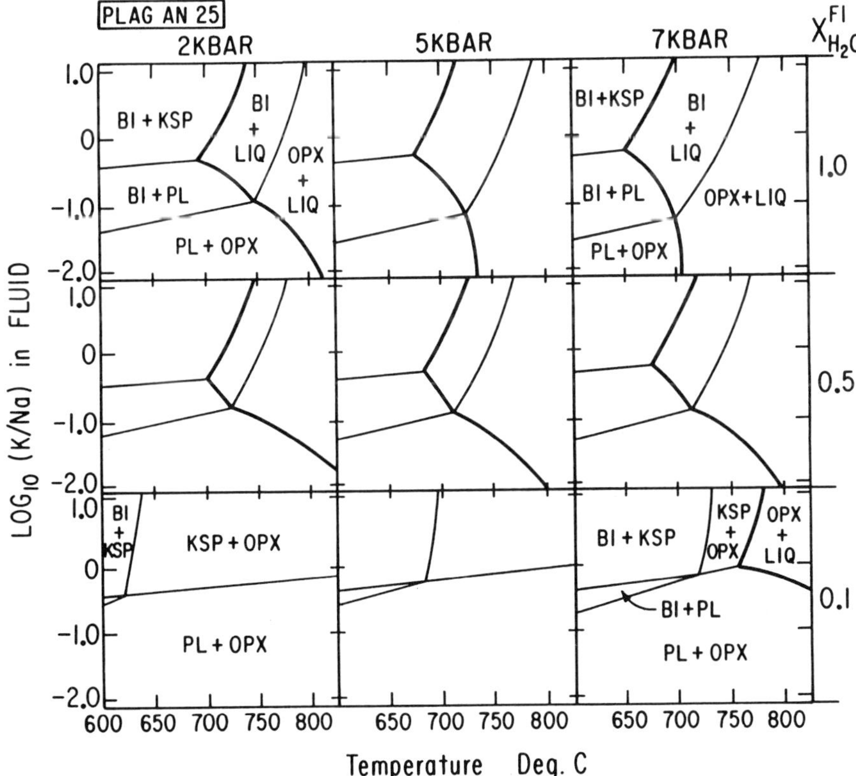

Figure 7.7 Fluid alkali ratio of various assemblages in the SiO_2-saturated portion of the KMASH-CO_2 system versus temperature, for various total pressures and H_2O mole fractions. BI, biotite (i.e. phlogopite); PL, plagioclase. Bold curves are solidus equilibria. Subsolidus curves are calculated as if the solutions are 2M chlorides, from data summarized in the text. They are corrected slightly for an assumed 25 mole percent of $CaAl_2Si_2O_8$ (anorthite) in the plagioclase. The most important feature is the low melting minimum for two-feldspar equilibrium, even at elevated CO_2 contents of the vapour.

Morey–Schreinemakers rule provides additional constraints on the locations of the melting curves. These considerations, together with the calculated position of Reaction (7.7), serve to fix the topologies rather rigidly, with only minor possible uncertainties in the curvatures of the melting reactions. Equilibrium subsolidus and melting equilibria for a variety of constant-*P* and constant-H_2O mole fractions, all at plagioclase compositions of An_{25}, are shown in Figure 7.7.

The diagrams for H_2O mole fractions of 0.5 and 1.0 show two cusps on the solidi, which may serve as important control points for rock melting. The melting minimum involves two feldspars and corresponds to formation of simple granitic or syenitic liquids. The higher-temperature cusp involves orthopyroxene, and may have a rôle in limited amounts of melting of the deeper crust under low-H_2O conditions. The large effect of the K : Na ratio of the fluids on the solidus indicates that K-enriched portions of the middle crust may undergo melting under some thermal régimes even if the lower crust remains solid. At very low H_2O contents of the fluids, the subsolidus field of biotite is greatly contracted and the solidus is elevated in temperature.

It must be emphasized that the foregoing analysis does not consider amphibole stability. A simple trondhjemite with An_{25} plagioclase might well have a small amount of calcic amphibole. The field evidence at Kabbaldurga indicates that this would break down to an orthopyroxene-bearing assemblage under some subsolidus conditions of low-H_2O mole fraction in a vapour phase; that reaction is not modelled in Figure 7.7. Ferromagnesian amphibole stability is also a possibility in some of the P–T–X_{H_2O} space considered here, but is not considered in the present simple phase analysis, which concentrates only on melting relations of biotite and orthopyroxene-bearing assemblages.

7.3 Interpretation of Closepet–charnockite relations

7.3.1 Effects of variable K activity

The melting minimum for saturation of both feldspars in a quartz-excess system allows for the possibility of mid-crustal melting by solutions rising through the crust, as shown in Figure 7.8. Representative conditions of 5–7 kbar and an H_2O mole fraction of 0.5 can be used to illustrate the possible effects over a range of crustal depth. H_2O–CO_2 solutions, initially undersaturated in K as they are introduced into the crust, will break down biotite at deeper levels, liberating K and H_2O, thus producing subsolidus charnockite formation. It is expected, therefore, that there will be an initial K activity gradient, as well as a temperature gradient, suggested in Figure 7.8 (crosses). Rising solutions would be expected to bring in heat, so that the initial evolution of temperature and solution composition would be as shown by the dotted arrows. At some higher level, potassium would be deposited because of the

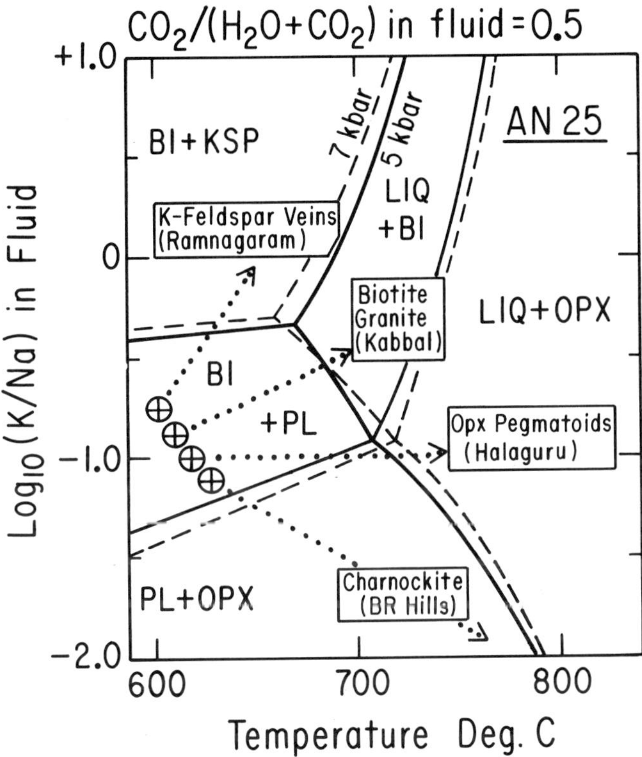

Figure 7.8 Univariant equilibrium curves from Figure 7.7, showing how K-depletion and charnockitization in the lower crust could be brought about by entry of hot fluids initially undersaturated in K, and how K could be deposited down a temperature and pressure gradient in the middle and upper crust, giving rise to mid-crustal melting and metasomatic veining, as observed at the indicated localities in the Closepet – BR Hills zone. The array of X-points suggests the initial K-activity and T gradient in the crust.

reduced solubility at lower pressure. Emplacement of K would occur first in plagioclase solid solution, and then in a melt phase as temperatures increase. This process, which is embodied in Reaction (7.12), is essentially K-infiltration melting, as suggested by chemical studies of migmatites by Olsen (1985) and Johannes (1988). At some intermediate level, K is neither removed nor emplaced, but the solutions continue to transfer heat upwards. The resulting horizontal vector could penetrate the field of orthopyroxene-liquid assemblages. This process may have formed the orthopyroxene pegmatoids of the

Halaguru banded charnockites. At upper crustal levels, increase of K activity and temperature may avoid the field of liquid, with metasomatic conversion of plagioclase to K-feldspar, as observed in the southern Closepet area (Divakara Rao *et al.* 1972). The metasomatic vein sequence in the southern Closepet area is very complex because of changing water and solute activities with decreasing temperature in the late stages of a fluid-streaming process. Carbonate replacement veins of greenschist grade in the Chitradurga area (Chadwick *et al.* 1981) and the quartz–carbonate gold-bearing veins of low amphibolite facies in the Kolar area (Hamilton & Hodgson 1986) may be upper-level expressions of the passage of fluids through the crust, associated with deeper charnockite formation (Cameron 1988).

The two isobaric invariant points will be preferred locations of vapour-present melting because of the buffering capacities of rocks with abundant plagioclase and biotite. Liquids which form at deeper levels and lower K activities seek the invariant point biotite–orthopyroxene–plagioclase–liquid. Because of the very large buffering capacity of plagioclase in trondhjemitic rock, pore fluids under cooler, H_2O-richer conditions would follow the line of Reaction (7.6), at least to some extent, which would result in melts formed at the invariant point sanidine–plagioclase–biotite–quartz–liquid. A further rise of temperature could be strongly limited by thermal buffering, as suggested by Hodges *et al.* (1988): the large amount of heat necessary to perform major rock melting would act as a brake on temperature. Very extensive melting would inevitably result in diapiric uprise of magma.

7.3.2 Change of fluid $H_2O:CO_2$ ratio with time

The character of anatectic liquid undoubtedly changes greatly along the univariant KMASH–CO_2 *P–T* melting curve. It has been inferred that both CO_2 and MgO increase coherently in the liquid with increasing pressure along the curve. As noted by Wendlandt (1981), the liquid must change from granitic (normative feldspar : normative quartz = 1) to syenitic with lower SiO_2 contents at higher pressures along the melting curve. At some elevated pressures the melt becomes lamprophyric because of the increase of dissolved CO_2 at high pressures in silicate liquids, increase of mafic components, and depolymerization of the liquid. It is shown in Figure 7.6 that at pressures above 5 kbar, appropriate for the southern Closepet region, vapours extremely rich in CO_2, of X_{H_2O} approaching 0.1, would be in equilibrium with the KMASH–CO_2 univariant liquid. This liquid could not be strictly granitic, but must be low in SiO_2 and fairly mafic. These considerations lead to the conclusion that Closepet Ganite, if it was ever largely molten (which seems probable), resulted from infiltration of relatively H_2O-rich solutions, and the qualitative conclusion is in accord with the lack of orthopyroxene coexisting with the granitic liquids in the southern Closepet area. Subsequent infiltrated fluids were poorer in H_2O and richer in CO_2, as documented by the relatively late appearance of charnockite overprinting both grey gneiss and granite, and by the lack of clear-cut K_2O emplacement in charnockitization and more nearly isochemical alteration (Hansen *et al.* 1987), suggesting lower solubilities in the fluid.

The deduced sequence of early H_2O-richer solutions followed by late CO_2-richer solutions has the potential of explaining some otherwise enigmatic phenomena, namely the lack of alteration of much of the orthopyroxene in the pegmatoids of the Halaguru banded gneisses, and the CO_2-rich fluid inclusions of granitic leucosomes as well as charnockitic veins. When the anatectic melts froze, H_2O released should have altered the orthopyroxene to a hydrate (biotite or amphibole) under equilibrium conditions, and would have been captured as fluid inclusions by crystallizing quartz. Wendlandt (1981) suggested a possible explanation of orthopyroxene-bearing migmatites by increase, with time, of CO_2 activity (and decrease of H_2O activity) of fluids coexisting with melts. A change of composition in the vapour phase towards CO_2-richer compositions will inevitably freeze the liquids, even though the temperature may continue to rise. Thus, the vapours coexisting with the last melts could be in equilibrium with orthopyroxene. Also, the last fluids trapped as the melts freeze could be very rich in CO_2. Post-consolidation entrapment of dry fluids could result in almost pure CO_2 inclusions. The sequence of fluids from early H_2O-rich to late CO_2-rich is a petrological deduction without explanation, inasmuch as the ultimate sources and mechanisms of fluid influx are unknown.

7.3.3 The Closepet zone as a crustal shear belt

The Closepet – Kabbal – Halaguru – BR Hills zone has been pictured as a metasomatized crustal shear zone with a depth-dependent alteration sequence which includes deep-crustal metasomatic charnockites, low-P_{H_2O} orthopyroxene-bearing migmatites at higher levels in the lower crust, metasomatic and anatectic K-rich granite containing biotite and amphibole and therefore reflecting more H_2O-rich conditions at mid-crustal levels, and a complex of high-temperature to low-temperature replacement veins deposited at upper-crustal levels. The average width of the belt in its southern portion near the incoming of orthopyroxene is 15–20 km, which is a typical width for deep-crustal megashear belts (Hanmer 1988).

The inferred aspect of the late Archaean crust during the metasomatic event in southern India is shown schematically in Figure 7.9. Hot fluids of deep origin destabilized biotite and amphibole in the deep crust, transforming low-K gneisses to charnockitic rocks and moving released K and H_2O upwards. In addition, the diagram suggests that there were pre-existing deep-crustal charnockites, perhaps resulting from an earlier metamorphic cycle. Fluids initially undersaturated in K could have destabilizised biotite and amphibole even at relatively high H_2O activity, according to Figure 7.7. The fluids would

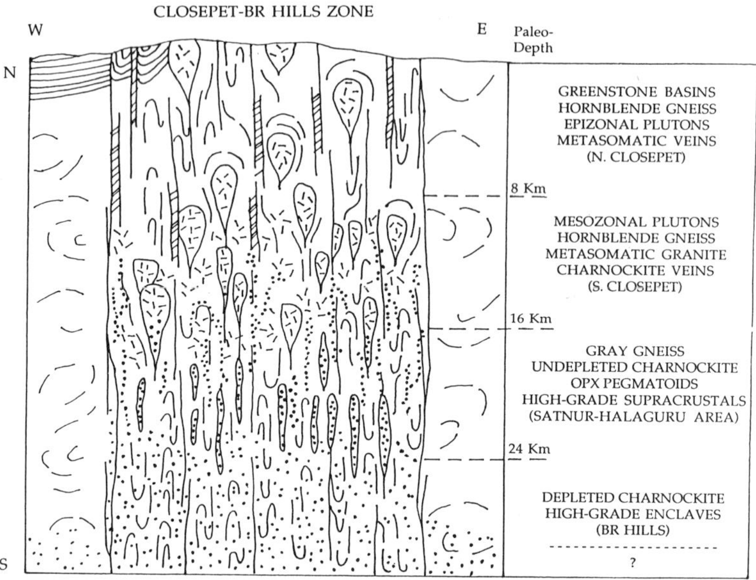

Figure 7.9 Schematic megashear zone proposed to account for the extreme elongation of the Closepet Granite and the succession of lithologies encountered southward in the Closepet – BR Hills zone of the Dharwar Craton, inferred to correspond to a depth-zone profile. The horizontal scale equals the vertical scale. The suggested charnockitic roots of the craton outside of the shear zone imply that an older generation of granulites, pre-dating 2.5 Ga, exists at some depth.

eventually come into alkali exchange equilibrium with the crustal rocks. Because of the decrease of K_2O solubility with decreasing fluid pressure and temperature, it is inevitable that K_2O was emplaced somewhere in the middle or upper crust. In the earlier, cooler stages, this could have been in the form of K-exchange of feldspars. As the fluids transported heat upwards, the K_2O could have been emplaced into a granitic melt instead. This mechanism, which depends on the establishment of a K activity gradient in the crust, could explain the generation of Closepet Granite at mid-crustal, rather than deep crustal, levels. Locally, more extensive melting resulted in the formation of small plutons which rose some distance into the upper crust.

At some level between deep-crustal granulite facies metamorphism and mid-crustal granite formation, there must have existed a level where K was neither greatly depleted nor enriched. Limited melting could have occurred in this zone, with the formation of orthopyroxene-bearing pegmatoids such as those found in the Satnur–Halaguru area, as suggested in Figure 7.8.

Late CO_2-richer fluids moved the orthopyroxene horizon somewhat higher in the crust. This could have occurred during the waning stages of the fluid event, under decreasing temperatures. The drier fluids effected freezing of the pegmatoids and granites and created the subsolidus charnockitic veins at localities such as Kabbal. During a cooler period the K-feldspar–carbonate veins could have been emplaced and, at higher levels, quartz–carbonate veins. Those emplaced into greenstone units may have precipitated gold deposits such as those of the Kolar area, perhaps by reduction of the solutions as they contacted Fe^{2+} silicates or sulphides.

7.4 Possible applications to other terrains

7.4.1 Metasomatic shear belts

Many examples exist in Precambrian gneiss terrains of profoundly metasomatized deep-crustal shear zones, although none have been reported with extensive development of discrete granite plutons as in the Closepet Zone. The Nordre Strømfjord linear belt of south-west Greenland is a zone of profound left-lateral shearing, 5–25 km wide, in the 1700–1800 Ma Nagsuggtoqudian Mobile Belt. The displacement across the zone is thought to be of the order of 120 km (Sørensen 1983). The metamorphic grade within the shear zone varies from greenschist facies in the east to granulite facies in the western extremity, and the width of the belt increases with increasing metamorphic grade, in the manner typical of through-crustal megashear zones (Hanmer 1988). Composite gneiss sampling across the high-grade portion of the Nordre Strømfjord belt shows systematic enrichment of K_2O and SiO_2 within the zone compared to gneisses outside the zone. Some of this granitization results from swarms of granitic pegmatoids (W.E. Glassley, personal communication, 1988) and some is metasomatic. The overall metamorphic grade of the zone corresponds generally to that of the surrounding terrain, but generally contains more biotite and less orthopyroxene, which may be a reflection of higher K activity rather than of different P–T conditions (Sørensen 1983).

Other metasomatized zones which may be comparable are the Laxfordian zones of the Scourie granulite facies terrain of north-west Scotland (Beach 1976) and the Athabasca Mobile Zone of the western Canada subsurface basement (Burwash & Krupička 1970). The former are retrogressive in character and perhaps 1 Ga younger than the granulite facies metamorphism. Consistent right-lateral offsets of the order of 20 km occur in the Laxford zones. The Athabasca Mobile Zone is represented mainly by drill cores which show K_2O enrichment correlated with the degree of shearing (Burwash & Krupička 1970). Metasomatic introduction of K is shown petrographically in the form of K-feldspar replacement of plagioclase and biotite replacement of amphibole.

Low-grade metasomatic megashears involving carbonate replacement occur in Hercynian schists in the Tien Shan, Central Asia (Baratov *et al.* 1984) and in the late Archaean Abitibi greenstone belts of southern Canada (Kerrich *et al.* 1987). It is possible that these upper-level deposits document the CO_2 component of fluids postulated to have moved through the crust. Numerous deep fault zones of Soviet Asia containing carbonatites and alkali metasomatites, described by Lapin & Ploshko (1988), may result from similar processes.

7.4.2 Orthopyroxene and CO_2-rich inclusions in migmatites

Orthopyroxene-bearing leucosomes in the progressive metamorphic terrain of Namaqualand, South Africa, were explained by Waters (1988) as the result of dehydration melting. According to this detailed and persuasive analysis, H_2O activity was lowered to the level of orthopyroxene stability by absorption of resident H_2O into anatectic melts. A problem raised by Waters is the relative lack of alteration of the orthopyroxene: when the granitic melts froze, the released H_2O should have back-reacted the orthopyroxene to biotite or amphibole under equilibrium conditions. He suggested that the H_2O had limited contact with orthopyroxene because of strong channelization and perhaps because of armouring by quartz and feldspar. Another possibility which is suggested, by analogy with the southern Closepet area, is influx of late CO_2-rich fluids. Following the suggestion of Wendlandt (1981), it is possible that the last liquids to freeze were CO_2-rich and actually in equilibrium with orthopyroxene, and that later post-consolidation fluids were even poorer in H_2O.

Olsen (1987) pointed to the problem of CO_2-rich inclusions in the Proterozoic migmatites of the Front Range, Colorado. They are much more numerous in leucosomes than in host rocks, suggesting a genetic link with the anatexis, and yet are seemingly inappropriate in their high CO_2 contents for equilibrium with granitic melts. This paradox is recognized as a general problem of migmatites by Hollister (1988), who refers to the CO_2-rich inclusions as 'enigmatic'. The deduction for southern India of an influx of late CO_2-rich fluids which froze the melts may be an alternative explanation of CO_2-rich inclusions in migmatites in the Front Range and elsewhere.

7.4.3 Possible sources of fluids

It is difficult to envisage a crustal source of CO_2 large enough to account for the massive metasomatism of the lower and middle crust postulated for the Closepet – Kabbal – BR Hills zone of southern India. It is unlikely that there are large quantities of limestone in the deepest crust which could have been mobilized in metamorphism to provide the necessary CO_2 flux. A recent study of gneiss–marble contact interactions in the Adirondack Mountains high-grade terrain (McLelland *et al.* 1988) suggests that the marbles are very refractory in the absence of infiltrating fluids and yield insignificant quantities of CO_2. A recent proposal whereby deep-crustal marble bodies may sink through partially melted enclosing gneisses and into basaltic underplates and thus decarbonate (Wickham 1988) is an ingenious mechanism, but no evidence of the requisite underplate in the form of gravity anomalies below the high-grade terrain (Subrahmanyam 1978) or gabbro bodies syngenetic with the Closepet Granite have been reported. The lack of evidence of mafic magmatism in the Closepet zone also raises a major problem of the heat source for metamorphism and melting.

The most probable source of fluids seems to be the upper mantle. A deep-seated source is consonant with the great length scale of the Closepet Granite outcrop which must, to some extent, reflect the vertical dimensions involved. The right-lateral shears manifest in the Closepet zone are, logically, avenues for the upward migration of fluids, as postulated in general for the Dharwar Craton banded gneiss ('high finite strain') zones of Drury & Holt (1980). Possible sources for abundant metasomatizing fluids are carbonated-hydrated mantle 'metasomes', which have been proposed as the source regions of lamprophyres and alkali basalts (Middlemost *et al.* 1988), asthenospheric liquids, which are likely to be carbonatitic with a high H_2O content (Wallace & Green 1988) or devolatilizing slabs subducted below the continental margin, as pictured for the western United States in the late Mesozoic by Hoisch (1987a) and Peacock (1987a). Volatile production from subducting slabs in the Precambrian might have been different from that of the Phanerozoic because of different chemical environments of sedimentation and seafloor alteration. In particular, the probable greater abundance of ferrous carbonate deposited in reducing ancient oceans might have resulted in greater amounts of slab-derived CO_2, because of the lower thermal stability of ferrous carbonate, than calcium carbonate.

A change with time in a metamorphic sequence from more H_2O-rich infiltrating fluids to more CO_2-rich fluids could arise from a number of factors connected with change of the angle, position, and velocity of the subducting slab. If, for example, the subduction rate decreased over time, the subduction angle might become steeper, so that the more refractory components of the subducted volatiles, namely the carbonates, would become major contributors to the volatile production underneath a given part of the continental margin. Progressive deepening of shear zones during a waning metamorphic cycle could have tapped asthenospheric fluids, bringing more CO_2 into the rising fluids.

The possibility that mantle-derived fluids can bring much of the heat for metamorphism and partial melting is a topic of current debate. Some calculations of heat advection suggest that slab-derived fluids can be important in the thermal budget of a metamorphic terrain (Hoisch 1987b), whereas other calculations have discounted this possibility, maintaining that such heating would be feasible only in thin zones adjacent to the fluid channels (Peacock 1987b). The possibility that substantial heating of the lower and middle crust could be caused by advecting fluids from the mantle would seem to remain open for shear zones of the sort postulated for the Closepet – BR Hills zone.

7.4.4 Younger shear zone granites possibly analogous to the Closepet

Several occurrences have been reported of granite bodies which have close spatial and temporal relations with large-scale transcurrent faults. These tend to be alkali-rich granites, typically emplaced tens of millions of years after major collisional orogenies (Sylvester 1989). Some of the Pan African alkaline granites of the Najd transcurrent zone in the Western Arabian Shield have chemistries quite similar to some of the Closepet Granites (Table 7.1; compare with the Tumkur–KD granites). The Najd faults may have several hundred kilometres of left-lateral offset and arc considered by Stern (1985) to be intracontinental transform faults. The reasons for the post-collisional emplacement and the heat source for the granites are major problems. Mantle-derived fluids as a source of halogens and CO_2 in crustal melting were suggested by Harris (1982). Structural and chemical similarities of the Closepet Granite with some of the Najd Zone granites suggest that processes of island-arc accretion and continental-margin reactivation, well documented for the Saudi Arabian Pan African Shield, may have operated also in the late Archaean of southern India.

Acknowledgements

The author's experimental research, including some of the phase equilibria described here, is supported by a U.S. National Science Foundation grant, no. EAR 87–07156. His experience with southern India, from which the ideas herein were largely drawn, was sponsored by an NSF grant, no. INT 82–19140. Valuable conversations with George Brimhall, Clark Friend, Nigel Harris, B. Mahabaleswar, Wally Pitcher, John J. W. Rogers, Paul Sylvester, and Steve Wickham are acknowledged.

References

Allen, P., K. C. Condie & G. P. Bowling 1986. Geochemical characteristics and possible origins of the Closepet Batholith, South India. *Journal of Geology* **94**, 283–99.

Baker, A. J. & A. E. Fallick 1988. Evidence for CO_2 infiltration in granulite facies marbles from Lofoten–Vesteralen, Norway. *Earth and Planetary Science Letters* **91**, 132–40.

Balasubrahmanyam, M. N. 1978. Geochronology and geochemistry of Archaean tonalitic gneisses and granites of South Kanara District, Karnataka State, India. In *The origin and evolution of Archaean continental crust*, B. F. Windley & S. M. Naqvi (eds), 59–77. Amsterdam: Elsevier.

Baratov, R. B., N. A. Gnutenko & V. N. Kuzemko 1984. Regional carbonitization connected with the epi-Hercynian tectonogenesis in the southern Tien Shan. *Doklady Akademii Nauk SSSR* **274**, 124–6.

Beach, A. 1976. The interrelations of fluid transport, deformation, geochemistry and heat flow in early Proterozoic shear zones in the Lewisian complex. *Philosophical Transactions of the Royal Society of London* **A280**, 569–604.

Berman, R. G. 1988. Internally-consistent thermodynamic data for minerals in the system Na_2O–K_2O–CaO–MgO–FeO–Fe_2O_3–Al_2O_3–SiO_2–TiO_2–H_2O–CO_2. *Journal of Petrology* **29**, 445–522.

Bohlen, S. R., A. L. Boettcher, V. J. Wall & J. D. Clemens 1983. Stability of phlogopite–quartz and sanidine–quartz: a model for melting in the lower crust. *Contributions to Mineralogy and Petrology* **83**, 270–7.

Burnham, C. W., J. R. Holloway & N. F. Davis 1969. *Thermodynamic properties of water to 1,000°C and 10,000 bars*. Geological Society of America, Special Paper 132.

Burwash, R. A. & J. Krupička 1970. Cratonic reactivation in the Precambrian basement of western Canada. Part II: Metasomatism and isostasy. *Canadian Journal of Earth Sciences* **7**, 1275–94.

Cameron, E. M. 1988. Archaean gold: relation to granulite formation and redox zoning in the crust. *Geology* **16**, 109–12.

Chadwick, B., M. Ramakrishnan & M. N. Viswanatha 1981. The stratigraphy and structure of the Chitradurga Region: an illustration of cover–basement interaction in the late Archaean evolution of the Karnataka Craton, Southern India. *Precambian Research* **16**, 31–54.

Clemens, J. D., S. Circone, A. Navrotsky & P. F. McMillan 1987. Phlogopite: high-temperature solution calorimetry, thermodynamic properties, Al–Si and stacking disorder, and phase equilibria. *Geochimica et Cosmochimica Acta* **51**, 2569–78.

Devaraju, T. C. & M. S. Sadashivaiah 1969. The charnockites of Satnur–Halaguru area, Mysore State. *Indian Mineralogist* **10**, 67–88.

Divakara Rao, V., U. Aswathanarayana & M. N. Qureshy 1972. Trace element geochemistry of the Closepet Granite, Mysore State, India. *Mineralogical Magazine* **38**, 678–86.

Drury, S. A. & R. W. Holt 1980. The tectonic framework of the South Indian Craton: a reconnaissance involving LANDSAT imagery. *Tectonophyics* **65**, T1–15.

Eggler, D. H. & A. A. Kadik 1979. The system $NaAlSi_3O_8$–H_2O–CO_2 to 20 kbar pressure: I. Compositional and thermodynamic relations of liquids and vapors coexisting with albite. *American Mineralogist* **64**, 1036–48.

Friend, C. R. L. 1983. The link between charnockite formation and granite production: evidence from Kabbaldurga, Karnataka, Southern India. In *Migmatites, melting and metamorphism*, M. P. Atherton & C. D. Gribble (eds), 264–76. Nantwich, UK: Shiva.

Friend, C. R. L. 1984. The origins of the Closepet Granites and the implications for the crustal evolution of Southern Karnataka. *Journal of the Geological Society of India* **25**, 73–84.

Friend, C. R. L. 1985. Evidence for fluid pathways through Archaean crust and the generation of the Closepet Granite, Karnataka, South India. *Precambrian Research* **27**, 239–50.

Fyfe, W. S. 1973. The generation of batholiths. *Tectonophysics* **17**, 273–83.

Gopalakrishna, D., E. C. Hansen & R. C. Newton 1986. The southern high-grade margin of the Dharwar Craton. *Journal of Geology* **94**, 247–60.

Grew, E. S. & W. I. Manton 1984. Age of allanite from Kabbaldurga Quarry, Karnataka. *Journal of the Geological Society of India* **25**, 193–5.

Hamilton, J. V. & C. J. Hodgson 1986. Mineralization and structure of the Kolar Gold Field, India. In *Gold '86*, A. J. MacDonald (ed.), 270–83. Willowdale, Ontario: Konsult International Inc.

Hanmer, S. 1988. Great Slave Lake Shear Zone, Canadian Shield: reconstructed vertical profile of a crustal-scale fault zone. *Tectonophysics* **149**, 245–64.

Hansen, E. C., A. S. Janardhan, R. C. Newton, W. K. B. N. Prame & G. R. Ravindra Kumar 1987. Arrested charnockite formation in southern India and Sri Lanka. *Contributions to Mineralogy and Petrology* **96**, 225–44.

Hansen, E. C., R. C. Newton & A. S. Janardhan 1984. Fluid inclusions in rocks from the amphibolite-facies gneiss to charnockite progression in southern Karnataka, India: direct evidence concerning the fluids of granulite metamorphism. *Journal of Metamorphic Geology* **2**, 249–64.

Harris, N. B. W. 1982. The petrogenesis of alkaline intrusives from Arabia and northeast Africa and their implications for within-plate magmatism. *Tectonophysics* **83**, 243–58.

Harris, N. B. W. & S. Jayaram 1982. Metamorphism of cordierite gneisses from the Bangalore region of the Indian Archean. *Lithos* **15**, 89–98.

Harrison, T. M., E. B. Watson & R. P. Rapp 1986. Does anatexis deplete the lower crust in heat-producing elements? Implications from experimental studies. *EOS, Transactions, American Geophysical Union* **67**, 386.

Helgeson, H. C. 1969. Thermodynamics of hydrothermal systems at elevated temperatures and pressures. *American Journal of Science* **267**, 729–804.

Hodges, K. V., P. LeFort & A. Pêcher 1988. Possible thermal buffering by crustal anatexis in collisional orogens. *Geology* **16**, 707–10.

Hoisch, T. D. 1987a. Heat transport by fluids during late Cretaceous regional metamorphism in the Big Maria Mountains, southeastern California. *Geological Society of America Bulletin* **99**, 549–53.

Hoisch, T. D. 1987b. Heat transport by fluids during channelized flow and thermal consequences for regional metamorphism. *Geological Society of America, Abstracts with Programs* **19**, 704.

Holland, T. J. B. & R. Powell 1985. An internally consistent thermodynamic dataset with uncertainties and correlations: 2. Data and results. *Journal of Metamorphic Geology* **3**, 343–70.

Hollister, L. S. 1988. On the origin of CO_2-rich fluid inclusions in migmatites. *Journal of Metamorphic Geology* **6**, 467–74.

Holloway, J. R. 1973. The system pargasite–H_2O–CO_2: a model for melting of a hydrous mineral with a mixed-volatile fluid – I. Experimental results to 8 kbar. *Geochimica et Cosmochimica Acta* **37**, 651–66.

Holtz, F., A. Ebadi, P. Barbey, W. Johannes & M. Pichavant 1988. Phase relations in the Qz–An–Or system at 2 and 5 kbar: the effect of a_{H_2O}. *Terra Cognita* **8**, 66.

Hutchison, W. W. 1970. Metamorphic framework and plutonic styles in the Prince Rupert region of the Central Coast Mountains, British Columbia. *Canadian Journal of Earth Sciences* **7**, 376–405.

Iiyama, J. T. 1965. Influence des anions sur les équilibres d'exchange d'ions Na–K dans les feldspaths alcalines à 600°C sous une pression de 1000 bars. *Bulletin de la Société Française de Minéralogie et de Cristallographie* **88**, 618–22.

Jackson, N. J., J. N. Walsh & E. Pegram 1984. Geology, geochemistry and petrogenesis of late Precambrian granitoids in the Central Hijaz Region of the Arabian Shield. *Contributions to Mineralogy and Petrology* **87**, 205–19.

Johannes, W. 1988. What controls partial melting in migmatites? *Journal of Metamorphic Geology* **6**, 451–66.

Kerrick, R., B. J. Fryer, R. W. King, L. M. Willmore & E. van Hees 1987. Crustal outgassing and LILE enrichment in major lithosphere structures, Archaean Abitibi greenstone belt: evidence on the source reservoir from strontium and carbon isotope tracers. *Contributions to Mineralogy and Petrology* **97**, 156–68.

Kerrick, D. M. & G. K. Jacobs 1981. A modified Redlich–Kwong equation for H_2O, CO_2, and H_2O–CO_2 mixtures at elevated pressures and temperatures. *American Journal of Science* **281**, 735–67.

Lamb, R. C., P. C. Smalley & D. Field 1986. *P–T* conditions for the Arendal granulites, southern Norway: implications for the roles of *P*, *T* and CO_2 in deep crustal LILE-depletion. *Journal of Metamorphic Geology* **4**, 143–60.

Lapin, A. V. & V. V. Ploshko 1988. Rock-association and morphological types of carbonatite and their geotectonic environments. *International Geology Review* **30**, 390–6.
Luth, W. C. 1963. *The system $KAlSiO_4$–Mg_2SiO_4–H_2O from 500 to 3000 bars and 800 to 1200 degrees, and its petrologic significance.* PhD thesis, Pennsylvania State University.

Mahabaleswar, B. & C. Naganna 1981. Geothermometry of Karnataka charnockites. *Bulletin de Minéralogie* **104**, 848–55.
Mahabaleswar, B., I. R. Vasant Kumar & C. R. L. Friend 1986. Geochemistry of the Archaean gneiss complex and associated rocks of the Kanakapura area, Karnataka, South India. *Journal of the Geological Society of India* **27**, 282–97.
McLelland, J., W. M. Hunt & E. C. Hansen 1988. The relationship between metamorphic charnockite and marble near Speculator, Central Adirondack Mountains, New York. *Journal of Geology* **96**, 455–68.
McLelland, J. M. & J. Husain 1986. Nature and timing of anatexis in the eastern and southern Adirondack Highlands. *Journal of Geology* **94**, 17–25.
Middlemost, E. A. K., D. K. Paul & I. R. Fletcher 1988. Geochemistry and mineralogy of the minette–lamproite association from the Indian Gondwanas. *Lithos* **22**, 31–42.
Monrad, J. R. 1983. Evolution of sialic terranes in the vicinity of the Holenarsipur belt, Hassan District, Karnataka, India. In *Precambrian of South India*, S. M. Naqvi & J. J. W. Rogers (eds), 343–64. Geological Society of India, Memoir 4.

Nesbitt, H. W. 1980. Genesis of the New Quebec and Adirondack granulites: evidence for their production by partial melting. *Contributions to Mineralogy and Petrology* **72**, 303–10.
Newton, R. C., T. V. Charlu & O. J. Kleppa 1980. Thermochemistry of the high structural state plagioclases. *Geochimica et Cosmochimica Acta* **44**, 933–41.

Olsen, S. N. 1985. Mass balance in migmatites. In *Migmatites*, J. R. Ashworth (ed.), 145–79. Glasgow: Blackie.
Olsen, S. N. 1987. The composition and role of the fluid in migmatites: a fluid inclusion study of the Front Range rocks. *Contributions to Mineralogy and Petrology* **96**, 104–20.
Orville, P. M. 1963. Alkali ion exchange between vapor and feldspar phases. *American Journal of Science* **261**, 201–37.

Peacock, S. M. 1987a. Serpentinization and infiltration metasomatism in the Trinity peridotite, Klamath province, northern California: implications for subduction zones. *Contributions to Mineralogy and Petrology* **95**, 55–70.
Peacock, S. M. 1987b. Thermal effects of metamorphic fluids in subduction zones. *Geology* **15**, 1057–60.
Peterson, J. W. & R. C. Newton 1987. Reversed biotite and quartz melting reactions. *EOS, Transactions, American Geophysical Union* **68**, 451.
Peterson, J. W. & R. C. Newton 1988a. Experimental constraints on the vapor-absent melting of phlogopite + quartz. *EOS, Transactions, American Geophysical Union* **69**, 498.
Peterson, J. W. & R. C. Newton 1988b. Experimental *P–T* constraints on the phlogopite–quartz–sanidine–enstatite–vapor–liquid invariant point. *Geological Society of America, Abstracts with Programs* **20**, A190.
Peterson, J. W. & R. C. Newton 1989. Reversed experiments on biotite–quartz–feldspar melting in the system KMASH: implications for crustal anatexis. *Journal of Geology* **97**, 465–85.
Peucat, J. J., P. H. Vidal, J. Bernard-Griffiths & K. C. Condie 1987. Sr, Nd, and Pb systems across the amphibolite to granulite facies transition in southern India. *Terra Cognita* **7**, 333.
Phillips, G. N. 1980. Water activity changes across an amphibolite–granulite facies transition, Broken Hill, Australia. *Contributions to Mineralogy and Petrology* **75**, 377–86.
Pichamuthu, C. S. 1965. Regional metamorphism and charnockitization in Mysore State, India. *Indian Mineralogist* **6**, 119–26.

Pichamuthu, C. S. & R. Srinivasan 1984. The Dharwar Craton. *Indian National Science Academy, Perspective Report Series* **7**, 3–34.
Powell, R. 1983. Processes in granulite-facies metamorphism. In *Migmatites, melting and metamorphism*, M. P. Atherton & C. D. Gribble (eds), 127–39. Nantwich, UK: Shiva.
Pride, C. & G. K. Muecke 1980. Rare earth element geochemistry of the Scourian Complex, N.W. Scotland – evidence for the granite–granulite link. *Contributions to Mineralogy and Petrology* **73**, 403–12.
Puziewicz, J. & W. Johannes 1988. Phase equilibria and compositions of Fe–Mg–Al minerals and melts in water-saturated peraluminous granitic systems. *Contributions to Mineralogy and Petrology* **100**, 156–68.

Quensel, P. 1952. The charnockite series of the Varberg district on the southwest coast of Sweden. *Arkiv för Mineralogi och Geologi* **1**, 229–332.

Raase, P., M. Raith, D. Ackermand & R. K. Lal 1986. Progressive metamorphism of mafic rocks from greenschist to granulite facies in the Dharwar Craton of South India. *Journal of Geology* **94**, 261–82.
Radhakrishna, B. P. 1956. *The Closepet Granite of Mysore State, India*. Bangalore: Mysore Geological Association Special Publication, 110 pp.
Radhakrishna, B. P. 1958. On the nature of certain brick-red zones in Closepet granite. *Mysore Geological Department Records* **48**, 61–73.
Raith, M., C. Hengst, B. Nagel, A. Bhattacharya & C. Srikantappa 1988. Metamorphic conditions in the Nilgiri granulite terrane and the adjacent Moyar and Bhavani shear zones: a reinterpretation. *Journal of the Geological Society of India* **31**, 112–13.
Rama Rao, B. 1940. The Archaean Complex of Mysore. *Mysore Geological Department Bulletin* **17**, 1–101.
Rogers, J. J. W. & J. K. Greenberg 1981. Trace elements in continental margin magmatism. III. Alkali granites and their relation to cratonization. *Geological Society of America Bulletin* **92**(II), 57–93.
Rogers, J. J. W., E. J. Callahan, K. O. Dennen, P. D. Fullagar, P. T. Stroh & L. F. Wood 1986. Chemical evolution of Peninsular Gneiss in the western Dharwar Craton, southern India. *Journal of Geology* **94**, 233–46.
Rubie, D. C. & W. D. Gunter 1983. The role of speciation in alkaline igneous fluids during fenite metasomatism. *Contributions to Mineralogy and Petrology* **82**, 165–75.

Schmid, R. 1978. Are the metapelites of the Ivrea–Verbano Zone restites? *Memorie di Scienze Geologiche* **33**, 67–9.
Sørensen, K. 1983. Growth and dynamics of the Nordre Strømfjord shear zone. *Journal of Geophysical Research* **88**, 3419–38.
Stähle, H. J., M. Raith, S. Hoernes & A. Delfs 1987. Element mobility during incipient granulite formation at Kabbaldurga, southern India. *Journal of Petrology* **28**, 803–34.
Stern, R. J. 1985. The Najd fault system, Saudi Arabia and Egypt: a late Precambrian rift-related transform system? *Tectonics* **4**, 497–511.
Subrahmanyam, C. 1978. On the relation of gravity anomalies to geotectonics of the Precambrian terrains of the South Indian Shield. *Journal of the Geological Society of India* **19**, 251–63.
Subramaniam, A. P. 1967. Charnockites and granulites of southern India: a review. *Dansk Geologisk Forening Meddelelser* **17**, 473–93.
Suryanarayana, K. V. 1960. The Closepet Granite and associated rocks. *Indian Mineralogist* **1**, 86–100.
Sylvester, P. 1989. Post-collisional alkaline granites. *Journal of Geology* **97**, 261–80.

Taylor, P. N., S. Moorbath, B. Chadwick, M. Ramakrishnan & M. N. Viswanatha 1984. Petrography, chemistry, and isotopic ages of Peninsular gneiss, Dharwar acid volcanic rocks, and the Chitradurga granite with special reference to the late Archean evolution of the Karnataka craton, southern India. *Precambrian Research* **23**, 349–75.

Touret, J. 1985. Fluid regime in southern Norway: the record of fluid inclusions. In *The deep Proterozoic crust in the North Atlantic provinces*, A. C. Tobi & J. L. R. Touret (eds), 517–49. Dordrecht: Reidel.

Touret, J. & Y. Bottinga 1979. Equation d'état pour le CO_2; application aux inclusions carboniques. *Bulletin de Minéralogie* **102**, 577–83.

Valley, J. W. & J. R. O'Neil 1984. Fluid heterogeneity during granulite facies metamorphism in the Adirondacks: stable isotope evidence. *Contributions to Mineralogy and Petrology* **85**, 158–73.

Wallace, M. E. & D. H. Green 1988. An experimental determination of primary carbonatite magma composition. *Nature* **335**, 343–6.

Waters, D. J. 1988. Partial melting and the formation of granulite facies assemblages in Namaqualand, South Africa. *Journal of Metamorphic Geology* **6**, 387–404.

Weaver, B. L. 1980. Rare-earth element geochemistry of Madras granulites. *Contributions to Mineralogy and Petrology* **71**, 271–9.

Weaver, B. L. & J. Tarney 1983. Elemental depletion in Archaean granulite-facies rocks. In *Migmatites, melting and metamorphism*, M. P. Atherton & C. D. Gribble (eds), 250–63. Nantwich, UK: Shiva.

Wendlandt, R. F. 1981. Influence of CO_2 on melting of model granulite facies assemblages: a model for the genesis of charnockites. *American Mineralogist* **66**, 1164–74.

Wickham, S. M. 1987. The segregation and emplacement of granitic magmas. *Journal of the Geological Society of London* **144**, 281–97.

Wickham, S. M. 1988. Underplating, anatexis, and assimilation of metacarbonate: a possible source for large CO_2 fluxes in the deep crust. *Journal of the Geological Society of India* **31**, 162.

Wood, B. J. 1976. The reaction phlogopite + quartz = enstatite + sanidine + H_2O. *Progress in experimental petrology. Natural Environment Research Council (Great Britain)* **3**, 17–20.

CHAPTER EIGHT

Fluid–rock interaction in the north-west Adirondack Mountains, New York State

Ian Cartwright & John W. Valley

8.1 Introduction

One of the principal tasks in metamorphic petrology is to discern the timing and extent of fluid–rock interaction during metamorphism. Fluids, if present in large volumes, will control a variety of metamorphic and metasomatic processes, including: the transport of chemical components and heat; the promotion or prevention of anatexis; and the genesis of economic deposits. The presence of fluids will also affect the rheology of the crust and its geophysical properties. Fluid-rich zones have been invoked as the cause of seismic reflections in the lower crust (e.g. Brown *et al.* 1983, Matthews 1986, Mooney & Brocher 1987) and regions of enhanced electrical conductivity (Shankland & Ander 1983).

The rôle of fluids during regional metamorphism is the subject of much debate. Permeability estimates for lower-crustal rocks range from 10^{-17} m^2 (Brace *et al.* 1968, Etheridge *et al.* 1983, McKenzie 1984) to 10^{-24} m^2 (Rutter & Brodie 1985); these low permeabilities inhibit fluid convection and imply that fluid flow in the deep crust is a 'one-pass' process (Walther & Orville 1982, Wood & Walther 1986). Given this restriction, and the absence of deep-seated reservoirs of comparable volumes to those that contain surface-derived fluids, it might be expected that medium to high grade metamorphism occurred with low fluid : rock ratios.

Nevertheless, several studies have concluded that many mid- or lower-crustal terrains interacted with large volumes of fluid during regional metamorphism. Taylor (1969) appealed to the ingress of fluids to elevate $\delta^{18}O$ values of the Marcy anorthosite in the Adirondack Mountains from a 'typical igneous value' of 6–7‰ to >9‰. Wickham & Taylor (1985) invoked the influx of ocean water to depths of 12–15 km to promote large-scale crustal melting in the Pyrenees. Ferry (1983, 1986) concluded that several calc-silicate rock units in New England were infiltrated by up to five rock volumes of fluid. Etheridge *et al.* (1983) proposed the occurrence of large-scale convection of fluids within the crust. Criss & Fleck (1986) invoked high fluid : rock ratios to explain ^{18}O-depleted rocks in the Eocene plutons of Idaho. Additionally, one commonly invoked model for the formation of granulite facies lithologies envisages the dehydration of these rocks to be due to the passage of CO_2-rich fluids through the crust (Janardhan *et al.* 1979, Newton *et al.* 1980).

It is evident that documenting the location, composition, and behaviour of fluids during metamorphism is crucial to our understanding of lower-crustal processes. In this chapter we summarize studies from the amphibolite–granulite facies transition zone of the Adirondack Mountains, New York State that utilize stable isotope and petrological data to document fluid–rock interaction in this area, and we review these results in the context of similar studies from elsewhere in the Adirondacks.

8.1.1 Adirondack geology

The Adirondack Mountains of New York State are a southeastern extension of the Grenville province of Canada. A simplified geological map of part of the Adirondacks is presented in Figure 8.1. The North-west Lowlands comprise mainly supracrustal rocks, including marble layers that are commonly >200 m thick, and biotite–quartz–plagioclase psammitic lithologies, together with syenitic to granitic plutonic bodies. By contrast, the Adirondack Highlands are dominated by meta-igneous lithologies, including massif-type anorthosites. U–Pb zircon studies indicate that the metaplutonic rocks in both the Lowlands and Highlands were emplaced between 1320 and 1050 Ma (McLelland *et al.* 1988). The Highlands underwent high-grade metamorphism in the Grenville event at 1050–1100 Ma (Silver 1969, McLelland & Isachsen 1986, Chiarenzelli *et al.* 1987, McLelland *et al.* 1988). Many authors (e.g. Bohlen *et al.* 1985) assumed that metamorphism in the Lowlands also occurred at this time. However, McLelland *et al.* (1988) interpret the lack of 1050–1100 Ma zircon ages from the Lowlands as indicating that this area did not undergo regional metamorphism at the same time as the Highlands. Grant *et al.* (1986) interpret a Rb–Sr whole-rock isochron of 1154 Ma from supracrustals in the Lowlands as dating the peak of Grenville metamorphism in that area. These authors proposed that regional metamorphism in the Lowlands occurred much earlier (~1150 Ma) and that the Carthage–Colton mylonite zone (Fig. 8.1) represents a major crustal suture which juxtaposes two terrains with different geological histories.

Metamorphic temperatures and pressures are extremely well characterized from the application of several geobarometers, geothermometers, and mineral equilibria to a wide variety of lithologies. Peak conditions systematically increased from 600°C at 0.65 GPa in the extreme north-west to 800°C at 0.7–0.8 GPa in the central Highlands (Bohlen *et al.* 1980, 1985). The

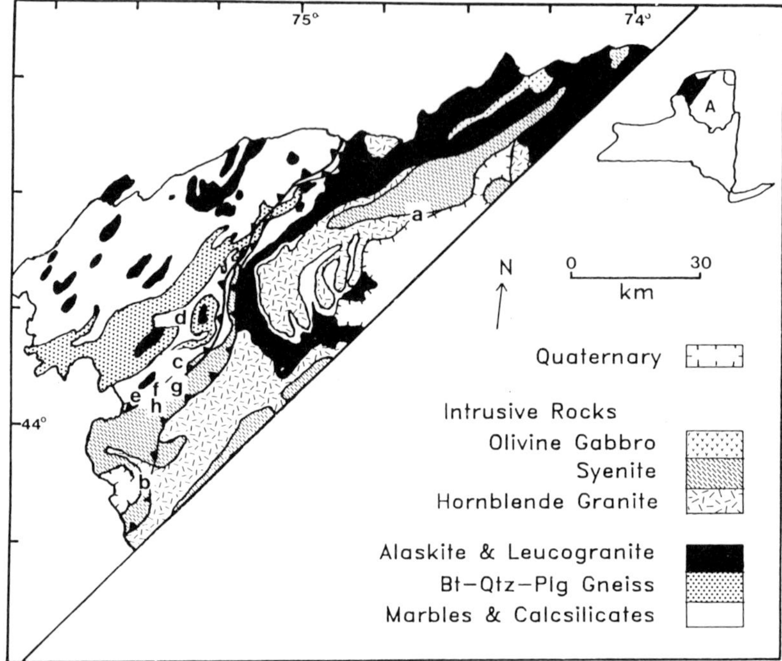

Figure 8.1 Simplified geological map of the north-west Adirondacks, showing localities discussed in the text. The inset shows the location of the study area in respect to the Adirondacks (A) and New York State. The hachured line shows the northern and southern boundaries of the Cathage–Colton mylonite zone.

general behaviour of fluids in the Adirondacks was recently summarized by Valley *et al.* (1989), who emphasized the heterogeneous nature of fluid–rock interaction in this terrain and the local control of fluid compositions.

Fluid movements and fluid–rock interaction in the north-west Adirondacks appear to have occurred in a number of spatially and temporally distinct episodes that may be broadly grouped into events that occurred prior to, or during, the regional metamorphism and those which post-date it.

8.2 Early fluid-related events

8.2.1 *Incursion of surface-derived fluids into metasyenites*

The Diana and Stark complexes are syenitic and granitic lithologies that were intruded at 1150–1160 Ma (McLelland *et al.* 1988), 100–150 Ma prior to the regional metamorphic event in the Highlands. Where unaffected by later shearing (see below), they typically have whole-rock $\delta^{18}O$ values of 9‰–10‰

(Hazelwood *et al.* 1987, Cartwright & Valley 1989a). However, at two localities (**a** and **b** in Fig. 8.1) whole-rock $\delta^{18}O$ values of 4–6‰ are recorded (Table 8.1). Igneous lithologies with whole-rock $\delta^{18}O$ values less than those of the mantle (~5.5‰; Kyser 1986, Taylor & Sheppard 1986) have either interacted with low-$\delta^{18}O$ surface-derived fluids or have assimilated rocks that have been previously altered by such fluids. Both low-$\delta^{18}O$ localities are at the margins of the complexes, and probably document the ingress of surface fluids into these rocks. Lithologies at **a** and **b** preserve peak-metamorphic mineralogies (including charnockitic orthopyroxene–clinopyroxene assemblages) with high-temperature ^{18}O fractionations (Fig. 8.2), indicating that the fluid infiltration occurred at, or prior to, the peak of regional metamorphism. Since it is unlikely that large volumes of hydrostatically pressured surface-derived fluids would penetrate below the brittle–ductile transition (where fluids are under lithostatic pressure), these low-$\delta^{18}O$ rocks most plausibly document a period of fluid–rock interaction that occurred at shallow levels prior to the Grenville event. Circulation of surface-derived fluids producing ^{18}O-depleted lithologies commonly accompanies high-level igneous intrusions (e.g. Criss & Taylor 1986). A similar origin for low $\delta^{18}O$ rocks at the margins of the Marcy anorthosite was proposed by Valley & O'Neil (1982) and Valley (1985). The volume of rock affected by these fluids is limited; localities **a** and **b** are within 20 m of the margins of these bodies, and at both the degree of depletion varies markedly on a scale of a few square metres (Table 8.1, Fig. 8.2). No low-$\delta^{18}O$

Table 8.1 $\delta^{18}O$ values of syenites from localities **a** and **b**.

	$\delta^{18}O$		(‰)			
Sample	Loc	WR	Ksp	Qtz	Other	Notes
87SK5P	a(1)	4.0	3.9	4.8	Hbl 2.3	Localities a(1) and a(2) are <10 m apart
87SK5T	a(1)	4.2	3.8	4.9	Hbl 2.6	
87SK5G	a(1)	4.1	4.1	5.0	Cpx 2.8	
87SK1P	a(2)	6.1	6.2	7.1	Hbl 4.2	
87SK1T	a(2)	5.9	6.0	7.3	Hbl 4.0	
87SK1G	a(2)	6.1	6.1	6.9	Cpx 4.2	
85CR12	b	9.5	9.1	10.8	Cpx 7.3	Samples collected within 20m^2
85CR13	b	5.8	5.8	7.6	Cpx 4.0	
85CR14	b	9.7	9.2	10.9	Hbl 7.6	

Abbreviations: Cpx = clinopyroxene; Hbl = hornblende; Ksp = alkali feldspar; Qtz = quartz; WR = whole rock.

Samples 87SK5G and 87SK1G are from pyroxene-breaing "charnockitization" zones (Newton & Hansen 1983) and were collected from within 10 cm of the hornblende bearing samples at a(1) and a(2).

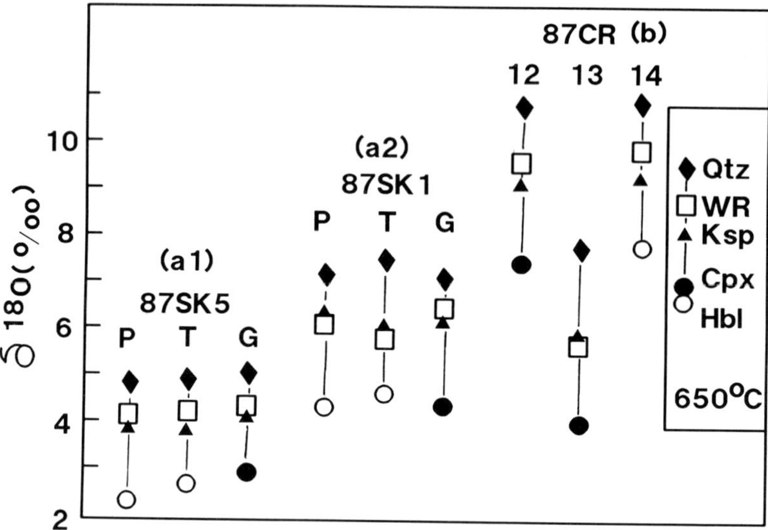

Figure 8.2 Whole-rock and mineral $\delta^{18}O$ values of samples from localities **a** and **b** in Figure 8.1 (data from Table 8.1). The close similarity between the observed isotopic fractionations and those expected at peak-metamorphic temperatures (shown in inset, data from O'Neil & Taylor 1967, Bottinga & Javoy 1973, 1975, Friedman & O'Neil 1977) implies that these minerals equilibrated close to the peak of metamorphism. Localities **a**(1) and **a**(2) are <10 m apart, and the samples from locality **b** were collected within 20 m^2, indicating the localized nature of the areas of low $\delta^{18}O$ values. The Cpx-bearing samples from **a**(1) and **a**(2) are from 'charnockitization' zones, and were collected within 10 cm of the Hbl-bearing gneisses from these localities.

lithologies have been located within the centres of the complexes, and fluids most likely were channelled along the contacts between these lithologies and the surrounding rocks.

By contrast, on the basis of locally higher geothermometry results (up to 880°C) from metasediments adjacent to the Diana complex, Powers & Bohlen (1985) argue that the Diana and Stark rocks were intruded at depth during the regional metamorphism. Small volumes of meteoric fluid may penetrate to depth along fault zones (McCaig 1988) and subsequently be channelled towards the surface along structural discontinuities, such as the contact of a meta-igneous body, resulting in ^{18}O depletions at depth. However, it is also possible that the higher temperatures of Powers & Bohlen (1985) reflect a pre-regional metamorphic contact aureole; an interpretation which is more consistent with the geochronology data discussed above.

8.2.2 Chemical transport at contacts between marbles and granites or syenites

The contacts between meta-igneous rocks and metasediments represent original chemical discontinuities which are extremely sensitive to fluid–rock inter-

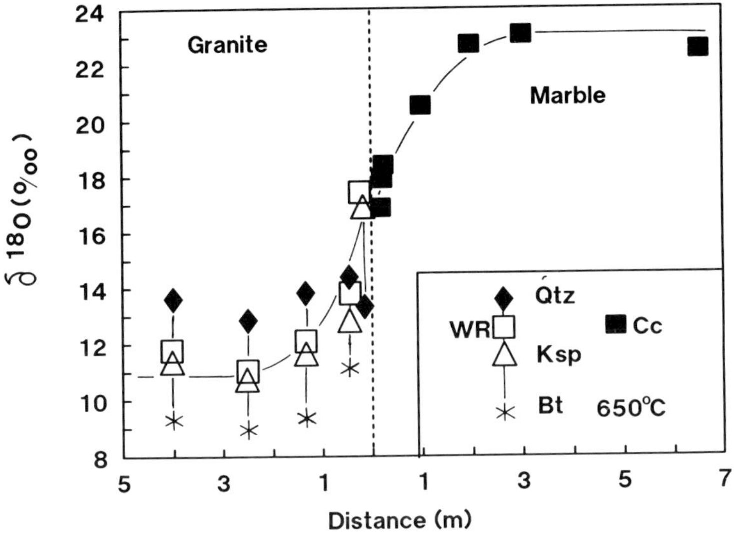

Figure 8.3 Oxygen isotope results from a traverse between granite and marble at locality **c** in Figure 8.1 (data from Cartwright & Valley 1988b, 1989b). The inset shows calculated fractionation between quartz, K-feldspar, whole-rock, and biotite at the peak-metamorphic temperature of 650°C (data sources as for Fig. 8.2). There is only one isotopic reversal, and all phases generally show high-temperature isotopic fractionations. Lines are best-fit to the calcite data within the marbles and the whole-rock values of the meta-igneous rocks ($\delta^{18}O$ of calcite is 1–2‰ heavier than that of the granite whole-rock at 650°C). Note that the isotopic profiles are developed in both marbles and meta-igneous lithologies, are convex towards the contact, and are bisected by the contact.

actions. In Figure 8.3 is shown a stable isotope profile from a traverse across the contact between a granite and a marble in the north-west Adirondacks, from locality **c** in Figure 8.1 (Cartwright & Valley 1988b, 1989b). The marbles >3 m from the contact have $\delta^{18}O$ calcite values of 22–24‰, which is typical for Adirondack marbles (Valley & O'Neil 1984). Granite at similar distances from the contacts has a whole-rock $\delta^{18}O$ value of 11–12‰, which is in the range recorded for Grenville granites by Shieh (1985) and Wu & Kerrich (1986). Minerals in the granite generally show a high-temperature fractionation of ^{18}O indicating that they equilibrated at, or close to, peak-metamorphic temperatures (~650°C). At the contact, isotopic values from both the granite and marble converge to a value of 17‰. Similar profiles were recorded at two other localities at the contacts between marbles and another granite pluton (locality **d**) and the Diana complex (locality **e**). Additionally, steep isotopic gradients were recorded at several other contacts between marbles and meta-igneous lithologies (Valley & O'Neil 1984, Valley *et al.* 1989).

The granitic and syenitic rocks were emplaced at, or prior to, the peak of the Grenville event (McLelland *et al.* 1988). The minerals in the granite at this and the other localities generally have high-temperature isotopic fractionations (except for a single reversal between qtz and ksp in the sample immediately adjacent to the contact; Fig. 8.3), which is probably most consistent with the profiles being formed at, or prior to, the peak of regional metamorphism.

It is shown in Figure 8.3 that oxygen isotope profiles are developed in both the marble and the granite which are convex to, and bisected by, the contact. The granite is plutonic and, upon intrusion, devolatilization of the marble, resulting in a depletion of ^{18}O close to the contact, would undoubtedly have occurred. However, such devolatilization can account for <2‰ of the observed variation in $\delta^{18}O$ values, and was not significant in forming the profile (Valley 1986). The observed profile has a similar shape to profiles resulting from non-steady-state diffusion of a chemical component between two semi-infinite media (Fig. 8.4a; Crank 1975, Fletcher & Hofmann 1974). Metre-scale diffusion of oxygen isotopes during metamorphism will only occur via a fluid phase, as diffusion coefficients at 600°C for oxygen in minerals are extremely sluggish, typically 10^{-18}–10^{-22} $m^2 s^{-1}$ (Cole & Ohmoto 1986), as against 10^{-7}–10^{-8} $m^2 s^{-1}$ for oxygen in a fluid phase (Bickle & McKenzie 1987). Cart-

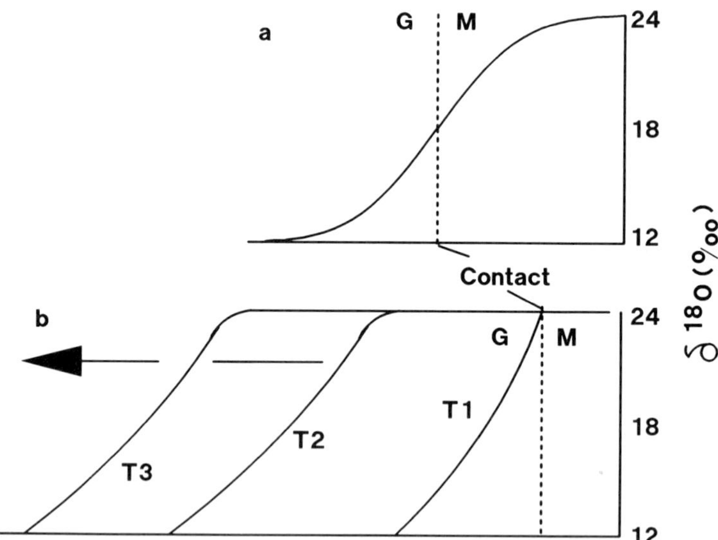

Figure 8.4 Chemical profiles formed by: (a) diffusion between two semi-infinite bodies; and (b) infiltration of a fluid from one lithology into another, where the component is transported by the infiltrating fluid and by diffusion within that fluid (calculated from Fletcher & Hofmann 1974). The $\delta^{18}O$ scale is used to model the contact between the granite and marble described in the text; in both cases the granite (G) lies to the left of the contact and the marble (M) to the right. There is a close similarity between the diffusion profile of (a) and that of Figure 8.3. By contrast the profile formed by infiltration (b) is only developed in the lithology into which the fluid is infiltrating. The distance scale is arbitrary: typically, the distance over which the infiltration profiles are formed is much greater than the distances of diffusional mass transport.

wright & Valley (1988b, 1989b) present details of the modelling of isotopic transport by diffusion of oxygen isotopes through an intergranular fluid phase at these contacts and conclude that, for reasonable porosities of 10^{-3}–10^{-4}, it would only take between 1 and 500 ka to form the isotopic profile of Figure 8.3. Isotopic profiles resulting from infiltration of a fluid are distinct from those due to diffusion (Fig. 8.4b), only being developed in the lithology into which the fluid is infiltrating, and are dissimilar to the observed isotopic profiles in the Adirondack lithologies. Hence, it is proposed that the profiles were formed largely by diffusion and that infiltration was not significant in the formation of these profiles. However, other scenarios, such as fluid flow in a broad zone parallel to the contact, are possible: distinguishing between this and diffusion requires profiles in other elements to be identified.

The profiles are most likely to have been formed either during the emplacement of the igneous bodies or during prograde regional metamorphism, as these are the times when fluids from igneous crystallization or metamorphic reactions are most likely to have been present at these localities, and are also the periods when the most elevated temperatures would have been recorded. The profiles were probably formed in a transient event (<1 Ma) which may reflect the timescales over which emplacement, crystallization, and cooling of igneous rocks or devolatilization reactions occur.

It is apparent that, while the formation of these isotopic profiles requires the presence of a fluid phase, no significant volumes of pervasive fluids moved across the contacts between these lithologies. Fluids almost certainly migrated away from these localities; however, such movement probably took place in channels (possibly represented by veins or zones of recrystallization not sampled in this study). The contacts examined are all lateral boundaries; sampling across the upper contacts of the igneous rocks may yield evidence of infiltration of fluids, exsolved during crystallization, into the overlying rocks. It is also evident that no large-scale fluid movements affected these localities after these diffusion events, as this would have modified or obliterated the isotopic profiles.

8.2.3 Formation of massive wollastonite deposits

Wollastonite-bearing marbles and calc-silicates are commonly found at the contacts with syenites in the north-west Adirondacks, and are the result of contact metamorphism (Valley 1985). At most of these localities the wollastonite-bearing zone is a few centimetres wide; however, at locality **f** (Fig. 8.1) wollastonite is found over an area of several km^2, again close to the contact between marble and syenite. This occurrence (the Valentine deposit) is mined and the ore typically comprises >80% wollastonite (usually with calcite, but no quartz). At the contact of this deposit are altered syenitic or granitic rocks (possibly Diana complex lithologies) comprising alkali feldspar- and epidote-bearing assemblages. These high-variance mineral assemblages must have been formed by interaction with large volumes of water-dominated fluids. Wollastonite from similar massive deposits at the contact between anorthosite and marble in the Adirondack Highlands has $\delta^{18}O$ values as low as $-1.3‰$ (Valley & O'Neil 1982), indicating that meteoric fluids were responsible for their formation. At locality **f** wollastonite has a $\delta^{18}O$ of 4.3–8.8‰ (Valley 1985), significantly lower than the sparse wollastonite found elsewhere at the contact between marble and syenite (typically 12–16‰), but not diagnostic of a surface-derived fluid. The epidote–alkali feldspar rock from locality **f** has a whole-rock $\delta^{18}O$ of 10.5‰, again showing no evidence that meteoric fluids caused this alteration. However, the ultimate source of these fluids may have been meteoric waters, since it is possible (especially in the case of a water-rich fluid infiltrating a syenite) that the isotopic composition of a fluid will equilibrate by interacting with lower volumes of rock than the fluid is capable of hydrating. Alternatively, the alteration may have been caused by fluids from the crystallization of the syenite or expelled connate water from the surrounding metasediments. The Valentine wollastonite deposit is a spectacular example of high fluid : rock ratios. However, on a regional scale it is an isolated example and probably represents a zone through which fluids migrated.

8.3 Late fluid incursions

8.3.1 Fluid-rock interaction in shear zones

During cooling from the peak of regional metamorphism, fluids channelled along centimetre- to metre-wide shear zones in the Diana and Stark complexes (Fig. 8.1) caused local retrogression. Within the shears, biotite replaces the pyroxene- and amphibole-bearing mineralogies that were stable at the peak of metamorphism, and a new fine-grained fabric is formed that cross-cuts the granulite facies fabrics of these rocks. As shown by the analysis of oxygen isotopes across shear zones at localities **g** and **h** in Figure 8.1, the sheared lithologies show an elevation of whole-rock $\delta^{18}O$ values by up to 4.5‰ relative to their unsheared counterparts, from 9.2‰ to 13.5‰ (Fig. 8.5). There is a correlation between the degree of isotopic elevation and the amount of shearing, with the highest $\delta^{18}O$ values recorded in sample C87HA16B, which shows the strongest shear fabric.

Fractionation of ^{18}O between minerals in the unsheared rocks indicates that they equilibrated isotopically close to the peak of regional metamorphism (650–700°C). The $\delta^{18}O$ values of coexisting minerals in the retrogressed assemblages imply that the shearing took place at 450–550°C (100–200°C below peak-metamorphic temperatures). Most minerals in the retrogressed lithologies document the resetting, showing elevation of $\delta^{18}O$ values and higher fractionations in the sheared lithologies relative to the unsheared rocks.

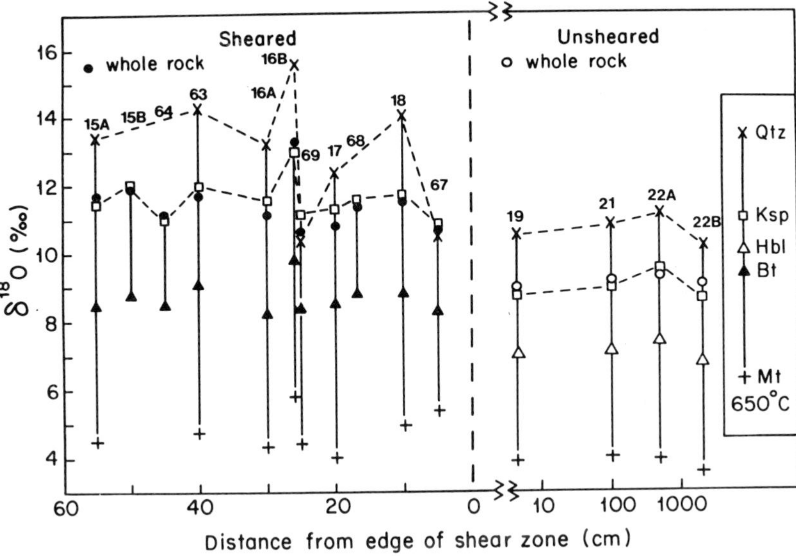

Figure 8.5 Whole-rock and mineral $\delta^{18}O$ values from traverses across three shear zones from localities **g** and **h** (Fig. 8.1), plotted against distance from the edge of the shear (data from Cartwright & Valley 1989b). The whole-rock $\delta^{18}O$ values of the sheared samples (solid circles) are uniformly higher than their unsheared counterparts (open circles). The inset shows the expected isotopic fractionations at 650°C (data sources as for Fig. 8.2). In the unsheared lithologies, the minerals have equilibrated isotopically at peak-metamorphic temperatures. The sheared rocks show isotopic resetting at lower temperatures; however, fractionations between quartz and the other minerals in the sheared rocks are lower than expected, and there are two isotopic reversals between quartz and K-feldspar. Sample numbers are prefixed by C87HA.

However, in two samples (C87HA64 & C87HA69) there is an isotopic reversal between quartz and K-feldspar. Cartwright & Valley (1988a, 1989a) document that, throughout the Diana and Stark complexes, all minerals in the unsheared rocks have concordant oxygen isotope fractionations but, in the sheared lithologies, quartz is often out of isotopic equilibrium with the other phases by up to 1.5‰. This variability in re-equilibration probably reflects differences in the degree of recrystallization during shearing. The two samples in Figure 8.5 that show isotopic reversals between quartz and K-feldspar do not have very strong shear fabrics and probably underwent only limited degrees of recrystallization. By contrast, sample C87HA16B, which is from a zone of high shear strain, underwent more extensive recrystallization and contains quartz which is isotopically concordant with the other minerals.

The fluids that ingressed via these shear zones were probably derived from the surrounding high-$\delta^{18}O$ metasediments; the adjacent marbles have whole-rock $\delta^{18}O$ values of 20–25‰ (Valley & O'Neil 1984), while the psammites have $\delta^{18}O$ values of 14–16‰ (Cartwright, unpublished data). In spite of the large elevations of $\delta^{18}O$ observed within the shears, fluid : rock ratios may have been as low as 0.3–0.4 (equivalent oxygen basis) if the fluids were derived solely from the surrounding marbles. For fluids derived wholly, or in part, from the psammites, fluid : rock ratios may still have been <1. The shear zones occupy much less than 1% of the current outcrop area of the Diana and Stark complexes, and the overall quantity of fluids that moved through these shear zones was probably vanishingly small compared with the volume of rock in this part of the terrain.

The retrogression within the shear zones also affected the Fe–Ti oxides in these rocks, and there is a good correlation between the degree of resetting of oxygen isotopes and magnetite–ilmenite temperatures and f_{O_2} values (Hazelwood 1987, Hazelwood *et al.* 1987, Cartwright & Valley 1988a, Cartwright *et al.* 1989).

8.3.2 Formation of late veins

Late veins are common throughout the north-west Adirondacks. Within the syenites at localities **g** and **h** in Figure 8.1 several vein types are present, including calcite, quartz, epidote–alkali feldspar, 'calc-silicate' (clinopyroxene–carbonate–wollastonite ± scapolite), sericite–calcite, and chlorite veins. These veins range from <1 mm to several centimetres in width; all cross-cut the peak-metamorphic fabric of the syenites, and some are observed to cut the shear zones described above. It is apparent that these veins document fluid movements within these rocks which occurred after the peak of Grenville metamorphism. In a study of the Marcy Anorthosite in the Adirondack Highlands (Fig. 8.1), Morrison & Valley (1988) showed that extensive microveins (<0.5 mm wide) which are visible under cathodoluminescence but are not detectable by transmitted light microscopy, are present. The Marcy lithologies also contained numerous millimetre-sized calcite–sericite–chlorite–quartz veins which are similar to some of those observed in the Diana and Stark lithologies.

At present, the timing, conditions, and regional significance of the fluid events responsible for the formation of these veins are not known. The vein-forming events may have occurred during the waning stages of Grenvillian metamorphism or may document later Phanerozoic fluid movements (Whitney & Davin 1987). Morrison & Valley (1988) estimate that the ~15 000 km^3 Marcy anorthosite, which is considered to be largely unaltered, contains in excess of 80 km^3 of hydrothermally produced calcite, indicating that these late fluid–rock interaction events warrant much further study.

At locality **e** (Fig. 8.1) very extensive but localized retrograde hydration is apparent in marbles close to the contact with the Diana complex, ~20 m from where the steep oxygen isotope profile was recorded. As in the case of the Valentine Wollastonite deposit (discussed earlier) this locality probably represents a region through which fluids were channelled.

8.4 Discussion

The evidence summarized in this chapter is that fluid infiltration is documented by lithologies at several localities within the north-west Adirondacks, and that the infiltration of fluids or chemical transport through fluids was important in shaping the mineralogical, chemical, and isotopic character of these rocks. However, this work illustrates the fact that fluid movements appear to have occurred in a number of spatially and temporally distinct episodes. Preservation of the isotopic signatures of the early low-$\delta^{18}O$ fluid infiltration and the lack of wholesale isotopic homogenization within the terrain implies that at no stage during the regional Grenvillian metamorphism did large volumes of pervasive fluids infiltrate these lithologies. Likewise, fluid exchange between adjacent lithologies seems to have been extremely limited, with diffusion dominating chemical transport at the contacts between marbles and meta-igneous lithologies. Where fluid movements can be documented, they were for the most part channelled (for example, the fluids associated with the late shear zones and veins). Only the fluids associated with the massive wollastonite deposits and the late retrograde alteration at locality **e** appear to have infiltrated pervasively (albeit over limited lateral distances), and even these may have been channelled along structural discontinuities between meta-igneous and metasedimentary lithologies. Both of these localities lie close to the northern margin of the Colton–Carthage mylonite, and it is possible that this structure acted as a channel for fluid movement.

8.4.1 Fluid sources

Part of the description of fluid–rock interaction must be a consideration of fluid sources. The low-$\delta^{18}O$ fluids were of surficial origin, and the fluids through which diffusion occurred at the contacts between marbles and the metaplutonic lithologies were almost certainly derived from either crystallization of the igneous lithology or devolatilization of the marble. In both these examples, the fluid–rock interaction occurred due to fluids being produced or circulated as a response to igneous intrusion.

The late fluids that ingressed into the metasyenites along the shear zones and veins probably equilibrated with the surrounding supracrustal rocks; however, the reasons for fluid being produced at these times is less obvious. Fluids may be produced during cooling by the crystallization of melts formed at the peak of metamorphism (e.g. Cartwright 1988); this may explain the fluids that were associated with the shear zones at $\sim 500°C$, but is unlikely to account for lower-temperature fluids. Most retrograde reactions are rehydration and recarbonation reactions which tend to consume fluid. Also, as the retrograde assemblages usually have larger volumes than their unaltered precursors, porosity is decreased during cooling, inhibiting fluid migration. However, it is generally not possible to remove a mixed volatile fluid phase entirely by retrograde reaction, and the late fluids may have migrated, perhaps due to fracturing in response to tectonic activity.

8.4.2 Metamorphic fluids in the Adirondacks

Overall, our work documents the complexity and heterogeneity of fluid–rock interaction in the Adirondacks and is consistent with the findings of a number of other studies of the behaviour of fluids in this terrain, which were recently summarized by Valley *et al.* (1989). Quantitative estimates of peak-metamorphic values of a_{H_2O} in the semi-pelitic biotite–quartz–plagioclase gneisses in the north-west Lowlands range from 0.1 to 0.4 (Powers & Bohlen 1985; Edwards & Essene 1988), and show no simple relationship to metamorphic temperatures or the distance from the amphibolite to granulite facies transition (Valley *et al.* 1989, Fig. 1). Likewise, a_{H_2O} and a_{CO_2} values calculated from marble equilibria are heterogeneous throughout the Lowlands and the Highlands (Valley & Essene 1980, Valley *et al.* 1983, Valley 1985) and demonstrate local control of fluid compositions. Valley (1985) and Lamb & Valley (1985) demonstrated that certain calc-silicate and meta-igneous lithologies in the Highlands document both low a_{H_2O} and a_{CO_2} values at the peak of Grenville metamorphism, implying that they were fluid-absent at that time. The presence of Cl-rich amphiboles in the Marcy anorthosite and orthogneisses from the Highlands also indicates that these lithologies were metamorphosed under fluid-absent conditions (Bohlen & Essene 1978, Morrison 1988).

There is evidence for metamorphism and fluid infiltration prior to Grenville event elsewhere in the Adirondacks. Wollastonite skarns adjacent to the Marcy anorthosite show depletion in $\delta^{18}O$ (down to $-1.3‰$) and, like the low-$\delta^{18}O$ values at the margins of the Diana complex syenites, are interpreted as representing the influx of surface-derived fluids during anorthosite emplacement at shallow levels (Valley & O'Neil 1982, 1984). McLelland & Husain (1986) proposed that partial melting in pelitic rocks of the central Adirondacks occurred prior to the peak of regional metamorphism, as a response to igneous intrusion. Sanidinite facies minerals in the Cascade Slide xenolith in the Highlands are also considered to be the product of contact metamorphism prior to the Grenville event (Valley 1985).

In general, these studies indicate that volatile activities during metamorphism were highly variable between lithologies and were locally controlled. The stable isotope and chemical character of most lithologies was largely formed prior to the peak of metamorphism and reflects the original sedimentary or igneous processes or early hydrothermal events in the upper crust. Certainly, there is no evidence that significant volumes of pervasive fluid were present in the terrain, either during the peak of metamorphism or at any other stage (as was envisaged by Taylor 1969). Many lithologies may have been fluid-absent during the Grenville event.

Aside from the surface-derived fluids, there is no need to invoke the influx of externally derived fluids at any stage of metamorphism. However, from similarities with rocks in southern India (e.g. Janardhan *et al.* 1979, Newton *et al.* 1980), Newton & Hansen (1983) attributed the formation of centimetre-wide zones of orthopyroxene-bearing syenite (charnockite) within dominantly hornblende-bearing lithologies at locality **a** (Fig. 8.1) to dehydration resulting from the passage of CO_2-rich fluids through the crust (Newton & Hansen 1983). The amphibole- and pyroxene-bearing lithologies at this locality preserve low $\delta^{18}O$ values (Table 8.1). Fluids associated with charnockitization are usually considered to originate from the mantle or from devolatilization of marbles in the lower crust. CO_2 in equilibrium with the mantle ($\delta^{18}O \approx 5.5‰$) would have a $\delta^{18}O$ of 10–11‰; CO_2 from the devolatilization of marbles would have a higher $\delta^{18}O$ value (up to 30‰). The preservation of $\delta^{18}O$ values depleted before regional metamorphism at these localities and the identical $\delta^{18}O$ values of both the pyroxene- and hornblende-bearing syenites (Table 8.1) is inconsistent with the formation of these textures by this process. It is possible the dehydration in the charnockite zones was caused by anatexis and melt escape. This interpretation is consistent with the presence of rims of amphibole around some orthopyroxenes in samples from this locality, which could have been formed by retrogression due to water exsolved from crystallizing melts during cooling (e.g. Cartwright 1988).

CO_2-rich fluid inclusions are found in many Adirondack granulites (Lamb *et al.* 1987, Morrison & Valley 1988). These fluid inclusions usually satisfy the criteria for primary inclusions trapping peak-metamorphic fluid. However, they often contain fluids that could not be in equilibrium with the peak-metamorphic mineralogies observed in the rocks, and must have been formed after the peak of metamorphism (Lamb *et al.* 1987). Morrison & Valley (1988) also demonstrated that apparently primary fluid inclusions were formed by the same fluids as minerals in late or post-metamorphic veins. It is evident that the fluid inclusions provide valuable information on the nature of metamorphic fluids, but that the infiltration events recorded may not be peak-metamorphic.

The metamorphism of this terrain appears to have occurred largely in response to magmatic processes. Many of the pre-regional metamorphic fluid-infiltration events were driven by the intrusion of igneous plutons, and the heat source for the regional Grenville metamorphism was probably from the emplacement of granitic lithologies (McLelland *et al.* 1988). There is abundant evidence for anatexis in both the psammitic rocks of the north-west Lowlands and in certain of the meta-igneous rocks in the Highlands; this partial melting probably accounts for the low and variable a_{H_2O} values throughout the terrain. The polymetamorphism apparent in this terrain resulted in many rocks being devolatilized prior to the regional metamorphism. In many cases the temperatures experienced during early contact metamorphism were comparable with, or higher than, those attained during the regional event. This resulted in lower volumes of fluids being generated during regional metamorphism, and led to certain lithologies being fluid-absent at that stage (Valley *et al.* 1989).

Acknowledgements

Funding for this research was provided by NATO Postdoctoral Fellowship GT8/F/86/GS/1 and a New York State Geological Survey Honorarium (to I.C.) and NSF grant EAR85–08102 and GRI grant 5086–260–1425 (to J.W.V.). We would like to thank K. Baker for assistance with the stable isotope analyses and S. Smith, A. Thawley, and W. Edwards for help with the figures. Discussions with W. Lamb and J. Morrison and helpful comments by A. Baker helped to improve the text.

References

Bickle, M. J. & D. McKenzie 1987. The transport of heat and matter by fluids during metamorphism. *Contributions to Mineralogy and Petrology* **95**, 384–92.

Bohlen, S. R. & E. J. Essene 1978. The significance of metamorphic fluorite in the Adirondacks. *Geochimica et Cosmochimica Acta* **42**, 1669–78.

Bohlen, S. R., E. J. Essene & K. S. Hoffman 1980. Update on feldspar and oxide thermometry in the Adirondack mountains, New York. *Geological Society of America Bulletin* **91**, 110–13.

Bohlen, S. R., J. W. Valley & E. J. Essene 1985. Metamorphism in the Adirondacks. I. Petrology, pressure and temperature. *Journal of Petrology* **26**, 971–92.

Bottinga, Y. & M. Javoy 1973. Comments on oxygen isotope geothermometry. *Earth and Planetary Science Letters* **20**, 250–65.

Bottinga, Y. & M. Javoy 1975. Oxygen isotope partitioning among the minerals in igneous and metamorphic rocks. *Reviews of Geophysics and Space Physics* **13**, 401–18.

Brace, W. F., J. B. Walsh & W. T. Frangos 1968. Permeability of granite under high pressure. *Journal of Geophysical Research* **73**, 2225–36.

Brown, L., C. Ando, S. Klemperer, J. Oliver, S. Kaufman, B. Czuchra, T. Walsh & Y. W. Isachsen 1983. Adirondack–Appalachian crustal structure: The COCORP Northeast Traverse. *Geological Society of America Bulletin* **94**, 1173–84.

Cartwright, I. 1988. Crystallization of melts, pegmatite intrusion and the Inverian retrogression of the Scourian complex, north-west Scotland. *Journal of Metamorphic Geology* **6**, 77–93.

Cartwright, I., A. M. Hazelwood & J. W. Valley 1989. Metamorphism of Diana and Stark complexes, Adirondack Mountains, New York. II: Oxygen fugacity results. Submitted to *Contributions to Mineralogy and Petrology*.

Cartwright, I. & J. W. Valley 1988a. Retrograde fluids and shearing in the Diana and Stark complexes, Adirondacks, New York: stable isotope and petrologic considerations. *EOS, Transactions, American Geophysical Union* **69**, 508.

Cartwright, I. & J. W. Valley 1988b. Diffusion and local fluid infiltration in the NW Adirondacks, New York. *Geological Society of America, Abstracts with Programs* **20**, 45.

Cartwright, I. & J. W. Valley 1989a. Metamorphism of Diana and Stark complexes, Adirondack Mountains, New York. I: Stable isotope results. Submitted to *Contributions to Mineralogy and Petrology*.

Cartwright, I. & J. W. Valley 1989b. Steep isotopic gradients, local fluid infiltration, and diffusion in the NW Adirondack Mountains, New York. Submitted to *Contributions to Mineralogy and Petrology*.

Chiarenzelli, J. R., M. E. Bickford, J. M. McLelland, Y. W. Isachsen & P. R. Whitney 1987. Early igneous history of the Adirondack Mountains as revealed by U/Pb ages. *Geological Society of America, Abstracts with Programs* **19**, 619.

Cole, D. R. & H. Ohmoto 1986. Kinetics of isotopic exchange at elevated temperatures and pressures. In *Stable isotopes in high temperature geological processes*, J. W. Valley, H. P. Taylor & J. R. O'Neil (eds), 41–90. Mineralogical Society of America, Reviews in Mineralogy 16.

Crank, J. R. 1975. *The mathematics of diffusion*, 2nd edn. Oxford: Oxford University Press.

Criss, R. E. & R. J. Fleck 1986. Petrogenesis, geochronology, and hydrothermal systems of the northern Idaho batholith region based on $^{18}O/^{16}O$, $^{87}Sr/^{86}Sr$, K–Ar, and $^{40}Ar/^{39}Ar$ studies. In Geology of the Blue Mountains region of Oregon, Idaho and Washington: the Idaho Batholith and its border zone, T. L. Vallier & H. C. Brooks (eds), U.S. Geological Survey Professional Paper **1436**, 95–137.

Criss, R. E. & H. P. Taylor 1986. Meteoric-hydrothermal systems. In *Stable isotopes in high temperature geological processes*, J. W. Valley, H. P. Taylor & J. R. O'Neil (eds), 373–424. Mineralogical Society of America, Reviews in Mineralogy 16.

Edwards, R. L. & E. J. Essene 1988. Pressure, temperature and C–O–H fluid fugacities across the amphibolite–granulite transition, NW Adirondack Mountains, NY. *Journal of Petrology* **29**, 39–72.

Etheridge, M. A., V. J. Wall & R. H. Vernon 1983. The role of the fluid phase during regional metamorphism and deformation. *Journal of Metamorphic Geology* **1**, 205–26.

Ferry, J. M. 1983. Applications of the reaction progress variable in metamorphic petrology. *Journal of Petrology* **24**, 343–76.

Ferry, J. M. 1986. Reaction progress: a monitor of fluid–rock interaction during metamorphic and hydrothermal events. In *Fluid–rock interactions during metamorphism*, J. V. Walther & B. J. Wood (eds), 60–88. New York: Springer.

Fletcher, R. C. & A. W. Hofmann 1974. Simple models of diffusion and combined diffusion–infiltration metasomatism. In *Geochemical transport and kinetics*, A. W. Hofmann, B. J. Giletti, H. S. Yoder Jr & R. A. Yund (eds), 243–59. Washington: Carnegie Institute.

Friedman, I. & J. R. O'Neil 1977. Compilation of stable isotope fractionation factors of geochemical interest. *U.S. Geological Survey, Professional Paper*, 440–KK.

Grant, N. K., R. J. Lepak, T. M. Maher, M. R. Hudson & J. D. Carl 1986. Geochronological framework for the Grenville rocks of the Adirondack mountains. *Geological Society of America, Abstracts with Programs* **18**, 620.

Hazelwood, A. M. 1987. *The role of metamorphic fluids in the amphibolite to granulite facies transition, Adirondack Mountains, New York*. Unpublished MS thesis, University of Wisconsin–Madison, 70 pp.

Hazelwood, A. M., I. Cartwright & J. W. Valley 1987. Oxygen isotope and f_{O_2} results from the Diana and Stark Complexes, Adirondacks, New York. *Geological Society of America, Abstracts with Programs* **19**, 697.

Janardhan, A. S., R. C. Newton & J. V. Smith 1979. Ancient crustal metamorphism at low P_{H_2O}: charnockite formation at Kabbaldurga, S. India. *Nature* **278**, 511–14.

Kyser, T. K. 1986. Stable isotope variations in the mantle. In *Stable isotopes in high temperature geological processes*, J. W. Valley, H. P. Taylor & J. R. O'Neil (eds), 141–64. Mineralogical Society of America, Reviews in Mineralogy 16.

Lamb, W. M. & J. W. Valley 1985. C–O–H fluid calculations and granulite genesis. In *The deep Proterozoic crust in the North Atlantic provinces*, A. C. Tobi & J. L. R. Touret (eds), 119–31. Dordrecht: Reidel.

Lamb, W. M., J. W. Valley & P. E. Brown 1987. Post-metamorphic CO_2-rich fluid inclusions in granulites. *Contributions to Mineralogy and Petrology* **96**, 485–95.

Matthews, D. H. 1986. Seismic reflections from the lower crust around Britain. In *The nature of the lower continental crust*, J. B. Dawson, D. A. Carswell, J. Hall & K. H. Wedepohl (eds), 11–22. Geological Society of London, Special Publication 24.

McCaig, A. M. 1988. Deep fluid circulation in fault zones. *Geology* **16**, 867–70.

McKenzie, D. 1984. The generation and compaction of partially molten rock. *Journal of Petrology* **25**, 713–65.

McLelland, J., J. Chiarenzelli, P. Whitney & Y. Isachsen 1988. U–Pb zircon geochronology of the Adirondack Mountains and implications for their geologic evolution. *Geology* **16**, 920–4.

McLelland, J. M. & J. Husain 1986. Nature and timing of anatexis in the eastern and southern Adirondack Highlands. *Journal of Geology* **94**, 17–25.

McLelland, J. M. & Y. W. Isachsen 1986. Synthesis of geology of the Adirondack mountains, New York, and their tectonic setting within the southwestern Grenville province. In *The Grenville Province*, J. M. Moore & A. J. Baer (eds), 75–94. Geological Association of Canada, Special Paper 31.

Mooney, W. D. & T. M. Brocher 1987. Coincident seismic reflection/refraction studies of the continental lithosphere: a global review. *Reviews of Geophysics* **25**, 723–42.

Morrison, J. 1988. *Petrology and stable isotope geochemistry of the Marcy Anorthosite Massif, Adirondack Mountains, N.Y.* PhD thesis. University of Wisconsin-Madison.

Morrison, J. & J. W. Valley 1988. Post-granulite facies fluid infiltration in the Adirondack Mountains. *Geology* **16**, 513–16.

Newton, R. C. & E. C. Hansen 1983. The origin of Proterozoic and late Archaean charnockites – evidence from field relations and experimental petrology. *Geological Society of America, Memoir* 161, 167–78.

Newton, R. C., J. V. Smith & B. F. Windley 1980. Carbonic metamorphism, granulites and crustal growth. *Nature* **288**, 45–50.

O'Neil, J. R. & H. P. Taylor 1967. The oxygen isotope and cation exchange chemistry of feldspars. *American Mineralogist* **52**, 1414–37.

Powers, R. E. & S. R. Bohlen 1985. The role of synmetamorphic igneous rocks in the metamorphism and partial melting of metasediment, N.W. Adirondacks. *Contributions to Mineralogy and Petrology* **90**, 401–9.

Rutter, E. H. & K. H. Brodie 1985. The permeation of water into hydrating shear zones. In *Metamorphic reactions: kinetics, textures and deformation*, A. B. Thompson & D. C. Rubie (eds), 242–50. Advances in Physical Geochemistry 4. New York: Springer.

Shankland, T. J. & M. E. Ander 1983. Electrical conductivity, temperatures, and fluids in the lower crust. *Journal of Geophysical Research* **88**, 9475–84.

Shieh, Y.-N. 1985. High-^{18}O granitic plutons from the Frontenac Axis, Grenville Province of Ontario, Canada. *Geochimica et Cosmochimica Acta* **49**, 117–23.

Silver, L. T. 1969. A geochronological investigation of the anorthosite complex, Adirondack mountains, New York. In *Origin of anorthosite and related rocks*, Y. W. Isachsen (ed.), 233–51. New York State Museum and Science Survey, Memoir 18.

Taylor, H. P. 1969. Oxygen isotope studies of anorthosites, with particular reference to the origin of bodies in the Adirondack Mountains, New York. In *Origin of anorthosites and related rocks*, Y. W. Isachsen (ed.), 111–34. New York State Museum and Science Service, Memoir 18.

Taylor, H. P. & S. M. F. Sheppard 1986. Igneous rocks: I. Processes of isotopic fractionation and isotope systematics. In *Stable isotopes in high temperature geological processes*, J. W. Valley, H. P. Taylor & J. R. O'Neil (eds), 227–72. Mineralogical Society of America, Reviews in Mineralogy 16.

Valley, J. W. 1985. Polymetamorphism in the Adirondacks: wollastonite at contacts of shallowly intruded anorthosite. In *The deep Proterozoic crust in the North Atlantic provinces*, A. C. Tobi & J. L. R. Touret (eds), 217–35. Dordrecht: Reidel.

Valley, J. W. 1986. Stable isotope geochemistry of metamorphic rocks. In *Stable isotopes in high temperature geological processes*, J. W. Valley, H. P. Taylor & J. R. O'Neil (eds), 445–90. Mineralogical Society of America, Reviews in Mineralogy 16.

Valley, J. W., S. R. Bohlen, E. J. Essene & W. Lamb 1989. Metamorphism in the Adirondacks. II. The role of fluids: Submitted to *Journal of Petrology*.

Valley, J. W. & E. J. Essene 1980. Calc-silicate reactions in Adirondack marbles: the role of fluids and solid solutions. *Geological Society of America Bulletin* **91**, I 114–17, II 720–815.

Valley, J. W., J. McLelland, E. J. Essene & W. Lamb 1983. Metamorphic fluids in the deep crust: evidence from the Adirondacks. *Nature* **301**, 226–8.

Valley, J. W. & J. R. O'Neil 1982. Oxygen isotope evidence for shallow emplacement of Adirondack anorthosite. *Nature* **300**, 497–500.

Valley, J. W. & J. R. O'Neil 1984. Fluid heterogeneity during granulite facies metamorphism in the Adirondacks: stable isotope evidence. *Contributions to Mineralogy and Petrology* **85**, 158–73.

Walther, J. V. & P. M. Orville 1982. Volatile production and transport in regional metamorphism. *Contributions to Mineralogy and Petrology* **79**, 252–7.

Whitney, P. R. & M. T. Davin 1987. Taconic deformation and metasomatism in Proterozoic rocks of the easternmost Adirondacks. *Geology* **15**, 500–3.

Wickham, S. M. & H. P. Taylor 1985. Stable isotope evidence for large-scale seawater infiltration in a regional metamorphic terrane, the Trois Seigneurs Massif, Pyrenees, France. *Contributions to Mineralogy and Petrology* **91**, 122–37.

Wood, B. J. & J. V. Walther 1986. Fluid flow during metamorphism and its implications for fluid–rock ratios. In *Fluid–rock interactions during metamorphism*, J. V. Walther & B. J. Wood (eds), 89–108. Advances in Physical Geochemistry 5. New York: Springer.

Wu, T.-W. & R. Kerrich 1986. Combined oxygen isotope-compositional studies of some granitoids from the Grenville province of Ontario, Canada: implications for source regions. *Canadian Journal of Earth Sciences* **23**, 1412–32.

CHAPTER NINE

Progressive reactions and melting in the Acadian metamorphic high of central Massachusetts and southwestern New Hampshire, USA

John C. Schumacher, Kurt T. Hollocher, Peter Robinson & Robert J. Tracy

9.1 Introduction

The purpose of this chapter is to describe briefly and summarize changes in the mineral assemblages and mineral compositions that are observed in various rock types along the metamorphic field gradient in the metamorphic high of central Massachusetts and southwestern New Hampshire, USA. The progressive metamorphism of rocks of basaltic composition is discussed here more extensively than for rocks of common pelitic compositions, because detailed descriptions of the progressive metamorphism of basaltic rocks at these metamorphic conditions are less common. This chapter is aimed, in part, at non-petrologists and postgraduate students. As a consequence, some basic subjects are discussed, including some general comments concerning *P–T–t* (pressure–temperature–time) paths and metamorphic field gradients. However, some aspects should interest petrologists looking for meatier topics.

9.2 *P–T–t* paths and metamorphic field gradients

Metamorphic field gradients (MFGs) are recognizable changes in metamorphic conditions (for example, sequences of isotherms, isobars, or mineral isograds) across a specified part of a terrane; whereas *P–T–t* paths express the pressure–temperature histories of rocks (outcrops from a restricted area) within the terrane. Each distinguishable part of an MFG will have its own *P–T–t* path, and each location along the MFG represents a unique point of pressure, temperature, and time. This means that the feature used to define the MFG did not necessarily form exactly contemporaneously across the terrane. The similarity among the paths will depend on tectonic complexity and diversity in the area. In multiply metamorphosed terranes, MFGs from unrelated metamorphic events may appear to be 'blended' together. Consequently, without further supporting geological data, there is a danger of using unrelated data to determine the MFG and of attributing unrelated *P–T–t* paths to the same metamorphic event.

9.2.1 Metamorphic field gradients

MFGs, as used here, can be equated with the terms *P–T arrays* (England & Richardson 1977), *piezothermic arrays* (Richardson & England 1979), and *sets of metamorphic zones* (Harte & Dempster 1987). England & Richardson (1977) have pointed out that prograde mineral assemblages most probably represent the temperature of maximum entropy (Ts_{max}), which may be lower than the true maximum temperature (T_{max}) experienced by the rock. This can occur if rocks are simultaneously cooled and uplifted such that they cross from the high- to low-pressure side of reactions with positive *P–T* slopes: this situation would show the largest discrepancies between Ts_{max} and T_{max}. Nevertheless, a loose correlation of prograde mineral assemblages with a regional variation in thermal maxima can be justified on the basis of reaction kinetics. Mineral reaction rates are very temperature dependent, and these reaction rates increase rapidly with increasing temperature. As a result, low-grade assemblages tend to react out quickly as the temperature increases. Cooling after the temperature maximum tends to quench further mineral reaction, thereby freezing the maximum temperature assemblage into the rock. Preservation of the maximum temperature assemblage is especially efficient if the assemblages developed via prograde devolatilization reactions, which become effectively irreversible if the released H_2O (or CO_2) is lost from the system and if the rock remains dry during cooling.

Using the depth–temperature correlation (geothermal gradients) and crustal thickening models, England & Richardson (1977) and England & Thompson (1984) have accounted for the origin of many observed MFGs. On the other hand, Harte & Dempster (1987) have emphasized the possible additional important contribution that lateral cooling at depth along major tectonic boundaries may provide to the development of steep MFGs (closely spaced metamorphic zones). The work cited above has provided tremendous insight into the significance of MFGs and the factors controlling them but, clearly, much of the detail of factors and processes leading to the development of MFGs remains to be uncovered.

9.2.2 *P–T–t paths*

If the real trace of a rock's *P–T–t* path on a *P–T* diagram could be obtained, it would reflect features of the local magmatic and tectonic histories. The approximate trace (what a petrologist derives) of a rock's *P–T–t* path on a *P–T* diagram is limited by the ability to fix the pressure, temperature, and time points along the path accurately. Constraints on a rock's *P–T–t* path can be either relative or absolute and can be attained, for example, through: the determination of reaction sequences using inclusions in porphyroblasts (e.g. Thompson *et al.* 1977); application of geothermometry, geobarometry, and geochronology to metamorphic rocks (e.g. Sanders *et al.* 1987); correlation of regional and local kinematic sequences with metamorphic assemblages (e.g. Robinson 1979); or analysis and interpretation of chemical zoning in minerals (e.g. Spear & Selverstone 1983).

The potential complexity of *P–T–t* paths makes generalization about their forms difficult. For example, Chamberlain & Karabinos (1987) have shown that, in structural terranes dominated by folding and thrusting, details of *P–T–t* paths can vary dramatically within a single metamorphic belt. This is due to heat redistribution caused by folding of the isotherms and by differential cooling rates that are dependent on the rock's position in the structure following deformation. However, from models of geological processes and from examples of metamorphic terranes (England & Thompson 1984, Hollister 1979, Thompson & England 1984 – also references therein), *P–T–t* paths can be simplified and broadly classified as curved-clockwise or curved-counterclockwise trajectories in *P–T* space.

9.2.3 *An example of P–T–t path and MFG*

In Figure 9.1 the examples of a *metamorphic field gradient* and contrasting *P–T–t* paths are modelled after the current understanding of metamorphism in central Massachusetts. The intent is not to imply that all metamorphic terranes and all MFGs form in this manner, but rather to provide an example that shows diverse features. Numerous examples of MFGs developed in terranes which do not show contrasting *P–T–t* paths can be found in the literature (e.g. Thompson & England 1984, Harte & Dempster 1987).

For the example in Figure 9.1A, rocks following a clockwise *P–T–t* path experience the pressure maximum at point C prior to the thermal maximum at point D. For rocks following a counterclockwise *P–T–t* path, the reverse is true; the temperature maximum at point A precedes the pressure maximum at point B. Another way to represent the *P–T–t* history of a rock is to use the analogy of a two-pen chart recorder printout on which variations in *P* and *T* are recorded over time (Figs 9.1B & C). The time interval covering the entire *P–T–t* path (beginning to end of the chart) may be completely or partly known, or even unknown. Likewise, the time span between specific points on

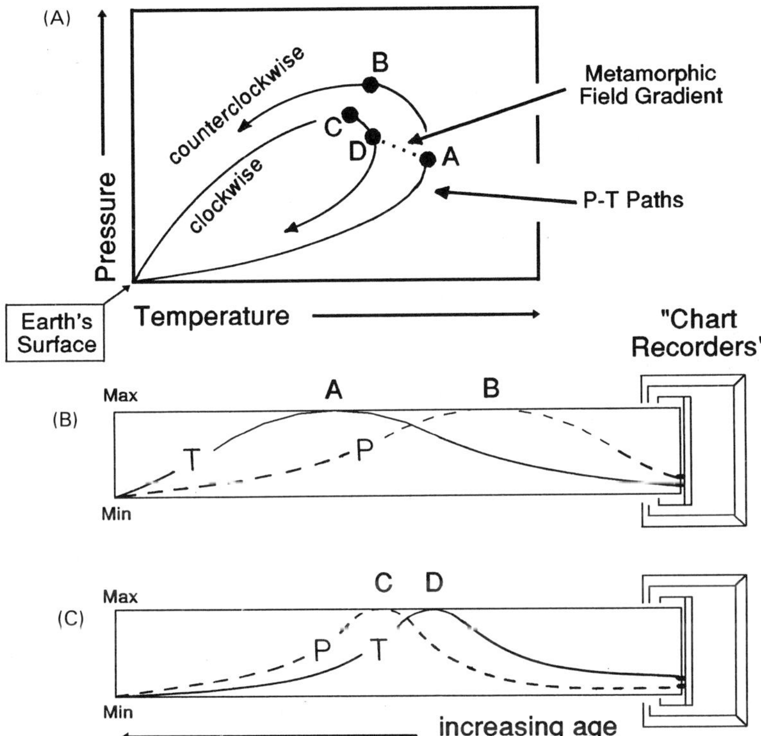

Figure 9.1 (A) Two contrasting types of *P–T–t* path, one clockwise and the other counterclockwise. Points B and C are the pressure maxima and points A and D are the temperature maxima of the respective paths. In this example, both paths were generated by the same orogenic event in geographically separated areas. The metamorphic field gradient (dotted line) is assumed to be independently determined. (B) A counterclockwise *P–T–t* path as it would look if it were recorded on a two-pen chart recorder; points A and B represent *P* and *T* as in part (A). (C) A clockwise *P–T–t* path as it would look if it were recorded on a two-pen chart recorder; points C and D represent *P* and *T* as in part (A). 'Min' and 'Max' indicate relative values of *P* and *T*. Note that in parts (B) and (C), if only the relative timing of the pressure and temperature maxima are known, then only their relative positions on the charts are known.

a *P–T–t* path may be constrained by age determinations, or only the relative ages of points on the *P–T–t* path may be determined. The timing of thermal and pressure maxima for the two *P–T–t* paths may be quite different (points A and B, Fig. 9.1B) or nearly coeval (points C and D, Fig. 9.1C).

An example of a possible *metamorphic field gradient* is also shown in Figure 9.1A. Assuming that the two *P–T–t* paths (a) are representative of the 'low- and high-temperature' ends of the MFG, (b) formed during the same regional metamorphic event, and (c) preserve prograde assemblages representing T_{max}, the MFG would be the approximate trace of the *T* maxima along line D–A of the two *P–T–t* paths. Note that the rocks from intervening areas will have followed their own unique paths that may or not closely resemble either of the paths on Figure 9.1A.

9.2.4 *Rôles of P–T–t paths and MFGs*

There are two broad uses of P–T–t information. (a) In essence, modelling of P–T evolution (e.g. Sleep 1979, England & Richardson 1977, England & Thompson 1984, Harte & Dempster 1987) or examining P–T–t paths from areas with independently known structural and geological histories serve as 'calibrations' of P–T–t information. (b) This 'calibration' transforms P–T–t information into a useful tool for making inferences about the tectonic processes operating in terranes where little is known about the geology.

During mapping, the essential first step in the investigation of any little-studied or unstudied metamorphic terrane, the presence or absence of an MFG is one of the initial important observations obtained. It is at the 'ends' of MFGs that the greatest differences in P–T–t paths are likely to be found; consequently, establishing the MFG helps to target areas for more detailed study.

Additionally, a well established MFG provides a reference frame for petrological research. A set of mineral isograds provides a relative measure of changing metamorphic conditions. If the immediate goal is a comparison of prograde metamorphic effects in different common bulk compositions (for example, metamorphosed basalts and shales), then it makes little difference whether the MFG records T_{max} or $T_{S_{max}}$; the sequence of reactions will be the same as would be found along a temperature interval at constant pressure. It is this type of use, comparing reaction sequences in different rock types, that is emphasized in the present chapter.

9.3 Background and setting

Numerous workers (Robinson 1963, 1967; Tracy *et al.* 1976; Robinson *et al.* 1982b; Robinson *et al.* 1986 and references therein) have provided the structural, stratigraphic, and analytical database that is necessary in order to understand the details of the metamorphism in central Massachusetts (Fig. 9.2). The T and P estimates, discussed extensively below, are based mainly on A. B. Thompson's (1976a,b) calibration of Fe–Mg exchange in biotite–garnet and the garnet–cordierite–sillimanite–quartz assemblage. Deformation and metamorphism of the region occurred during the Devonian Acadian Orogeny and affected stratified and intrusive rocks of Late Precambrian, Ordovician, Silurian, and Lower Devonian ages. The generalized sequence of deformation included: (1) west-directed fold nappes; (2) west-directed thrust nappes that truncate the axial surfaces of the fold nappes and carried high-temperature rocks over cooler rocks; (3) local and regional east-directed backfolding of the

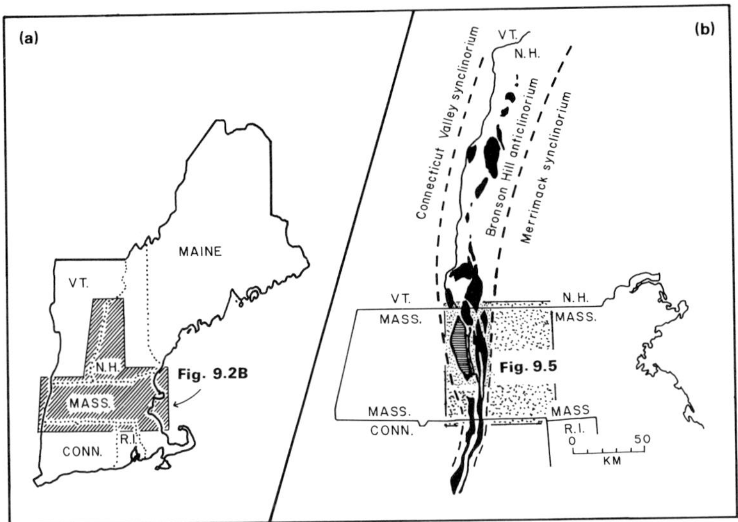

Figure 9.2 Maps showing the location of the subject area in the northeastern United States. Abbreviations: CONN, Connecticut; MASS, Massachusetts; N.H., New Hampshire; R.I., Rhode Island; VT, Vermont. (a) Map showing the general location of part (b) (shaded area) in New England. (b) Location map showing the approximate location of the three major tectonic subdivisions in the area; the Connecticut Valley synclinorium, the Bronson Hill anticlinorium, and the Merrimack synclinorium. The distribution of gneiss domes within the Bronson Hill anticlinorium is also shown. The stippled area shows the location of Figure 9.5.

axial surfaces of (1) and thrusts of (2); (4) development of mylonites in the late stages of eastward backfolding; and (5) deformation accompanying the gravity-driven rise of gneiss domes in the Bronson Hill anticlinorium (Fig. 9.2) (Robinson *et al.* 1986).

In Figure 9.3 are shown the *P–T–t* paths for the Bronson Hill anticlinorium (west) and the Merrimack synclinorium (east) (Robinson *et al.* 1982b, 1986; Schumacher *et al.* 1989). The triangular tie lines in Figure 9.3 correlate the sequence of deformational events along the two *P–T–t* paths. The *P–T–t* paths in Figure 9.3 were derived using many pieces of evidence, but the original basis for the two paths is relatively simple to understand. Rocks in the Bronson Hill anticlinorium that were metamorphosed only during the Acadian Orogeny contain kyanite or sillimanite as the common Al-silicate mineral. Andalusite or pseudomorphs after andalusite have not been found; this suggests that these rocks passed first through the kyanite stability field during prograde metamorphism, and subsequent increasing temperature drove some rocks into the sillimanite field, avoiding the andalusite field completely. The counterclockwise *P–T–t* path for the Merrimack synclinorium is indicated by the extensive occurrence of sillimanite pseudomorphs after andalusite; this suggests a path passing early through the andalusite stability field and later during increasing temperature and pressure into the sillimanite field. The

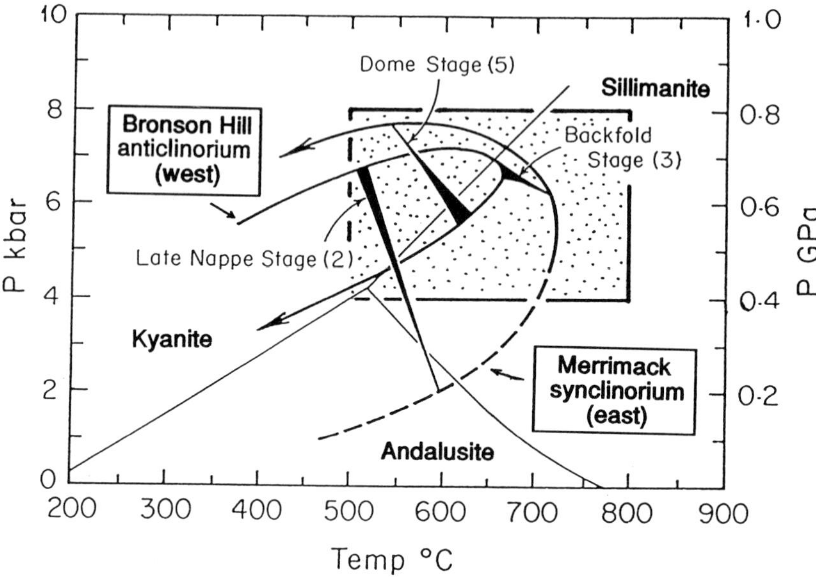

Figure 9.3 General *P–T–t* trajectories for the Bronson Hill anticlinorium and the Merrimack synclinorium in central Massachusetts and southwestern New Hampshire. The thick ends of the tie lines lie on the Bronson Hill trajectory, and the thin ends lie on the Merrimack synclinorium trajectory. The stippled area is the *P–T* region shown in Figures 9.9, 11, 14 & 18. Modified after Schumacher *et al.* (1989, Fig. 4); reproduced by permission of the Geological Society of London from 'Acadian metamorphism in central Massachusetts and southwestern New Hampshire' by J. C. Schumacher *et al.*, in *Evolution of metamorphic belts* (1989).

counterclockwise nature is further supported by *P–T* estimates placing these rocks well within the sillimanite stability field at pressures above the Al-silicate triple point of Holdaway (1971), shown on Figure 9.3. The lack of development of kyanite in the rocks following the counterclockwise path most probably reflects the notorious metastable persistence of kyanite and sillimanite in each others' stability fields or the lack of Al-silicate-forming reactions along the retrograde part of this path within the kyanite stability field.

Numerous sheet-like intrusions preceding and accompanying the early nappe formation, followed by tectonic thickening that accompanied continued nappe formation and refolding, generally explain the counterclockwise *P–T–t* path in the eastern part of the terrane (Robinson *et al.* 1982b). The early intrusive activity is absent in the western part of the terrane. Heating and tectonic thickening accompanying nappe formation and refolding were followed by uplift that was caused by the rise of the gneiss domes; these events account for the clockwise *P–T–t* path in the western part of the terrane (Robinson *et al.* 1982b).

9.3.1 Bulk chemistry of common rocks

The discussion below concerns the progressive metamorphism of pelitic rocks that were originally shales, basaltic rocks, low-Ca high-Al mafic rocks, and Ca-enriched basaltic rocks. The basaltic rocks represent tuffs, flows, or mafic intrusions (Hollocher 1985; J. C. Schumacher 1983, 1988). The low-Ca, high-Al mafic rocks are believed to represent hydrothermally altered mafic volcanics (J. C. Schumacher 1988), and the high-Ca amphibolites, at least in part, are thought to be basaltic volcanics that were contaminated with carbonate prior to metamorphism (R. Schumacher 1986).

The geochemical differences between these four rock types are shown in Figures 9.4A–D, where the tetrahedra represent the AKCM chemical system [Al_2O_3–Na_2O–K_2O] (A)–$KAlO_2$ (K)–CaO (C)–MgO (M), as projected from quartz, H_2O, and the albite component of plagioclase. The locations of end-member compositions of important metamorphic minerals, and the approximate composition volumes (delineated by the tie lines) in which the four different rock types are inscribed, are shown. The composition volumes show the volume of composition space (possible rock compositions) that could have metamorphic assemblages similar to the four rock types of interest. The most important features are: (a) potential assemblages of the pelitic volume would contain either muscovite or K-feldspar + Al-silicate; (b) the assemblages of the high-Al, low-Ca mafic rocks (i) have biotite as the only K-bearing phase and (ii) form from bulk compositions with $A:(A+C)$ ratios greater than 0.5 (peraluminous); (c) assemblages from the basaltic compositions (amphibolites) (i) have biotite as the only K-bearing phase and (ii) form from bulk compositions with $A:(A+C)$ ratios less than 0.5 (subaluminous), but here restricted to Ca contents less than the Plag–Trem–Bio plane; (d) the assemblages of higher Ca basalts and carbonate-contaminated basalts (i) have biotite as the only K-bearing phase and (ii) here form from bulk compositions with Ca contents between the Plag–Trem–Bio and Zo–Cpx–Bio planes.

The composition volumes do not represent the actual composition range of each rock type, which is generally restricted to a small space within the respective volume. In Figure 9.4A, most pelitic rocks would plot near the K–M–A face of the tetrahedron within the composition volume. Most low-Ca, high-Al mafic rocks plot near the M–A edge, or near the C–M–A face of the tetrahedron within the composition volume of Figure 9.4B. Most amphibolites plot near the C–M–A face of the tetrahedron within the composition volume of Figure 9.4C. Most high-Ca amphibolites plot near the C–M–A face of the tetrahedron within the composition volume of Figure 9.4D. Mechanical mixing of different rocks with different bulk compositions, possible in a volcanosedimentary environment, and various chemical alteration processes, permit a wide range of potential bulk compositions.

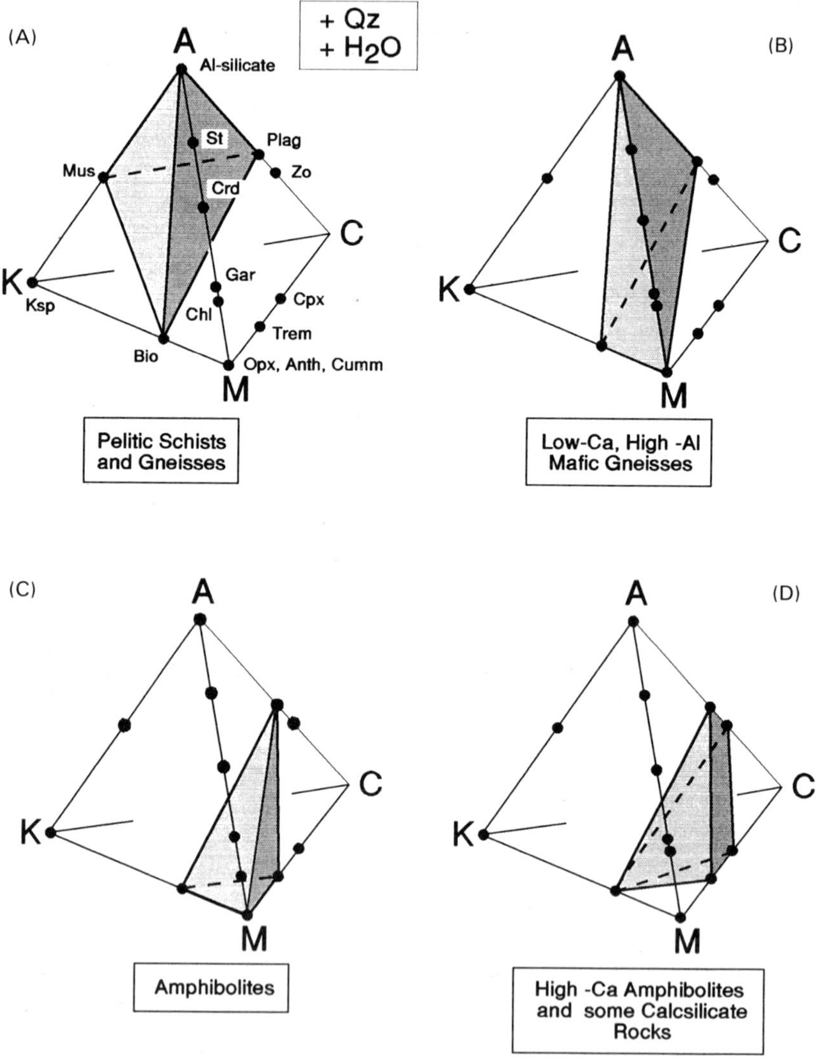

Figure 9.4 Projections from quartz, H_2O, and the albite component of plagioclase into $KAlO_2$–CaO–[MgO + FeO]–[Al_2O_3–Na_2O–K_2O] (KCMA) tetrahedra. Volumes inscribed within the tetrahedra indicate bulk compositions that may show the metamorphic mineralogy common in (A) pelitic rocks, (B) low-Ca, high-Al mafic rocks, (C) amphibolites and mafic granulites, and (D) calcareous amphibolites and high-Ca mafic rocks. Common representatives of these rock types occupy restricted regions within the volumes. Abbreviations: Anth, anthophyllite; Bio, biotite; Chl, chlorite; Cpx, clinopyroxene; Crd, cordierite; Cumm, cummingtonite; Gar, garnet; Ksp, K-feldspar, Mus, muscovite; Opx, orthopyroxene; Plag, plagioclase; St, staurolite; Trem, tremolite; Zo, zoisite.

9.4 The metamorphic zones of central Massachusetts

The central Massachusetts metamorphic high, east of the Connecticut Valley border fault, has been divided into six metamorphic zones that are based on mineral assemblages in pelitic schists (Fig. 9.5 and Tracy *et al.* 1976, Robinson *et al.* 1978). All of the common pelitic schists in central Massachusetts contain assemblages including quartz + garnet + biotite + plagioclase + graphite + ilmenite ± pyrrhotite. The assemblages that define the six metamorphic zones and the transitions forming the boundaries between zones are given in Table 9.1.

The occurrence of a zone index assemblage in any given outcrop in a zone is commonly dependent on the precise bulk composition of the rocks in question. For example, plagioclase in Zone IV commonly has a composition of An_{20-35}. At some localities, rocks with Na-rich plagioclase contain sillimanite + orthoclase but are muscovite-free, whereas rocks with more Ca-rich plagioclase may contain sillimanite + muscovite and be orthoclase-free (Tracy

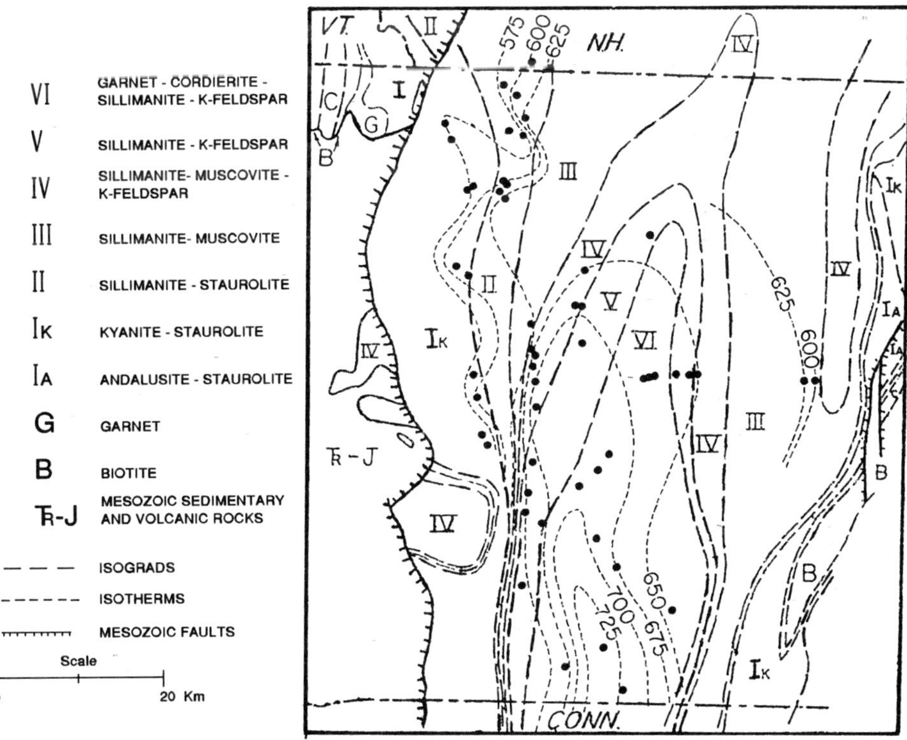

Figure 9.5 Contour map showing the positions of estimated isotherms of peak metamorphism based on studies of zoned garnets in pelitic schists (numbers), and the metamorphic zones (Roman numerals) (adapted from Robinson *et al.* 1982b).

Table 9.1 Definition of zones.

Zone	Zone index assemblage	Transitional reaction
Ik	Kyanite-staurolite-muscovite-garnet	
		Kyanite succeeded by sillimanite
II	Sillimanite-staurolite-muscovite-garnet	
		Disappearance of staurolite
III	Sillimanite-muscovite-garnet	
		Appearance of orthoclase
IV	Sillimanite-muscovite-orthoclase-garnet	
		Disappearance of muscovite
V	Sillimanite-orthoclase-garnet	
		Coexistence of garnet and cordierite
VI	Sillimanite-orthoclase-cordierite-garnet	

1975, 1978). Orthopyroxene–orthoclase–quartz assemblages in Zone VI, indicative of the granulite facies, have been recognized in mafic bulk compositions (Hollocher 1985).

9.5 Crossing isograds and isotherms

Since the metamorphic zones are generally based on highly temperature-sensitive dehydration reactions in pelitic rocks, isotherms that roughly parallel the isograds might be expected. However, in Figure 9.5 the isotherms are seen to cross metamorphic zone boundaries or isograds. If bulk compositions are assumed to be similar (here, fold repetition of the same stratigraphic units across the terrane supports this assumption), two causes for the isograd–isotherm crossings remain; variations in pressure or in fluid composition.

In a region undergoing metamorphism, a reaction that defines a metamorphic isograd will usually occur at higher temperature in areas that are at higher pressure (Fig. 9.6A, solid circles) due to the positive dP/dT slope of most metamorphic reactions. Alternatively, areas that have different metamorphic fluid compositions will allow a reaction to take place at different temperatures (Fig. 9.6B, solid diamonds): for example, the less H_2O in the metamorphic fluid, the lower the temperature at which the reaction will take

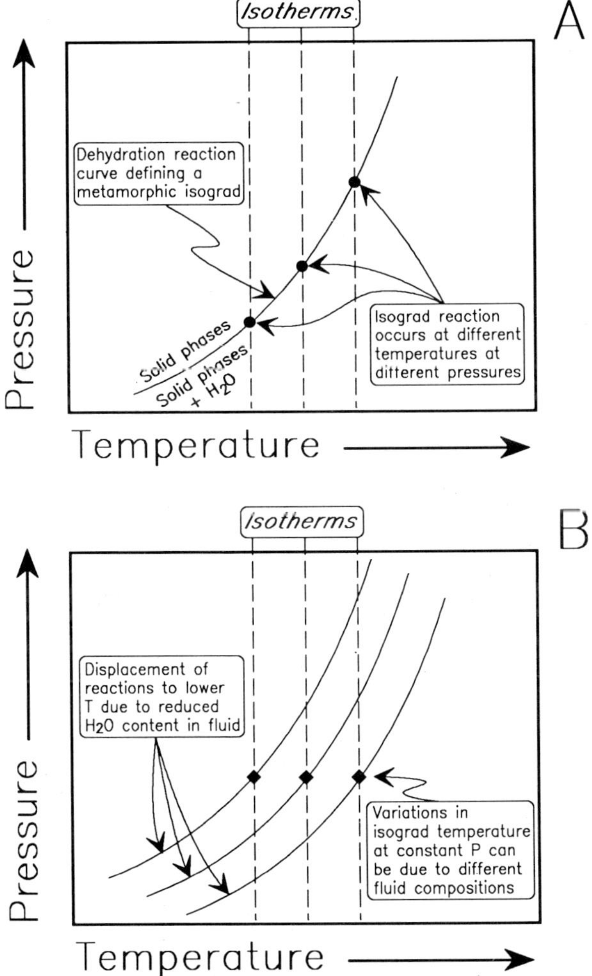

Figure 9.6 Generalized *P–T* diagrams that illustrate two possible origins of intersecting isotherms and isograds (zone boundaries). (A) The isograd reaction takes place at higher temperature in rocks at higher pressure. (B) Reduced X_{H_2O} in the metamorphic fluid phase displaces the isograd reaction to lower temperatures.

place. Therefore, rocks at the current erosion surface that were metamorphosed in regional pressure or fluid composition gradients will, alone or together, allow isograd boundaries to intersect isotherms. Large-scale fluid composition gradients are known, and appear to be possible at a scale of about 1–5 km (Carmichael 1970); however, a summary by Rice & Ferry (1981) suggests that variations in fluid compositions typically reflect local (internally buffered) differences in composition rather than regional (externally controlled) gradients. As a consequence, the best interpretation is that the intersecting isotherms and isograds probably reflect a pressure gradient across the terrane. With respect to *P–T–t* paths, this means that different parts of the metamorphosed terrane must have experienced maximum metamorphic temperatures at different pressures (Fig. 9.3).

9.6 Temperature versus TiO_2 content of minerals

The solubility of Ti in various silicate phases as a function of temperature has been examined by numerous workers (Guidotti 1970, 1984, Raase 1974, Tracy 1975). Evidence for high Ti solubility in various minerals can be seen in thin sections from higher-grade metamorphic terranes where biotites, clinoamphiboles, orthoamphiboles, and pyroxenes commonly contain oriented, exsolved rutile needles or ilmenite plates. Biotite, when viewed with a petrographic microscope, tends to change in colour from green or greenish-brown to brown, orange–brown, or red–brown with increasing grade. A discussion of

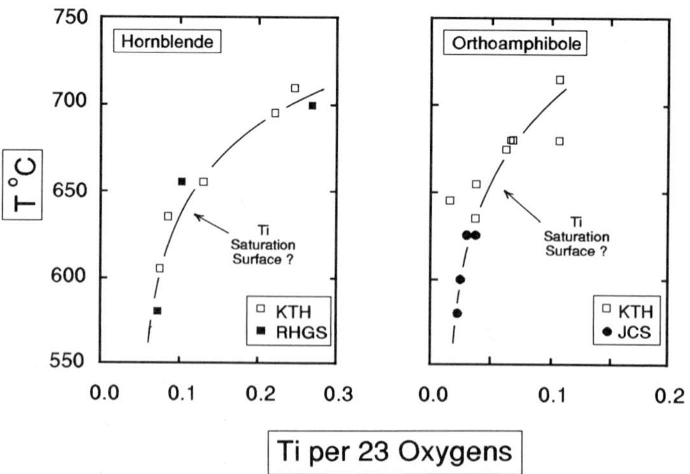

Figure 9.7 Plots of Ti versus temperature for hornblende and orthorhombic amphiboles (gedrite and anthophyllite) coexisting with ilmenite or rutile as the Ti-saturating phase. Open squares, data from Hollocher (1985) (KTH); solid squares, data from R. Schumacher (1986) (RHGS); solid circles, data from J. C. Schumacher (unpublished) (JCS).

equilibria controlling Ti in amphiboles can be found in Robinson *et al.* (1982a).

9.6.1 Amphiboles

The correlation between clinoamphibole TiO_2 contents and metamorphic grade in central Massachusetts has been discussed by Hollocher (1985) and R. Schumacher (1986) and a similar, but less drastic, correlation has also been found in orthoamphiboles. A plot of Ti content versus estimated metamorphic temperature for representative hornblende and orthoamphibole analyses from

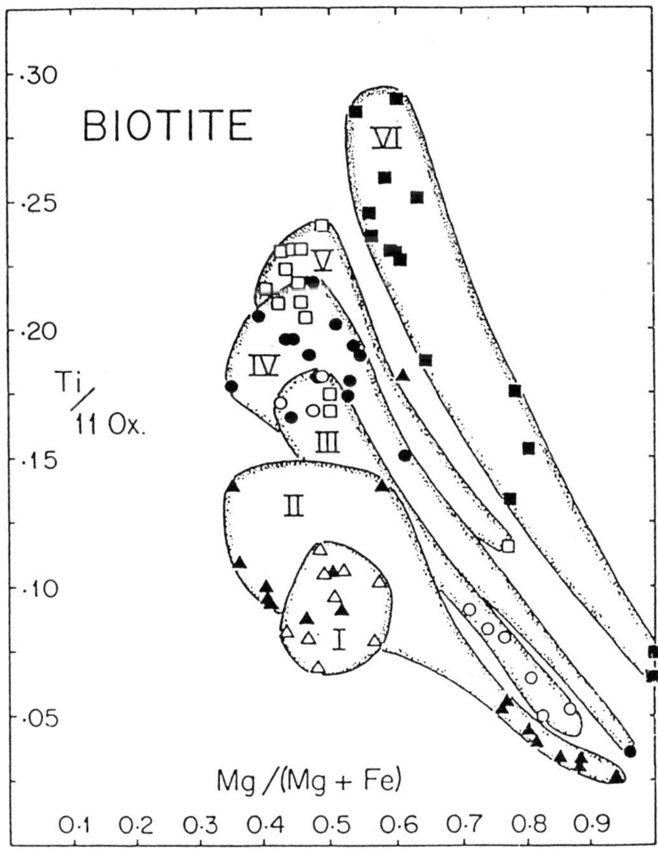

Figure 9.8 Plot of Mg/(Mg + Fe) versus Ti per 11 oxygens for biotites from central Massachusetts that coexist with Al-silicate, staurolite, or garnet as Al-saturating phases and ilmenite or rutile as Ti-saturating phases. Open triangles, Zone I; solid triangles, Zone II; open circles, Zone III; solid circles, Zone IV; open squares, Zone V; solid squares, Zone VI. Magnesian biotite compositions from Zones II and III are from the Small Falls Formation in Maine (Guidotti *et al.* 1977).

metamorphic zones I–VI is shown in Figure 9.7. In both hornblendes and orthoamphiboles, Ti contents almost triple between about 575 and 700°C and, at a given temperature, hornblende has two to three times the Ti content of orthoamphibole (Fig. 9.7). Tentative Ti saturation surfaces for both hornblendes and orthoamphiboles are also shown in Figure 9.7.

9.6.2 Biotite

The Ti contents of biotite coexisting with either rutile or ilmenite, grouped with respect to the metamorphic zone, are shown in Figure 9.8. Despite minor irregularities, the Ti content of biotites generally increases with metamorphic grade (temperature) for any given Fe : Mg ratio, but the positive correlation between Ti content and Fe : Mg ratio is striking. This confirms and expands the results of Guidotti *et al.* (1977) who studied the effects of the Fe–Mg content on Ti in biotites from the Smalls Falls Formation of Maine, and found that biotites from the same metamorphic grade show a drastic decrease in Ti content with increasing Mg content.

9.7 Prograde assemblages

The phase relations of different rock types are shown below on simplified and projected phase diagrams. Readers who are not familiar with projection techniques or specific types of projections are referred to Thompson (1957), Robinson & Jaffe (1969b), Greenwood (1975), and Spear *et al.* (1982) for additional information.

9.7.1 Prograde reactions in pelitic schists

The first appearance of sillimanite in pelitic schists marks the boundary between Zones I and II (Fig. 9.9); that is, it separates rocks in which the Al-silicate-forming reactions that involve Al–Fe–Mg (AFM) phases last produced either kyanite or sillimanite. Rocks which contain sillimanite replacing kyanite are rare; consequently, the sillimanite to kyanite polymorphic transition is not an important sillimanite-producing reaction. Additionally, although sillimanite pseudomorphs after andalusite are abundant in the Merrimack synclinorium, the transition from andalusite- to sillimanite-bearing is not present along this part of the metamorphic field gradient.

The Zone I to II transition is controlled by more than one AFM reaction, a problem that has been discussed in detail by Robinson (1963) and Hall (1970). The most likely reactions are the Fe–Mg continuous reactions:

$$\text{staurolite} + \text{muscovite} + \text{quartz} = \text{biotite} + \text{sillimanite} + H_2O \qquad (9.1)$$

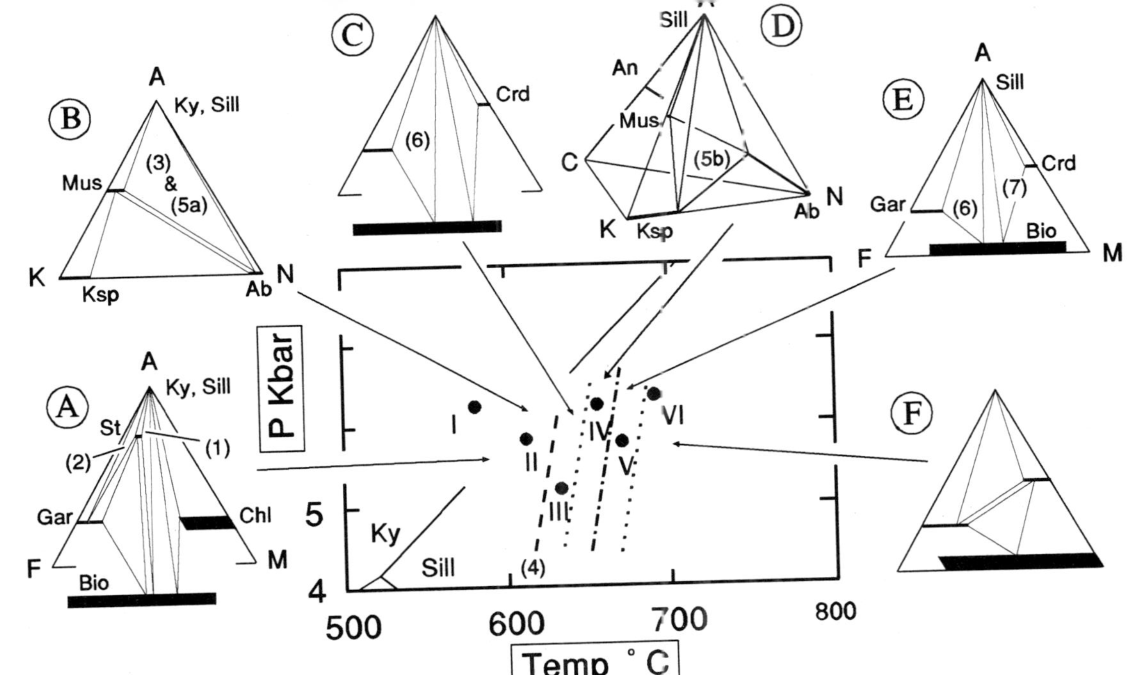

Figure 9.9 *P–T* diagram showing average conditions for the six metamorphic zones, with the phase relations for the pelitic rocks. Compatibility triangles (A) and (C) are projections from quartz, H_2O, and muscovite (Mus) onto an Al_2O_3–FeO–MgO plane; compatibility triangle (B) is a projection from quartz, H_2O, and anorthite onto an [Al_2O_3–K_2O–Na_2O–CaO]–$KAlO_2$ –$NaAlO_2$ plane; compatibility diagram (D) is a projection from quartz and H_2O into an [Al_2O_3–K_2O–Na_2O]–$KAlO_2$–$NaAlO_2$–CaO tetrahedron; compatibility triangles (E) and (F) are projections from quartz, H_2O, and K-feldspar (Ksp) onto an Al_2O_3–FeO–MgO plane. Numbers in parentheses correspond to reactions (prefixed 9.) in the text. Dashed, dotted, and dot–dashed lines are crudely located boundaries between the metamorphic zones. The dashed line is the staurolite-out isograd and the dot–dashed line is the muscovite-out isograd. Thin solid lines separate the aluminosilicate stability fields (Salje 1986). Abbreviations: Ab, albite; An, anorthite; Bio, biotite; Chl, chlorite; Crd, cordierite; Gar, garnet; Ksp, K-feldspar; Ky, kyanite; Mus, muscovite; Sill, sillimanite; St, staurolite.

and

$$\text{staurolite} + \text{muscovite} + \text{quartz} = \text{garnet} + \text{sillimanite} + H_2O \qquad (9.2)$$

(both apparent in triangle A on Fig. 9.9), and the K–Na continuous reaction:

$$\text{muscovite (Na component)} + \text{quartz} = \text{sillimanite} + \text{albite} + H_2O \quad (9.3)$$

(triangle B in Fig. 9.9). Minor variations in bulk composition determine which of these reactions is appropriate to specific samples.

The boundary between Zones II and III is defined by the last occurrence of staurolite, governed in general by the discontinuous reaction:

$$\text{staurolite} + \text{muscovite} + \text{quartz} = \text{biotite} + \text{garnet} + \text{sillimanite} + H_2O \quad (9.4)$$

by which staurolite in equilibrium with muscovite vanishes on its own projected composition between triangles A and C (Fig. 9.9). Although this reaction is discontinuous in the K_2O–FeO–MgO–Al_2O_3–SiO_2–H_2O (KFMASH) system, it is probably continuous (divariant or trivariant) over a limited temperature range due to the effects of ZnO in staurolite and CaO and MnO in garnet. This reaction, which is approximately located in P–T space on Figure 9.9 (heavy dashed line), defines the staurolite-out isograd by ending the stable coexistence of the three phases staurolite + muscovite + quartz. This isograd reaction is only applicable to rocks of pelitic composition, since staurolite is stable to much higher grades in muscovite- or quartz-free rocks.

The breakdown of muscovite controls the first appearance of sillimanite in some rocks, beginning in Zone II, and is also responsible for the first appearance of K-feldspar that marks the beginning of Zone IV. Tracy (1975, 1978) has studied this reaction extensively in Zones II–IV. In Zones II and III its generalized form is:

$$\begin{aligned}&\text{Na-bearing muscovite} + \text{quartz}\\ &\quad = \text{sillimanite} + \text{K-richer muscovite} + \text{Na-plagioclase} + H_2O \quad (9.5a)\end{aligned}$$

In Zone IV the generalized muscovite breakdown reaction is:

$$\begin{aligned}&\text{Na-bearing muscovite} + \text{plagioclase} + \text{quartz} = \text{sillimanite}\\ &\quad + \text{K-richer muscovite} + \text{Ca-richer plagioclase} + \text{K-feldspar} + H_2O \quad (9.5b)\end{aligned}$$

These two reactions are shown in triangle B and tetrahedron D in Figure 9.9. With increasing T, in the reaction assemblage the K : Na ratios of muscovite and K-feldspar increase and the Ca : Na ratio of plagioclase increases, which drives the reactions progressively through more K-rich and Ca-rich bulk compositions. Na from both muscovite and plagioclase form the Na component of K-feldspar. The Zone III/IV boundary, which is the first appearance of coexisting sillimanite + K-feldspar + muscovite, is seen in Na-richer schists that commonly contain plagioclase of about An_{20}. The Zone IV/V boundary, which is the last appearance of coexisting sillimanite + K-feldspar + muscovite, is seen in Ca-richer schists that commonly contain plagioclase of about An_{33}.

During and following the breakdown of muscovite, the dehydration of biotite via the continuous Fe–Mg reaction:

$$\text{sillimanite} + \text{FeMg biotite} + \text{quartz} = \text{garnet} + \text{K-feldspar} + \text{Mg-richer biotite} + H_2O \quad (9.6)$$

controls the Fe : Mg ratios of coexisting garnet and biotite in this assemblage (triangle E in Fig. 9.9). In Zones I–IV the pyrope contents of garnets reach maximum values of about 16–18%, but they increase to about 25% in Zone VI. Biotites coexisting with garnet show a sympathetic increase in Mg, attain values up to about $X_{Mg} = 0.50$ in Zones I–IV and change to about $X_{Mg} = 0.60$ in Zone VI. This increase of Mg in both garnet and biotite is largely due to reaction (9.6). Rarer Mg-rich bulk compositions experience a similar reaction:

$$\text{sillimanite} + \text{MgFe biotite} + \text{quartz} = \text{cordierite} + \text{K-feldspar} + \text{Fe-richer biotite} + H_2O \quad (9.7)$$

(triangle E in Fig. 9.9), but which drives the biotite to more Fe-rich compositions. In the pure KFMASH system, the two preceding continuous Fe–Mg reactions would meet, and the last tie line between sillimanite and biotite would break (triangle E in Fig. 9.9; and Fig. 9.10). The transition from Zone V to Zone VI is defined by the first occurrence of coexisting

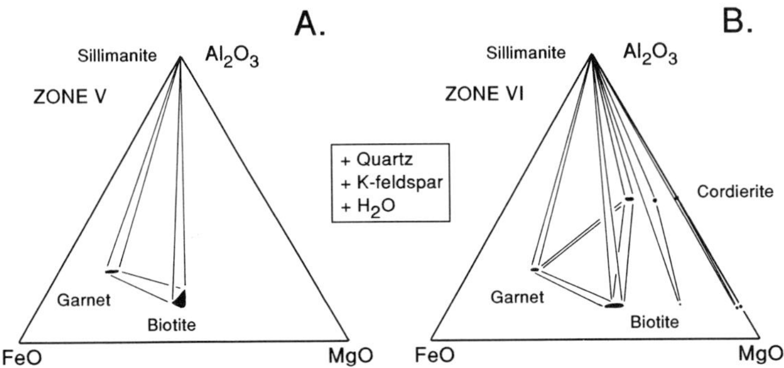

Figure 9.10 Compositions of minerals from pelitic rocks projected from quartz, H_2O, and K-feldspar onto an Al_2O_3–FeO–MgO plane: (A) Zone V assemblages and (B) Zone VI assemblages. Only garnet core compositions are represented. The extremely Mg-rich compositions are the result of sulphide–silicate reactions that depleted the Fe in the silicates (see Tracy & Robinson 1988).

sillimanite + garnet + cordierite + biotite (triangle F in Fig. 9.9; Fig. 9.10), by the general discontinuous reaction:

sillimanite + biotite + quartz = garnet + cordierite + K-feldspar + H_2O (9.8)

Due to the stabilizing effects of Mn and Ca in garnet, Ca in plagioclase, and Na in K-feldspar, the observed assemblage sillimanite + garnet + cordierite + biotite + orthoclase + plagioclase + quartz is at least trivariant.

9.7.2 Low-Ca, high-Al mafic rocks

In Ca-poor, Al-rich bulk compositions that are also generally K-poor, orthorhombic amphiboles commonly occur with other aluminous minerals such as garnet, staurolite, and aluminium silicates, and with sapphirine and corundum in quartz-free assemblages (Robinson & Jaffe 1969a,b, Schumacher 1983, Schumacher & Robinson 1987). The most abundant and well studied outcrops of this rock type are found in Zones I and II, but Robinson & Tracy (1979), Robinson *et al.* (1982a), and Hollocher (1985) have studied limited occurrences of this rock type from Zone IV. Robinson & Jaffe (1969b), Hudson & Harte (1985) and Spear & Rumble (1986) have given petrogenetic grids for quartz-bearing orthoamphibole rocks. One of the invariant points, common to all of these grids, is shown in Figure 9.11 and is relevant to quartz- and plagioclase-bearing assemblages. The ternaries A–E (Fig. 9.11) show possible two-phase assemblages and three-phase, continuous Fe–Mg reaction assemblages – remember that the projected phases are also present – in the relevant parts of P–T space. The numerous continuous reactions are not discussed here in detail, but the most important features of the diagram are described by the discontinuous reactions. The assemblage staurolite + cordierite + quartz + plagioclase (Fig. 9.11, ternary A) is stable on the low-P side of the discontinuous Fe–Mg reaction, which is a fuller version of Reaction (9.9b) below:

cordierite + staurolite + quartz + plagioclase = gedrite + sillimanite + H_2O

At higher P (Fig. 9.11, ternary B), gedrite + aluminosilicate are stable. With increasing T, staurolite will break down via the discontinuous Fe–Mg reaction:

staurolite + quartz + plagioclase = gedrite + garnet + sillimanite + H_2O

which gives the staurolite-free assemblages of ternary C (Fig. 9.11). At lower P (Fig. 9.11, ternary D), gedrite + sillimanite is unstable via the discontinuous Fe–Mg reaction:

gedrite + sillimanite + quartz = garnet + cordierite + plagioclase + H_2O

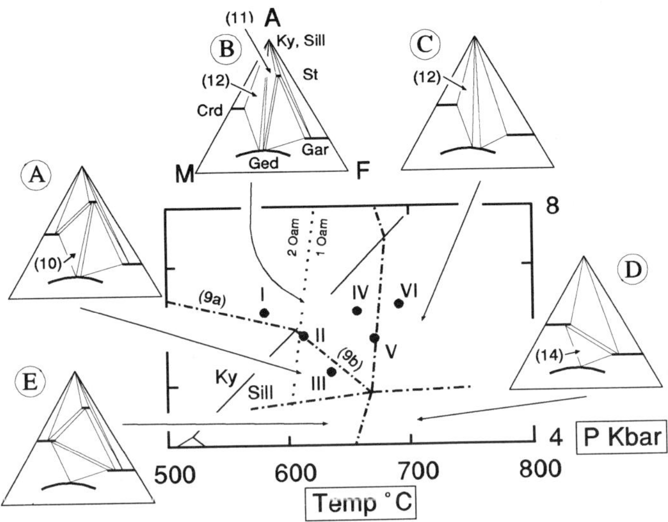

Figure 9.11 *P–T* diagram showing average conditions for the six metamorphic zones (after Robinson *et al.* 1982b) and showing the phase relations for low-Ca, high-Al mafic rocks containing cordierite and/or gedrite. Compatibility triangles (A)–(E) are projections from quartz, H_2O, and plagioclase onto an Al_2O_3–FeO–MgO plane; numbers in parentheses correspond to numbered reactions (prefixed 9.) in the text. Dot–dash lines are 'univariant' reactions after Spear & Rumble (1986; their invariant point has been slightly repositioned to fit better with observations in central Massachusetts); the heavy dotted line is the approximate crest of the orthoamphibole solvus; thin solid lines separate the aluminosilicate stability fields (Salje 1986). Abbreviations: Crd, cordierite; Gar, garnet; Ged, gedrite; Ky, kyanite; Sill, sillimanite; St, staurolite.

At lower *T* (Fig. 9.11, ternary E), staurolite + quartz is stabilized by the reaction:

$$\text{staurolite} + \text{quartz} = \text{cordierite} + \text{garnet} + \text{sillimanite} + H_2O$$

At slightly higher *P* (Fig. 9.11, ternary A), cordierite + garnet + plagioclase becomes unstable via the discontinuous Fe–Mg reaction:

$$\text{gedrite} + \text{staurolite} + \text{quartz} = \text{garnet} + \text{cordierite} + \text{plagioclase} + H_2O$$

For this limited part of *P–T* space, the lower-*P* assemblages will generally contain cordierite + staurolite or cordierite + garnet; relatively higher *P*'s are indicated by the stable gedrite + aluminosilicate assemblages.

In Zone Ik, rare coexisting orthoamphiboles are found (Robinson *et al.* 1982a); whereas in Zone II the orthoamphiboles are demonstrably hypersolvus, and only one orthoamphibole solid solution is stable (Robinson *et al.* 1971, 1982a, Spear 1980). The locations of the discontinuous reactions:

$$\text{cordierite} + \text{staurolite} + \text{quartz} = \text{gedrite} + \text{kyanite} + H_2O \qquad (9.9a)$$

$$\text{cordierite} + \text{staurolite} + \text{quartz} = \text{gedrite} + \text{sillimanite} + H_2O \quad (9.9b)$$

(Fig. 9.11, cordierite is on the low-*P* side of the reaction curves) have been slightly modified from the approximate *P–T* location given by Spear & Rumble (1986), as has the invariant point which is the highest-temperature invariant point emanating five reaction curves in Figure 13 of Spear & Rumble (1986). The reactions are placed near the average *P–T* conditions for Zones I and II in Figure 9.11 because assemblages representing the continuous Fe–Mg reactions:

$$\text{MgFe gedrite} + \text{FeMg staurolite} = \text{MgFe cordierite} + \text{Fe-richer gedrite} + \text{Fe-richer staurolite} \quad (9.10)$$

(Fig. 9.11A),

$$\text{Al-silicate} + \text{MgFe gedrite} = \text{staurolite} + \text{Mg-richer gedrite} \quad (9.11)$$

(Fig. 9.11B), and

$$\text{Al-silicate} + \text{MgFe gedrite} = \text{MgFe cordierite} + \text{Fe-richer gedrite} \quad (9.12)$$

(Fig. 9.11B) are all known from both Zones I and II. Since cordierite-forming reactions tend to be highly pressure dependent, the overlap of these continuous reaction assemblages in *P–T* space is a function of the local Fe : Mg ratio of the rock. This bulk compositional dependence has been shown by Robinson & Jaffe (1969a) and Schumacher & Robinson (1987) for Reaction (9.12).

For Zones IV and V, Robinson *et al.* (1982b) report the assemblages:

$$\text{sillimanite} + \text{staurolite} + \text{cordierite} + \text{garnet} \quad (9.13)$$

and

$$\text{gedrite} + \text{cordierite} + \text{garnet} \quad (9.14)$$

in high-Al, non-pelitic rocks. At first glance, these assemblages appear more appropriate to Figures 9.11D and E which are located at lower pressures than the *P–T* estimates for Zones IV and V. However, this apparent discrepancy might be explained by considering the possible stabilizing effects of MnO and ZnO. The relative X_{Mn} [Mn/(Mn + Zn + Fe + Mg)] of the relevant phases is garnet $\gg$ gedrite $\approx$ staurolite $>$ cordierite $\approx$ spinel, and the relative X_{Zn} [Zn/(Zn + Mn + Fe + Mg)] of the relevant phases is spinel $\gg$ staurolite $\gg$ gedrite $>$ garnet $\approx$ cordierite (e.g. Schumacher & Robinson 1987). Parts A, B, and C of Figure 9.12 correspond to triangles A, B, and C in Figure 9.11, and illustrate possible assemblages in Zn- and Mn-bearing bulk compositions. Assemblage (9.13) given above could be stable at higher pressure in rocks with

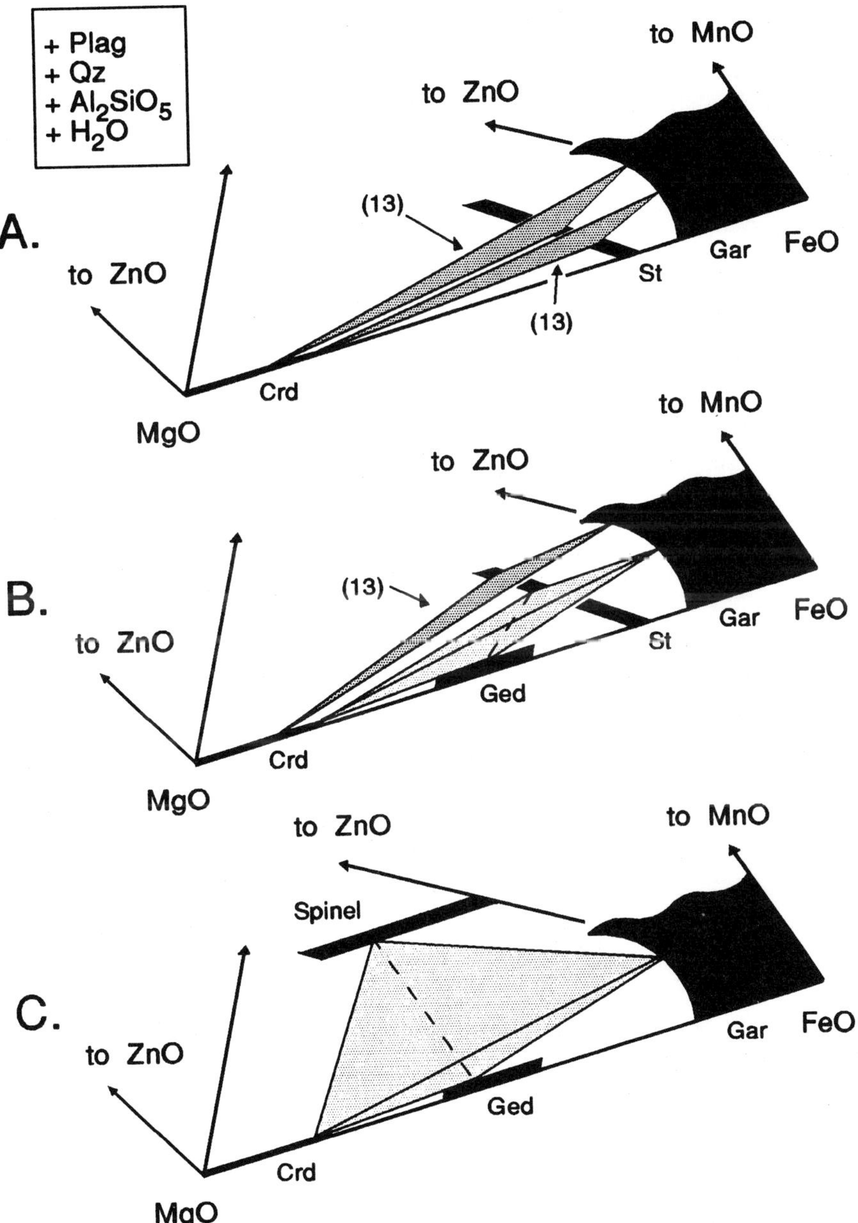

Figure 9.12 Projections from quartz, H_2O, aluminosilicate, and plagioclase into FeO–MgO–MnO–ZnO tetrahedra which show assemblages (shaded planes and a shaded volume) that could result from variations in the ZnO and MnO content of the bulk composition. Zn solid solution in minerals is exaggerated for clarity. Parts (A), (B), and (C) correspond to ternaries A, B, and C in Figure 9.11. The '(13)' corresponds to a numbered assemblage (9.13) in the text. Abbreviations: Crd, cordierite; Gar, garnet; Ged, gedrite; St, staurolite.

higher Mn and Zn contents (Figs 9.11B & 12B). Although not presented here, assemblage (9.14) can be shown to be equally valid at the $P-T$ conditions of Zones IV and V. These bulk compositions have not yet been reported from Zone VI, but spinel should occur in Zn-bearing assemblages following the breakdown of staurolite (Fig. 9.12C).

9.7.3 Basaltic rocks

The basaltic and high-Ca basaltic rock compositions are summarized in Figure 9.13 in plagioclase projection in the ACFM system (ACFM: $A = [Al + Fe^{3+} - Na + 2Ti]$, $F = [Fe^{2+} + Mn - Ti]$, $M = Mg$, $C = Ca$). Common tholeiitic basalts project between the Fe–Mg join and the hornblende field and within the hornblende field. High-Ca basaltic rocks project within or below the hornblende field. Prograde changes that are observed in these rocks (Figs 9.14 & 15C,D) from central Massachusetts have been studied by R. Schumacher (1986) and Hollocher (1985), and their major observations are summarized in Figure 9.14 and as follows. Epidote disappears under Zone

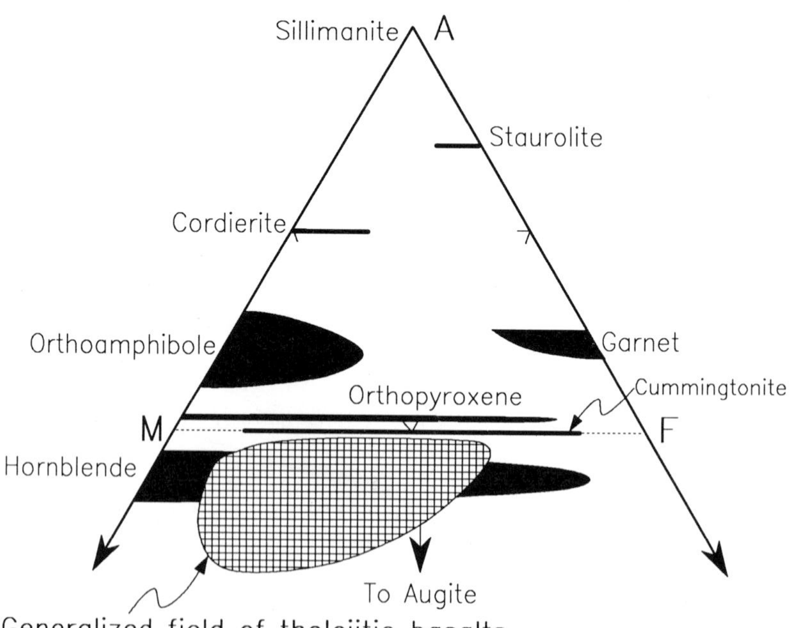

Figure 9.13 Projection from plagioclase in the Al_2O_3–FeO–MgO–Na_2O–CaO (AFMNC) system onto the AFM plane. Minerals related to important dehydration reactions in medium- and high-grade rocks of tholeiitic composition are shown (see Fig. 9.15). Composition fields for typical and Ca-rich tholeiitic basalts are also shown. $A = [((Al + Fe^{3+} - Na)/2) + Ti - Ca]$; $F = [Fe^{2+} + Mn - Ti]$; $M = Mg$.

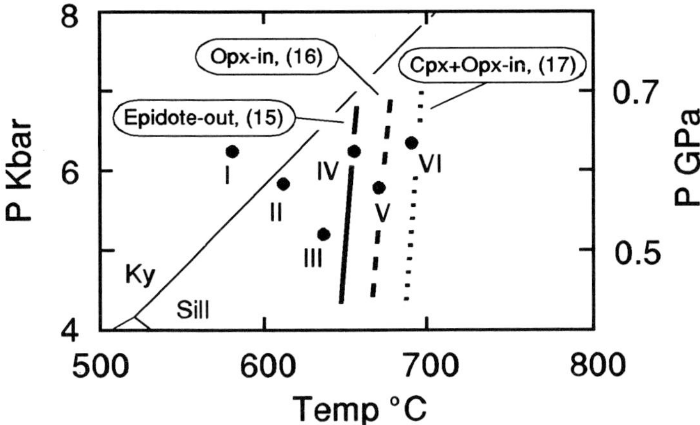

Figure 9.14 *P–T* diagram showing the average conditions for the six metamorphic zones (after Robinson *et al.* 1982b), and showing the approximate location of the last appearance of epidote in Ca-enriched basaltic compositions (solid line, R. Schumacher 1986), and the first appearance of orthopyroxene-bearing assemblages in Zone V (dashed line, first appearance of orthopyroxene) and Zone VI (dotted line, first appearance of orthopyroxene + clinopyroxene) in basaltic compositions (Hollocher 1985). Numbers in parentheses correspond to numbered reactions (prefixed 9.) in the text.

IV conditions (sillimanite–muscovite–orthoclase–garnet grade) via the generalized reaction:

$$\text{epidote} + \text{hornblende} + \text{quartz} = \text{plagioclase} + \text{clinopyroxene} + H_2O \quad (9.15)$$

(Fig. 9.14; R. Schumacher 1986). Within Zone V (sillimanite–orthoclase–garnet grade), orthopyroxene first appears as a breakdown product of cummingtonite by the generalized reaction:

$$\text{cummingtonite} + \text{hornblende} + \text{garnet} + \text{quartz} = \text{orthopyroxene} + \text{plagioclase} + H_2O \quad (9.16)$$

(Fig. 9.14; Hollocher 1985). In Zone VI (sillimanite–orthoclase–cordierite–garnet grade), orthopyroxene–clinopyroxene assemblages are produced by the breakdown of hornblende by the generalized reaction:

$$\text{hornblende} + \text{quartz} = \text{clinopyroxene} + \text{orthopyroxene} + \text{plagioclase} + H_2O \quad (9.17)$$

(Fig. 9.14; Hollocher 1985).

The phase relations in metamorphosed basaltic rocks have been determined in detail by first analysing minerals from a wide range of metamorphosed mafic rock compositions throughout central Massachusetts, and then tying

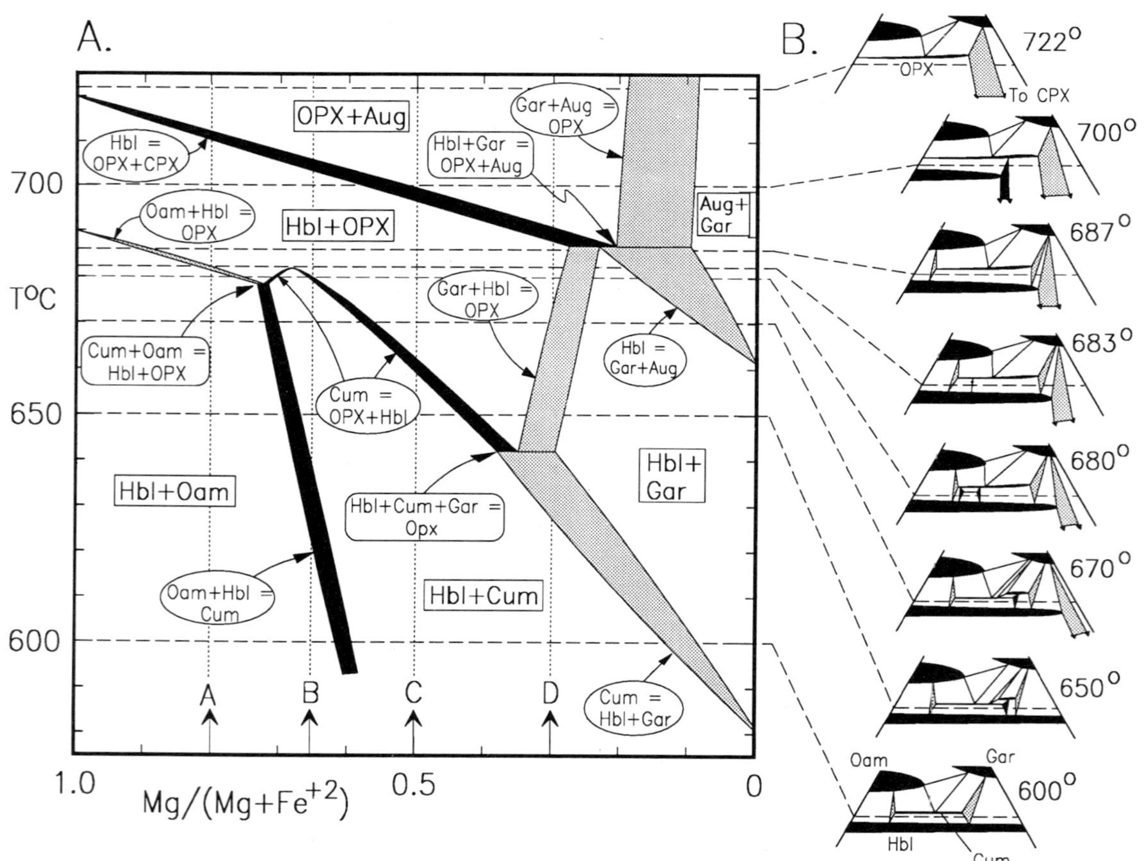
A.
OPX+Aug
Hbl =
OPX+CPX
Gar+Aug =
OPX
Hbl+Gar =
OPX+Aug
Oam+Hbl =
OPX
Hbl+OPX
Aug+
Gar
Gar+Hbl =
OPX
Cum+Oam =
Hbl+OPX
Hbl =
Gar+Aug
Cum =
OPX+Hbl
Hbl+Oam
Hbl+
Gar
Hbl+Cum+Gar =
Opx
Oam+Hbl =
Cum
Hbl+Cum
Cum =
Hbl+Gar
A
B
C
D
700
T°C
650
600
1.0
0.5
0
Mg/(Mg+Fe+2)
B.
722°
OPX
To CPX
700°
687°
683°
680°
670°
650°
Oam
Gar
600°
Hbl
Cum

those rocks to metamorphic temperature estimates (Fig. 9.5). Since all rocks examined were apparently metamorphosed at pressures near 6 kbar (e.g. Fig. 9.15), to a first approximation the mineral assemblages have equilibrated under isobaric conditions. Using the assumption of an isobaric metamorphic field gradient in conjunction with isotherms from Figure 9.5 and the mineral compositions in many mafic rock assemblages, some amphibole dehydration reactions were constrained in T–X_{Mg} space. Other reactions were constrained based on the occurrence of the requisite assemblages and reasonable estimates of the correct phase relations.

For the purpose of the following discussion, remember that the phase relations in plagioclase projection require the presence of the phases from which the projection is made, including quartz and plagioclase, for the projection to be valid. Figure 9.15 was constructed assuming that the pressures in the field areas were all roughly 6 kbar, and that the fluid composition was constant or varied smoothly with metamorphic temperature.

Rock A in Figure 9.15A represents a magnesian basalt (picrite). The temperatures pertain only to Figure 9.15 and are accurate to no better than ±50°C, although the temperatures are correct relative to one another. At 600°C rock A has a hornblende–orthoamphibole assemblage. With increasing temperature this assemblage remains stable, although mineral compositions change slightly. At about 682°C, the reaction orthoamphibole + hornblende = Opx (+plagioclase, +H_2O) takes place, represented as a narrow reaction loop in Figure 9.15A. In Figure 9.15B this reaction loop is seen as a three-phase field that moves towards more magnesian compositions with increasing temperature. The orthoamphibole + hornblende = orthopyroxene three-phase field is a continuous reaction, and orthoamphibole in rock A reacts out when this three-phase field moves past the rock's bulk composition at 684°C. At higher temperatures the hornblende–orthopyroxene assemblage in rock A persists, although minor changes in mineral composition occur. At about 710°C the three-phase hornblende-out continuous reaction hornblende = orthopyroxene + clinopyroxene (+plagioclase, +H_2O) in Figure 9.15B, and a reaction

Figure 9.15 T–X_{Mg} diagram (A) showing estimated phase relations in mafic rocks from central Massachusetts at about 6 kbar (modified after Figure 42 of Hollocher 1985). The T–X_{Mg} plane of section lies between the Fe–Mg join and hornblende in plagioclase projection (B), at an A′ value of about 0.05. The three-phase Fe–Mg reaction loops in (A) (elliptical labels) represent three-phase fields in the plagioclase projections of (B), corresponding to three-phase continuous reactions. Two-phase fields (rectangular labels) separate the reaction loops. Four-phase discontinuous reactions (rounded rectangular labels) are horizontal lines in (A), from which diverge three or four reaction loops. The three-phase loops are approximately constrained by actual mineral compositions and temperature estimates based on garnet–biotite geothermometry in associated schists (Fig. 9.5). Solid black, best constrained loops and three-phase fields; dotted, less well constrained reactions. Although the reaction loops are topologically reasonable, the absolute temperatures are probably accurate to no better than 50°C. The temperatures pertaining to this figure in the text are similarly inaccurate, although consistent relative to one another.

loop in Figure 9.15A, intersect the bulk composition of rock A. At 713°C the three-phase field passes over rock A, and at higher temperatures rock A contains the anhydrous orthopyroxene–augite assemblage of the pyroxene granulite facies.

To give an indication of how sensitive assemblages and reactions are to bulk composition, notice in Figure 9.15B (687°C) that if the bulk composition of rock A projected above the Fe–Mg join, between orthopyroxene and orthoamphibole, then hornblende would react out first rather than orthoamphibole. Also, if rock A projected substantially below the Fe–Mg join in the hornblende field, then this rock would be a common hornblende–plagioclase amphibolite without other amphiboles or pyroxene, until hornblende finally broke down at high temperature.

Rock B in Figure 9.15A is more iron-rich than rock A, and it is intersected by the three-phase field representing the reaction orthoamphibole + hornblende = cummingtonite (+plagioclase, +H_2O) at about 619°C, resulting in the three-amphibole assemblage orthoamphibole–cummingtonite–hornblende. This reaction is not a very strong dehydration reaction (small entropy of reaction), so this three-phase field moves rather slowly towards more magnesian compositions with increasing temperature. At about 640°C the three-phase reaction field finally moves past the bulk composition of rock B, resulting in a two-phase, hornblende–cummingtonite assemblage. Further increases in temperature have no effect on the assemblage until, at 679°C, rock B is intersected by the three-phase field representing the cummingtonite dehydration reaction cummingtonite (+plagioclase) = hornblende + orthopyroxene (+H_2O), which reacts out all cummingtonite. At higher temperatures the new hornblende–orthopyroxene two-phase assemblage remains unchanged until the hornblende dehydration reaction hornblende = orthopyroxene + clinopyroxene (+ plagioclase, +H_2O) passes over the bulk composition of rock B, resulting in the same granulite facies assemblage as in rock A, but at a slightly lower temperature.

Rock C is more iron-rich than rock B so it does not intersect an orthoamphibole stability field. Notice in Figure 9.15B that if rock C is projected above the Fe–Mg join, above orthopyroxene and cummingtonite, then orthoamphibole would be a likely phase. The assemblages in Figure 9.15A proceed from hornblende–cummingonite, to hornblende–cummingtonite–orthopyroxene, to hornblende–orthopyroxene, to hornblende–augite–orthopyroxene, and finally to the granulite facies augite–orthopyroxene assemblage as various three-phase continuous reaction fields pass over the bulk composition of rock C.

Rock D is substantially more iron-rich than common tholeiitic basalts, but such peculiar compositions can be found relatively easily because of the ready field identification of almandine garnet. Rock D starts with a hornblende–cummingtonite two-phase assemblage that gives way to the cummingtonite (+plagioclase) = hornblende + garnet (+H_2O) reaction assemblage at about 627°C, as this three-phase field passes over the bulk composition of rock D. This assemblage persists to about 641°C when orthopyroxene first appears in plagioclase projection by the four-phase discontinuous reaction cummingtonite + garnet + hornblende = orthopyroxene (+ plagioclase, + H_2O). The bulk composition of rock D was chosen to lie in the garnet + hornblende = orthopyroxene (+ plagioclase, + H_2O) three-phase field above 641°C after orthopyroxene appears. At about 674°C this three-phase field has moved towards more iron-rich compositions, leaving rock D with a hornblende–orthopyroxene assemblage. At about 688°C, rock D is intersected by the hornblende-out, three-phase field, and finally, at about 692°C, all hornblende has reacted out, leaving rock D with the augite–orthopyroxene granulite facies assemblage at a temperature about 20°C cooler than the more magnesian rock A.

This discussion illustrates how bulk composition as well as temperature are critically important for determining the mineral assemblages actually found in a rock. It is also shown in Figure 9.15A that, in the part of the AFM plagioclase projection dominated by tholeiitic basalts, three-phase fields representing continuous metamorphic reactions tend to occupy little area in T–X space compared to the wide and monotonous two-phase fields. This is why finding reaction assemblages in rocks in commonly difficult.

9.8 Melting

9.8.1 Mafic rocks

At metamorphic temperatures (about 600–750°C), rocks can begin to melt during metamorphism via two general mechanisms. One mechanism involves either the presence of a flux, typically an H_2O-rich fluid phase, or the addition of such a flux into a rock from outside resulting in a reaction:

solid phases + H_2O = melt ± new solid phases

In the second mechanism, melting occurs in the absence of a fluid phase (H_2O-rich phase) by a process known as fluid-absent or dehydration melting:

hydrous mineral ± other solid phases
= hydrous melt ± new anhydrous or less-hydrous solid phases

Fluid-absent reactions need not involve hydrous phases and can take place in H_2O-free systems:

anhydrous solid phases = anhydrous melt ± new anhydrous solid phases

Hollocher, in Robinson *et al.* (1986), has reported veins, anastomosing

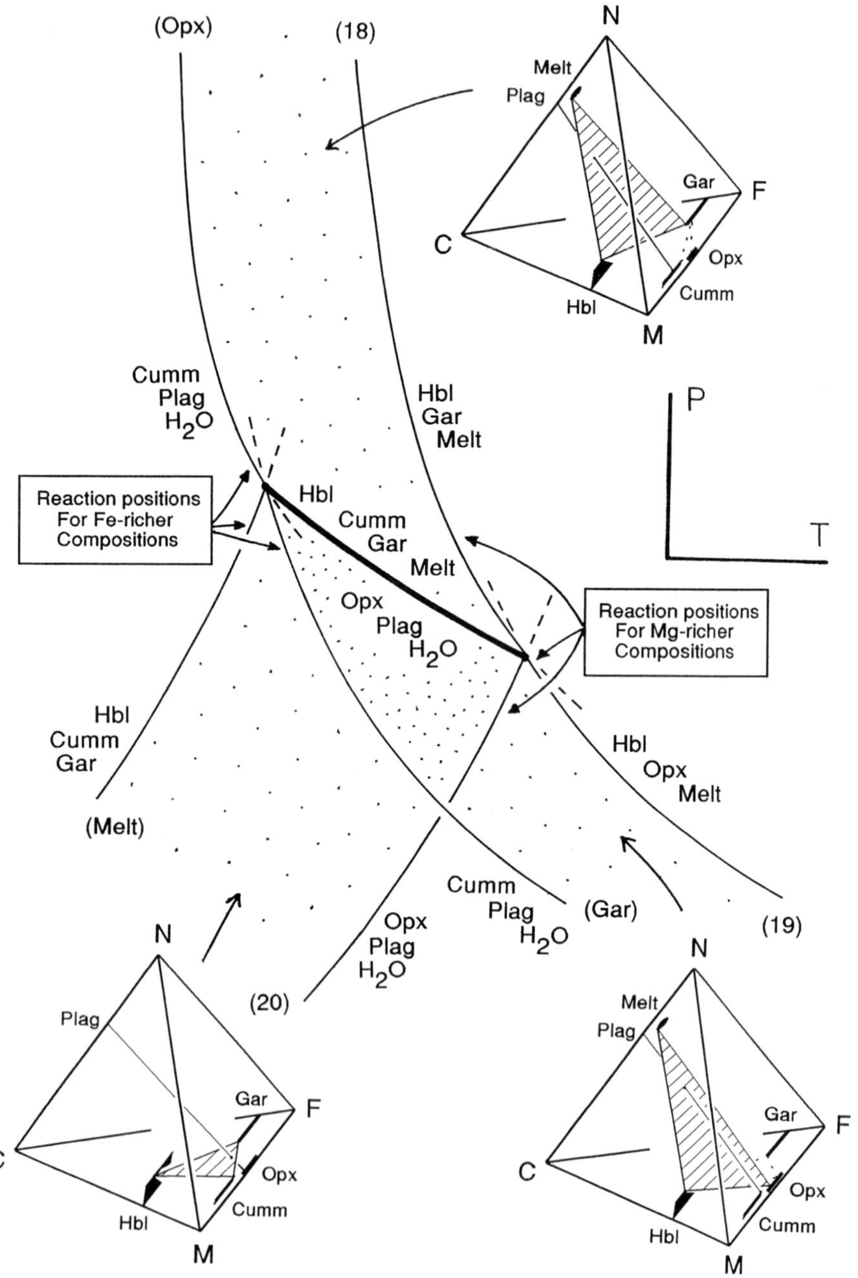
(Opx)
(18)
N
Melt
Plag
Gar
F
C
Opx
Cumm
Hbl
M
Cumm
Plag
H_2O
Hbl
Gar
Melt
P
T
Reaction positions
For Fe-richer
Compositions
Hbl
Cumm
Gar
Melt
Opx
Plag
H_2O
Reaction positions
For Mg-richer
Compositions
Hbl
Cumm
Gar
(Melt)
Hbl
Opx
Melt
Cumm
Plag
H_2O
(Gar)
(19)
Opx
Plag
H_2O
(20)
N
Plag
Gar
F
C
Opx
Hbl
Cumm
M
N
Melt
Plag
Gar
F
C
Opx
Cumm
Hbl
M

veins, small dykes, and pods of coarse-grained tonalitic rocks in amphibolite from Zone V. Two melting reactions involving the breakdown of cummingtonite have been suggested by Hollocher (Robinson *et al.* 1986). These were originally interpreted as fluid-absent reactions but it is now thought likely that melting involved the presence of a fluid phase, since the estimated P–T conditions at which these tonalitic melts developed coincide with the H_2O-saturated tonalite melting curve (Wyllie 1977). These reactions in simplified form are:

$$\text{cummingtonite} + \text{plagioclase} + H_2O = \text{hornblende} + \text{garnet} + \text{melt} \quad (9.18)$$

(metamorphic $T = 655°C$), and

$$\text{cummingtonite} + \text{plagioclase} + H_2O = \text{hornblende} + \text{orthopyroxene} + \text{melt} \quad (9.19)$$

(metamorphic $T = 670°C$). These melting reactions can be combined with Hollocher's (1985) reaction by which orthopyroxene first appears in ACFM plagioclase projection:

$$\text{cummingtonite} + \text{hornblende} + \text{garnet} + \text{quartz} = \text{orthopyroxene} + \text{plagioclase} + H_2O \quad (9.20)$$

to show some possible metamorphic phase relations of Zone V mafic rocks that involve silicate melt. On a P–T diagram (Fig. 9.16) both the Mg and Fe analogues of these three reactions intersect at points: the assemblage defined by each point is actually divariant due to the additional effects of Ca–Na solid solution in participating phases. In Figure 9.16 the reaction assemblages are schematically represented in projections from quartz and H_2O into Na_2O–CaO–FeO–MgO (NCFM) tetrahedra. Both melting reactions can occur within a restricted range of pressures.

In some parts of Zone VI, fluid-absent melting during hornblende breakdown appears to have caused a second, distinct episode of partial melting in

---

Figure 9.16 Hypothetical phase diagram for divariant melting reactions in cummingtonite-bearing amphibolites derived from a subsolidus reaction described by Hollocher (1985) and two melting reactions from Robinson *et al.* (1986). Thin lines show the relative positions of analogous sets of Fe and Mg reactions; reaction numbers from the text are shown in parentheses at the ends of the Mg reactions; the phases absent from the reactions are shown in parentheses at the ends of the Fe reactions. Several possible reaction curves have been left off for clarity. Stippled areas define the P–T areas of the divariant reactions; the more heavily stippled area shows the overlap of the two divariant reactions. The heavy line is a possible univariant Fe–Mg reaction. Na_2O–CaO–FeO–MgO (NCFM) tetrahedra schematically show reaction assemblages in the stippled fields. Abbreviations: Cumm, cummingtonite; Gar, garnet; Hbl, hornblende; Opx, orthopyroxene; Plag, plagioclase.

mafic rocks. A simplified version of the reaction postulated by Hollocher (in Robinson *et al.* 1986) is:

hornblende + quartz = orthopyroxene + clinopyroxene + plagioclase + melt (9.21)

(metamorphic $T = 705°C$).

9.8.2 Pelitic rocks

Tracy (1975, 1978) has discussed in detail the muscovite breakdown reactions and associated partial melting in Zone IV, and has pointed out the important rôle of Ca in both types of reactions. The sequence of reactions encountered with rising T for pelitic compositions at the pressures appropriate for central Massachusetts is illustrated in Figure 9.17. Melting begins in Na-rich bulk compositions by the reaction:

Na-bearing muscovite + Na-bearing plagioclase + quartz + H_2O
= sillimanite + K-richer muscovite + Ca-richer plagioclase + melt (9.22)

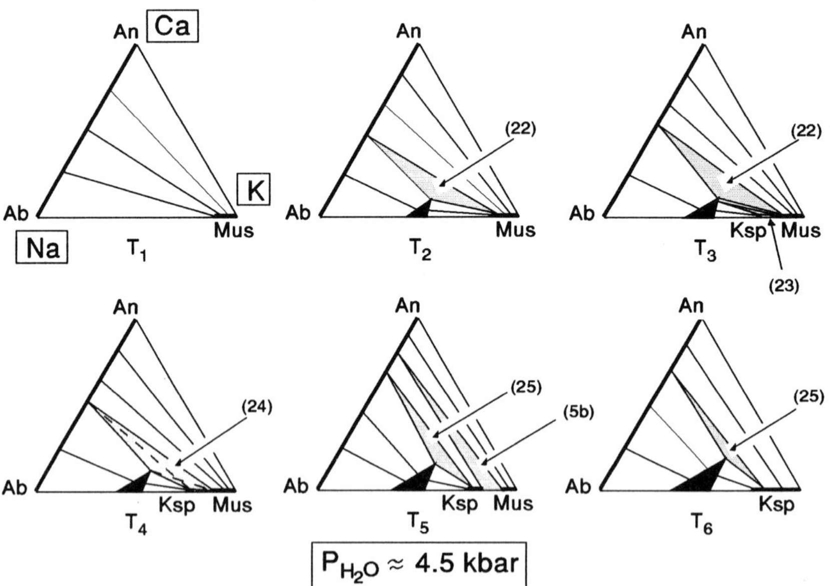

Figure 9.17 Projections from sillimanite, quartz, and H_2O onto the K–Na–Ca plane of a K–Na–Ca–Al tetrahedron that show muscovite dehydration and melting reactions with increasing temperature from T_1 to T_6 (from Tracy 1978). The roughly triangular, solid black areas are the melt compositions; heavy lines represent solid solution in phases; thin lines are tie lines; shaded areas show continuous and discontinuous reactions; numbers in parentheses correspond to numbered reactions (prefixed g.) in the text. Abbreviations: Ab, albite; An, anorthite; Ksp, K-feldspar, Mus, muscovite.

(Fig. 9.17, T_2). As melting progresses with increasing temperature, both the melt and plagioclase become more Ca-rich (Tracy 1978). At higher values of T in K_2O- and Na_2O-rich rocks, K-feldspar appears by the reaction:

$$\text{Na-bearing muscovite} + \text{quartz} + \text{melt} = \text{Na-bearing K-feldspar} + \text{sillimanite} \quad (9.23)$$

(Fig. 9.17, T_3), again with the melt becoming richer in Ca as T increases. Higher values of T lead to the discontinuous reaction:

$$\text{muscovite} + \text{quartz} + \text{melt} = \text{K-feldspar} + \text{plagioclase} + \text{sillimanite} + H_2O \quad (9.24)$$

(Fig. 9.17, T_4) (Tracy 1978). At higher values of T in Ca-rich rocks, continued muscovite breakdown occurs independent of melting via reaction (9.5b) (Fig. 9.17, T_5; see also Fig. 9.9, tetrahedron D) (Tracy 1978) and, in Na-rich rocks melting would continue via the granite melting reaction:

$$\text{Na-bearing plagioclase} + \text{Na-bearing K-feldspar} + \text{quartz} + H_2O = \text{melt} + \text{Ca-richer plagioclase} + \text{K-richer K-feldspar} \quad (9.25)$$

(Fig. 9.17, T_5 & T_6).

Fluid-absent melting involving the breakdown of biotite has been suggested for some garnet–cordierite gneisses from Zone VI (Robinson *et al.* 1986). In one sample studied in detail, the granite melting curves (K-feldspar + albite + quartz + H_2O = melt) and the stability curves for the six-phase assemblage quartz + orthoclase + sillimanite + garnet + biotite + cordierite from Holdaway & Lee (1977) were used to estimate P, T, and the X_{H_2O} in a fictive metamorphic fluid. A pressure of 5.2 kbar, temperature of 705°C, and an X_{H_2O} of 0.38 (fictive) were obtained: these pressures and temperatures are in reasonable agreement with other metamorphic estimates of 6.2 kbar (Tracy *et al.* 1976) and 685°C (Thompson 1976b).

The approximate P–T locations of the various melting episodes are summarized in Figure 9.18. With increasing T, melt textures appear first in pelitic rocks in Zone IV by melting reactions involving muscovite breakdown. Melting in Ca-richer pelitic compositions is delayed to higher temperatures (Fig. 9.18, stippled). Melting by the breakdown of cummingtonite in mafic rocks (Fig. 9.18, black) occurs at about Zone V conditions, which is above the stability of muscovite in pelitic rocks. Fluid-absent melting in pelitic rocks in Zone VI (Fig. 9.18, diagonal cross-hatching) constitutes a second and separate phase of melting experienced by pelitic rocks. Fluid-absent melting in mafic rocks took place in Zone VI during hornblende breakdown, producing a second distinct phase of melt textures (Fig. 9.18, axis-parallel cross hatching).

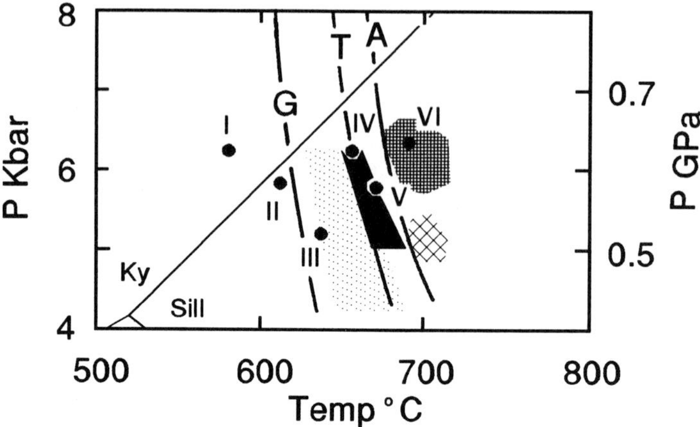

Figure 9.18 *P–T* diagram showing average conditions for the six metamorphic zones (after Robinson *et al.* 1982b), and showing the approximate fields of melting in various bulk compositions. Stippled area, melting in pelitic rocks accompanying dehydration of muscovite (Tracy 1978); black-filled area, melting in cummingtonite-bearing amphibolites that accompany cummingtonite dehydration (Hollocher in Robinson *et al.* 1986); diagonal cross-hatched area, possible fluid-absent melting in pelitic rocks caused by biotite dehydration; horizontal cross-hatched area, possible fluid-absent melting in mafic rocks caused by hornblende dehydration. Water-saturated melting curves for granite (G), tonalite (T), and amphibolite (A) (from Wyllie 1977) are given for comparison.

9.9 Summary

The minerals and textures found in metamorphic rocks formed as a result of complex interactions between *P*, *T*, time, rock composition, metamorphic fluid composition, and reaction kinetics. In general, rocks follow *P–T–t* paths during metamorphism that approximate the clockwise or counterclockwise trajectories shown in Figure 9.3. Because reactions between minerals are strongly temperature dependent, the dominant mineral assemblages commonly found in rocks that have undergone a single episode of metamorphism are those assemblages that formed under the highest temperature conditions that the rock experienced along its *P–T–t* path. Lower-temperature assemblages tend to react away quickly with rising temperature, and cooling of the partially or wholly dehydrated prograde assemblage following the highest-temperature conditions quickly quenches most reactions. Nonetheless, relict textures, minerals, minor retrograde reactions, partial melting, and mineral zoning can be used to establish *P–T–t* paths for certain areas, which can then be combined to reconstruct the dynamic history of a metamorphic belt.

In southwestern New Hampshire and central Massachusetts, the lower-grade Bronson Hill terrane (west) followed a clockwise *P–T–t* path, while the higher-grade Merrimack terrane (east) followed a counterclockwise *P–T–t* path. Geothermometry suggests that the temperature maxima (metamorphic field gradient) span about 150°C. A variety of metamorphic rock types, the equivalents of originally stratified sediments and volcanics, are found across this terrane. As a result, changes in mineral assemblages and mineral chemistry, as well as the advent of partial melting along the same *P–T* gradient can be correlated among the rock types. A firmer understanding of these relationships will enable better estimates of metamorphic conditions in regions with sparser outcrops or more monotonous rock types.

Acknowledgements

Much of the past and continuing research concerning metamorphism, stratigraphy, and structure in central Massachusetts and southwestern New Hampshire has been and is being undertaken by workers not listed as authors; some, but not all, of these workers are listed in the references, and their contributions are gratefully acknowledged. We would like to thank B. Harte and G. T. R. Droop for thorough and helpful reviews. Over the past 20 years, the research summarized here has been supported by various National Science Foundation Grants to P. Robinson and co-investigators.

References

Carmichael, D. M. 1970. Intersecting isograds in the Whetstone Lake area, Ontario. *Journal of Petrology* **11**, 147–81.

Chamberlain, C. P. & P. Karabinos 1987. Influence of deformation on pressure–temperature paths of metamorphism. *Geology* **15**, 42–4.

England, P. C. & S. W. Richardson 1977. The influence of erosion upon the mineral facies of rocks from different metamorphic environments. *Journal of the Geological Society of London* **134**, 201–13.

England, P. C. & A. B. Thompson 1984. Pressure–temperature–time paths of regional metamorphism I. Heat transfer during the evolution of regions of thickened continental crust. *Journal of Petrology* **25**, 894–928.

Greenwood, H. J. 1975. Thermodynamically valid projections of extensive phase relationships. *American Mineralogist* **60**, 1–8.

Guidotti, C. V. 1970. The mineralogy and petrology of the transition from the lower to the upper sillimanite Zone in the Oquossoc area, Maine. *Journal of Petrology* **11**, 277–336.

Guidotti, C. V. 1984. Micas in metamorphic rocks. In *Micas*, S. W. Bailey (ed.), 357–467. Mineralogical Society of America, Reviews in Mineralogy 13.

Guidotti, C. V., J. T. Cheney & S. Guggenheim 1977. Distribution of titanium between coexisting muscovite and biotite in pelitic schists from northwestern Maine. *American Mineralogist* **62**, 438–48.

Hall, D. J. 1970. *Compositional variations in biotites and garnets from kyanite and sillimanite zone mica schists, Orange area, Massachusetts and New Hampshire*. Contribution no. 4 (MS thesis), Department of Geology and Geography, University of Massachusetts, Amherst, 110pp.

Harte, B. & T. J. Dempster 1987. Regional metamorphic zones: tectonic controls. *Philosophical Transactions of the Royal Society of London* **A321**, 105–27.

Holdaway, M. J. 1971. Stability of andalusite and the aluminum silicate phase diagram. *American Journal of Science* **271**, 97–131.

Holdaway, M. J. & S. M. Lee 1977. Fe–Mg cordierite stability in high-grade pelitic rocks based on experimental, theoretical, and natural observations. *Contributions to Mineralogy and Petrology* **63**, 175–98.

Hollister, L. S. 1979. Metamorphism and crustal displacements: new insights. *Episodes* IUGS, 1979, 3–8.

Hollocher, K. T. 1985. *Geochemistry of metamorphosed volcanic rocks in Middle Ordovician Partridge Formation, and amphibole dehydration reactions in the high-grade metamorphic zones of central Massachusetts.* Contribution no. 56 (PhD thesis), Department of Geology and Geography, University of Massachusetts, Amherst, 275 pp.

Hudson, N. F. C. & B. Harte 1985. K_2O-poor, aluminous assemblages from the Buchan Dalradian, and the variety of orthoamphibole assemblages in aluminous bulk compositions in the amphibolite facies. *American Journal of Science* **285**, 224–66.

Raase, P. 1974, Al and Ti contents of hornblende, indicators of pressure and temperature of regional metamorphism. *Contributions to Mineralogy and Petrology* **45**, 231–6.

Rice, J. M. & J. M. Ferry 1981. Buffering, infiltration, and the control of intensive variables during metamorphism. In *Characterization of metamorphism through mineral equilibria*, J. M. Ferry (ed.), 263–326. Mineralogical Society of America, Reviews in Mineralogy 10.

Richardson, S. W. & P. C. England 1979. Metamorphic consequences of crustal eclogite production in overthrust orogenic zones. *Earth and Planetary Science Letters* **42**, 183–90.

Robinson, P. 1963. *Gneiss domes of the Orange area, west-central Massachusetts and New Hampshire.* PhD dissertation, Harvard University.

Robinson, P. 1967. Gneiss domes and recumbent folds of the Orange area, west-central Massachusetts. In *New England Intercollegiate Geology Conference, 59th Annual Meeting, Guidebook for field trips in the Connecticut Valley of Massachusetts*, 12–47.

Robinson, P. 1979. Bronson Hill anticlinorium and Merrimack synclinorium in central Massachusetts. In *The Caledonides in the USA: geological excursions in the northeast Appalachians*, J. W. Skeehan, S. J. & P. H. Osberg (eds), 126–50. IGCP Project 27, Caledonide Orogen. Weston Observatory, Weston, Massachusetts.

Robinson, P. & H. W. Jaffe 1969a. Aluminous enclaves in gedrite–cordierite gneiss from southwestern New Hampshire. *American Journal of Science* **267**, 389–421.

Robinson, P. & H. W. Jaffe 1969b. Chemographic exploration of amphibole assemblages from central Massachusetts and southwestern New Hampshire. In *Pyroxenes and amphiboles: crystal chemistry and phase petrology*, J. J. Papike (ed.), Mineralogical Society of America Special Paper **2**, 251–74.

Robinson, P., M. Ross & H. W. Jaffe 1971. Composition of the anthophyllite–gedrite series, comparisons of gedrite and hornblende, and the anthophyllite–gedrite solvus. *American Mineralogist* **56**, 1005–41.

Robinson, P., F. S. Spear, J. C. Schumacher, J. Laird, C. Klein, B. W. Evans, & B. L. Doolan 1982a. Phase relations of metamorphic amphiboles: natural occurrence and theory. In *Amphiboles: petrology and experimental phase relations*, D. R. Veblen & P. H. Ribbe (eds), 1–227. Mineralogical Society of America, Reviews in Mineralogy 9B.

Robinson, P. & R. J. Tracy 1979. Gedrite–cordierite, gedrite–orthopyroxene, and cummingtonite–orthopyroxene assemblages from ferromagnesian gneisses, sillimanite–K-feldspar zone, central Massachusetts. *Geological Society of America, Abstracts with Programs* **11**, 504.

Robinson, P., R. J. Tracy, K. T. Hollocher, & C. W. Dietsch 1982b. High grade Acadian regional metamorphism in south-central Massachusetts. In R. A. Joesten & S. S. Quarrier (eds),

Guidebook for fieldtrips in Connecticut and south-central Massachusetts: State Geological and Natural History Survey of Connecticut Guidebook 5, 289–339.

Robinson, P., R. J. Tracy, K. T. Hollocher, J. C. Schumacher & H. N. Berry IV 1986. The central Massachusetts metamorphic high. In P. Robinson & D. C. Elbert (eds) *Regional metamorphism and metamorphic phase relations in northwestern and central New England, Field Trip Guidebook (B-5), 14th General Meeting, International Mineralogical Association*, 195–234. Stanford, California, USA.

Robinson, P., R. J. Tracy & R. T. Tucker 1978. The Acadian (Devonian) metamorphic high of central Massachusetts. *Geological Society of America, Abstracts with Programs* **10**, 479–80.

Salje, E. 1986. Heat capacities and entropies of andalusite and sillimanite: the influence of fibrolitization on the phase diagram of the Al_2SiO_5 polymorphs. *American Mineralogist* **71**, 1366–71.

Sanders, I. S., J. S. Daly & G. R. Davies 1987. Late Proterozoic high-pressure granulite facies metamorphism in the north-east Ox inlier, north-west Ireland. *Journal of Metamorphic Geology* **5**, 69–85.

Schumacher, J. C. 1983. *Stratigraphic, geochemical and petrologic studies of the Ammonoosuc Volcanics, north-central Massachusetts and southwestern New Hampshire*. PhD dissertation, University of Massachusetts, Amherst, 237 pp.

Schumacher, J. C. 1988. Stratigraphy and geochemistry of the Ammonoosuc Volcanics, central Massachusetts and southwestern New Hampshire. *American Journal of Science* **288**, 619 63.

Schumacher, J. C. & P. Robinson 1987. Mineral chemistry and metasomatic growth of aluminous enclaves in gedrite–cordierite gneiss, southwestern New Hampshire, USA. *Journal of Petrology* **28**, 1033 73.

Schumacher, J. C., R. Schumacher & P. Robinson 1989. Acadian metamorphism in central Massachusetts and southwestern New Hampshire: evidence for contrasting *P–T* trajectories. In *Evolution of metamorphic belts*, J. S. Daly, R. A. Cliff & B. W. D. Yardley (eds), Geological Society of London, Special Publication, in press.

Schumacher, R. 1986. *Petrology and geochemistry of epidote- and clinopyroxene-bearing amphibolites and calc-silicate rocks from central Massachusetts, USA*. PhD dissertation, University of Bonn, FRG, 273 pp.

Sleep, N. H. 1979. A thermal constraint on the duration of folding with reference to Acadian geology, New England (USA). *Journal of Geology* **87**, 583–9.

Spear, F. S. 1980. The gedrite–anthophyllite solvus and the composition limits of orthoamphibole from the Post Pond Volcanics, Vermont. *American Mineralogist* **65**, 1103–18.

Spear, F. S., J. M. Ferry, & D. Rumble III 1982. Analytical formulation of phase equilibria: the Gibbs' method. In *Characterization of metamorphism through mineral equilibria*, J. M. Ferry (ed.), 105–52. Mineralogical Society of America, Reviews in Mineralogy 10.

Spear, F. S. & D. Rumble III 1986. Pressure, temperature, and structural evolution of the Orfordville belt, west-central New Hampshire. *Journal of Petrology* **27**, 1071–94.

Spear, F. S. & J. Selverstone 1983. Quantitative *P–T* paths from zoned minerals: theory and tectonic applications. *Contributions to Mineralogy and Petrology* **83**, 348–57.

Thompson, A. B. 1976a. Mineral reactions in pelitic rocks: I. Prediction of *P–T–X* (Fe–Mg) phase relations. *American Journal of Science* **276**, 401–24.

Thompson, A. B. 1976b. Mineral reactions in pelitic rocks: II. Calculation of some *P–T–X* (Fe–Mg) phase relations. *American Journal of Science* **276**, 425–54.

Thompson, A. B. & P. C. England 1984. Pressure–temperature–time paths of regional metamorphism II. Their inference and interpretation using mineral assemblages in metamorphic rocks. *Journal of Petrology* **25**, 929–55.

Thompson, A. B., P. T. Lyttle, & J. B. Thompson Jr 1977. Mineral reactions and A–Na–K and

A–F–M facies types in the Gassetts schist, Vermont. *American Journal of Science* **277**, 1124–51.

Thompson, J. B., Jr 1957. The graphical analysis of mineral assemblages in pelitic schists. *American Mineralogist* **42**, 842–58.

Tracy, R. J. 1975. *High grade metamorphic reactions and partial melting in pelitic schist, Quabbin Reservoir area, Massachusetts.* Contribution No. 20 (PhD thesis), Department of Geology and Geography, University of Massachusetts, Amherst, 127 pp.

Tracy, R. J. 1978. High grade metamorphic reactions and partial melting in pelitic schist, west-central Massachusetts. *American Journal of Science* **278**, 150–78.

Tracy, R. J. & P. Robinson 1988. Silicate–sulfide–oxide–fluid reactions in granulite-grade pelitic rocks, central Massachusetts. *American Journal of Science* **288-A**, 45–74.

Tracy, R. J., P. Robinson, & A. B. Thompson 1976. Garnet composition and zoning in the determination of temperature and pressure of metamorphism, central Massachusetts. *American Mineralogist* **61**, 762–75.

Wyllie, P. J. 1977. Crustal anatexis: an experimental review. *Tectonophysics* **43**, 41–71.

CHAPTER TEN

Granulite facies metamorphism of metabasic and intermediate rocks in the Highland Series of Sri Lanka

Renate Schumacher, Volker Schenk, Peter Raase & P. W. Vitanage

10.1 Introduction: why Sri Lanka?

The island of Sri Lanka is located to the south-east of India and is formed predominantly of Precambrian rocks that occur in several major terrains: the granulite facies Highland Series in the central part of the island, from which several authors separated a South West Group (e.g. Cooray 1965; Katz 1971, 1972); and the West and East Vijayan, both of which are predominantly composed of migmatitic granitic gneisses of middle to upper amphibolite facies conditions (Cooray 1962, Katz 1971, Oliver 1985) and border the Highland Series (Fig. 10.1). Due to its variety of rock types, which include granitic to basic meta-igneous rocks and metasediments (quartzites, metapelites, calc-silicate rocks, marbles), and due to the juxtaposition of amphibolite and granulite facies terranes in a relatively small area, Sri Lanka has been and still is the subject of petrological, geochronological, and structural interest.

This petrological investigation of the Highland Series should give answers to the following questions:

(1) Were the Highland Series rocks subjected to pressures and temperatures characteristic of the lower crust?
(2) Are there any metamorphic gradients within the Highland Series?
(3) What kind of reaction textures occur and what information do they give about the *P–T–t* path?

The rock types most likely to help resolve these questions are the metapelites (Raase & Schenk 1988), charnockites, and garnet–pyroxene-bearing intermediate to basic rocks (Schumacher & Schenk 1988). This chapter focuses

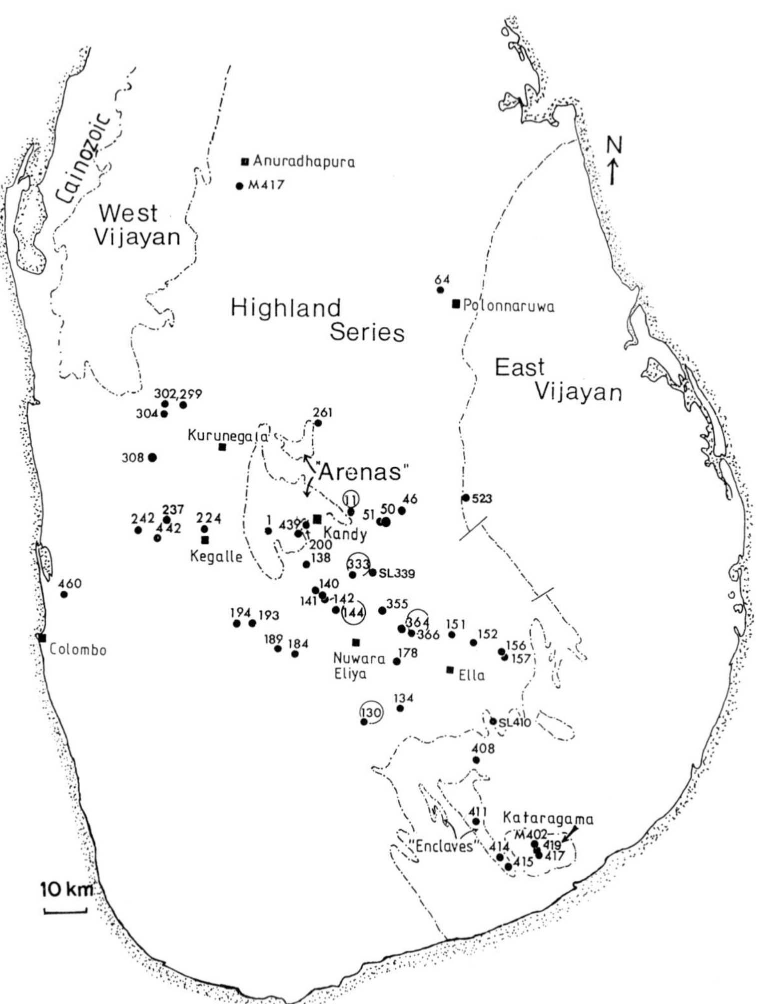

Figure 10.1 Simplified map of the lithological–tectonic subdivisions of Sri Lanka, with the three major Precambrian units, the West Vijayan, East Vijayan, and Highland Series, the Kataragama area, and the 'Arena' area around Kandy, based on the Geological Map of Sri Lanka (1982). In addition, sample localities of samples discussed in the text are shown. Sample numbers without a prefix are prefixed with a K in the text, figures, and tables.

on the petrology and *P–T* evolution of the intermediate to basic garnet–pyroxene-bearing granulite facies rocks.

10.2 Previous and recent work

Early work on Sri Lanka was directed toward distinguishing the main metamorphic terranes, which are the Highland Series, the South West Group, the

East Vijayan, and the West Vijayan, and determining their major lithological differences (e.g. Adams 1929; Cooray 1962, 1978; Hapuarachchi 1975). In particular, the relation between the Highland Series and the Vijayan was subject of discussions; one group of scientists regarded the Vijayan as the older basement of the Highland Series (e.g. Coates 1935; Katz 1971, 1972; Munasinghe & Dissanayake 1979, 1980a,b), while the other group interpreted the Vijayan as younger (Vitanage 1959) or as the product of retrograde metamorphism of the Highland Series (Cooray 1978, Crawford & Oliver 1969, Berger 1973, Berger & Jayasinghe 1976). The boundary of the Highland Series and the East Vijayan is, in part, sheared and tectonized (Vitanage 1959, Sandiford *et al.* 1988) and is the location of serpentinite bodies (Dissanayake & Van Riel 1978). Gravity measurements support the fault-character of the Highland Series/East Vijayan boundary (Hatherton *et al.* 1975). One plate tectonic model suggests that the boundary represents a convergent plate suture (Munasinghe & Dissanayake 1979, 1980a, 1982). On the basis of differences that are predominantly lithological, Cooray (1965) separated the South West Group from the Highland Series. According to Katz (1972) the South West Group is characterized by the occurrence of thick quartzite units, the presence of wollastonite in the marbles, and the occurrence of cordierite and of coexisting andalusite–sillimanite. He suggested that the Highland Series was subjected to an intermediate-pressure type of metamorphism, while the South West Group appeared to have experienced a lower-pressure metamorphism (Katz 1971, 1972). Perera (1984) described coexisting cordierite–garnet assemblages from the Highland Series and the South West Group, and pointed out the chemical control on this assemblage rather than a low-pressure origin. He thus doubted the existence of a paired metamorphic belt as suggested by Katz (1971, 1972). Similarities also exist in metamorphic evolution (Welberts & Raith 1988, Schumacher & Schenk 1988). Therefore it is questionable whether the separation of a South West Group is meaningful.

Ages of formation of the various rock types and the time of metamorphism have also been a matter of debate (Crawford 1969, Crawford & Oliver 1969, Kröner *et al.* 1987, Milisenda *et al.* 1988, Hölzl *et al.* 1988). A recent ion microprobe study, on single zircons from a quartzite and a garnet–biotite gneiss from the northern Highland Series near Polonnaruwa, and from a garnet–biotite gneiss from the southeastern part of the Highland Series, suggests that the high-grade metamorphism took place at 1100 Ma, and that later events occurred between 1000 and 500 Ma (Kröner *et al.* 1987). Recent age determinations with the U–Pb, Sm–Nd, and Rb–Sr methods indicate that granite plutonism in the Highland Series occurred at 560–600 Ma, and that associated *in situ* charnockite formation must be younger (Hölzl *et al.* 1988). Zircons of an orthogneiss from south of Ella (sample SL 410, Fig. 10.1) and zircons and monazites of a paragneiss from near Nuwara Eliya (sample SL 339, Fig. 10.1) suggest that the high-temperature metamorphism ended at *c.* 560 Ma (Hölzl *et al.* 1988). Rb–Sr biotite-whole-rock ages of several samples range from 405 to 470 Ma (Hölzl *et al.* 1988), and represent either a later regional event or slow cooling after tectonic uplift from the lower-crustal level (Schenk *et al.* 1988).

The difference between the Highland Series (including the South West Group) and the orthogneisses of the Vijayan Complex is supported by the results of Nd isotopic studies which give Nd model ages of 2.2–3.0 Ga for the Highland Series and the South West Group, and 1.0–1.8 Ga for the East and West Vijayan (Milisenda *et al.* 1988). According to these results the Vijayan Complex represents a separate younger-crustal terrane, and is not the product of retrograde metamorphism and migmatization of former Highland Series rocks. In addition, the Nd studies suggest that the South West Group and the Highland Series have similar crustal residence ages.

Due to the occurrence of orthopyroxene–plagioclase symplectites around garnet porphyroblasts (Coates 1935, Cooray 1961), two major periods of metamorphism have been distinguished within the Highland Series (Katz 1971, Hapuarachchi 1975, Jayawardena & Carswell 1976, Cooray 1978, Sandiford *et al.* 1988). The characteristic symplectite reaction texture, which suggests re-equilibration at relatively low pressures, is interpreted as the result of the erosion of a tectonically thickened crust or a separate Pan African thermal perturbation (Sandiford *et al.* 1988). In both cases, Sandiford *et al.* (1988) see in the symplectite stage a near-isothermal (uplift) event which followed the prograde metamorphism in the lower-crustal level. Reaction textures indicating isobaric cooling in the lower crust prior to tectonic uplift (Schenk *et al.* 1988, Schumacher & Schenk 1988) contradict Sandiford *et al.*'s (1988) interpretation.

10.3 Rock types

Within all three Precambrian terranes, the Highland Series, the East and the West Vijayan, granitic to dioritic orthogneisses constitute the common rock types. The West Vijayan is composed chiefly of granitoid gneisses and migmatites, with inclusions of biotite and hornblende gneisses, locally containing clinopyroxene (Cooray 1978). The majority of the samples were collected along a north-west–south-east profile about 200 km in length, extending from the West Vijayan via Kurunegala, Kandy and Nuwara Eliya to Kataragama and along an east–west profile between Colombo and Kandy (locations shown in Fig. 10.1). So far, no orthopyroxene has been found in the West Vijayan area, as it is delineated in this chapter on the basis of the Geological Map of Sri Lanka (1982). In contrast, granulite facies assemblages with garnet and orthopyroxene occur in 'enclaves' within the East Vijayan to the north-west of Kataragama (Fig. 10.1, samples K 408, 411 & 415), the tectonic significance of which so far has not been resolved. Orthopyroxene-bearing assemblages have been found at locality K 523 (Fig. 10.1) in the East Vijayan. The granulite facies Kataragama area (Fig. 10.1) is interpreted as a Highland Series nappe resting on the East Vijayan (Kröner *et al.* 1987, Sandiford *et al.* 1988).

In the north-west part of the Highland Series around Kurunegala (Fig. 10.1) orthopyroxene-bearing gneisses were subjected to a late-stage orthopyroxene formation which is discordant to the regional foliation, whereas massive charnockites have been found in most exposures in the south-east part. Metasediments are closely associated with the charnockitic gneisses (e.g. Cooray 1978).

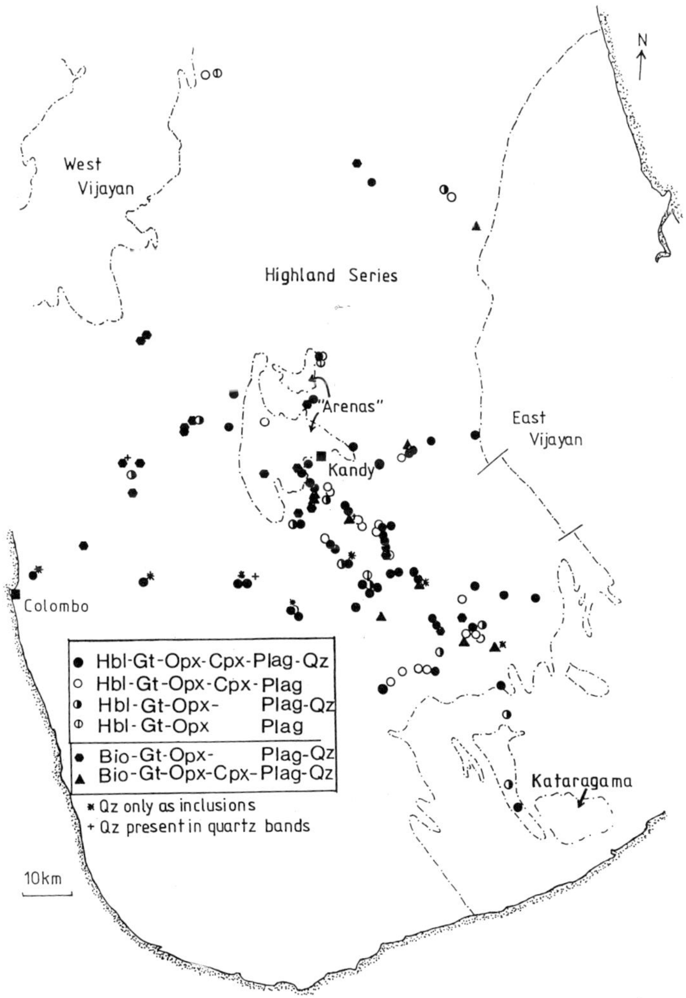

Figure 10.2 Distribution of biotite–garnet–orthopyroxene and hornblende–garnet–orthopyroxene-bearing assemblages in metabasic and intermediate rocks from the Highland Series granulite facies 'enclaves' within the East Vijayan and the Kataragama area, which is interpreted as a Highland Series nappe.

Within the granulite facies Highland Series, metabasites and hornblende-bearing gneisses are common in the central part and in the south-east (Fig. 10.2), whereas metapelites, marbles, and quartzites are subordinate. In the west and north-west parts of the Highland Series, metabasites are extremely rare. The 'Arenas' form an area with open synformal folds (e.g. Berger & Jayasinghe 1976, Vitanage 1985, Sandiford *et al.* 1988) and with hornblende gneisses as the predominant rock type.

Mineral assemblages (for abbreviations see Table 10.1) in the tonalitic to basic rocks of the granulite facies Highland Series terrane including the 'Arena' area (Fig. 10.1) are listed in Table 10.2.

Opaque phases are ilmenite, magnetite, and less commonly, Fe-sulphide. Apatite, zircon, monazite, and allanite may occur as accessory phases. The distribution of the garnet- and orthopyroxene-bearing mineral assemblages within the Highland Series is shown in Figure 10.2. Hornblende–garnet–orthopyroxene–clinopyroxene–quartz-bearing assemblages are common in the south-east part of the Highland Series and are very rare in the north and west parts and in the 'Arena' near Kandy, indicating a decrease in metamorphic grade towards the west (Fig. 10.2). Hornblende-free metabasites and intermediate rocks are less common than hornblende-bearing metabasites, but occur also in the north and in the 'Arena' area (Fig. 10.2).

Table 10.1 Abbreviations used in the text and figures.

Amph	=	amphibole	Qz	=	quartz
And	=	andalusite	R	=	rim
Bio	=	biotite	Sec	=	secondary
C	=	core	Sil	=	sillimanite
C/R	=	core-rim transition area	Sph	=	titanite
			Sympl.	=	symplectite
Chl	=	chlorite	T	=	temperature
Cord	=	cordierite	Vij	=	Vijayan
Cpx	=	clinopyroxene	W	=	west
E	=	east	X_{al}	=	almandine component in garnet
Gross	=	grossular			
Gt	=	garnet	X_{an}	=	anorthite component in plagioclase
Hbl	=	hornblende			
Ilm	=	ilmenite	X_{en}	=	enstatite component in orthopyroxene
K	=	Kataragama			
Kf	=	K-feldspar	X_{gr}	=	grossular component in garnet
Ky	=	kyanite			
Mt	=	magnetite	X_{py}	=	pyrope component in garnet
Opx	=	orthopyroxene			
P	=	pressure			
Plag	=	plagioclase			

Table 10.2 Mineral assemblages in the tonalitic to basic metamorphosed rocks of the Highland Series.

Hbl	- Gt	- Opx	- Cpx	- Plag	- Qz	± Kf	± Bio	± Opaques
Hbl	- Gt	- Opx	- Cpx	- Plag		± Kf	± Bio	± Opaques
Hbl	- Gt	- Opx		- Plag	- Qz	± Kf	± Bio	± Opaques
Hbl	- Gt	- Opx		- Plag		± Kf	± Bio	± Opaques
Hbl	- Gt		- Cpx	- Plag	- Qz	± Kf	± Bio	± Opaques
Hbl	- Gt		- Cpx	- Plag		± Kf	± Bio	± Opaques
Hbl	- Gt			- Plag	- Qz	± Kf	± Bio	± Opaques
Hbl	- Gt			- Plag		± Kf	± Bio	± Opaques
Hbl		- Opx	- Cpx	- Plag	- Qz	± Kf	± Bio	± Opaques
Hbl		- Opx	- Cpx	- Plag		± Kf	± Bio	± Opaques
Hbl		- Opx		- Plag	- Qz	± Kf	± Bio	± Opaques
Hbl			- Cpx	- Plag	- Qz	± Kf	± Bio	± Opaques
Hbl			- Cpx	- Plag		± Kf	± Bio	± Opaques
Hbl				- Plag	- Qz	± Kf	± Bio	± Opaques
Hbl				- Plag		± Kf	± Bio	± Opaques
	Gt	- Opx	± Cpx	- Plag	- Qz	± Kf	± Bio	± Opaques
		Opx	- Cpx	- Plag	- Qz	± Kf	± Bio	± Opaques

10.4 Reaction textures

Within a large number of metabasites, porphyroblasts of orthopyroxene, clinopyroxene, and plagioclase up to several mm in size belong to the highest temperature mineral assemblage. In the Mg-richer metabasites the orthopyroxenes may have fine (100) clinopyroxene lamellae and, more commonly, the clinopyroxenes may exhibit several generations of fine exsolution lamellae parallel to '100'. Part of these high-temperature pyroxenes recrystallized to a finer-grained matrix. At a few localities (Anuradhapura, to the west of Kataragama and in the Kataragama area, samples M 417, K 414 & K 419 in Fig. 10.1) Fe-rich orthopyroxenes (inverted pigeonites) in rocks of granitic composition exhibit coarse clinopyroxene-exsolution lamellae parallel to '001'. The host as well as the coarse exsolution lamellae contain finer exsolution lamellae parallel to '100'.

During further cooling of the orthopyroxene-bearing metabasites, reaction rims of garnet and clinopyroxene or garnet–quartz symplectites and smaller amounts of clinopyroxene formed around orthopyroxene and plagioclase porphyroblasts (Figs 10.3A & B). These reaction rims display a variety of grain sizes ranging from several mm (e.g. sample K 333) to several μm (e.g. GK-364, Fig. 3B). In general, fine-grained clinopyroxene occurs next to garnet–quartz intergrowths around orthopyroxene; less common are clinopyroxene rims around orthopyroxene (Fig. 10.3A) or clinopyroxene–garnet symplectites. In some samples with large garnets that may have been part of the high-temperature assemblage, secondary garnet overgrowths occur (Fig. 10.3C; 'garnet growth' in Tables 10.3 & 4). These reaction textures indicate that a garnet–clinopyroxene-forming reaction has occurred, which can be represented by the simplified reaction:

$$2 \text{ orthopyroxene} + 1 \text{ anorthite} = 1 \text{ garnet} + 1 \text{ clinopyroxene} + 1 \text{ quartz}$$

The larger molar volume of garnet compared to that of clinopyroxene is mirrored by the predominance of modal garnet compared to clinopyroxene in the reaction rims.

After the garnet-forming reaction the garnet–clinopyroxene assemblages became unstable again, as indicated by orthopyroxene–plagioclase rims and orthopyroxene–plagioclase ± magnetite symplectites around garnet porphyroblasts (Fig. 10.3D) and around fine-grained garnet rims (Fig. 10.3B). In addition, orthopyroxene rims around clinopyroxene are common in metabasites throughout a large part of the Highland Series. So far, these reaction textures have been found in the central, eastern, and southeastern parts of the Highland Series, but not in the western part of the Highland Series between Kegalle and Columbo (Fig. 10.1) and only in one sample from the 'Arena' area (sample K 439–2, Fig. 10.1). Both the earlier garnet-forming and the later garnet-breakdown reactions can be observed in only a few metabasite samples, but these are distributed over a large area which covers the central, east, and south-east Highland Series (e.g. samples K 11, K 130, K 144, K 333 & K 364, circled sample numbers in Fig. 10.1).

Different types of garnet- and pyroxene-forming reactions involve the breakdown of Ti-rich hornblende porphyroblasts or biotites or both.

Figure 10.3 (A) Photomicrograph of garnet–quartz symplectites and thin clinopyroxene rim around orthopyroxene porphyroblast in a metatonalite (sample K 366). Nicols at an oblique angle. (B) Photomicrograph of fine-grained garnet–quartz intergrowth around plagioclase porphyroblast in a metabasite (plane-polarized light). Narrow reaction rims of new plagioclase occur next to garnet, and those of orthopyroxene occur around clinopyroxene porphyroblast (sample GK 364). (C) Photomicrograph of a garnet porphyroblast with garnet–quartz overgrowth (sample K 366). Nicols at an oblique angle. (D) Photomicrograph of orthopyroxene–plagioclase symplectites around garnet porphyroblasts and of orthopyroxene rims around clinopyroxene in a metabasite (sample K 144, plane-polarized light).

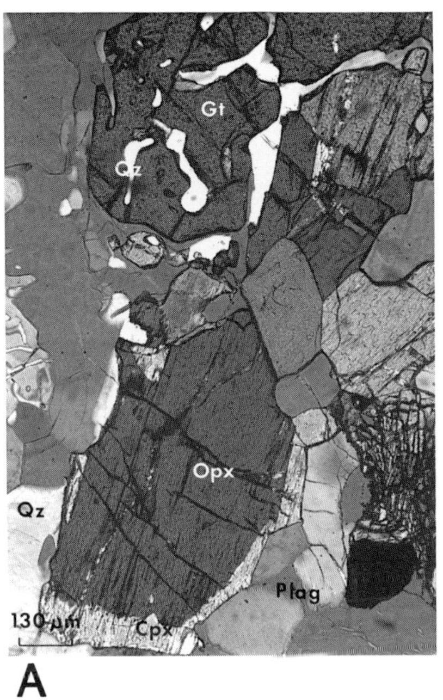
Gt
Qz
Opx
Qz
Plag
Cpx
130 μm
A

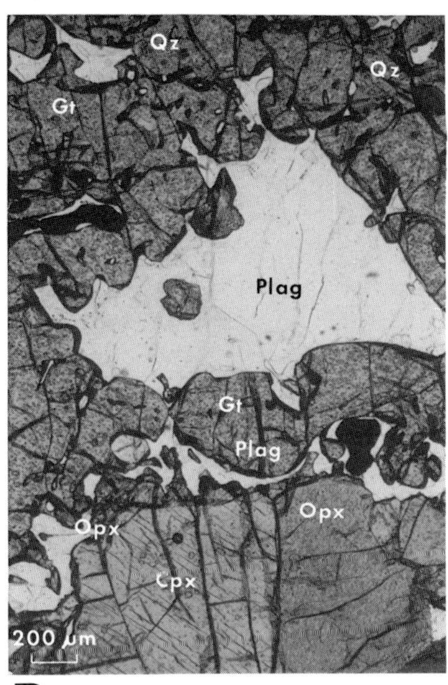
Qz
Qz
Gt
Plag
Gt
Plag
Opx
Opx
Cpx
200 μm
B

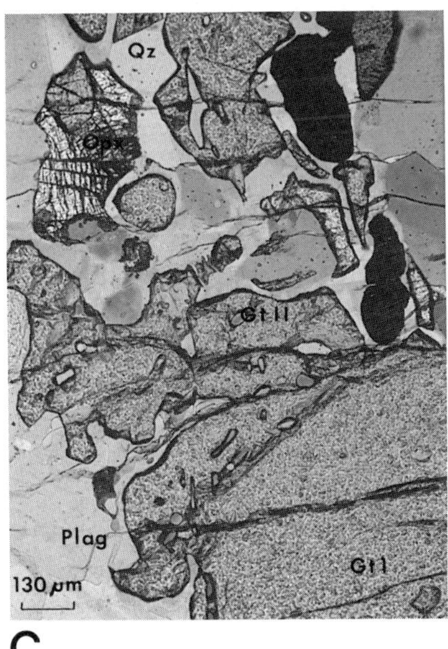
Qz
Opx
Gt II
Plag
Gt I
130 μm
C

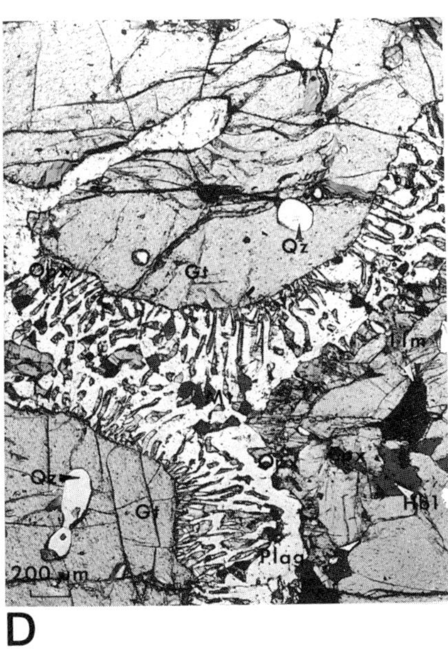
Qz
Gt
Ilm
Qz
Gt
Plag
Hbl
200 μm
D

Table 10.3 Microprobe analyses of minerals from metabasic to intermediate rocks from Sri Lanka. Analyses are recalculated on the basis of the following stoichiometric assumptions: garnet – 8 cations, 12 oxygens; orthopyroxene – 4 cations, 6 oxygens; clinopyroxene – 4 cations, 6 oxygens; plagioclase – 8 oxygens; ilmenite – 2 cations, 3 oxygens. Localities of samples are given in Figure 10.1. 'Gt growth' refers to the metamorphic stage 4; 'sympl.' refers to the symplectitic stage 5 discussed in the text.

Sample	K 1-5			K 50-1				K 130-1				K 130-1	
Analysis No.	Gt 14	Opx 31	Plag 38	Gt 43	Opx 54	Plag 57	Cpx 47	Gt 10	Cpx 24	Opx 2	Plag 20	Gt 16	Cpx 21
Remarks	Cores			Cores				Cores				Gt growth	
SiO_2	38.71	51.24	55.46	37.75	49.71	59.26	50.45	38.10	50.67	50.41	51.18	38.06	50.37
TiO_2	0.02	0.08	-	0.05	0.11	-	0.24	0.06	0.32	0.09	-	0.10	0.38
Al_2O_3	21.35	2.43	28.43	20.52	1.04	26.34	1.80	20.89	3.35	1.65	29.52	21.34	3.51
Cr_2O_3	-	0.03	-	-	0.04	0.04	-	-	0.02	-	0.05	0.02	-
FeO	25.55	23.64	0.10	30.46	35.71	0.17	19.49	26.89	10.67	28.88	0.15	27.87	13.72
MnO	3.83	1.42	-	0.93	0.46	0.03	0.26	0.79	0.17	0.39	-	1.21	0.13
MgO	7.21	19.94	-	3.50	12.53	-	9.70	6.16	11.71	17.58	-	4.82	11.78
CaO	3.45	0.30	10.51	6.97	0.86	8.59	18.33	6.80	22.07	0.64	13.80	6.78	19.26
Na_2O	-	-	5.30	-	-	6.22	0.29	-	0.45	0.02	3.65	-	0.47
K_2O	-	-	0.21	-	-	0.57	-	-	-	-	0.21	-	-
Sum	100.12	99.08	100.01	100.18	100.46	101.22	100.56	99.69	99.43	99.66	98.56	100.20	99.62
Structural formulae													
Si	3.008	1.950	2.496	2.993	1.971	2.621	1.942	2.977	1.904	1.942	2.364	2.983	1.910
Al^{IV}	-	0.050	1.507	-	0.029	1.373	0.058	-	0.096	0.058	1.607	-	0.090
Al	1.955	0.060	-	1.917	0.020	-	0.024	1.923	0.067	0.017	-	1.971	0.067
Ti	0.001	0.003	-	0.003	0.003	-	0.007	0.004	0.012	0.003	-	0.006	0.011
Cr	-	-	-	-	0.001	0.001	-	-	-	-	0.002	0.001	-
Fe^{3+}	0.026	-	-	0.092	0.002	-	0.043	0.116	0.045	0.038	-	0.051	0.036
Mg	0.835	1.122	-	0.414	0.740	-	0.556	0.717	0.637	1.010	-	0.563	0.666
Fe^{2+}	1.635	0.759	0.004	1.928	1.181	0.006	0.586	1.641	0.304	0.893	0.006	1.776	0.399
Mn	0.252	0.014	-	0.062	0.015	0.001	0.008	0.052	0.005	0.013	-	0.080	0.004
Ca	0.287	0.041	0.507	0.592	0.037	0.407	0.756	0.569	0.893	0.026	0.683	0.569	0.783
Na	-	0.001	0.463	-	-	0.533	0.022	-	0.038	0.001	0.327	-	0.035
K	-	-	0.012	-	-	0.032	-	-	-	-	0.012	-	-
Sum Cations	7.999	4.000	4.989	8.001	3.999	4.974	4.002	7.999	4.001	4.001	5.001	8.000	4.001

Sample	K 130-1			K 134-1				K 134-1			K 144-4	
Analysis No.	Gt 50	Opx 55	Plag 46	Gt 46	Opx 81	Plag 65	Cpx 68	Gt 18	Cpx 34	Ilm 1	Opx 70	Cpx+
Remarks	Rim	Sympl.*		Cores				Hbl breakdown			Cores	
SiO_2	37.78	50.84	49.52	38.30	50.47	58.94	51.21	37.91	51.90	-	50.43	49.90
TiO_2	0.08	0.06	0.01	0.18	0.10	0.02	0.38	-	0.19	52.22	0.10	0.44
Al_2O_3	20.91	1.09	32.08	21.33	0.82	26.93	2.14	21.34	1.56	0.04	1.91	3.64
Cr_2O_3	-	-	0.06	-	0.03	0.02	-	0.02	0.01	-	0.02	0.01
FeO	28.32	29.43	0.28	27.82	31.59	0.06	13.55	29.20	11.81	46.02	27.57	12.73
MnO	1.14	0.42	0.06	0.81	0.38	0.02	-	1.22	0.19	0.36	0.36	0.21
MgO	4.53	17.01	-	5.35	16.16	-	11.88	4.02	12.00	0.99	18.13	11.33
CaO	6.87	0.51	15.39	7.04	0.66	8.72	19.94	7.01	22.24	0.12	0.72	21.42
Na_2O	-	-	2.72	-	0.01	6.36	0.42	-	0.33	-	0.03	0.40
K_2O	-	-	0.14	-	-	0.27	-	-	0.01	-	-	-
Sum	99.63	99.36	100.26	100.83	100.22	101.34	99.52	100.72	100.24	99.75	99.27	100.08
Structural formulae												
Si	2.985	1.973	2.260	2.974	1.958	2.602	1.947	2.972	1.953	-	1.937	1.884
Al^{IV}	-	0.027	1.725	-	0.037	1.401	0.053	-	0.047	-	0.063	0.116
Al	1.947	0.023	-	1.951	-	-	0.043	1.972	0.022	0.001	0.013	0.046
Ti	0.005	0.002	-	0.011	0.003	0.001	0.011	-	0.004	0.985	0.004	0.012
Cr	-	-	0.002	-	0.001	0.001	-	0.001	0.001	-	0.001	-
Fe^{3+}	0.073	-	-	0.080	0.041	-	0.019	0.083	0.021	0.028	0.045	0.075
Mg	0.534	0.984	-	0.619	0.935	-	0.673	0.470	0.684	0.037	1.052	0.638
Fe^{2+}	1.798	0.956	0.011	1.726	0.985	0.002	0.412	1.832	0.342	0.937	0.839	0.327
Mn	0.076	0.014	0.002	0.053	0.012	0.001	-	0.081	0.004	0.008	0.011	0.007
Ca	0.582	0.021	0.753	0.585	0.027	0.412	0.811	0.589	0.904	0.003	0.031	0.866
Na	-	-	0.240	-	0.001	0.536	0.031	-	0.018	-	0.004	0.029
K	-	-	0.008	-	-	0.020	-	-	-	-	-	-
Sum Cations	8.000	4.000	5.001	7.999	4.000	4.976	4.000	8.000	4.000	1.999	4.000	4.000

* Sympl. = symplectites

+ reintegrated composition (lamellae and host)

Table 10.3 (*Continued*)

Sample	K 144-4				K 144-4				K 200			K 200		
Analysis No.	Gt 47	Opx 73	Plag63	Cpx 14	Gt 31	Opx 19	Plag30	Cpx 6	Gt 11	Opx 60	Plag 32	Gt 24	Opx 25	Plag 19
Remarks	Gt growth				Rim	Sympl.*		Rim	Cores			Rim	Sympl.*	
SiO_2	38.60	50.47	56.73	49.49	38.11	50.54	46.94	50.87	37.97	49.52	54.00	37.98	50.50	52.89
TiO_2	0.05	0.06	0.03	0.44	0.10	0.05	-	0.31	0.13	0.11	0.01	0.08	0.04	-
Al_2O_3	21.02	1.34	26.67	3.55	20.75	1.43	33.11	2.76	20.91	1.13	28.24	20.50	1.24	28.93
Cr_2O_3	-	0.04	0.03	-	0.03	0.01	-	0.04	-	-	-	-	0.07	-
FeO	26.14	28.01	0.20	13.45	28.23	28.85	0.42	12.11	24.28	31.97	0.26	28.98	31.58	0.35
MnO	0.73	0.39	0.03	0.21	1.62	0.51	0.02	0.19	0.49	0.74	-	0.92	0.53	-
MgO	7.03	18.59	0.01	11.15	4.83	17.45	-	11.61	4.25	15.63	-	3.55	15.93	-
CaO	6.70	0.54	9.65	20.72	6.81	0.51	17.37	21.87	11.87	0.68	11.70	7.82	0.62	12.45
Na_2O	-	0.03	5.55	0.13	-	-	1.68	0.46	-	-	4.54	-	0.01	4.25
K_2O	-	-	0.72	-	-	-	0.10	-	-	-	0.35	-	-	0.29
Sum	100.27	99.47	99.62	99.14	100.48	99.35	99.64	100.22	99.90	99.78	99.10	99.83	100.52	99.16
Structural formulae														
Si	2.982	1.942	2.563	1.894	2.985	1.956	2.169	1.914	2.963	1.936	2.464	3.012	1.955	2.420
Al^{IV}	-	0.058	1.420	0.106	-	0.044	1.804	0.086	-	0.052	1.519	-	0.045	1.560
Al	1.914	0.002	-	0.054	1.915	0.021	-	0.036	1.924	-	-	1.916	0.012	-
Ti	0.003	0.002	0.001	0.013	0.006	0.001	-	0.009	0.008	0.003	-	0.005	0.001	-
Cr	-	0.001	0.001	-	0.002	-	-	0.001	-	-	-	-	0.002	-
Fe^{3+}	0.116	0.053	-	0.035	0.101	0.021	-	0.065	0.133	0.070	-	0.051	0.030	-
Mg	0.810	1.061	-	0.636	0.564	1.006	-	0.651	0.495	0.911	-	0.420	0.919	-
Fe^{2+}	1.573	0.844	0.008	0.395	1.748	0.913	0.016	0.317	1.452	0.975	0.010	1.871	0.992	0.013
Mn	0.048	0.013	0.001	0.007	0.107	0.017	0.001	0.006	0.032	0.025	-	0.062	0.017	-
Ca	0.555	0.022	0.467	0.850	0.571	0.021	0.860	0.882	0.993	0.028	0.572	0.664	0.026	0.610
Na	-	0.002	0.486	0.010	-	-	0.151	0.034	-	-	0.402	-	0.001	0.377
K	-	-	0.041	-	-	-	0.006	-	-	-	0.020	-	-	0.017
Sum Cations	8.001	4.000	4.988	4.000	7.999	4.000	5.007	4.001	8.000	4.000	4.987	8.001	4.000	4.997

* Sympl. = symplectites

Sample	K 242-2			K 242-2			K 302			K 333-4 (Qz free)			
Analysis No.	Gt 17	Opx 12	Plag 9	Gt 30	Opx 29	Plag 28	Gt 23	Opx 20	Plag 27	Gt 68	Opx 50	Gt 8	Cpx 41
Remarks	Cores			Fine-grained			Cores			Gt growth		Gt growth	
SiO_2	37.42	49.52	60.63	37.78	49.69	61.23	37.35	50.38	59.86	39.09	51.14	39.81	49.87
TiO_2	-	0.08	-	0.02	0.10	0.01	0.04	0.09	-	0.10	0.07	0.04	0.39
Al_2O_3	21.25	2.21	25.23	21.15	2.43	24.50	20.84	1.63	25.67	21.42	3.97	21.93	4.66
Cr_2O_3	0.27	0.08	-	0.16	0.12	-	0.06	0.01	-	0.08	0.08	0.08	0.04
FeO	31.66	31.63	0.15	33.21	31.05	0.11	32.19	31.74	0.21	20.23	19.02	19.52	8.12
MnO	1.16	0.32	0.02	1.26	0.36	-	1.98	0.54	-	0.88	0.30	0.77	0.22
MgO	6.12	15.61	-	5.06	16.13	-	4.44	15.67	-	10.40	24.74	11.22	13.48
CaO	1.84	0.17	6.78	1.91	0.20	6.20	2.43	0.30	7.61	7.20	0.34	7.10	22.70
Na_2O	-	0.02	7.22	-	0.03	7.30	-	-	6.76	-	0.01	-	0.51
K_2O	-	-	0.43	-	-	0.52	-	-	0.43	-	-	-	-
Sum	99.72	99.64	100.46	100.55	100.11	99.87	99.33	100.36	100.54	99.40	99.67	100.47	99.99
Structural formulae													
Si	2.957	1.933	2.687	2.984	1.924	2.722	2.994	1.955	2.655	2.968	1.870	2.975	1.846
Al^{IV}	-	0.067	1.318	-	0.076	1.284	-	0.045	1.342	-	0.130	-	0.154
Al	1.979	0.035	-	1.969	0.037	-	1.968	0.030	-	1.917	0.041	1.931	0.049
Ti	-	0.002	-	0.001	0.003	-	0.002	0.003	-	0.006	0.002	0.002	0.011
Cr	0.017	0.002	-	0.010	0.003	-	0.004	-	-	0.005	0.002	0.005	0.001
Fe^{3+}	0.090	0.026	-	0.050	0.022	-	0.036	0.009	-	0.130	0.083	0.111	0.120
Mg	0.721	0.908	-	0.596	0.935	-	0.530	0.906	-	1.177	1.349	1.250	0.744
Fe^{2+}	2.003	1.006	0.006	2.144	0.982	0.004	2.123	1.021	0.008	1.154	0.499	1.109	0.132
Mn	0.078	0.011	0.001	0.084	0.011	-	0.134	0.018	-	0.057	0.010	0.049	0.007
Ca	0.156	0.007	0.322	0.162	0.007	0.295	0.209	0.012	0.362	0.586	0.013	0.568	0.900
Na	-	0.002	0.620	-	0.001	0.629	-	-	0.582	-	0.001	-	0.037
K	-	-	0.024	-	-	0.029	-	-	0.025	-	-	-	-
Sum Cations	8.001	3.999	4.978	8.000	4.001	4.963	8.000	3.999	4.974	8.000	4.000	8.000	4.001

Table 10.3 *(Continued)*

Sample	K 333-4 (Qz free)		K 355-1				K 408			K 460-1		
Analysis No.	Gt 30	Opx 29	Gt 10	Opx 11	Plag 7	Cpx 9	Gt 29	Opx 27	Plag 28	Gt 50	Opx 9	Plag 22
Remarks	Rim	Sympl.*	Gt growth				Rim	Sympl.*		Cores		
SiO_2	39.39	52.13	38.44	51.25	59.02	51.65	38.77	51.59	52.50	37.77	49.61	57.19
TiO_2	0.01	0.01	0.10	0.08	0.02	0.26	0.04	0.06	-	-	0.12	-
Al_2O_3	21.51	3.44	20.91	1.34	25.60	2.00	21.14	1.33	30.35	21.25	3.17	25.40
Cr_2O_3	0.04	0.03	-	-	0.02	-	0.04	0.05	0.03	0.07	0.02	-
FeO	22.12	17.88	28.40	29.54	0.13	12.56	26.03	25.39	0.35	33.40	28.74	1.54
MnO	1.79	0.51	1.21	0.37	0.04	0.14	2.52	0.76	-	1.02	0.72	-
MgO	9.51	25.45	4.95	16.81	-	12.16	5.82	19.69	-	5.41	16.48	-
CaO	5.99	0.30	6.72	0.75	8.69	21.44	5.73	0.37	13.16	2.02	0.26	8.11
Na_2O	-	-	-	0.02	6.07	0.32	-	-	3.90	-	0.03	6.70
K_2O	-	-	-	-	0.46	-	-	-	0.16	-	-	0.39
Sum	100.36	99.75	100.73	100.16	100.05	100.53	100.09	99.24	100.45	100.94	99.15	99.33
	-											
Structural formulae												
Si	2.990	1.898	3.001	1.975	2.638	1.956	3.027	1.968	2.373	2.966	1.927	2.599
Al^{IV}	-	0.102	-	0.025	1.348	0.044	-	0.032	1.617	-	0.073	1.361
Al	1.924	0.046	1.924	0.036	-	0.041	1.945	0.028	-	1.966	0.072	-
Ti	0.001	-	0.001	0.002	0.001	0.007	0.002	0.002	-	-	0.004	-
Cr	0.002	0.001	-	-	0.001	-	0.002	0.002	0.001	0.004	0.001	-
Fe^{3+}	0.093	0.055	0.071	-	-	0.016	-	-	-	0.098	-	-
Mg	1.076	1.381	0.576	0.965	-	0.681	0.677	1.120	-	0.633	0.954	-
Fe^{2+}	1.312	0.489	1.784	0.952	0.005	0.351	1.700	0.810	0.013	2.095	0.934	0.059
Mn	0.115	0.016	0.080	0.012	0.001	0.006	0.167	0.025	-	0.068	0.024	-
Ca	0.488	0.012	0.562	0.031	0.416	0.870	0.479	0.015	0.637	0.170	0.011	0.395
Na	-	-	-	0.001	0.526	0.028	-	-	0.342	-	0.002	0.590
K	-	-	-	-	0.026	-	-	-	0.009	-	-	0.023
Sum Cations	8.001	4.000	7.999	3.999	4.962	4.000	7.999	4.002	4.992	8.000	4.002	5.027

* Sympl. = symplectites

Possible reactions are:

$$\text{Ti–Hbl} + \text{Plag} + \text{Qz} = \text{Gt} + \text{Cpx} + \text{Kf} + \text{Ilm} + H_2O$$
$$\text{Ti–Bio} + \text{Plag} + \text{Qz} = \text{Gt} + \text{Opx} + \text{Kf} + \text{Ilm} + H_2O$$

(Hollocher 1985)

$$\text{Ti–Hbl} + \text{Qz} = \text{Gt} + \text{Opx} + \text{Cpx} + \text{Ilm} + H_2O$$

(modified after Robinson *et al.* 1982)

The occurrence of thin rims of K-feldspar around the plagioclase, which itself forms rims around garnet, suggests that these reactions took place after the orthopyroxene + plagioclase ± magnetite-forming stage; that is, along the retrograde part of the *P–T–t* path. This implies that these late dehydration reactions would be the result of a decrease in H_2O activity or a decrease in pressure. However, it is unlikely that a dehydration reaction would occur while temperature decreased, and even more unlikely that the denser parageneses with garnet and clinopyroxene could be formed during pressure decrease. The problem of prograde or retrograde breakdown of hornblende remains unresolved.

10.5 Mineral chemistry

Mineral analyses have been obtained by wavelength-dispersive electron microprobe analysis, using a Camebax microprobe, at the University of Kiel, with a 15 kV acceleration potential and a take-off angle of 40°. Corrections were carried out with the PAP correction program (Pouchou & Pichoir 1984). Representative microprobe analyses and examples of chemical parameters used for *P–T* estimates are presented in Tables 10.3 and 10.4, respectively. Ferric iron estimates from microprobe analyses of garnet, orthopyroxene, clinopyroxene, and the opaque phases are based on a fixed number of oxygens (feldspar, hornblende) or cations and oxygens (garnet, pyroxene, and ilmenite; Table 10.3). In most samples, ilmenite is the only matrix oxide phase present, and magnetite occurs as a reaction product. This general lack of magnetite in the matrix and the relatively low Fe^{3+} content in ilmenite (e.g. Table 10.3) suggest that the hornblendes have only low Fe^{3+} contents.

The garnet–clinopyroxene-forming reaction spans a temperature (or pressure) interval due to the effects of Fe–Mg and Na–Ca solid solution in the phases involved in the reactions. Despite the apparent *P–T* interval, no distinctive zoning has been found in the garnet from originally garnet-free, two-pyroxene metabasites, in which only reaction rims of garnet and clinopyroxene formed around orthopyroxene (e.g. sample K 140, Table 10.3). In contrast, garnets which show the breakdown to orthopyroxene and plagioclase display zoning towards the rims, which in some cases can be very strong; presumably,

Table 10.4 Mineral chemical data of individual minerals used for geothermometry (Harley 1984, Ellis & Green 1979) and geobarometry (Newton & Perkins 1982). Mineral analyses are given in Table 10.3.

Sample	K 1-5 Cores	K 50-1 Cores	K 130-1 Cores	K 130-1 Gt growth *	K 130-1 Sympl. **	K 134-1 Cores	K 134-1 Cores	K 134-1 Hbl breakdown	K 144-4 Gt growth *	K 144-4 Sympl. **
Harley 1984										
X_{gr}	0.104	0.202	0.194		0.200	0.200			0.189	0.198
$(Fe/Mg)_{Gt}$	1.958	4.657	2.289		3.367	2.788			1.943	3.099
$(Fe/Mg)_{Opx}$	0.676	1.596	0.884		0.972	1.053			1.257	0.908
K_D	2.896	2.918	2.589		3.464	2.646			1.546	3.798
P=5 kbar : T= (°C)	709	738	798		660	789			829	666
P=8 kbar : T= (°C)	726	755	816		675	806			848	682
Ellis & Green 1979										
$(Fe/Mg)_{Cpx}$		1.054	0.477	0.599			0.612	0.500		0.487
$(Fe/Mg)_{Gt}$		4.657	2.289	3.155			3.898	3.898		3.099
K_D		4.418	4.799	5.267			6.369	7.796		6.363
P=5 kbar : T= (°C)		822	789	763			810	666		712
P=8 kbar : T= (°C)		831	799	772			820	675		721
Newton & Perkins 1982										
X_{py}	0.278	0.138	0.241		0.179	0.211***			0.271	0.189
X_{al}	0.543	0.644	0.551		0.601	0.589			0.527	0.585
X_{gr}	0.095	0.198	0.191		0.195	0.200			0.186	0.191
X_{an}	0.516	0.419	0.668		0.752	0.426			0.470	0.846
X_{en}	0.596	0.385	0.531		0.507	0.460			0.443	0.551
T=600°C: P= (kbar)	5.73	6.87	7.67		6.34	8.4 (max)			8.5	6.2
T=800°C: P= (kbar)	6.32	7.77	8.26		6.39	9.6 (max)			9.6	6.2

* Gt growth = Garnet forming stage 4

** Sympl. = symplectite stage 5

*** Qz-free

Sample	K 200 Cores	K 200 Sympl. **	K 242-2 Cores	K 242-2 fine-grained	K 302 Cores	K 333-4 Gt growth	K 333-4 Gt growth	K 333-4 Sympl. **	K 355-1 Gt growth*	K 408 Sympl. **	K 460-1 Cores
Harley 1984											
X_{gr}	0.338	0.225	0.054	0.056	0.073	0.201		0.170	0.192	0.168	0.059
$(Fe/Mg)_{Gt}$	2.933	4.455	2.778	3.597	4.006	0.980		1.219	3.097	2.511	3.310
$(Fe/Mg)_{Opx}$	1.070	1.079	1.108	1.050	1.127	0.370		0.354	0.987	0.723	0.979
K_D	2.741	4.129	2.507	3.426	3.555	2.649		3.444	3.138	3.473	3.381
P=5 kbar : T= (°C)	818	599	764	619	610	788		653	701	649	625
P=8 kbar : T= (°C)	836	614	783	635	625	805		668	715	664	641
Ellis & Green 1979											
$(Fe/Mg)_{Cpx}$							0.177		0.515		
$(Fe/Mg)_{Gt}$							0.887		3.097		
K_D							5.011		6.014		
P=5 kbar : T= (°C)							776		723		
P=8 kbar : T= (°C)							786		732		
Newton & Perkins 1982						***	***	***			
X_{py}	0.167	0.139	0.244	0.200	0.177				0.192	0.224	0.213
X_{al}	0.489	0.620	0.677	0.718	0.709				0.594	0.562	0.706
X_{gr}	0.334	0.220	0.053	0.054	0.070				0.187	0.158	0.057
X_{an}	0.575	0.608	0.333	0.310	0.374				0.430	0.645	0.392
X_{en}	0.483	0.481	0.474	0.488	0.470				0.503	0.580	0.505
T=600°C: P= (kbar)	8.6	5.6	5.1	3.9	3.6				7.0	6.1	3.7
T=800°C: P= (kbar)	9.3	5.8	5.9	4.5	4.1				8.0	6.4	4.1

** Sympl. = symplectite stage 5

plateau compositions are preserved in the cores. However, the whole grain may be re-equilibrated (Lasaga 1983), so that the temperatures and pressures deduced from the core compositions represent minimum values. Two contrasting zoning profiles are shown in Figures 10.4 and 10.5. The garnets from sample K 200 from a locality to the south-east of Kandy (Fig. 10.4) and sample K 333 from the Nuwara Eliya area (Fig. 10.5) both show an increase in Fe and Mn content towards the rim. In the garnet from sample K 200 the Ca-content decreases towards the edge, whereas no Ca-zoning occurs in the garnet from sample K 333. This difference could be explained by the amounts of clinopyroxene in the rock. In sample K 200 only relics of clinopyroxene occur as inclusions in garnet, so that the garnet and clinopyroxene consuming reaction

garnet + clinopyroxene + quartz = orthopyroxene + plagioclase

could have stopped, and further formation of orthopyroxene and plagioclase

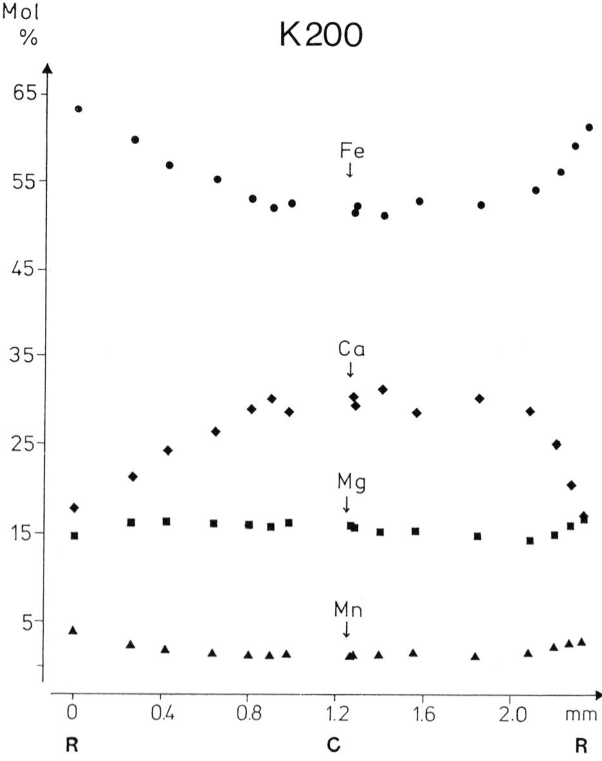

Figure 10.4 Compositional microprobe profile through garnet porphyroblast (sample K 200) which coexists with accessory clinopyroxene and has reacted to form orthopyroxene–plagioclase symplectites at the rims.

could have continued by using the grossular component of the garnet, because the matrix clinopyroxene has been consumed. On the other hand, the abundant matrix clinopyroxene of sample K 333 can react, as indicated by the orthopyroxene rims around the clinopyroxene porphyroblasts (Fig. 10.3D). In contrast to the garnets, the exsolved orthopyroxene and clinopyroxene porphyroblasts generally show no distinct zoning; however, core compositions of clinopyroxene may show higher Al_2O_3 contents. The difference in Al_2O_3 between core and rim may be up to 2 wt% (for example, 4.63 wt% Al_2O_3 in the core and 2.99 wt% Al_2O_3 in the rim of a clinopyroxene in sample K 333). The fineness of the exsolution lamellae in pyroxenes from metabasites precluded microprobe analysis in most cases. The compositions of the exsolved hosts are identical to those of the recrystallized grains within the same rock.

The majority of the matrix plagioclase has an An content of 40–60%. In contrast, the plagioclases that occur in the orthopyroxene–plagioclase coronas around garnet are always more calcic in composition (An_{70-90}) (Table 10.3).

From the experimental data of Green & Ringwood (1967) it can be concluded that the growth of garnet and clinopyroxene from orthopyroxene and

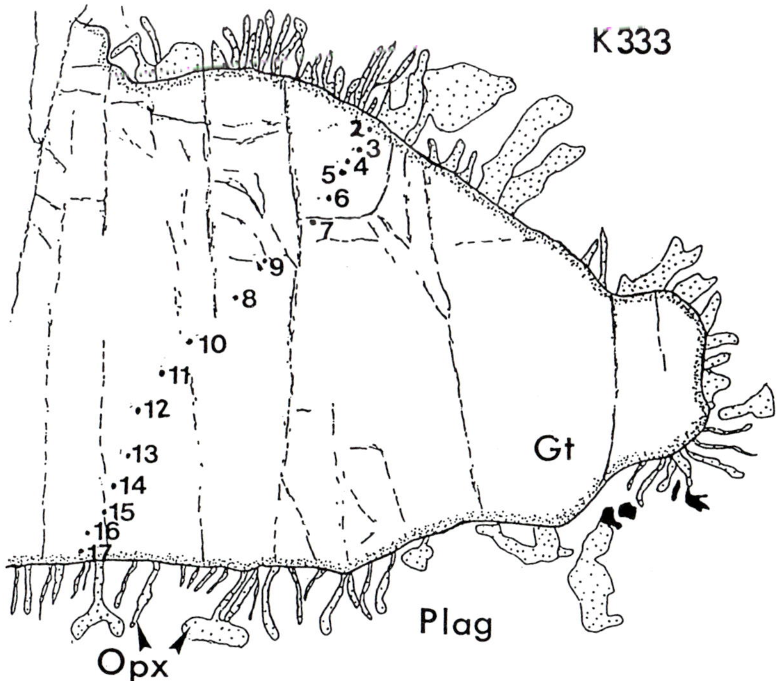

Figure 10.5 Compositional microprobe profile through garnet porphyroblast (sample K 333) which coexists with clinopyroxene and has reacted to form orthopyroxene–plagioclase–magnetite symplectites at the rims.

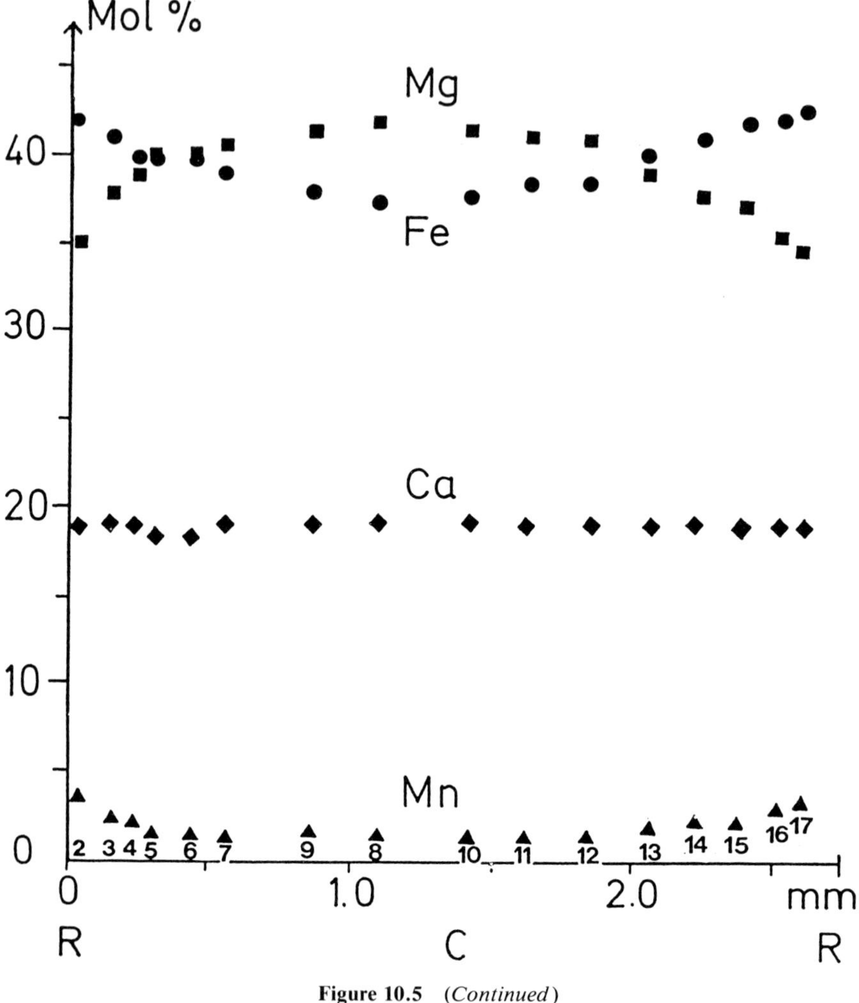

Figure 10.5 (*Continued*)

plagioclase with falling temperatures or increasing pressures involves a continuous reaction, in which the Fe–Mg minerals should become Mg-richer. In Figure 10.6 the changes in mineral composition of sample K 144 are presented in a projection from anorthite and albite, ilmenite, and quartz (the inset in Fig. 10.6 shows the projection procedure). The Ab + An projection is only a simplified method to represent the phase relations graphically; however, the Na and Fe^{3+} contents of the projected mineral phases are very low. The primary two-pyroxene assemblage (core of homogeneous Opx, re-integrated Cpx host and lamellae) gives way to the more Mg-rich three-phase assemblage: Opx with intermediate composition between core and rim (C/R) + exsolved Cpx (C/R) + Gt core (C) (+ Plag, + Qz) (Fig. 10.6). During the breakdown of

Cpx
70
90
1/2 Ca
46
E' Highland
Series
Amph.
Gt
51
64
50;R
50
70
Opx
Cpx
70
90
50
Ca
50
50
Mg
Fe
Central Highland
Series
138-5
25
Gt
189-10
194
184-2
193
← Mg
50
70
90
Fe 2+
Opx

A.

B.

Cpx
70
90
50
146-4
SE' Highland
Series
Gt
142
144-4
333-4
130
134
Opx
50
←Mg
70
90
50
Central Highland
Series
25
Gt
261-6
151
224-2
138-4
50
Opx
70
90
Fe 2+

C.

D.

1/2 Ca
90
50
Arenas (1-5) and W'
Highland Series
25
• 242-2
• 442
• 460
Gt
1-5
299-6
304-2
302-2
308-4
50
Opx
70
Cpx
70
90
50
SE' Highland
Series
25
Gt
355
M402
156
366
141
140
178
157
152
50
70
90
Opx
Fe 2+

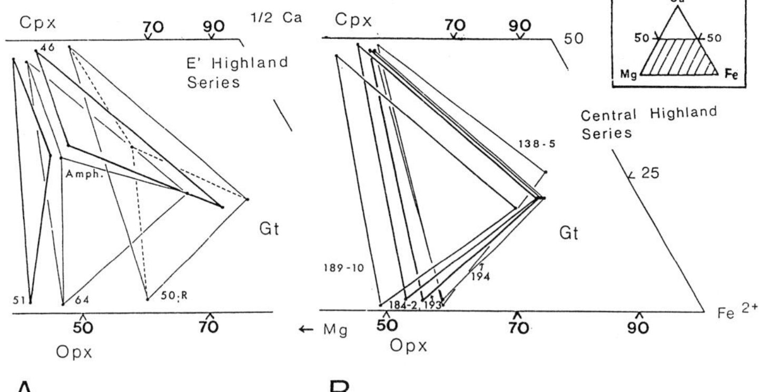

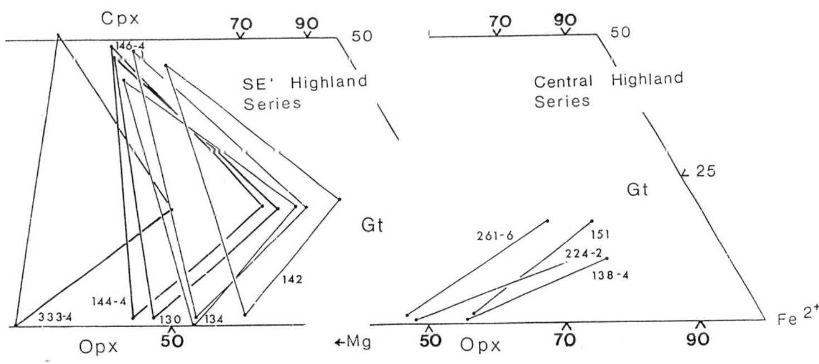

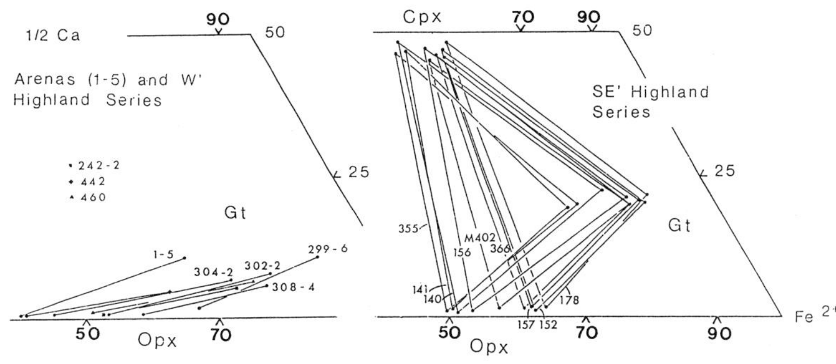

E.

F.

lines in Fig. 10.7C) is similar to the range in composition of the garnet + clinopyroxene from the south-east Highland Series that form coronas around orthopyroxene (Fig. 10.7F).

(2) In Figure 10.7C the most Mg-rich compositions are of the samples K 144–4 and K 333–4 from the south-east Highland Series, in which garnet–clinopyroxene coronas around orthopyroxene show breakdown to orthopyroxene–plagioclase symplectites. This may indicate that in rocks that show the breakdown of garnet + clinopyroxene to orthopyroxene + plagioclase, the matrix clinopyroxene, orthopyroxene, and, if still present, garnet compositions of samples from one specific area of the Highland Series are generally Mg-rich (e.g. heavy tie lines in Fig. 10.7C) compared to the mineral compositions of rocks with no reaction textures (light tie lines in Fig. 10.7C).

(3) The grossular contents of the cores of garnets from garnet–clinopyroxene–orthopyroxene-bearing metabasites (Figs 10.7A–C & F) display a narrow range (with the one exception of sample K 138–5, $X_{Gross} = 0.17–0.22$).

(4) In contrast to the two-pyroxene-bearing assemblage, garnet–orthopyroxene-bearing rocks without clinopyroxene as a Ca-buffering phase contain Fe-richer garnets with highly variable grossular contents (Figs 10.7D & E), $X_{Gross} = 0.04–0.17$, similar to the garnets from Enderby Land, Antarctica (Harley 1985). In samples from the 'Arena' area and the western part of the Highland Series, tie lines between garnet–orthopyroxene are flatter, thus indicating lower temperatures; a similar observation has been made by Harley (1985) on his samples from Enderby Land.

10.6 Metamorphic conditions

The regional distribution of the pressures from the garnet porphyroblast stage in the pyroxene granulites shows that pressures of 7.3–9.0 kbar have occurred in the central and south-east parts of the Highland Series, and a continuous decrease in pressure to about 3.5 kbar can be observed from Kandy towards Colombo in the west (Fig. 10.8A). These pressures were obtained from core compositions of garnet porphyroblasts, matrix plagioclase and exsolved orthopyroxene in quartz-bearing rocks (Newton & Perkins 1982), and the results agree well with the pressure estimates on the metapelites by application of the Newton & Haselton (1981) geobarometer (Raase & Schenk 1988).

The corresponding temperatures show a similar regional distribution with temperatures of 760–840°C in the central and south-east parts of the Highland Series, and a decrease in temperatures to about 600°C from Kandy towards Colombo in the west (Fig. 10.8A; orthopyroxene–garnet thermometer of Harley 1984).

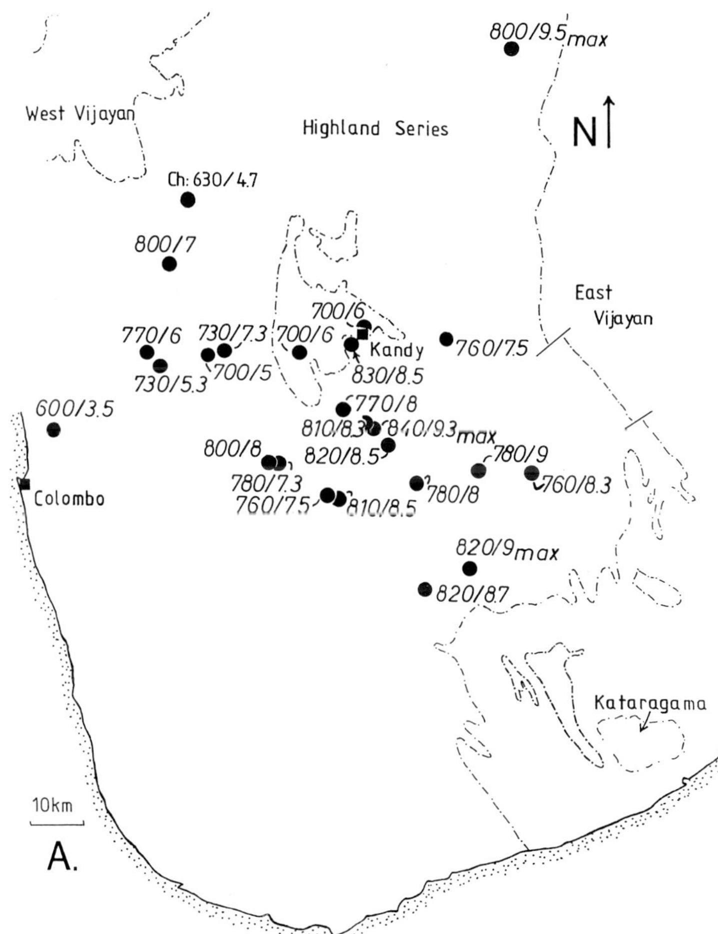

Figure 10.8 (A) Distribution of average temperatures and pressures within the Highland Series, obtained from core compositions of garnet, exsolved orthopyroxene, and plagioclase in metabasic and intermediate rocks and one charnockite (Ch) (thermometer of Harley 1984, barometer of Newton & Perkins 1982); max, maximum pressures obtained from quartz-free assemblages. (B) Distribution of average temperatures within the Highland Series, obtained from garnet–clinopyroxene porphyroblasts, fine-grained garnet–quartz–clinopyroxene rims and symplectites (s), and garnet–quartz overgrowths on garnet porphyroblasts (thermometer of Ellis & Green 1979). Corresponding pressures were obtained from exsolved orthopyroxene compositions, garnet, and matrix plagioclase (barometer of Newton & Perkins 1982). (C) Distribution of average temperatures and pressures during the symplectitic stage within the Highland Series: garnet–clinopyroxene breakdown to orthopyroxene–plagioclase symplectites and rims (thermometer of Harley 1984, barometer of Newton & Perkins 1982).

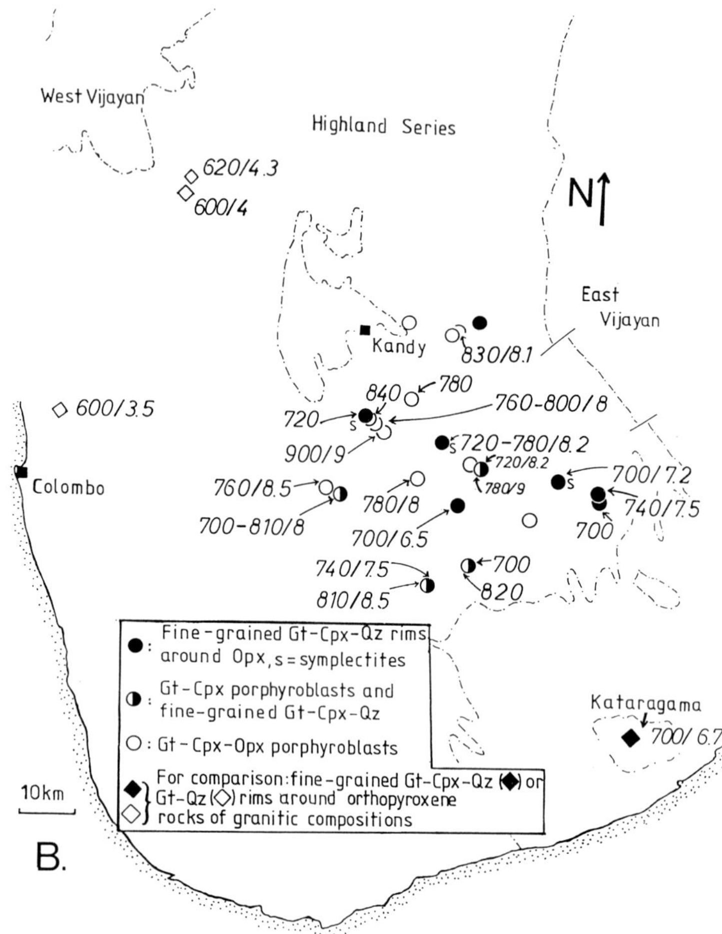

Figure 10.8 (*Continued*)

In comparison, thermometry on garnet and clinopyroxene (Ellis & Green 1979) gives a large temperature interval of 700–900°C for metabasites from the east and south-east parts of the Highland Series (Fig. 10.8B). The highest temperatures of 760–900°C are obtained from porphyroblasts of garnet and clinopyroxene, while lower temperatures are obtained from finer-grained and symplectitic garnet–clinopyroxene. In Figure 10.8B temperatures determined from samples with porphyroblasts (open circles), fine-grained garnet–clinopyroxene or symplectites (solid circles) or both textures (half-solid circles) are distinguished. From four samples with garnet–clinopyroxene porphyroblasts as well as fine-grained garnet–clinopyroxene, a large temperature range could be deduced, which for one sample was 700–820°C (Fig. 10.8B). Temperatures obtained from late-stage unzoned garnet and clinopyroxene may be better constrained than temperatures obtained from rim compositions of a matrix orthopyroxene and the newly formed garnet. However, low temperatures may be the result of re-equilibration of very fine-grained reaction products.

The pressures obtained from the cores of garnet porphyroblasts, matrix plagioclase, and orthopyroxene (Newton & Perkins 1982), coexisting with clinopyroxene porphyroblasts, are 8–9 kbar in samples from the south-east part of the Highland Series (Fig. 10.8B). For the pressure estimates of the finer-grained garnet–clinopyroxene ± quartz reaction rims around matrix orthopyroxene, rim compositions of matrix plagioclase and of matrix orthopyroxene have been used in addition to the composition of the newly formed

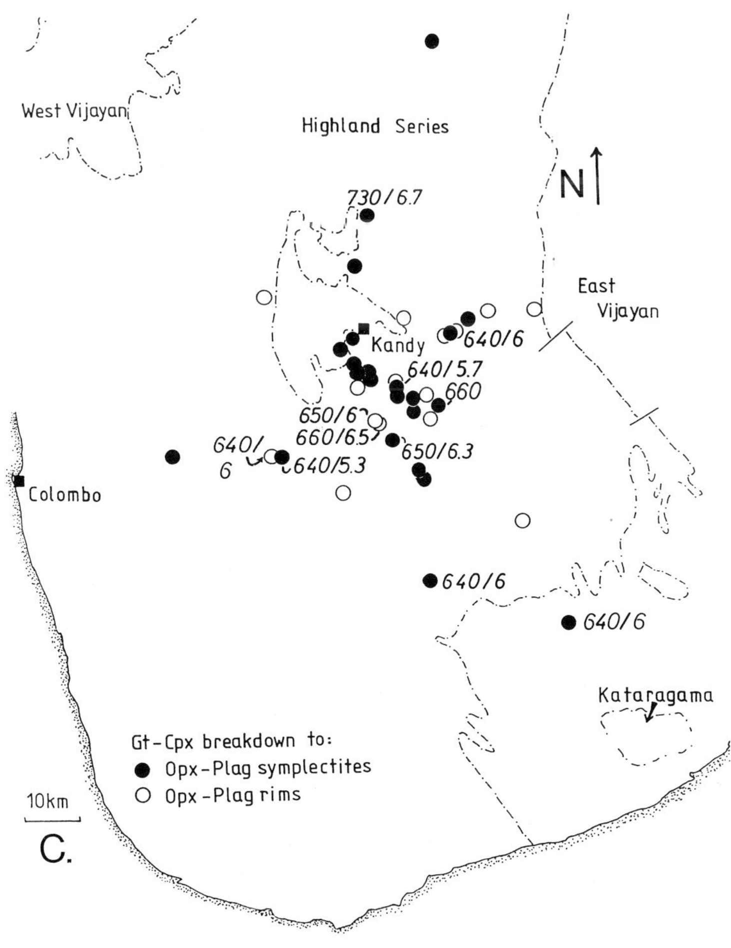

Figure 10.8 (*Continued*)

unzoned garnet. These pressures are 6.5–8.2 kbar, with the majority of the pressures around 7.5 kbar. The low pressure of 6.5 kbar (Fig. 10.8B) may be due to the re-equilibration of the fine-grained garnets.

Temperatures (Harley 1984) of about 650°C and pressures (Newton & Perkins 1982) of 5.3–6.3 kbar characterize the formation of orthopyroxene and plagioclase (symplectite stage) from garnet, clinopyroxene, and quartz in the east, central, and south-east parts of the Highland Series (Fig. 10.8C). The metamorphic conditions have been determined from garnet rim compositions and the compositions of new plagioclase and orthopyroxene.

So far, the 'Arena' area has been regarded as an amphibolite facies terrane within the granulite facies Highland Series (Sandiford *et al.* 1988). Our observation of charnockite formation discordant to the regional foliation and the (rare) occurrence of garnet–orthopyroxene suggests a granulite facies event in the *P–T* history of the Arena rocks.

10.7 Ti in amphiboles

The Ti content of amphiboles that coexist with Ti-rich phases (e.g. titanite, ilmenite, and rutile) can be used as a rough temperature indicator (e.g. Raase 1974). The evidence from natural assemblages has been experimentally confirmed by Spear (1981) at oxygen fugacities of the QFM buffer at 1 kbar (small dots connected by solid line in Fig. 10.9) and 3 kbar (small squares connected by solid line in Fig. 10.9), respectively.

Attention was paid to the textural relations of the Ti-rich hornblendes from the Highland Series and their Ti contents. The majority of the amphiboles appear to have formed as primary matrix minerals. This is supported by the high Ti contents, which are plotted in Figure 10.9 as the number of cations in the hornblende formula on the basis of 23 oxygens. For comparison, secondary hornblendes which occur within orthopyroxene–plagioclase symplectites in two Highland Series samples have lower Ti contents of about 0.1 cations (half-solid circles labelled 'sec' in Fig. 10.9). The temperatures of the Highland Series samples in Figure 10.9 have been obtained from the cores of garnet and orthopyroxene porphyroblasts (Harley 1984). Some hornblendes coexist with ilmenite, while others coexist with ilmenite and magnetite, which in most cases has been formed in the late-stage orthopyroxene–plagioclase–magnetite symplectites (Fig. 10.9; symbols given in the legend). The data fall well within the maximum and minimum limits given by Raase (1974) (dashed line in Fig. 10.9). The point labelled 'K', with about 0.25 cations Ti, is from a granitic rock (sample K 417) from the Kataragama nappe (Fig. 10.1). This hornblende indicates that the temperatures were similar to, or only a little lower than, those of the major part of the Highland Series. The Ti content of the sample labelled 'Vij' comes from an East Vijayan hornblende that coexists with

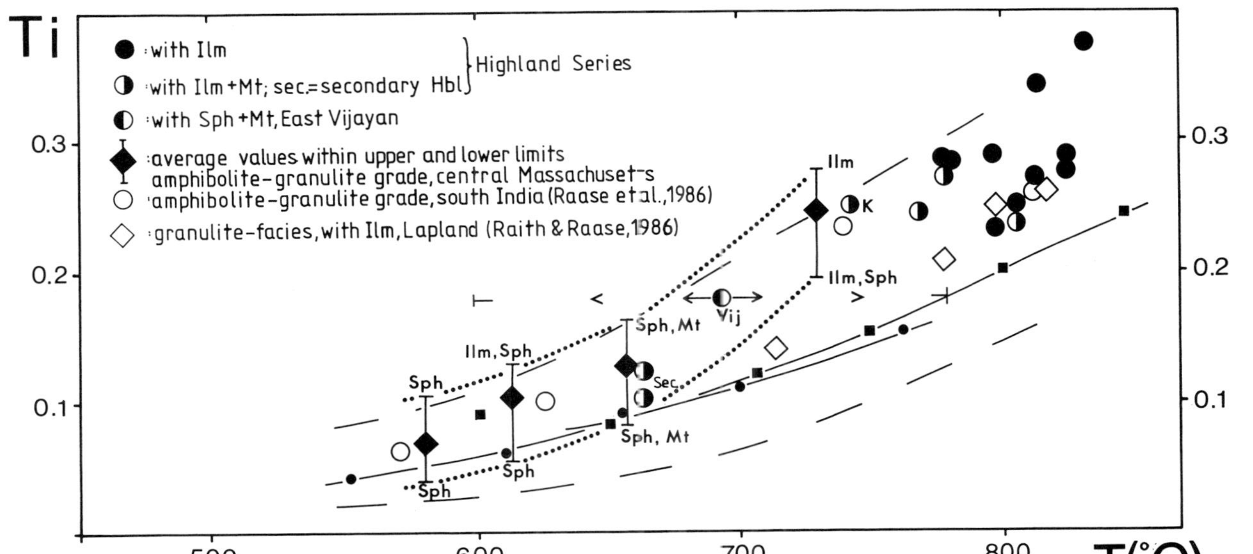

Figure 10.9 Plot of Ti contents of amphiboles (cations per 23 oxygen formula) versus temperature of metamorphism. Small dots connected by lines and small squares connected by lines show the experimental data of Spear (1981) for 1 kbar and 3 kbar, respectively, at QFM buffer conditions; long dashed curves show the maximum and minimum Ti contents from Raase (1974). Temperatures for the Highland Series: values obtained by the Harley (1984) thermometer. The explanation of the symbols is given in the figure. Abbreviations are given in Table 10.1.

titanite (sample K 523 in Fig. 10.1). Since, apart from the orthopyroxene-bearing assemblages in the 'enclaves' within the East Vijayan, no primary fresh orthopyroxene has yet been found in these Vijayan rocks, the temperature estimate was obtained from garnet–biotite pairs (Ferry & Spear 1978). The calculated temperatures show a large spread between 600 and 780°C, as indicated by arrows in Figure 10.9, and are interpreted as minimum and maximum values that are unrealistic actual temperatures. An average temperature of about 690°C appears to be realistic for the Vijayan rocks in respect to the Ti content. Secondary hornblende from two samples occurs in orthopyroxene–plagioclase symplectites around garnet, from which a temperature of 660°C has been deduced (thermometer of Harley 1984). The Ti content of about 0.1 cations per formula unit based on 23 oxygens agrees well with the temperature results. For comparison, Ti contents of hornblendes from various amphibolite facies and granulite facies terranes are shown in Figure 10.9. Data from amphibolite facies to lower granulite facies hornblendes that coexist with epidote and/or clinopyroxene and one or several Ti phases (rutile, ilmenite, and titanite) from central Massachusetts (Schumacher 1986) are also given (Fig. 10.9; brackets connected by dots show upper and lower Ti limits, solid diamonds represent the average values). They exceed the maximum limit of Raase (1974) at temperatures above 650°C. In addition, mean values for granulite and amphibolite facies zones from South India (Raase *et al.* 1986, open circles in Fig. 10.9) and mean values from data given by Raith & Raase (1986) for granulite facies rocks of Finnish Lapland (open squares in Fig. 10.9) are given.

A comparison of these results supports the temperature dependence of the Ti content in amphiboles, and suggests that the granulite facies hornblendes of Sri Lanka are of primary origin and have undergone metamorphism at temperatures comparable with those estimated from the granulite facies terrane of southern India (Raith *et al.* 1983) or (as can be seen from extremely high Ti values of over 0.3 cations per 23 oxygens) even exceed those temperatures.

10.8 *P–T–t* path

Several *P–T* stages can be deduced from the textural and *P–T* data of the basic to intermediate garnet–pyroxene granulites and associated metapelites in the central, eastern, and southeastern parts of the Highland Series:

(1) In sillimanite-bearing metapelites, early kyanite inclusions in garnet from localities south of Kandy, south of Ella, and from the Kataragama area (Fig. 10.1) indicate that the prograde *P–T–t* path went through the kyanite stability field and then entered the sillimanite field (Raase & Schenk 1988, stage 1).
(2) The highest temperatures were estimated from pre-exsolution composi-

tions of orthopyroxene and clinopyroxene (thermometer after Lindsley 1983) in metabasic and charnockitic rocks and exceed 900°C (Schenk *et al.* 1988). The estimated pre-exsolution compositions of charnockite pyroxenes with coarse exsolution lamellae lie in the pigeonite field. It remains uncertain whether the pigeonites have formed during prograde metamorphism or during magmatic cooling of a charnockite intrusion. Due to the absence of garnet, a pressure determination cannot be obtained.

(3) The third stage is characterized by the exsolution of the large pyroxenes and their recrystallization.

(4) At stage 4, garnet and clinopyroxene form from orthopyroxene and plagioclase. According to the garnet–orthopyroxene and garnet–clinopyroxene thermometers (Ellis & Green 1979, Harley 1984) the temperatures obtained from core compositions of porphyroblasts are about 820°C at pressures of about 8 kbar (barometer of Newton & Perkins 1982; Fig. 10.10). Temperatures obtained from garnet overgrowths on garnet porphyroblasts and symplectitic garnet–clinopyroxene ± quartz

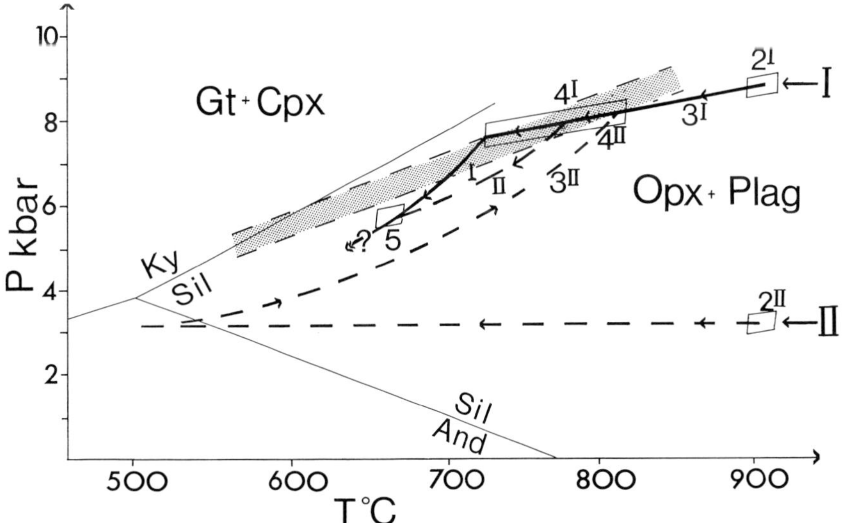

Figure 10.10 Two possible *P–T–t* paths, I and II, of the central, eastern and southeastern parts of the Highland Series, based on reaction textures and *P–T* estimates deduced from the metabasic to intermediate rocks (stages 2–5). *P–T–t* path I (modified from Schenk *et al.* 1988) is the preferred *P–T–t* path for the Highland Series rocks. Stage 1 (not shown) is represented by early (prograde) kyanite inclusions in garnets from metapelites associated with the metabasic rocks, and stage 6 represents retrograde growth of andalusite at the expense of cordierite in metapelites. The stippled field separates garnet + clinopyroxene (higher *P*) from the orthopyroxene + plagioclase (lower *P*) stability field (Green & Ringwood 1967). Aluminium silicate triple point after Holdaway (1971). The *P–T* boxes drawn at stages 2, 4, and 5 represent the range of calculated values and give no estimate of calibration errors.

coronas around orthopyroxene give a range down to about 700°C. Pressure estimates are about 7.5 kbar (Newton & Perkins 1982).

(5) This stage (Fig. 10.10) is characterized by the instability of garnet and clinopyroxene and the formation of a second generation of orthopyroxene and plagioclase ± magnetite in the form of symplectites or rims. The pressures are estimated to be 5.5–6 kbar at about 650°C (Fig. 10.10) (geobarometer of Newton & Perkins 1982, thermometer of Harley 1984). Textures may indicate that another, later formation of garnet, pyroxene, ilmenite, and K-feldspar from hornblende and biotite occurred at pressures of about 5.5 kbar (question mark in Fig. 10.10) or could have formed during prograde metamorphism and then re-equilibrated.

(6) At this stage, formation of retrograde andalusite at the expense of cordierite in metapelites indicates that pressures of less than 4 kbar must have been reached at temperatures of about 550°C (Raase & Schenk 1988).

10.9 Discussion

The textures of the metabasic to intermediate rocks can be interpreted differently, leading to two contrasting *P–T–t* paths, one of which is characterized by garnet growth due to isobaric cooling in the lower crust (I), while the other involves prograde growth of garnet due to burial (II).

Alternative I (solid line in Fig. 10.10). The very-high-temperature stage 2 may represent either metamorphic or magmatic temperatures at an unknown depth. Assuming that a mafic intrusion into the lower crust occurred (stage 2^{I} in Fig. 10.10) and cooled slowly at almost constant pressures while the pyroxenes exsolved and recrystallized (stage 3^{I} in Fig. 10.10), the garnet–clinopyroxene field would be reached. Garnet would then form over an almost isobaric temperature interval of about 840–720°C at pressures of about 8–7.5 kbar before decompression would lead to the orthopyroxene–plagioclase symplectite stage 5 (Fig. 10.10). The kyanite inclusions in garnets (stage 1) reflect burial of sediments to lower crustal depths.

Alternative II (dashed line in Fig. 10.10). Since the pressure of stage 2 cannot be determined, a mafic intrusion could have occurred at shallower crustal depths (stage 2^{II} in Fig. 10.10), cooled down to the conditions of a normal geotherm and then, at an unknown time, the crust could have been buried by tectonic processes. In this case, prograde growth of garnet followed by later decompression (stage 5) could have been the result of an anticlockwise *P–T–t* path. The temperature range deduced for stage 4 with garnet–clinopyroxene thermometry (Ellis & Green 1979) would be in part the result of retrograde re-equilibration of garnet rims and of the finer-grained garnet–clinopyroxene symplectites, and in part it could be the result of isobaric cooling after peak metamorphism. This *P–T–t* path is likely to be characteristic of a magmatically thickened and heated crust such as that expected in a continental arc environment (Bohlen 1987).

The common occurrence of two generations of garnet suggests that the finer-grained garnet–clinopyroxene symplectites and garnet overgrowths on porphyroblasts formed due to almost isobaric cooling rather than at increasing pressures. Therefore, even though the highest-temperature part of the *P–T–t* path remains obscure, the almost isobaric cooling from temperatures of at least 840°C to about 720°C represents the preferred interpretation for the garnet-forming stage. This interpretation of the *P–T–t* path contrasts with the almost isothermal decompression model deduced by Sandiford *et al.* (1988) on metabasites from the central and eastern part of the Highland Series.

P–T–t paths which show isobaric cooling have been recognized in several other granulite facies terranes such as the Napier Complex of Enderby Land, Antarctica (Ellis 1980, Ellis & Green 1985, Harley 1985, Sandiford 1985), the Labwor Hills of Uganda (Sandiford *et al.* 1987), or the Arunta Block of central Australia (Warren 1983, Hensen & Warren 1988). Several models could explain the thermal perturbation at very high temperatures, which is followed by cooling at almost constant pressures:

(a) Tectonic crustal doubling (England & Thompson 1984, Ellis 1987)
(b) Magmatic underplating (e.g. Wells 1980, Bohlen 1987)
(c) Extensional tectonics (Sandiford & Powell 1986).

All of these have the common feature that the final excavation of the metamorphic terrain represents a separate event that is unrelated to prograde metamorphism.

The orthopyroxene–plagioclase symplectite stage 5 may be the consequence of either decompression following the prograde equilibrium or a second thermal perturbation during the Pan African event (Sandiford *et al.* 1988).

Acknowledgements

We wish to thank J. C. Schumacher for helpful comments on the manuscript, and D. Lattard and J. C. Schumacher for support during fieldwork as well as for providing computer programs. J. S. Daly and S. L. Harley are thanked for their reviews which helped to improve the manuscript. Mrs A. Weinkauf drafted several figures, Mr H. Unger prepared the thin sections, and D. Ackermand and B. Mader assisted with microprobe analyses: their help is gratefully acknowledged. We are indebted to Rani Vitanage (Kandy) who kindly provided maternal care, food, and accommodation between hot days in the field. We thank the Deutsche Forschungsgemeinschaft for financial support of this research, which is part of the Lower Crust program. All members of the Sri Lanka consortium contributed to lively discussions and through exchange of sample material.

References

Adams, F. D. 1929. The geology of Ceylon. *Canadian Journal of Research* **1**, 425–511.

Berger, A. R. 1973. The Precambrian metamorphic rocks of Ceylon: a critique of a radical interpretation. *Geologische Rundschau* **62**, 342–7.

Berger, A. R. & N. R. Jayasinghe 1976. Precambrian structure and chronology in the Highland Series of Sri Lanka. *Precambrian Research* **3**, 559–76.

Bohlen, S. R. 1987. Pressure–temperature–time paths and a tectonic model for the evolution of granulites. *Journal of Geology* **95**, 617–32.

Coates, J. S. 1935. The geology of Ceylon. *Ceylon Journal of Science*, Section B **19**, 101–87.

Cooray, P. G. 1961. *The geology of the area around Rangala.* Memoirs of the Geological Survey of Ceylon **2**, 138 pp.

Cooray, P. G. 1962. Charnockites and their associated gneisses in the Precambrian of Ceylon. *Quarterly Journal of the Geological Society of London* **118**, 239–73.

Cooray, P. G. 1965. *Geology of the country around Aluthgama.* Memoirs of the Geological Survey of Ceylon **3**, 1–111.

Cooray, P. G. 1978. Geology of Sri Lanka. In *Proceedings of the 3rd Regional Conference on Geology and Mineral Resources of SE Asia.* Bangkok, Thailand, 701–10.

Crawford, A. R. 1969. India, Ceylon and Pakistan: new age data and comparisons with Australia. *Nature* **223**, 380–4.

Crawford, A. R. & R. L. Oliver 1969. *The Precambrian geochronology of Ceylon.* Special Publication of the Geological Society of Australia **2**, 283–306.

Dissanayake, C. B. & B. J. Van Riel 1978. Petrology and geochemistry of a recently discovered nickeliferous serpentinite in Sri Lanka. *Journal of the Geological Society of India* **19**, 467–71.

Ellis, D. J. 1980. Osumilite–sapphirine–quartz granulites from Enderby Land, Antarctica: *P–T* conditions of metamorphism, implication for garnet–cordierite equilibria and the evolution of the deep crust. *Contributions to Mineralogy and Petrology* **74**, 201–10.

Ellis, D. J. 1987. Origin and evolution of granulites in normal and thickened crusts. *Geology* **15**, 167–70.

Ellis, D. J. & D. H. Green 1979. An experimental study of the effect of Ca upon garnet–clinopyroxene Fe–Mg exchange equilibria. *Contributions to Mineralogy and Petrology* **71**, 13–22.

Ellis, D. J. & D. H. Green 1985. Garnet-forming reactions in mafic granulites from Enderby Land, Antarctica – implications for geothermometry and geobarometry. *Journal of Petrology* **26**, 633–62.

England, P. C. & A. B. Thompson 1984. Pressure–temperature–time paths of regional metamorphism I. Heat transfer during the evolution of regions of thickened continental crust. *Journal of Petrology* **25**, 894–928.

Ferry, J. M. & F. S. Spear 1978. Experimental calibration of the partitioning of Fe and Mg between biotite and garnet. *Contributions to Mineralogy and Petrology* **66**, 113–17.

Geological Map of Sri Lanka 1982. Geological Survey Dept. of Sri Lanka, Colombo, Scale: 8 miles to 1 inch.

Green, D. H. & A. E. Ringwood 1967. An experimental investigation of the gabbro to eclogite transformation and its petrological applications. *Geochimica et Cosmochimica Acta* **31**, 767–833.

Hapuarachchi, D. J. A. C. 1975. *The granulite facies of Sri Lanka*. Geological Survey of Sri Lanka, Professional Paper **4**, 29 pp.
Harley, S. L. 1984. An experimental study of the partitioning of Fe and Mg between garnet and orthopyroxene. *Contributions to Mineralogy and Petrology* **86**, 359–73.
Harley, S. L. 1985. Garnet–orthopyroxene bearing granulites from Enderby Land, Antarctica: metamorphic pressure–temperature–time evolution of the Archaean Napier Complex. *Journal of Petrology* **26**, 819–56.
Hatherton, T., D. B. Patiarachchi & V. V. C. Ranasinghe 1975. *Gravity map of Sri Lanka*, 1 : 1 000 000. Geological Survey of Sri Lanka, Professional Paper, **3**.
Hensen, B. J. & R. G. Warren 1988. Isobaric cooling and locally controlled fluid activities in granulites from the Arunta Block, central Australia. *Terra Cognita* **8**, 246.
Holdaway, M. J. 1971. Stability of andalusite and the aluminum silicate phase diagram. *American Journal of Science* **271**, 97–131.
Hölzl, S., H. Köhler & W. Todt 1988. Datierungen an Magmatiten und Metamorphiten Sri Lankas. *Fortschritte der Mineralogie* **66**, 68.
Hollocher, K. T. 1985. *Geochemistry of metamorphosed volcanic rocks in the Middle Ordovician Partridge Formation, and amphibole dehydration reactions in the high-grade metamorphic zone of central Massachusetts*. Contribution no. 56, PhD thesis, University of Massachusetts, Amherst.

Jayawardena, D. E. de S. & D. A. Carswell 1976. The geochemistry of 'charnockites' and their constituent ferromagnesian minerals from the Precambrian of southeast Sri Lanka (Ceylon). *Mineralogical Magazine* **40**, 541–54.

Katz, M. B. 1971. Precambrian metamorphic rocks of Ceylon. *Geologische Rundschau* **60**, 1523–49.
Katz, M. B. 1972. Facies series of the high grade metamorphic rocks of the Ceylon Precambrian. *Proceedings, Section 2, 24th International Geological Congress*, 43–51.
Kröner, A., I. S. Williams, W. Compston, N. Baur, P. W. Vitanage & L. R. K. Perera 1987. Zircon ion microprobe dating of high-grade rocks in Sri Lanka. *Journal of Geology* **95**, 775–91.

Lasaga, A. C. 1983. Geospeedometry: an extension of geothermometry. In *Kinetics and equilibrium in mineral reactions*, S. K. Saxena (ed.), 81–114. New York: Springer.
Lindsley, D. J. 1983. Pyroxene thermometry. *American Mineralogist* **68**, 477–93.

Milisenda, C. C., T. C. Liew, A. W. Hofmann & A. Kröner 1988. Isotopic mapping of age provinces in Precambrian high-grade terrains: Sri Lanka. *Journal of Geology* **96**, 608–15.
Munasinghe, T. & C. B. Dissanayake 1979. Is the Highland – Eastern Vijayan boundary in Sri Lanka a possible mineralised belt? *Economic Geology* **74**, 1495–6.
Munasinghe, T. & C. B. Dissanayake 1980a. Is the Highland – Eastern Vijayan boundary in Sri Lanka a possible mineralised belt? – a reply. *Economic Geology* **75**, 775–7.
Munasinghe, T. & C. B. Dissanayake 1980b. Pink granites in the Highland Series of Sri Lanka – a case study. *Journal of the Geological Society of India* **21**, 446–52.
Munasinghe, T. & C. B. Dissanayake 1982. A plate tectonic model for the geologic evolution of Sri Lanka. *Journal of the Geological Society of India* **23**, 369–80.

Newton, R. C. & H. T. Haselton 1981. Thermodynamics of the garnet–plagioclase–Al_2SiO_5–quartz geobarometer. In *Thermodynamics of minerals and melts*, R. C. Newton, A. Navrotsky & B. J. Wood (eds), 131–47. New York: Springer.
Newton, R. C. & D. Perkins 1982. Thermodynamic calibration of geobarometers based on the assemblages garnet–plagioclase–orthopyroxene–(clinopyroxene)–quartz. *American Mineralogist* **67**, 203–22.

Oliver, R. L. 1985. A comparison of the basement geology of Sri Lanka and Southern Eyre Peninsula, South Australia. In *Recent advances in the geology of Sri Lanka*, C. B. Dissanayake & P. G. Cooray (eds), 57–62. Centre International pour la Formation et les Echanges Géologiques. Publication occasionelle no. 6.

Perera, L. R. K. 1984. Coexisting cordierite–almandine: a key to the metamorphic history of Sri Lanka. *Precambrian Research* **25**, 349–64.

Pouchou, J. L. & F. Pichoir 1984. A new model for quantitative X-ray microanalysis, Part I. Application to the analysis of homogeneous samples. *Recherche Aerospatiale* **3**, 13–38.

Raase, P. 1974. Al and Ti contents of hornblende, indicators of pressure and temperature of regional metamorphism. *Contributions to Mineralogy and Petrology* **45**, 231–6.

Raase, P. & V. Schenk 1988. Petrologie granulitfazieller Metapelite Sri Lankas. *Fortschritte der Mineralogie* **66**, 130.

Raase, P., M. Raith, D. Ackermand & R. K. Lal 1986. Progressive metamorphism of mafic rocks from greenschist to granulite facies in the Dharwar Craton of South India. *Journal of Geology* **94**, 261–82.

Raith, M. & P. Raase 1986. High grade metamorphism in the granulite belt of Finnish Lapland. In *The nature of the lower continental crust*, J. B. Dawson, D. A. Carswell, J. Hall & K. H. Wedepohl (eds), 283–95. Geological Society of London, Special Publication 24.

Raith, M., P. Raase, D. Ackermand & R. K. Lal 1983. Regional geothermobarometry in the granulite facies terrane of South India. *Transactions of The Royal Society of Edinburgh: Earth Sciences* **73**, 221–44.

Robinson, P., F. S. Spear, J. C. Schumacher, J. Laird, C. Klein, B. W. Evans & B. L. Doolan 1982. Phase relations of metamorphic amphiboles: natural occurrence and theory. In *Amphiboles: petrology and experimental phase relations*, D. R. Veblen & P. H. Ribbe (eds), 181–227. Mineralogical Society of America, Reviews in Mineralogy 9B.

Sandiford, M. 1985. The origin of retrograde shear zones in the Napier Complex: implications for the tectonic evolution of Enderby Land, Antarctica. *Journal of Structural Geology* **7**, 477–88.

Sandiford, M., F. B. Neall & R. Powell 1987. Metamorphic evolution of aluminous granulites from Labwor Hills, Uganda. *Contributions to Mineralogy and Petrology* **95**, 217–25.

Sandiford, M. & R. Powell 1986. Deep crustal metamorphism during continental extension: modern and ancient examples. *Earth and Planetary Science Letters* **79**, 151–8.

Sandiford, M., R. Powell, S. F. Martin & L. R. K. Perera 1988. Thermal and baric evolution of garnet granulites from Sri Lanka. *Journal of Metamorphic Geology* **6**, 351–64.

Schenk, V., P. Raase & R. Schumacher 1988. Very high temperatures and isobaric cooling before tectonic unlift in the Highland Series of Sri Lanka. *Terra Cognita* **8**, 265.

Schumacher, R. 1986. *Petrology and geochemistry of epidote- and clinopyroxene-bearing amphibolites and calc-silicate rocks from central Massachusetts, USA*. Unpublished PhD thesis, University of Bonn, FRG.

Schumacher, R. & V. Schenk 1988. Phasenbeziehungen und Metamorphoseverlauf der Metabasite der Highland Series Sri Lankas. *Fortschritte der Mineralogie* **66**, 146.

Spear, F. S. 1981. An experimental study of hornblende stability and compositional variability in amphibolite. *American Journal of Science* **281**, 697–734.

Vitanage, P. W. 1959. Geology of the country around Polonnaruwa. *Memoirs of the Geological Survey of Ceylon* **1**, 75 pp.

Vitanage, P. W. 1985. The geology, structure and tectonics of Sri Lanka and South India. In *Recent advances in the geology of Sri Lanka*, C. B. Dissanayake & P. G. Cooray (eds), 57–62. Centre International pour la Formation et les Echanges Géologiques. Publication occasionelle no. 6.

Warren, R. G. 1983. Metamorphic and tectonic evolution of granulites, Arunta Block, Central Australia. *Nature* **305**, 300–3.
Welberts, S. & M. Raith 1988. Metamorphe Entwicklung der Granat–Pyroxen–Granulite im SW-Teil Sri Lankas. *Fortschritte der Mineralogie* **66**, 165.
Wells, P. R. A. 1980. Thermal models for the magmatic accretion and subsequent metamorphism of continental crust. *Earth and Planetary Science Letters* **46**, 253–65.

CHAPTER ELEVEN

Local, mid-crustal granulite facies metamorphism and melting: an example in the Mount Stafford area, central Australia

R. H. Vernon, G. L. Clarke & W. J. Collins

11.1 Introduction

This chapter describes amphibolite to granulite facies regional metamorphism and partial melting that occurred at remarkably low pressure (around 2–4.5 kbar) in a section of Proterozoic crust at *c.* 1820 Ma ago (Collins *et al.* 1989b), and discusses processes responsible for unusually high temperatures at such shallow-crustal pressures. A relatively small area (about 260 km^2) of low-pressure regional metamorphic rocks (mainly metapelites and metapsammites, with subordinate mafic rocks) occurs in the Mount Stafford region, at the northwestern end of the Anmatjira Range, central Australia (Fig. 11.1). The rocks belong to the Lander Rock Beds of the Proterozoic Arunta Block, and were deposited about 1870 Ma ago, on the basis of correlation with similar rocks of that age in the Warramunga Group of the Devonport Basin (Blake & Page 1988). Primary sedimentary features can be recognized throughout the Mount Stafford area, even in the highest-grade rocks, and the metamorphic isograds appear to cut the near-flat-lying sediments.

The rocks we describe are unusual, in that they are very-low-pressure granulites. This is well demonstrated by the occurrence of interlayered andalusite-bearing migmatites and two pyroxene mafic granofelses. To place the rocks into a metamorphic and spatial perspective, we first review some general features of granulite facies terrains, and then summarize briefly the regional geology of the Arunta Block.

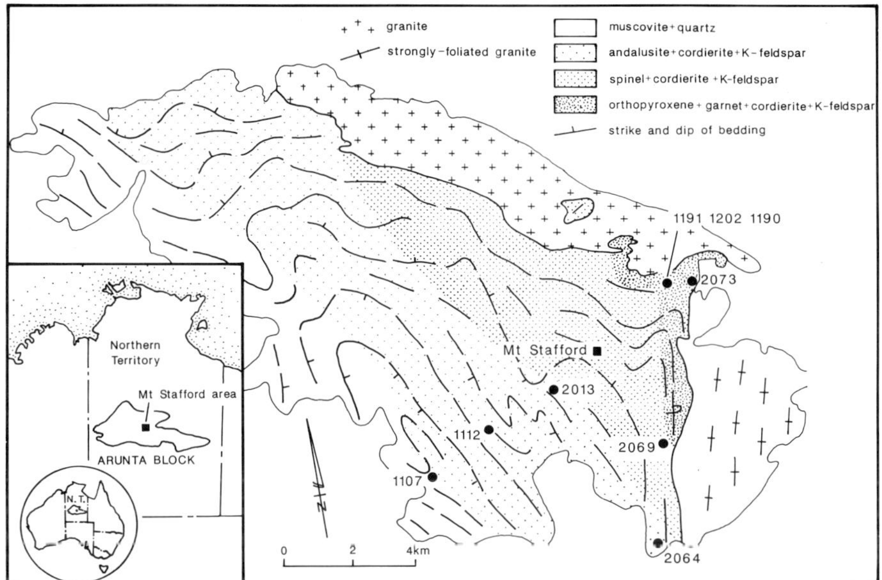

Figure 11.1 Map of the Mount Stafford area, showing bedding trends, metamorphic zones, granitoid intrusions, and the main sample localities referred to in this chapter. The inset shows the position of Mount Stafford and the Arunta Block in the Australian continent.

11.2 What causes granulite facies metamorphism?

Regional, high-grade metamorphic rocks are generally thought to to represent the exposed roots of old mountain belts, metamorphism being the result of burial during crustal overthickening. Granulite facies rocks preserve evidence of metamorphic conditions that were very hot and, most probably, of low water activity. Such gneisses are particularly common in metamorphosed Precambrian terrains and, in places, they are presumed to comprise rocks that have been exhumed from the very base of the continental crust. However, the following features of granulite facies terrains imply that the cause of granulite facies metamorphism must involve more than simple crustal overthickening.

First, granulite facies conditions require a gross perturbation of the normal continental geotherm. In the rocks we describe herein, the extent of this perturbation at mid-crustal levels is well illustrated by the syn-deformational overprinting of andalusite by sillimanite. Second, many high-grade Precambrian rocks preserve evidence of prolonged residence in the mid to lower crust, *after* the effects of metamorphism and deformation. For example, following high-grade metamorphism, the Archaean Napier Complex (Sheraton *et al.* 1980) in Enderby Land, Antarctica apparently remained near the base of the crust for half the history of the Earth (Sandiford 1985). Third, the timing of peak metamorphism with respect to deformation in the high-grade Precambrian terrains is different from that in Phanerozoic collisional zones: in many Precambrian terrains, deformation continued during cooling at constant or slightly increasing crustal thickness (Phillips & Wall 1981, Warren 1983, Clarke *et al.* 1987).

The crustal stability that apparently postdates metamorphism in many granulite facies terrains contrasts with the isothermal, rapid uplift of metamorphic rocks from the roots of many Phanerozoic mountain chains (e.g. Selverstone *et al.* 1984), where the effects of erosion, extension or both seem to terminate metamorphism (England & Richardson 1977, England 1987). Implicit in inferring crustal stability after granulite facies metamorphism and deformation, is that the synmetamorphic deformation did not result in an overthickened crust. The rocks in these terrains were probably returned to the Earth's surface by later events that were causally unrelated to the metamorphism. Thermal perturbations required for granulite facies metamorphism are difficult to establish by just an overthickening of the crust (Thompson & England 1984), and in any event would only occur late in the evolution of such a mountain belt during the isothermal decompression of deeply buried rocks. Although the remains of an eroded mountain chain may cool isobarically from a late thermal peak of metamorphism (Thompson & England 1984, Ellis 1987), the timing of deformation and peak metamorphism in many granulite facies terrains precludes such a tectonic setting. Thus, an external source of heat must be involved in granulite facies metamorphism. The contribution of heat to a terrain by advection, in the form of felsic or mafic magmas, is often presented as a direct cause of granulite facies metamorphism. However, the magmas commonly do not appear to be abundant enough (Wells 1980), or they are not synchronous with the peak of metamorphism. Moreover, the tectonic setting of the thermal perturbation that produced the magmas, and thus the metamorphism, then needs to be explained. Hence the general cause of granulite facies metamorphism in many terrains must be external to the rocks that we observe, and most probably external to the crust. However, magmas can be responsible for the local development of granulite facies rocks in thermally perturbed regions, as discussed later in this chapter.

11.3 The Arunta Block

The Arunta Block (Noakes 1953) is an extensive outcrop area of complexly deformed, mostly high-grade Proterozoic metasediments and orthogneisses that have been separated into three tectonostratigraphic divisions (Stewart *et al.* 1984). Sm–Nd model ages (Windrum & McCulloch 1986) allow a maximum crust formation age of 2000 Ma for the rocks, assuming no isotopic source mixing. The Mount Stafford area is part of the Anmatjira–Reynolds

Range region in the Arunta Block (Stewart 1981), which was affected by four temporally distinct metamorphic and deformation events between 1870 and 1580 Ma ago (Collins *et al.* 1989a,b; Clarke *et al.* 1990). The rocks described in this chapter preserve the effects of the first metamorphic and deformational event in the region.

Subsequent to the high-grade metamorphism(s), the Arunta Block remained at depth in the Earth's crust (e.g. Warren 1983), and the only constraint on its initial uplift is that its upper parts (amphibolite facies) must have been exposed prior to the deposition of Upper Proterozoic to Middle Palaeozoic sedimentary basins. The Late Devonian to Early Carboniferous Alice Springs Orogeny caused further basement uplift, folding, faulting, and thrusting of the sedimentary basins, and intense ductile faulting of the Early Proterozoic metamorphic basement (e.g. Collins & Teyssier 1989). This probably exposed the granulite facies rocks, because no undoubted unconformity between them and the cover rocks is known.

11.4 The Mount Stafford area

The metamorphic grade in the Mount Stafford area varies from low-pressure amphibolite facies in the south-west to low-pressure granulite facies in the north-east, over a lateral distance of only 10 km or so (Fig. 11.1), which indicates a steep local thermal gradient (about 75°C per km, as discussed later). Most of the metapelitic rocks show evidence of partial melting, which has produced nebulitic migmatites (terminology of Mehnert 1968); stromatic migmatites are typically absent. The following metamorphic zones (Fig. 11.1) have been delineated: (a) muscovite; (b) andalusite–cordierite–K-feldspar; (c) spinel–cordierite–K-feldspar; and (d) orthopyroxene–garnet–cordierite–K-feldspar.

Important features of the rocks are: (a) the common preservation or partial preservation of sedimentary mesostructures and igneous microstructures, even at highest metamorphic grades; (b) a relatively simple fold history; and (c) relatively fine grain sizes of some of the metasedimentary and meta-igneous rocks. In fact, many of the rocks resemble hornfelses, so much so that the metamorphism was assigned to the hornblende hornfels and pyroxene hornfels facies by Stewart (1981). This terminology is acceptable, on the basis of mineral assemblages, using the facies classification of Turner (1981) but, owing to the regional extent of the metamorphism, we prefer instead to use the terms low-pressure amphibolite facies and low-pressure granulite facies, respectively.

Deformed megacrystic granitoids occur in the north and north-east of the Mount Stafford area, generally adjacent to the highest-grade rocks (Fig. 11.1). However, the contacts cut across isograds (Fig. 11.1) and the granitoids have random xenoliths of metamorphic rocks similar to those of the Mount

Figure 11.2 Random xenoliths of metasedimentary rocks, similar to those of the Mount Stafford succession, in an intrusive, largely undeformed, megacrystic granitoid in the north of the Mount Stafford area (Fig. 11.1).

Stafford area (Fig. 11.2). Therefore, the metamorphic sequence cannot be regarded as a simple regional aureole, in the sense of White *et al.* (1974), as suggested by Stewart *et al.* (1980). However, the granitoids may have transported heat to the Mount Stafford area, and may have intruded through lower-grade zones that were originally arranged domally around a granitoid-rich core (e.g. den Tex 1963), as discussed at the end of the chapter.

Mineral symbols used in this chapter follow the recommendations of Kretz (1983), with the addition of As (aluminosilicate), East (eastonite), Fcrd (iron cordierite) and Mgts (Mg-Tschermak's molecule).

11.5 The structure of the Mount Stafford area

Three episodes of folding occurred during high-grade metamorphism in the Mount Stafford area. High-grade rocks in the Mount Weldon area, to the immediate south-east of the Mount Stafford area, and in the south-east Reynolds Range, to the south of the Mount Stafford area, have each been affected by two or three episodes of folding that have minor or no expression in the Mount Stafford rocks (Collins *et al.* 1989a). Moreover, U–Pb dating of zircon grains with the ion microprobe (Collins *et al.* 1989b) has shown that both the Mount Weldon and the Reynolds Range metamorphisms and accom-

Table 11.1 Metamorphic and folding episodes in the Anmatjira and Reynolds Ranges, central Australia.

Area	Metamorphism	Folding
Mount Stafford (Anmatjira Range)	M_1 (*c*.1820 Ma ago)	F_{1a},F_{1b},F_{1c}
Mount Weldon (Anmatjira Range)	M_2 (*c*.1760 Ma ago)	F_{2a},F_{2b},F_{2c}
SE Reynolds Range	M_3 (*c*.1730 Ma ago)	F_{3a},F_{3b}

panying fold sequences are temporally distinct from each other and from the Mount Stafford metamorphism and deformation. Therefore, we suggest a consistent scheme for metamorphic and folding events, in which the Mount Stafford metamorphism is labelled M_1 and the accompanying folding episodes are labelled F_{1a},F_{1b}, and F_{1c} (Table 11.1).

Most of the Mount Stafford area (Fig. 11.1) is characterized by shallowly

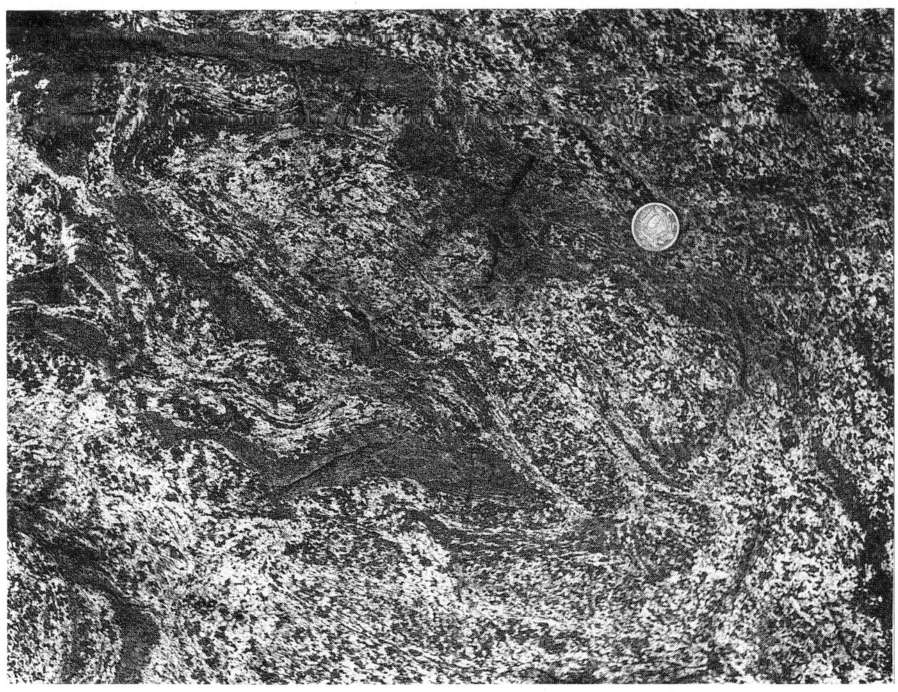

Figure 11.3 Outcrop at locality 2013 (Fig. 11.1) near the northern edge of the andalusite–cordierite–K-feldspar zone, showing recumbent, isoclinal folds (F_{1b}) outlined by dark, metapsammitic layers (S_0). Metapelitic layers show abundant dark porphyroblasts of altered cordierite, whereas more psammopelitic layers show a strong foliation (S_{1a}) that has been folded, along with S_0, by the F_{1b} folds. The F_{1b} folds have been gently folded by upright F_{1c} folds. The diameter of the coin is 23 mm (after Clarke *et al.* 1990, by permission of Blackwell Scientific Publications Ltd).

dipping bedding (S_0) and, in many rocks, tectonic foliations are weak, even in the high-grade rocks. Metamorphic isograds cut across stratigraphic units. Some of the highest-grade rocks (Fig. 11.1) show the most complete structural history. In some outcrops of these rocks (Fig. 11.3), a layer-parallel foliation (S_{1a}) is parallel to the axial surfaces of rootless, intrafolial F_{1a} folds, and is folded by west-verging, recumbent F_{1b} folds that have only a weak axial-

Figure 11.4 Open folds (F_{1c}) in metapelite (lighter) and metapsammitic (darker) beds in the andalusite–cordierite–K-feldspar zone. The metapelitic beds are mottled with altered cordierite porphyroblasts, and show a weak axial-surface foliation marked by weak alignment of leucosomes and weak elongation of porphyroblasts. The knife is 9 cm long.

Figure 11.5 Porphyroblasts of altered cordierite (dark) strongly elongate in S_{1a-1b} and crenulated to form a new foliation (S_{1c}) in the andalusite–cordierite–K-feldspar zone. The knife is 9 cm long.

surface foliation (S_{1b}). The folds are locally refolded by open, upright to reclined, north- to north-east-verging F_{1c} folds (Fig. 11.3). All of these structures are delineated by the high-grade mineral assemblages. Thus, we infer that metamorphism accompanied all three folding episodes. Since S_{1a} and S_{1b} are layer-parallel in most outcrops, they are generally indistinguishable.

In lower-grade rocks, the main foliation observed in outcrops and in thin sections is a commonly weak, upright foliation labelled S_{1c} because it is parallel to S_{1c} in the higher-grade rocks and locally is parallel to the axial surfaces of open, upright folds of similar orientation to F_{1c} folds in the higher-grade area (Fig. 11.4). Commonly, S_0 is also visible, together with an early tectonic foliation that varies from approximately parallel to distinctly oblique to S_0. Correlating with structures in the higher-grade rocks, this foliation could be either S_{1a} or S_{1b}, and we have no way of distinguishing between these alternatives.

S_{1a-1b} is more strongly developed in the andalusite–cordierite–K-feldspar zone than in the muscovite zone. The metamorphic minerals (especially matrix biotite and inclusions in andalusite) commonly are aligned in S_{1a-1b} although, in places, biotite grains and cordierite porphyroblasts are elongate in both S_{1a-1b} and S_{1c} (Fig. 11.5). Thus, we infer that the metamorphism occurred during all folding episodes in the andalusite–cordierite–K-feldspar zone, as in the higher-grade zones. However, the dominant foliation changes from S_{1c} in the lower-grade rocks to S_{1a-1b} in the higher-grade rocks, which suggests either that the higher-grade rocks reached their metamorphic peak earlier (with respect to the folding sequence) than the lower-grade rocks, or that S_{1c} was more strongly developed in the lower-grade rocks. The first interpretation is supported by the dominant alignment (where present) of leucosomes parallel to S_{1c} in the andalusite–cordierite–K-feldspar zone and parallel to S_{1a-1b} in the highest-grade rocks.

Sedimentary structures are common at all metamorphic grades (Figs 11.6–10). Cross-bedding (Figs 11.6 & 9) are graded bedding (Fig. 11.7) are well preserved in many places, even in the highest-grade rocks (Fig. 11.7), although tectonic boudinage has also occurred (Fig. 11.10). The abundance of sedimentary structures indicates that the Mount Stafford area, as a whole, has

Figure 11.6 Cross-bedding in metapsammitic rock in the andalusite–cordierite–K-feldspar zone. The coin diameter is 23 mm.

Figure 11.7 Graded bedding in the orthopyroxene–garnet–cordierite–K-feldspar zone. The pelitic parts of the beds are now relatively coarse-grained and spotted with cordierite porphyroblasts, whereas the more psammitic parts are relatively fine-grained and show residual sedimentary laminations. The knife is 9 cm long.

Figure 11.8 Metapelitic boudins in metapsammite in the andalusite–cordierite–K-feldspar zone. The metapelitic boudins are characterized by random, dark porphyroblasts of cordierite that have been largely to completely replaced by symplectite (see text). The knife is 9 cm long.

Figure 11.9 Metasedimentary breccia in the andalusite–cordierite–K-feldspar zone, containing various rock fragments, including metapelite with large, dark porphyroblasts of altered cordierite. The metapsammite at the top of the photograph shows cross-bedding. The diameter of the lens cap is 50 mm.

Figure 11.10 Distorted (?)brecciated structure in the orthopyroxene–garnet–cordierite–K-feldspar zone. Large, dark porphyroblasts of orthopyroxene and garnet are visible. The knife is 9 cm long.

not undergone very strong deformation prior to the three folding episodes now evident, although local high-strain zones may have been present, as discussed later.

In several places in all metamorphic zones (Fig. 11.1), sedimentary deformation has produced sharply defined blocks of metapelite and metapsammite (Figs 11.8–10), although these appear to have been distorted by tectonic boudinage in the highest-grade rocks (Fig. 11.10). At locality 1112 (Fig. 11.1), breccias (Figs 11.8 & 9) form part of an inferred sedimentary slump that continues, in apparent stratigraphic conformity, across the entire Mount Stafford area. We suggest a sedimentary origin for the fragmentation, in view of the observed continuity of S_{1a-1b} from blocks to matrix. These rocks show the contrasting response of different rock types to metamorphism and deformation. Blocks of metapelite have abundant leucosome and large porphyroblasts of cordierite, whereas the enclosing metapsammite commonly preserves bedding laminations (Figs 11.8 & 9). Both rock types have S_{1a-1b}, but in the metapelite S_{1a-1b} tends to be preserved mainly in the large cordierite porphyroblasts.

Most of the folding is relatively open (Fig. 11.1), especially F_{1b} and F_{1c}, although evidence of strong, multiple deformation occurs in metapelitic layers in some places (Fig. 11.3). Evidence of strong, repeated, penetrative deformation in metapelites, interbedded with some psammitic layers showing much less deformation (including beds with residual sedimentary structures) has been observed in several areas (e.g. Weiss & McIntyre 1957, Lister & Williams 1983), being related to strong partitioning of deformation between different rock types.

Apart from mylonitic foliations due to mid-Palaeozoic deformation (Collins & Teyssier 1989), the granitoids in the north of the Mount Stafford area (Fig. 11.1) are relatively weakly deformed and their contacts truncate S_{1a-1b} in the metasediments. The foliation in the more strongly deformed granitoids in the east of the area (Fig. 11.1) is similar in orientation to S_{1c} in the metasediments. However, it is also parallel to S_2, which is strongly developed in the adjacent Mount Weldon area. In view of the intensity of the foliation, as well as evidence (discussed later) of later metamorphic minerals aligned in a parallel foliation in the metasediments adjacent to the granitoids, we infer that this foliation is S_2.

11.6 Metamorphic (M_1) zones

11.6.1 The muscovite zone

The lowest-grade metasedimentary rocks in the Mount Stafford area (Fig. 11.1) are interbedded metasandstones and metashales occurring south of the Lander River. They consist of muscovite, biotite, and quartz, with or without minor tourmaline, ilmenite, and apatite. Quartzose rock fragments are visible in hand specimen, and detrital quartz is commonly observed in thin section. Sedimentary structures, including bedding (S_0), graded bedding and ripple cross-laminations, are well preserved, which indicate younging-upwards in all outcrops observed, although the reverse situation could also be expected, in view of the recumbent folding referred to previously. A single, near-vertical cleavage (S_{1c}) is parallel to the axial surfaces of upright, open to tight, south-east trending folds, which control the outcrop pattern of the Mount Stafford area (Fig. 11.1). In thin section, the cleavage is seen to be defined by foliae of fine-grained white mica that anastomose around detrital mineral and rock fragments. Mafic rocks in the zone have residual igneous microstructures masked, to various degrees, by growth of fibrous, blue–green amphibole.

A megacrystic granitoid has intruded the south-west portion of the muscovite zone. It is surrounded, for up to about a kilometre, by muscovite–biotite schists that are much coarser-grained than elsewhere in this zone, indicating a local contact effect of the intrusion.

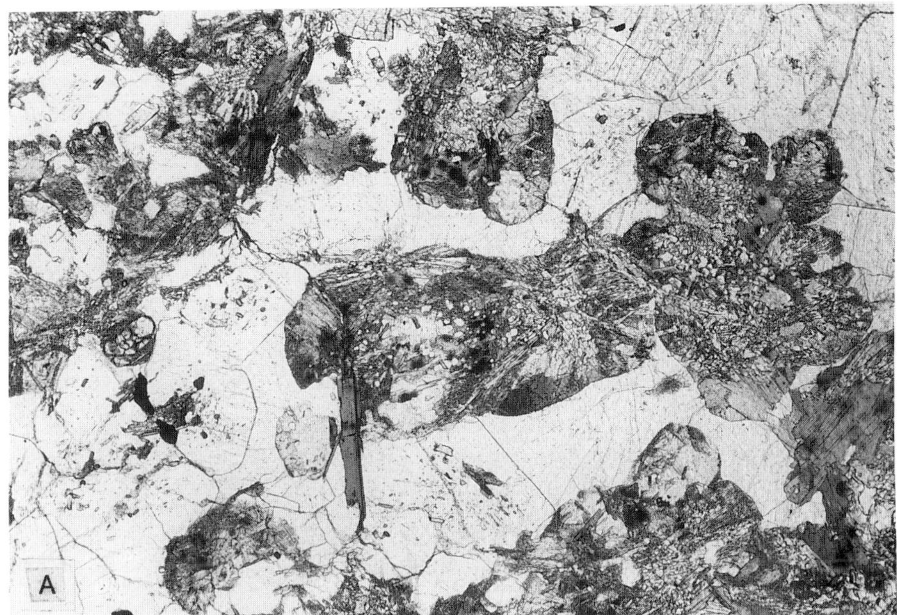

Figure 11.11 Metapelite in the andalusite–cordierite–K-feldspar zone, consisting of quartz, microcline–microperthite (with tartan twinning and small inclusions of biotite, quartz, and ilmenite), cordierite (completely replaced by random, symplectic aggregates of andalusite, biotite and quartz), and minor andalusite (top edge). The microcline has crystal faces against quartz (left and bottom). (A) Plane-polarized light; (B) crossed polars; base of photographs 4.4 mm.

Figure 11.12 Random, radiating, symplectic intergrowths of andalusite, biotite, and quartz that have partly replaced cordierite (centre and right) and microcline (left) in a metapelite of the andalusite–cordierite–K-feldspar zone. Crossed polars; base of photograph 1.75 mm.

11.6.2 The andalusite–cordierite–K-feldspar zone

Further north and north-east (Fig. 11.1), the metasedimentary rocks are less schistose (foliations generally being weak, although locally strong) and commonly show bedding and other sedimentary structures, such as cross-bedding (Fig. 11.6). Metapelitic rocks are characterized by abundant dark spots (Figs 11.4, 5, 8 & 9) that represent former cordierite porphyroblasts, now extensively to completely replaced by fine-grained, symplectic aggregates of biotite, andalusite, and quartz (Figs 11.11 & 12), discussed below in Section 11.8 on post-peak reactions. In many rocks, cordierite porphyroblasts are random (Fig. 11.8), but locally they are weakly (Fig. 11.4) to strongly (Fig. 11.5) elongate in S_{1a-1b}; these elongate porphyroblasts may be crenulated (Fig. 11.5), thus becoming elongate in S_{1c}. Evidently cordierite, and hence also the other minerals of the main assemblage in this zone, grew during F_{1a-1b} and F_{1c} folding.

The peak-metamorphic assemblage is andalusite + cordierite + K-feldspar + quartz + biotite + ilmenite ± either garnet or spinel ± tourmaline. The K-feldspar is microcline–microperthite (Fig. 11.12). Muscovite occurs only as a retrograde mineral, so that prograde metamorphic conditions exceeded the stability limit of quartz + muscovite. Plagioclase is typically absent, reflecting

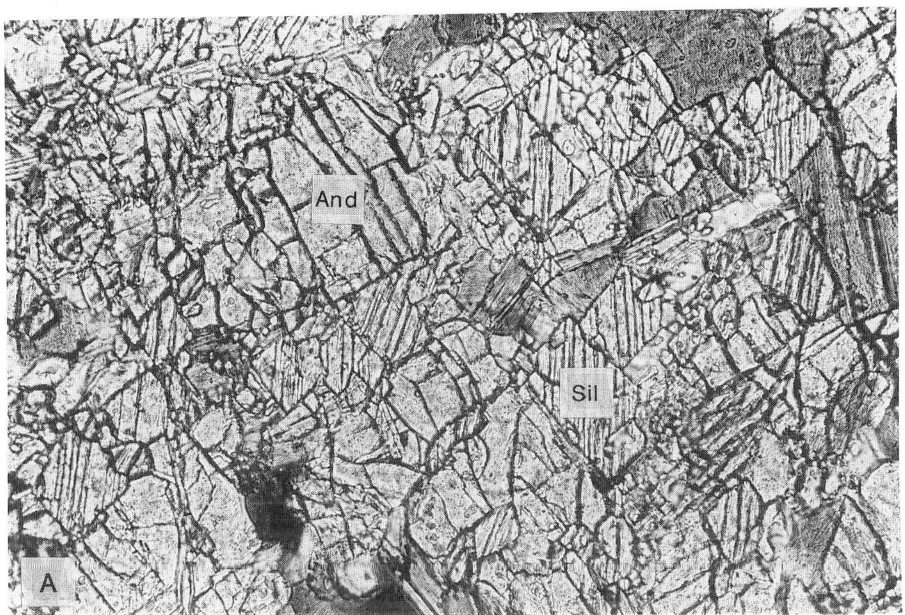

Figure 11.13 Prismatic sillimanite (Sil) pseudomorphing andalusite (And) in a metapelite of the andalusite–cordierite–K-feldspar zone. A coaxial relationship between the two minerals (e.g. Bosworth 1910, Vernon 1987) is indicated by the cleavage angles (A) and the simultaneous extinction (B). (A) Plane-polarized light; (B) crossed polars; base of photograph 0.70 mm.

very K-rich, Ca-poor bulk compositions. Cordierite typically has been completely pseudomorphed by random, fine-grained, symplectic aggregates of andalusite, biotite, and quartz (Fig. 11.11), and microcline has also been partly replaced (Fig. 11.12).

Some metapelites which occur in the higher-grade parts of this zone show prismatic sillimanite replacing andalusite and cordierite. Locally, the replacement of andalusite is coaxial (cf., e.g., Bosworth 1910, Vernon 1987), as shown in Figure 11.13. Late fibrous sillimanite is also common, as discussed later in Section 11.8 on post-peak reactions. Most of the sillimanite is random and so presumably post-deformational, but some is aligned in a new foliation. If this foliation is S_{1c}, it implies that the main assemblage (andalusite–cordierite–K-feldspar–biotite) grew during the development of either S_{1a} or S_{1b} in the rock concerned. The development of sillimanite reflects an increase in pressure, rather than temperature, as discussed in Section 11.11 on *P–T–t* paths. Prograde sillimanite is not observed in the Mount Stafford area, all sillimanite being due to post-peak reactions, as discussed in Section 11.8.

Many metapelitic rocks in this zone show evidence of partial melting, in the form of patchy leucosomes that are confined generally to metapelitic beds (Fig. 11.14), although very local migration of leucosome into adjacent metapsammitic beds may be evident. Melting in the Mount Stafford rocks is discussed more fully in Section 11.7.

Figure 11.14 Metasedimentary rocks in the andalusite–cordierite–K-feldspar zone, showing extensive, patchy (nebulitic) leucosome that is largely confined to coarser-grained metapelitic beds that are interbedded with finer-grained, laminated metapsammite.

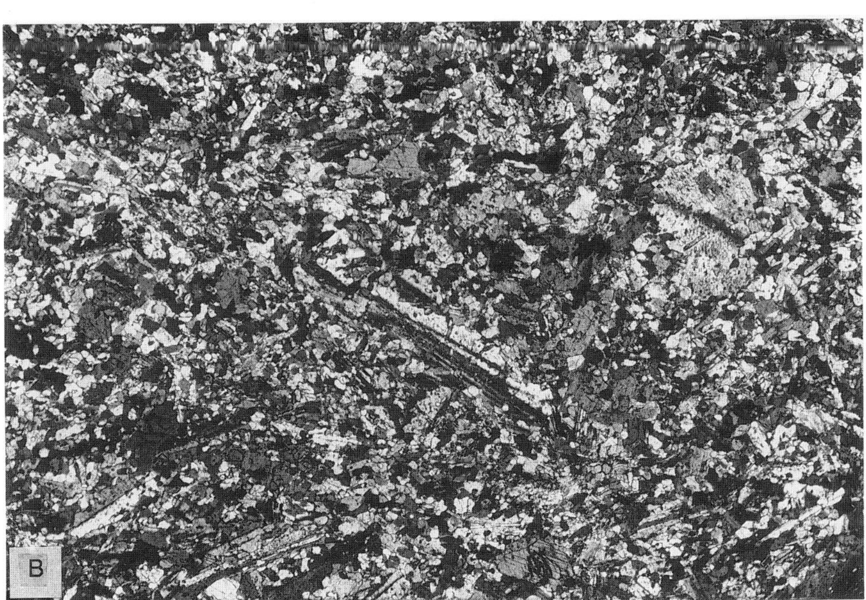

Figure 11.15 (A) Metadolerite of the andalusite–cordierite–K-feldspar zone, consisting of orthopyroxene, clinopyroxene, brown hornblende, plagioclase, and ilmenite. Although the rock is locally recrystallized to a granoblastic aggregate, residual ophitic microstructure is also present. (B) Metabasalt of the andalusite–cordierite–K-feldspar zone, consisting of the same minerals as the rock shown in (A), but with a much finer grain size and more extensive development of granoblastic microstructure. Residual phenocrysts of plagioclase are present. Crossed polars; base of photographs 4.4 mm.

Mafic rocks in the andalusite–cordierite–K-feldspar zone are massive to poorly foliated metagabbros and metadolerites, less commonly metabasalts (Fig. 11.15), containing orthopyroxene + clinopyroxene + brown hornblende + plagioclase + ilmenite ± red-brown biotite. Most are partly granoblastic, but ophitic or basaltic microstructures commonly are relatively well preserved (Fig. 11.15). Some of the mafic rocks are relatively fine-grained, despite their high metamorphic grade (Fig. 11.15). Mafic rocks in the higher-grade zones have the same mineral assemblage as in this zone.

11.6.3 The spinel–cordierite–K-feldspar zone

In this zone, metasedimentary rocks still show bedding, despite extensive partial melting (Fig. 11.16) and abundant spotting caused by large cordierite porphyroblasts that are dark grey, owing to extensive replacement by the symplectite of biotite, andalusite, and quartz (Fig. 11.16). Also present are nebulitic migmatites, in which sedimentary bedding has been disrupted and dispersed through leucosome (Fig. 11.17). Andalusite is typically absent, and prograde sillimanite has not been observed. Evidently, the pressure was too low for cordierite to react to garnet + sillimanite + quartz. The most common assemblage in metapelites in this zone is cordierite + K-feldspar + quartz + biotite + ilmenite ± spinel ± either garnet or orthopyroxene. The spinel is

Figure 11.16 Bedded metasedimentary rock in the spinel–cordierite–K-feldspar zone, showing relatively coarse-grained metapelitic beds, with abundant leucosome and dark porphyroblasts of altered cordierite, and finer-grained metapsammitic beds.

Figure 11.17 Nebulitic migmatite (schlieren migmatite) in the spinel–cordierite–K-feldspar zone, showing discontinuous, wispy relics of relatively fine-grained metapsammitic beds dispersed through coarser-grained leucosome.

restricted to small grains in the cores of some cordierite grains, suggesting that they are residual reactants of a reaction that removed spinel form the rest of the rock, as discussed in the next section.

11.6.4 The orthopyroxene-garnet-cordierite-K-feldspar zone

The highest-grade zone occurs in the east and north-east of the Mount Stafford area (Fig. 11.1), where many of the metasedimentary rocks are relatively coarse-grained granofelses (Fig. 11.10) and nebulitic migmatites, although some relatively fine-grained granofelses, resembling hornfelses, are also present (Fig. 11.7). In general, metapsammitic rocks tend to be finer-grained than metapelitic rocks (Fig. 11.7). Sedimentary structures, including graded bedding (Fig. 11.7), are preserved locally. However, much bedding has been tectonically boudinaged (Fig. 11.10), and in nebulites it has been mostly disrupted and occurs as discontinuous wisps in leucosome, as in the spinel–cordierite–K-feldspar zone (Fig. 11.17).

The most common assemblage in this zone is orthopyroxene + cordierite + K-feldspar + quartz + biotite + ilmenite ± garnet ± plagioclase. Spinel occurs mostly as inclusions in cordierite. Sillimanite is typically absent, for reasons discussed in the previous section, except where formed in late (post-peak) reactions, discussed later. The cordierite and K-feldspar occur as polygonal

Figure 11.18 Metapelite locally rich in microcline in the orthopyroxene–garnet–cordierite– K-feldspar zone. The microcline, which has small inclusions of quartz, biotite, and ilmenite, occurs mainly in polygonal aggregates of polygonal grains, except in leucosome, where it has crystal faces against quartz (right). Crossed polars; base of photograph 4.4 mm.

Figure 11.19 Metapelite locally rich in cordierite in the orthopyroxene–garnet–cordierite–K-feldspar zone. In places the cordierite occurs in polygonal aggregates, but it mainly has crystal faces against quartz. Minor plagioclase and K-feldspar are also present. Crossed polars; base of photograph 12 mm.

Figure 11.20 Metapelite in the orthopyroxene–garnet–cordierite–K-feldspar zone, showing relatively large, irregularly shaped porphyroblasts of orthopyroxene (Opx) dispersed through a matrix consisting mainly of K-feldspar, cordierite, biotite, and quartz. Crossed polars; base of photograph 4.4 mm.

(Fig. 11.18) to complexly intergrown aggregates in mesosomes, but commonly occur also as idioblastic grains against quartz in leucosomes (Figs 11.18 & 19), which are generally very pervasive in this zone. Orthopyroxene and garnet typically occur as relatively large, xenoblastic grains (Fig. 11.20), and biotite occurs as dispersed grains with (001) faces. The overall similarity in grain size, mutual inclusion, and absence of evidence of mutual reaction suggest that these minerals are all part of the same metamorphic assemblage. However, the restriction of spinel to cores in cordierite grains suggest that the main assemblage has been produced in reactions that destroyed spinel. As shown later, these reactions involved an increase in pressure, but a decrease in temperature. Because the minerals of the main assemblage delineate S_{1a-1b} and S_{1c} (where present), it appears that the thermal peak occurred early during F_{1a-1b}. Thus, F_{1c} and part of F_{1a-1b} occurred after the thermal peak of metamorphism in this zone.

11.7 Partial melting

Metapelites in the andalusite–cordierite–K-feldspar zone and higher-grade zones show abundant evidence of extensive partial melting (Figs 11.14, 16, 17

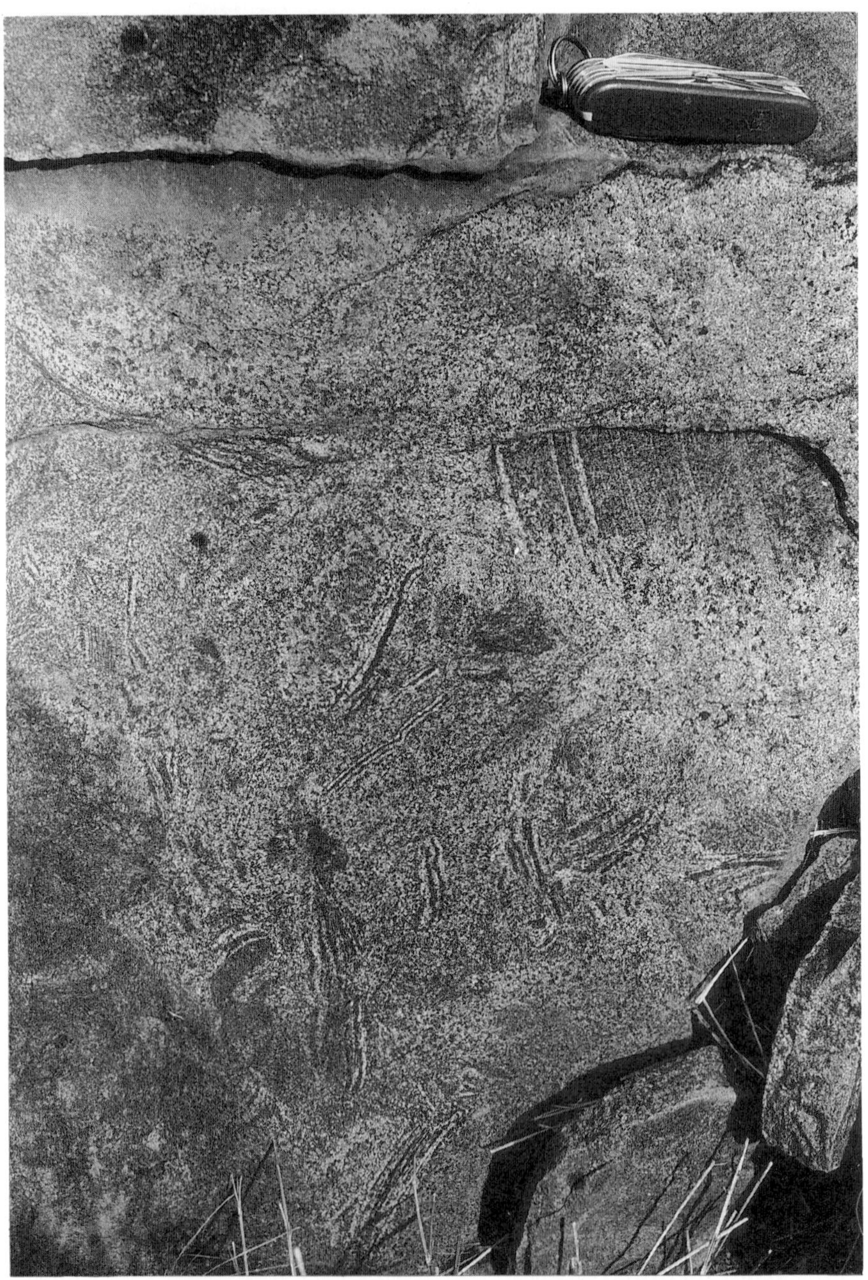

Figure 11.21 Partly melted metapelite in the orthopyroxene–garnet–cordierite–K-feldspar zone. Locally (e.g. at the top-right) melting has been complete in some layers, whereas others can be traced continuously into the otherwise completely melted rock. The high degree of melting indicates the local availability of water. The knife is 9 cm long.

& 21). Most leucosome occurs pervasively in metapelitic beds (Figs 11.14 & 16), although minor migration has involved penetration of adjacent metapsammitic beds and axial surfaces of minor folds, as well as the local formation of small pegmatitic pods (Vernon & Collins 1988). Overall, the leucosome has not moved out of the source rocks, as large bodies or veins of pegmatitic rock are absent. These features also suggests that no melt has been introduced from outside the area.

The resulting migmatites are typically not stromatic, using the terminology summarized by McLellan (1983). Instead, the leucosome tends to be patchy to pervasive, so that the most strongly melted rocks can be described as nebulites (Mehnert 1968) or 'schlieren migmatites' (McLellan 1983). The formation of granitic melts at the relatively low temperatures of the andalusite–cordierite–K-feldspar zone implies a high level of water activity (e.g. Johannes 1985), and the restriction of water-rich fluid inclusions to leucosomes in stromatic migmatites (e.g. Johannes 1985, p. 80) may suggest that leucosomes in at least some stromatic migmatites are due to influx of water into selected layers, which focuses melting in those layers. This process should apply especially to strongly foliated rocks, whereas the Mount Stafford rocks lacked a strong, tectonic, compositional foliation prior to high-grade metamorphism and melting. Therefore, the melting was unfocused, producing nebulitic, rather than stromatic migmatites. Melting is progressively more extensive with increasing metamorphic grade, and many of the higher-grade rocks are nebulites, with residual wisps and distorted layers of unmelted psammitic material (Fig. 11.17).

Minerals in leucosomes in the andalusite–cordierite–K-feldspar zone are quartz, K-feldspar, cordierite, andalusite, and minor biotite (Vernon & Collins 1988), so that the melts were strongly peraluminous. Andalusite locally occurs as large, pink, radiating crystals in pegmatitic pods (Vernon & Collins 1988), as shown in Figure 11.22. Orthopyroxene does not occur in the leucosomes, probably because water activities were too high or temperatures too low to stabilize orthopyroxene in hydrous peraluminous melts, namely around 800°C at 1–5 kbar and water activities of about 0.2–0.4 (Clemens & Wall 1981).

The microstructures of the leucosomes are typically igneous (Vernon & Collins 1988). Cordierite, K-feldspar, and andalusite have crystal faces against quartz (Figs 11.11, 18, 19 & 22), whereas in mesosomes these minerals are typically polygonal (Fig. 11.18). Biotite shows (001) faces in either situation, as is typical of biotite in both igneous and metamorphic rocks (e.g. Kretz 1966; Vernon 1968, 1976). Preservation of igneous microstructures is not common in migmatites, although it has been described (e.g. Wyborn 1977, Kenah & Hollister 1983). It requires absence of penetrative deformation during or after the melting, as this causes recrystallization which obliterates igneous microstructures. Such deformation is especially likely in stromatic migmatites, because focusing of fluid into selected layers is likely to promote deformation and recrystallization in those layers (Mawer *et al.* 1988). In contrast, unfocused

Figure 11.22 Locally abundant andalusite in leucosome in the andalusite–cordierite–K-feldspar zone. The andalusite mostly has crystal faces against quartz. Crossed polars; base of photograph 4.4 mm.

melting has not promoted selective deformation in the Mount Stafford migmatites, thereby preserving the igneous microstructures.

The presence of magmatic andalusite implies very low pressure during melting. In fact, commonly accepted curves for andalusite = sillimanite (Holdaway 1971) and water-saturated 'granite' melting (e.g. Tuttle & Bowen 1958, Merrill *et al.* 1970) leave little or no $P–T$ space for andalusite to coexist in a stable manner with water-saturated felsic melt. Boron and especially fluorine move the water-saturated solidus to lower temperatures (Manning & Pichavant 1983), and so make more $P–T$ space for andalusite. However, apart from very local tourmaline, no fluorine- or boron-bearing minerals have been found at Mount Stafford, and the biotite is not unusually rich in fluorine (Table 11.2). Furthermore, the typical absence of plagioclase from these rocks moves the water-saturated solvus to even higher temperatures than those of the hydrous 'granite' melting curve, which exacerbates the problem. In fact, the coexistence of K-feldspar, cordierite, andalusite, and silica-saturated melt (Vernon & Collins 1988) suggests that the melting reaction 'quartz + K-feldspar + cordierite + water = aluminosilicate + melt' may be applicable to the plagioclase-free rocks in the Mount Stafford area. According to the grid of Vielzeuf & Holloway (1988, Fig. 5), this melting curve occurs at about 720°C at 3.5 kbar, and about 760°C at 2 kbar. Therefore, we suggest that the andalusite–sillimanite curve of Richardson *et al.* (1969) is more appropriate

Table 11.2 Representative microprobe analyses of minerals in samples 1112, 1190f, 2073, and 2064 (localities in Fig. 11.1).

1112	Crd	Bt symp	Bt +sil	Bt matr	Bt in grt	Ksp	Grt core	Grt S_{1c}	Spl	Ilm	And S_{1a-1b}	Sil S_{1c}
SiO_2	47.11	34.85	34.54	34.05	34.50	65.06	37.01	37.38	0.00	0.00	37.04	36.61
TiO_2	0.00	2.62	4.05	3.69	2.61	0.00	0.00	0.00	0.11	53.33	0.00	0.00
Al_2O_3	32.10	30.97	18.09	18.33	19.40	18.74	20.68	21.24	59.41	0.06	62.33	64.47
FeO	12.05	18.30	22.10	21.17	16.80	0.10	37.94	38.75	40.07	46.88	0.19	0.10
MnO	0.23	0.07	0.00	0.00	0.00	0.00	0.71	1.42	0.11	0.32	0.00	0.00
MgO	5.53	4.66	7.82	9.20	11.92	0.00	3.52	2.47	1.67	0.13	0.00	0.00
CaO	0.00	0.00	0.00	0.00	0.07	0.00	0.38	0.55			0.00	0.00
Na_2O	0.18	0.05	0.10	0.08	0.12	1.36	0.08	0.00			0.00	0.00
K_2O	0.03	6.66	9.12	8.59	8.66	14.61	0.03	0.04			0.00	0.00
ZnO	0.00								0.76	0.00		
Total	97.23	98.18	95.82	95.11	94.08	99.87	100.35	101.85	102.13	100.72	99.56	101.18
Si	5.00	4.95	5.31	5.24	5.24	2.99	2.99	2.99	0.00	0.00	1.00	0.98
Ti	0.00	0.28	0.47	0.43	0.30	0.00	0.00	0.00	0.01	1.00	0.00	0.00
Al	4.01	5.18	3.27	3.32	3.47	1.02	1.97	2.00	7.89	0.00	1.99	2.03
Fe	1.07	2.17	2.84	2.72	2.14	0.00	2.56	2.59	3.78	0.98	0.00	0.00
Mn	0.02	0.01	0.00	0.00	0.00	0.00	0.05	0.10	0.01	0.01	0.00	0.00
Mg	0.87	0.99	1.79	2.11	2.71	0.00	0.43	0.29	0.28	0.01	0.00	0.00
Ca	0.00	0.00	0.00	0.00	0.00	0.00	0.03	0.05			0.00	0.00
Na	0.04	0.01	0.03	0.03	0.04	0.12	0.01	0.00			0.00	0.00
K	0.00	1.21	1.79	1.69	1.68	0.86	0.00	0.00			0.00	0.00
Zn	0.00								0.06	0.00		
Total	11.01	14.80	15.50	15.54	15.59	4.99	8.04	8.02	12.03	2.00	2.99	3.01
X_{Fe}	0.55	0.69	0.61	0.56	0.44		0.86	0.90	0.93	0.99		

1190f	Crd	Bt in grt	Bt matr	Bt co spl	Grt rim	Grt core	Spl in crd	Ilm	Opx core	Opx rim	Ksp
SiO_2	49.23	36.56	35.71	43.36	37.20	34.47	0.39	0.15	47.97	47.87	66.08
TiO_2	0.00	4.20	4.36	3.54	0.00	0.00	0.00	53.27	0.23	0.24	0.00
Al_2O_3	32.79	17.81	18.22	15.17	22.17	25.06	56.66	0.11	2.76	2.71	19.32
FeO	12.86	17.31	24.31	21.64	37.43	37.55	40.97	47.84	40.70	39.75	0.09
MnO	0.19	0.00	0.07	0.00	0.44	0.77	0.14	0.17	0.65	0.64	0.00
MgO	5.83	11.94	6.96	4.94	4.06	3.78	1.45	0.34	9.33	9.71	0.00
CaO	0.00	0.00	0.06	0.06	0.15	0.13			0.17	0.11	0.00
Na_2O	0.06	0.13	0.08	0.00	0.00	0.00			0.00	0.00	1.52
K_2O	0.00	9.59	8.73	7.88	0.00	0.04			0.03	0.05	14.08
ZnO							1.82				
Total	100.96	97.54	98.50	96.59	101.45	101.80	101.43	101.88	101.84	101.08	101.09
Si	5.03	5.38	5.36	6.38	2.95	2.74	0.05	0.00	1.92	1.92	2.99
Ti	0.00	0.47	0.49	0.39	0.00	0.00	0.00	0.99	0.01	0.01	0.00
Al	3.95	3.09	3.22	2.63	2.07	2.35	7.69	0.00	0.13	0.13	1.03
Fe	1.10	2.13	3.05	2.67	2.48	2.50	3.95	0.99	1.36	1.34	0.00
Mn	0.02	0.00	0.01	0.00	0.03	0.04	0.01	0.00	0.02	0.02	0.00
Mg	0.89	2.62	1.56	1.09	0.48	0.45	0.25	0.01	0.56	0.58	0.00
Ca	0.00	0.00	0.01	0.01	0.01	0.01			0.01	0.01	0.00
Na	0.01	0.04	0.03	0.00	0.00	0.00			0.00	0.00	0.13
K	0.00	1.80	1.67	1.48	0.00	0.00			0.00	0.00	0.82
Zn							0.16				
Total	11.00	15.53	15.40	14.65	8.02	8.09	12.11	1.99	4.01	4.01	4.97
X_{Fe}	0.55	0.45	0.66	0.71	0.84	0.85	0.94	0.99	0.71	0.70	

Table 11.2 (*Continued*)

2073	Grt rim	Grt core	Grt symp	Grt symp	Bt >crd	Ksp	Ilm in bt	Crd	Sil >crd	Plag
SiO_2	37.43	37.25	35.16	39.44	38.09	65.46	0.29	48.27	36.63	58.93
TiO_2	0.00	0.00	0.00	0.00	0.00	0.00	50.83	0.00	0.00	0.00
Al_2O_3	21.05	21.12	21.53	20.23	19.61	18.49	0.08	32.61	63.02	25.84
FeO	37.02	36.81	39.52	37.67	20.33	0.00	45.60	11.23	0.45	0.00
MnO	1.04	0.98	1.01	1.19	0.05	0.00	0.13	0.13	0.00	0.00
MgO	3.14	3.37	2.82	2.04	8.50	0.00	0.14	6.46	0.05	0.00
CaO	0.99	1.07	0.72	0.88	0.00	0.16	0.00	0.00	0.00	8.11
Na_2O	0.00	0.04	0.00	0.00	0.30	2.09	0.00	0.34	0.00	8.51
K_2O	0.00	0.05	0.00	0.06	8.61	13.52	0.00	0.00	0.00	0.22
ZnO										
Total	100.67	100.69	100.76	101.51	95.49	99.72	97.07	99.04	100.15	101.61
Si	3.00	2.99	2.87	3.13	5.73	3.00	0.01	5.00	0.99	2.61
Ti	0.00	0.00	0.00	0.00	0.00	0.00	0.99	0.00	0.00	0.00
Al	1.99	1.99	2.07	1.89	3.47	1.00	0.00	3.98	2.00	1.34
Fe	2.48	2.47	2.69	2.50	2.56	0.00	0.99	0.97	0.01	0.00
Mn	0.07	0.07	0.07	0.08	0.01	0.00	0.00	0.01	0.00	0.00
Mg	0.38	0.40	0.34	0.24	1.91	0.00	0.01	1.00	0.00	0.00
Ca	0.09	0.09	0.06	0.08	0.00	0.01	0.00	0.00	0.00	0.39
Na	0.00	0.01	0.00	0.00	0.09	0.19	0.00	0.07	0.00	0.73
K	0.00	0.01	0.00	0.00	1.65	0.79	0.00	0.00	0.00	0.01
Zn										
Total	8.01	8.03	8.10	7.92	15.42	4.99	2.00	11.03	3.00	5.08
X_{Fe}	0.87	0.86	0.89	0.91	0.57		0.99	0.49	1.00	

2064	Grt core	Grt rim	Bt in grt	Bt matrix	Bt >crd	Sil S_2	And S_{1a-1b}	Crd	Mag matr	Ksp
SiO_2	37.85	37.56	36.21	34.18	35.99	36.32	37.03	50.86	0.14	64.39
TiO_2	0.00	0.00	3.75	2.85	2.64	0.00	0.00	0.00	0.08	0.00
Al_2O_3	20.45	20.99	17.85	17.32	20.75	63.38	62.78	34.98	0.42	19.34
FeO	35.73	38.19	17.29	19.17	20.27	1.25	0.65	8.22	92.32	0.00
MnO	0.70	0.77	0.00	0.00	0.00	0.00	0.00	0.00	0.00	0.00
MgO	4.58	3.09	12.43	13.75	9.24	0.03	0.04	8.61	0.00	0.00
CaO	0.38	0.47	0.00	0.07	0.00	0.00	0.00		0.00	0.00
Na_2O	0.00	0.00	0.13	0.13	0.14	0.00	0.00		0.00	0.00
K_2O	0.03	0.00	8.91	7.62	9.31	0.00	0.00		0.00	14.56
ZnO	0.00	0.00	0.00					0.00		
Total	99.72	101.07	96.57	95.09	98.34	100.98	100.50	102.67	92.96	99.77
Si	3.03	3.01	5.37	5.19	5.30	0.98	1.00	5.00	0.02	2.97
Ti	0.00	0.00	0.42	0.33	0.29	0.00	0.00	0.00	0.01	0.00
Al	1.93	1.98	3.12	3.10	3.60	2.01	1.99	4.05	0.08	1.05
Fe	2.40	2.56	2.14	2.43	2.50	0.03	0.02	0.68	11.82	0.00
Mn	0.05	0.05	0.00	0.00	0.00	0.00	0.00	0.00	0.00	0.00
Mg	0.55	0.37	2.74	3.11	2.03	0.00	0.00	1.26	0.00	0.00
Ca	0.03	0.04	0.00	0.01	0.00	0.00	0.00		0.00	0.00
Na	0.00	0.00	0.04	0.04	0.04	0.00	0.00		0.00	0.13
K	0.00	0.00	1.68	1.48	1.75	0.00	0.00		0.00	0.86
Zn	0.00	0.00	0.00					0.00		
Total	7.99	8.01	15.51	15.69	15.51	3.02	3.01	10.99	11.93	5.01
X_{Fe}	0.81	0.87	0.44	0.44	0.55	1.00	1.00	0.35	1.00	

for these rocks than lower-pressure equilibria, such as that of Holdaway (1971). Some other studies on low-pressure metamorphic rocks have come to a similar conclusion (e.g. Vernon 1982), and the necessity for a P–T stability field of andalusite in felsic melts has been emphasized by Clarke *et al.* (1976). However, a recent detailed investigation of the problem by Kerrick & Speer (1988) suggests that the effect of boron and fluorine, together with proposed lowering of the solidus by excess Al in the melt, can account for magmatic andalusite, without recourse to the Richardson *et al.* (1969) curve. Nevertheless, as stated previously, the Mount Stafford rocks show little evidence for unusually high boron or fluorine. Moreover, although some of the andalusite in the Mount Stafford rocks is relatively high in Fe, as discussed later, no systematic distribution of Fe between andalusite and sillimanite occurs in the same rock, and the amount of impurity appears to be too small to modify the andalusite–sillimanite curve of Holdaway (1971) sufficiently to intersect the field of andalusite + felsic melt (Vielzeuf & Holloway 1988, Fig. 5), as noted by Kerrick & Speer (1988, p. 185) and as indicated by their Figure 2.

Of course, some of the melting at peak-metamorphic conditions in the higher-grade zones may have been water undersaturated, involving a reaction such as biotite + quartz + andalusite = cordierite + K-feldspar + melt. If so, temperatures would have been higher than for water-saturated melting, but this does not present a problem, because leucosomes in the higher grade zones do not contain andalusite.

In some of the highest-grade rocks, melting locally has been almost complete, and remnants of foliated mesosome pass laterally into massive leucosome, via a zone of nebulitic migmatite with a few relics of the metamorphic layering that are continuous with the foliation in the unmelted rock (Fig. 11.21). This appears to be due to complete, *in situ* melting, which we ascribe to local access of fluid, permitting water-saturated melting (e.g. Collins *et al.* 1989a,b).

Very locally, partial melting has promoted the growth of K-feldspar megacrysts, both in pegmatitic pods and in metapelite just ahead of a local melting 'front'. These megacrysts have simple twinning and are mesoscopically euhedral, although in thin section their margins are typically irregular. Euhedral, simply twinned megacrysts of K-feldspar typically grow in melts (Vernon 1986). Therefore, the K-feldspar megacrysts in these rocks probably grew in minor amounts of melt leucosome, which can generally be seen in the outcrop. Reasons for the development of megacrysts are unknown, this being a nucleation problem that is difficult to solve (Vernon 1986).

11.8 Post-peak reactions

The Mount Stafford rocks show abundant evidence of reactions that occurred after peak-metamorphic temperatures had been reached. Some of these reactions occurred during cooling, as part of the M_1 metamorphism, but others appear be due to the incipient effects of metamorphism M_2, which strongly affected the adjacent Mount Weldon area (Table 11.1). The products of these reactions are typically fine-grained and pseudomorphous, and commonly occur in symplectic intergrowths. The most common intergrowths are the random, symplectic aggregates of andalusite, biotite and quartz, mentioned previously. These intergrowths typically pseudomorph cordierite (Figs 11.11 & 12) and, to a small extent, K-feldspar (Fig. 11.12), especially in the andalusite–cordierite–K-feldspar zone, in which the replacement of cordierite is commonly complete (Fig. 11.11). The intergrowths are identical to those in the Cooma Complex, south-east Australia (Vernon 1978, Vernon & Pooley 1981). As the symplectites are generally random, they post-date S_{1c}.

As mentioned previously, melted material (leucosome) evidently did not escape from the rocks at Mount Stafford, so that water must have been released from the leucosomes as they crystallized during cooling. We infer that the water reacted with cordierite and some K-feldspar to form the symplectic aggregates of andalusite, biotite, and quartz. Some potassium from the melts may also have been involved in the reaction. Crystallization of abundant, pervasive leucosome material retained in the rocks may account for the unusually high proportion of symplectite in this area. However, the symplectite is less common in higher-grade rocks, probably because the water was released while

Figure 11.23 Random, fibrous to prismatic sillimanite replacing twinned cordierite. Crossed polars; base of photograph 4.4 mm.

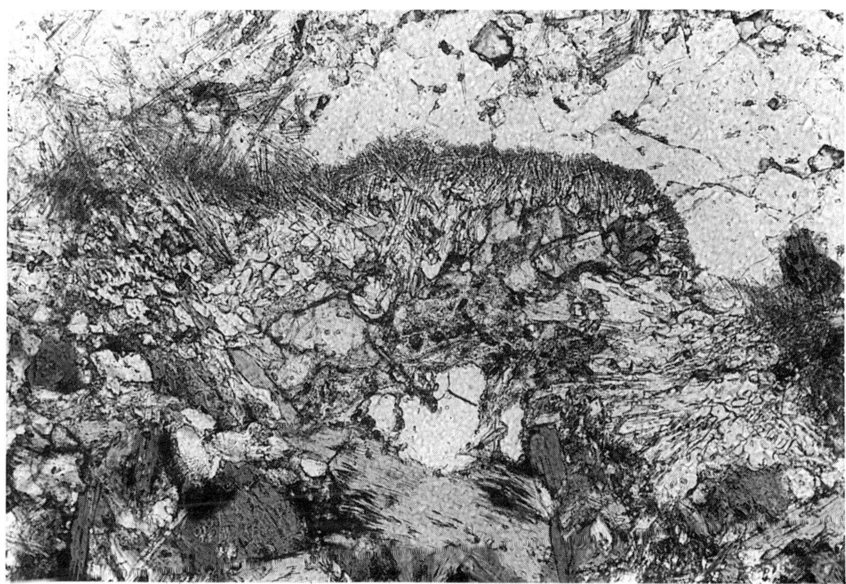

Figure 11.24 Late fibrous sillimanite occurring as a fringe on a symplectic intergrowth of andalusite, biotite, and quartz. Plane polarized light; base of photograph 4.4 mm.

the temperature was high enough to maintain stability of cordierite and K-feldspar.

Late sillimanite is common in the Mount Stafford area, mainly as fibrous aggregates replacing or rimming most or all of the prograde minerals, in an identical manner to fibrous sillimanite in the Cooma Complex, south-east Australia (Vernon 1979). Slightly coarser-grained sillimanite also randomly replaces cordierite in some rocks (Fig. 11.23). Microstructural relationships indicate that the fibrous sillimanite post-dates the symplectic aggregates of andalusite, biotite, and quartz (Fig. 11.24), so that it also post-dates crystallization of the leucosomes and the development of S_{1c}. Because the formation of andalusite-bearing symplectite in the higher-grade rocks involved cooling from the sillimanite *P–T* field into the andalusite field, the later growth of sillimanite (assuming that it grew in a stable manner) probably would have involved re-heating the rocks, and so implies a separate metamorphic event. The most likely event is metamorphism M_2 which, although it strongly affected the rocks of the adjacent Mount Weldon area, appears to have affected the Mount Stafford area only weakly. This interpretation is supported by the increased amount of sillimanite in the east of the Mount Stafford area, where the occurrence of abundant sillimanite appears to transgress the main pattern of M_1 isograds, and where some sillimanite is oriented in a new foliation. This foliation is parallel to S_2, which is strongly developed in the adjacent Mount Weldon area, and which is the main foliation in the deformed granitoids in the east of the Mount Stafford area (Fig. 11.1). The formation of late sillimanite in the Mount Stafford area evidently did not involve much deformation, except locally, where sillimanite foliae were produced, and in the east of the area. However, it did involve an increase in pressure, because cordierite was replaced by sillimanite and garnet in the east of the area.

11.9 Chemical compositions of minerals

Analyses were carried out on an ETEC electron microprobe, with an accelerating voltage of 15 kV, in the School of Earth Sciences, Macquarie University. Representative analyses of minerals in typical high-variance metapelitic assemblages are given in Table 11.2, and compositional ratios used for P–T calculations are given in Table 11.3. The main chemical features are discussed below.

Garnet has X_{Fe} ranging from 0.85 to 0.95, grossular and spessartine contents being consistently less than 3%. The garnet grains are typically zoned, rims having an X_{Fe} that is 0.02–0.06 higher than the cores. The rims of most zoned garnet grains are also slightly richer in spessartine than the cores, but the grossular content is constant. In sample 2073 (Table 11.2), relatively large garnet grains have $X_{Fe} = 0.86$, whereas symplectic garnet overgrowths have $X_{Fe} = 0.90$.

Biotite has variable X_{Fe}, ranging from 0.44 to 0.71, even in the same specimen. Except for one sample, oriented inclusions of biotite in andalusite, cordierite, garnet, and K-feldspar have lower Fe contents than biotite in the matrix lying in S_{1b} or S_{1c} (Table 11.2). We infer that these inclusions belong to the syn-S_{1a} assemblage. Biotite in the symplectic intergrowths of andalusite, biotite, and quartz (Figs 11.11 & 12) generally also is richer in Fe than these oriented biotite inclusions, except for the lower-grade samples. Values of the $TiO_2:(TiO_2 + Al_2O_3)$ ratio vary between 0.05 and 0.13, apparently unsystematically.

Cordierite has values of X_{Fe} ranging from 0.35 to 0.56, without zoning, even in the largest grains. *Orthopyroxene* has values of X_{Fe} varying from 0.69 to 0.74, also without zoning in either Fe or Al. *Spinel* is hercynite, with values of X_{Fe} between 0.92 and 0.95, so that it is always richer in Fe than coexisting garnet. The spinel may have up to 4.5% ZnO, but negligible Cr_2O_3.

The opaque minerals are *ilmenite*, in which the ilmenite end-member varies from 0.97 to 1.00, and *magnetite*, which is pure. *Andalusite* and *sillimanite* have negligible Mn, but Fe_2O_3 may reach more than 1%. No systematic variation in Fe_2O_3 content was observed between andalusite and sillimanite in the same rock. *K-feldspar* varies from Or_{70} to Or_{91}, commonly with slight variation in individual rocks. *Plagioclase* is An_{25} in sample 1140, An_{38} in 1190c, and An_{39} in 2073 (Table 11.2).

Table 11.3 Average pressure calculations on samples from the Mount Stafford area, using the approach of Powell & Holland (1988) with the expanded data set of Holland & Powell (1990).

(a) Average pressure calculations for rock 1190f

(i) the activity of mineral endmembers in the rock

	prp	alm	en	fs	mgts	phl	ann	east
a	0.00427	0.524	0.0827	0.442	0.0143	0.0150	0.0602	0.0110

	kfs	qtz	crd	fcrd
a	0.859	1.00	0.210	0.279

(ii) an independent set of reactions involving the mineral endmembers

(1) 2prp + 3qtz = 2en + crd
(2) 2alm + 3qtz = 2fs + fcrd
(3) 6en + 3fcrd = 4prp + 2alm + 9qtz
(4) 2en + 3fcrd = 2alm + 4mgts + 9qtz
(5) phl + crd = prp + east + 3qtz
(6) en + 2ann + 2crd = 2prp + 2alm + 2kfs + $2H_2O$
(7) 15fs + 6east + 6crd = 8prp + 10alm + 6kfs + $6H_2O$

(iii) calculated pressures for these reactions, for the given activity data and at fixed T = 750°C and a_{H_2O} = 0.3

	P	σ	dP/dT	ln K
1	3.8	1.93	-0.0033	4.367
2	2.3	0.34	0.0069	-1.617
3	3.0	1.34	0.0008	-4.332
4	2.9	1.58	-0.0014	-9.468
5	3.0	1.62	-0.0027	-4.206
6	2.5	1.45	-0.0128	-1.275
7	3.2	1.42	-0.0085	-2.354

(iv) average pressures calculated from the intercepts of the reactions

T °C	700	750	800
av P	2.10	2.50	3.00
σ	0.31	0.32	0.34
σ fit	0.70	0.80	1.10

(v) the effect of varying a_{H_2O} on the average pressure at 750°C

a_{H_2O}	$P \pm 2\sigma$	σ fit
0.1	2.3 ± 1.3	2.1
0.3	2.5 ± 0.6	0.8
0.5	2.5 ± 0.6	0.6
0.7	2.6 ± 0.7	0.6

Table 11.3 (*Continued*)

(b) Average pressure calculations for rock 1202a

(i) the activity of mineral endmembers in the rock

	prp	alm	en	fs	mgts	phl	ann	east
a	7.2e-5	0.520	0.0655	0.458	0.00737	0.0246	0.0611	0.00645

	kfs	qtz	crd	fcrd
a	0.772	1.00	0.181	0.319

(ii) an independent set of reactions involving the mineral endmembers

(1) 2alm + 3qtz = 2fs + fcrd
(2) 2alm + en + 3qtz = 3fs + crd
(3) 2ann + 6qtz = 3fs + 2kfs + $2H_2O$
(4) prp + 6ann + 6fcrd = 10alm + 3mgts + 6kfs + $6H_2O$
(5) ann + crd = alm + east + 3qtz
(6) 5alm + 3east + 9qtz = 2prp + 3ann + 3fcrd
(7) 2alm + 2phl + 3qtz = 2en + 2ann + crd

(iii) calculated pressures for these reactions, for the given activity data and at fixed T = 750°C and $a_{H_2O} = 0.5$

	P	σ	dP/dT	ln K
1	2.1	0.32	0.0067	-1.396
2	2.5	0.55	0.0062	-0.018
3	2.6	1.73	0.0352	2.730
4	2.9	1.48	-0.0063	10.340
5	1.5	1.07	0.0049	-1.193
6	5.1	1.30	0.0136	-12.485
7	4.8	1.54	0.0123	-4.033

(iv) average pressures calculated from the intercepts of the reactions

Using all mineral endmembers:

T °C	700	750	800
av P	1.8	2.1	2.4
σ	0.55	0.66	0.75
σ fit	2.0	2.2	2.4

excluding pyrope from the calculations:

T °C	700	750	800
av P	1.9	2.2	2.6
σ	0.33	0.36	0.41
σ fit	1.1	1.2	1.3

Table 11.3 *(Continued)*

(c) Average pressure calculations for the overprinting mineral assemblages in rock 1112m

(i) the activity of mineral endmembers in the rock

	prp	alm	phl	ann	east	crd	fcrd	kfs
a	6.2e-4	0.559	0.0195	0.0477	0.0191	0.203	0.288	0.834

	sill	qtz
a	1.00	1.00

(ii) an independent set of reactions involving the mineral endmembers

(1) 3crd = 2prp + 4sil + 5qtz
(2) 3fcrd = alm + 4sill + 5qtz
(3) 5east + 4crd = prp + 5phl + 12 sill
(4) east + 2crd = 2prp + kfs + 3sill + H_20
(5) 5ann + 6fcrd = 9alm + 5kfs + 3sill + $5H_20$

(iii) calculated pressures for these reactions, for the given activity data and at fixed T = 650°C and a_{H_2O} = 0.5

	P	σ	dP/dT	ln K
1	3.3	0.81	0.0007	-9.985
2	3.7	0.28	0.0076	2.571
3	6.4	1.35	0.0088	-0.902
4	3.7	1.24	-0.0078	-7.803
5	4.2	0.53	0.0002	16.541

(iv) average pressures calculated from the intercepts of the reactions

T °C	550	600	650	700
av P	3.2	3.5	3.9	4.3
σ	0.48	0.35	0.31	0.37
σ fit	2.1	1.5	1.3	1.5

(v) the effect of varying a_{H_2O} on the average pressure at 650°C

a_{H_2O}	$P \pm 2\sigma$	σ fit
0.1	3.57 ± 1.40	2.8
0.3	3.73 ± 0.76	1.5
0.5	3.92 ± 0.62	1.3
0.7	4.14 ± 0.76	1.5

Table 11.3 *(Continued)*

(d) Average pressure calculations for the overprinting mineral assemblages in sample 2064, excluding biotite

(i) the activity of mineral endmembers in the rock

	prp	alm	sill	qtz	crd	fcrd
a	0.00181	0.589	1.00	1.0	0.249	0.229

(ii) an independent set of reactions involving the mineral endmembers

(1) 3crd = 2prp + 4sil + 5qtz
(2) 3fcrd = 2alm + 4sill + 5qtz

(iii) calculated pressures for these reactions, for the given activity data and at fixed T = 650°C and $a_{H_2O} = 0.5$

	P	σ	dP/dT	ln K
1	4.2	0.75	0.0019	-8.458
2	4.2	0.32	0.0082	3.363

(iv) average pressures calculated from the intercepts of the reactions

T °C	550	600	650	700
av P	3.8	4.0	4.3	4.6
σ	0.28	0.29	0.30	0.31
σ fit	0.9	0.3	0.2	0.7

(v) the effect of varying a_{H_2O} on the average pressure at 650°C

a_{H_2O}	$P \pm 2\sigma$	σ fit
0.1	3.61 ± 0.6	0.22
0.3	3.97 ± 0.6	0.22
0.5	4.33 ± 0.6	0.21
0.7	4.65 ± 0.6	0.20

11.10 *P–T* grids for the mineral assemblages

A KFMASH grid (Fig. 11.25) has been developed as an extension of two relatively well known KFMASH grids, namely a lower-temperature grid not involving spinel (Grant 1985, Fig. 3.12) and a higher-temperature grid not involving biotite (Hensen 1986, Fig. 1b). Figure 11.25 is drawn for spinel X_{Fe} > garnet X_{Fe}, as observed for coexisting spinel and garnet in samples from the Mount Stafford area.

Extension of KFMASH into KFMASH–TiO_2–Fe_2O_3 (KFMASHTO) has been discussed by Clarke *et al.* (1989) and the grid, projected from cordierite, is shown in Figure 11.26. The KFMASH grid (Fig. 11.25) includes the [Opx] intersection, which is metastable with respect to KFASH equilibria in

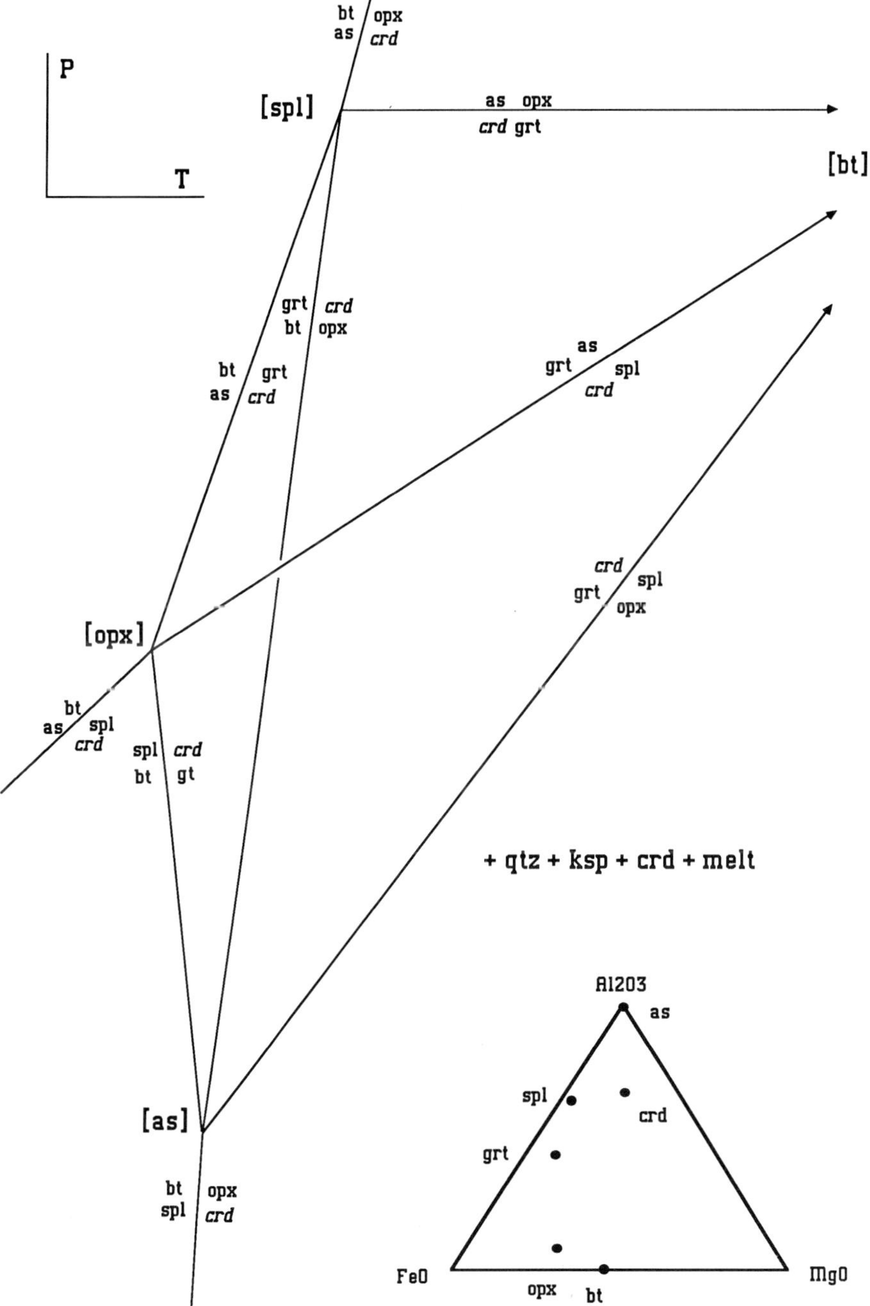

Figure 11.25 The K_2O–FeO–MgO–Al_2O_3–SiO_2–H_2O (KFMASH) *P–T* grid, projected from qtz, Kfs, melt, and crd, calculated for *P–T* conditions inferred to have affected metapelitic rocks in the Mount Stafford area (after Clarke *et al.* 1989, by permission of Blackwell Scientific Publications Ltd). The [opx] intersection is metastable in KFMASH, with respect to the H_2O-present solidus and to KFASH equilibria, but it is included because it is stable in the system KFMASHTO, as shown in Figure 11.26. *Crd* is picked out in italics because it is a constantly present phase, serving as a projection point.

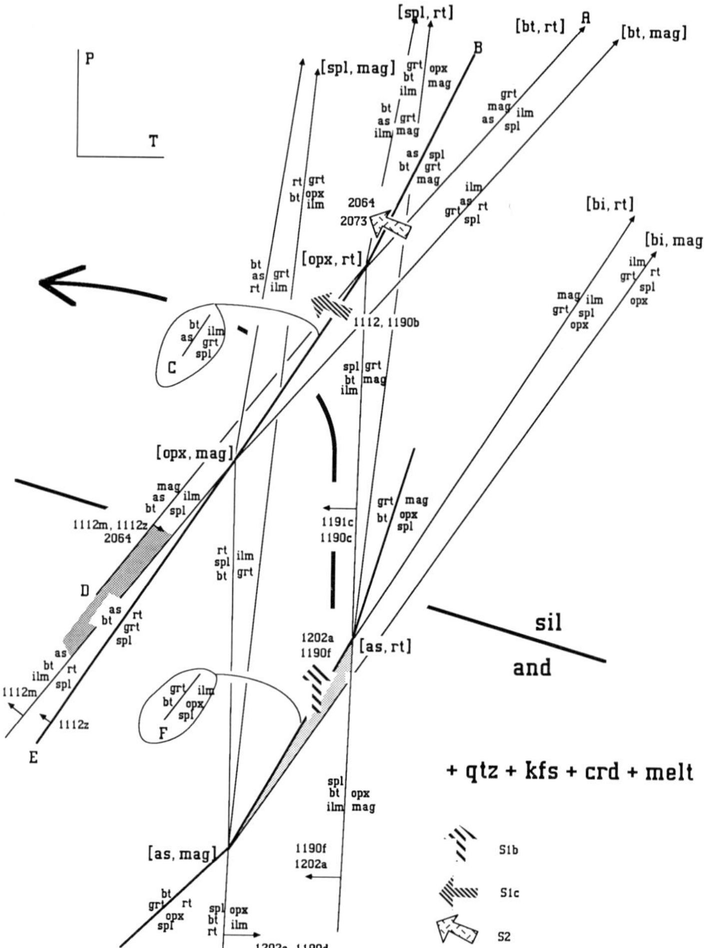

Figure 11.26 The KFMASH–TiO_2–O_2 (KFMASHTO) *P–T* grid, after Clarke *et al.* (1989, 1990, by permission of Blackwell Scientific Publications Ltd). The thicker lines are effectively the loci of the KFMASH intersections [opx] and [as] in Figure 11.25, as Fe_2O_3 and TiO_2 are added to the KFMASH system. The small numbered arrows refer to constraints on the peak-metamorphic conditions provided by the rock of that number. Samples 1112 and 2064 plot at a lower *P* and *T* than 1190 and 1202, which is consistent with field relationships (Fig. 11.1). The larger arrows and reactions A–F refer to overprinting assemblages discussed in the text. The discontinuous line indicates the inferred *P–T–t* path for rocks from the highest-grade parts of the area.

Figure 11.27 Metapelite locally rich in intergrown spinel, sillimanite, and twinned cordierite in the orthopyroxene–garnet–cordierite–K-feldspar zone. Crossed polars; base of photograph 1.75 mm.

KFMASH (Grant 1985), but appears to be stable in KFMASHTO (Clarke *et al.*, 1989). Microstructural evidence suggests that garnet formed during or after breakdown or cordierite in some of the Mount Stafford rocks, so that the assemblages may also be considered on a KFMASH grid projected from garnet, extended into KFMASHTO. However, this grid is almost identical to the projection from cordierite, so that Figure 11.26 serves to illustrate the relevant reactions.

The positions of peak-metamorphic conditions implied by individual assemblages are indicated in Figure 11.26 by small arrows, the large arrows representing part of the *P–T–t* path, as discussed below. The highest-grade assemblages occur at localities 1190 and nearby 1202 (Fig. 11.1). Samples 1190f and 1202a contain spinel and biotite, as part of a univariant assemblage in KFMASHTO. However, spinel is confined to cores of cordierite grains with spinel-free rims, which is consistent with growth of cordierite after spinel had become unstable with respect to quartz and feldspar in the matrix. Biotite is commonly intergrown with garnet and locally appears to have replaced orthopyroxene. As discussed previously, these observations suggest an early assemblage (presumably syn-F_{1a} or early syn-F_{1b}) which contains spinel and orthopyroxene, and a latter assemblage (presumably syn-S_{1b} and/or syn-S_{1c}) which contains biotite, this being consistent with the rocks cooling and reacting across line F in Figure 11.26. Sample 1190b contains local coarse-grained sillimanite intergrown with cordierite and spinel (Fig. 11.27), and ilmenite enclosed by coarse-grained biotite + magnetite. These relationships are consistent with the rock cooling and reacting across reaction D in Figure 11.26.

In sample 2073, which also occurs in the highest-grade zone in the east of the area (Fig. 11.1), cordierite is pseudomorphed by poorly oriented aggregates (syn-S_2) of sillimanite, biotite, magnetite, and garnet. In sample 2063a, which occurs near 2064, cordierite appears to have been replaced by garnet. These relationships are consistent with cooling and reaction across reactions A and B in Figure 11.26, at a higher pressure than that of the [Opx, Rt] invariant point, and probably at a higher pressure than that during the growth of late sillimanite in 1112, discussed below. Thus, as discussed previously, we suggest that the eastern part of the Mount Stafford area underwent continued burial after the main thermal peak of metamorphism, probably during F_2 folding.

11.11 *P–T–t* path

If these inferences are correct, the cooling path (discontinuous line in Figure 11.26) for rocks at localities 1190 and 1202 passes above the [Opx, Mag] invariant point, at higher pressures than at peak-temperature conditions and within the sillimanite field.

The peak assemblage for rocks at locality 1112 occurs on the grid at lower temperatures than those of 1190 and 1202, which is consistent with the occurrence of 1112 in the andalusite–cordierite–K-feldspar zone (Fig. 11.1). The rocks at 1112 contain abundant sillimanite, much of which occurs as fibrous radiating aggregates (Figs 11.23 & 24) and transgressive foliae (S_{1c}), but some of which is coarser-grained and prismatic (Figs 11.13 & 23). The change from stability of andalusite to stability of sillimanite was probably due to an increase in pressure, rather than temperature, because the rocks do not contain the higher-temperature assemblages of 1190 and 1202. Thus, the growth of sillimanite may have occurred at the same time as the reactions discussed above, in response to increasing pressure during cooling. The widespread replacement of cordierite by symplectic andalusite, biotite, and quartz, which predates the late fibrous sillimanite (Fig. 11.24), probably occurred in response to release of fluids from crystallizing melt leucosomes, as discussed previously. Nevertheless it is compatible with growth during increasing pressure, as the reaction 'cordierite + K-feldspar + water = andalusite + biotite + quartz' is relatively pressure sensitive (e.g. Hoffer 1976, Vielzeuf & Holloway 1988).

All the foregoing observations and interpretations involving post-peak reactions are consistent with an anticlockwise *P–T–t* path, which is typical of low-pressure metamorphic areas (e.g. Phillips & Wall 1981, Vernon 1982, Warren 1983, Clarke *et al.* 1987). However, whereas most workers have suggested that initial cooling in low-pressure granulite facies terrains is isobaric (e.g. Bohlen 1987), the Mount Stafford study shows that cooling may be accompanied by crustal thickening, reflected in higher-pressure late assemblages, as inferred for the Olary Block of the Proterozoic Willyama Complex, South Australia, by Clarke *et al.* (1987). This is consistent with abundant evidence of growth of metamorphic minerals during folding in low-pressure terrains (e.g. Zwart 1962, Bell & Rubenach 1983, Vernon 1988). This may perhaps prove to be typical of such terrains, where sufficient evidence is available for detailed analysis of post-peak assemblages to be carried out.

11.12 Peak-metamorphic temperatures

It is difficult to determine an accurate peak-metamorphic temperature for the Mount Stafford rocks, as calculations using available geothermometers gave inconsistent results. However, a peak temperature of about 750°C for the andalusite–cordierite–K-feldspar zone is suggested by the occurrence of melts with magmatic cordierite and andalusite (Vernon & Collins 1988), using the grid of Vielzeuf & Holloway (1988, Fig. 5). This temperature is consistent with the occurrence of two pyroxenes in rocks of mafic composition at relatively low water activity (Phillips 1980); lower water activities are more likely in mafic rocks than in metapelitic rocks (e.g. Phillips 1980). Peak temperatures for the higher-grade zones would have been higher than 750°C.

11.13 Peak-metamorphic pressure

Mineral assemblages described in the previous sections have considerable potential for the estimation of metamorphic pressures, using the average-pressure approach of Powell & Holland (1988), with the expanded internally consistent data set of Holland & Powell (1990). The approach involves the averaging of pressures calculated for each reaction in an independent set of reactions at a stipulated temperature, involving all the mineral end-members shared between an equilibrium mineral assemblage and the data set. The calculations maximize the information available in a mineral assemblage.

The application of the average-pressure approach to the Mount Stafford metapelites is complicated by the problem of re-equilibration during cooling from peak temperature, affecting mineral compositions by, for example, Fe–Mg exchange, even if the peak mineral assemblage is correctly identified. The problem is particularly acute with spinel, because exsolution of magnetite has occurred during cooling in some samples. Moreover, if biotite is included in the calculations, the activity of water is an additional variable. In rocks partly melted at peak conditions in the highest-grade zones, the water activity may be less or even much less than unity, except where melting has been very extensive to locally complete (Figs 11.17 & 21).

As a consequence of these potential difficulties, the most successful calculations were obtained on rocks without spinel, samples 1190f and 1202a giving the best pressure estimates (Table 11.3). A peak temperature of 750°C was selected for the calculations, on the basis of other mineralogical information (discussed previously), although the calculations are effectively insensitive to temperature. Peak pressures are 2.50 ± 0.60 (2σ) kbar for sample 1190f, and 2.24 ± 0.72 (2σ) kbar for sample 1202a, at 750°C and $a_{H_2O} = 0.3$, with a very small a_{H_2O} dependence, owing to solubility of water in cordierite. Pyrope was excluded, because of its small activity, the variability of its concentration, and on the basis of the diagnostic values discussed by Powell & Holland (1988).

Pressure calculations for the later mineral assemblages are very difficult, because of partial overprinting of the peak assemblage. However, samples 1112m and 2064 gave acceptable results (Table 11.3). A pressure for the inferred syn-S_{1c} sillimanite-bearing assemblage of 3.92 ± 0.62 (2σ) kbar, at 650°C and $a_{H_2O} = 0.5$, was obtained for sample 1112 m. The pressure depends strongly on water activity, uncertainties increasing considerably for much higher or much lower water activities. A pressure of 4.30 ± 0.60 (2σ) kbar, at 650°C and $a_{H_2O} = 0.5$, was obtained for the poorly oriented, syn-F_2, garnet–sillimanite aggregates of sample 2064. Biotite was excluded, on the basis of the diagnostics discussed by Powell & Holland (1988).

In summary, the average pressures calculated from samples 1190f and 1202a indicate that the peak-metamorphic pressure at Mount Stafford was between 1.5 and 3.1 kbar. Low pressures of this order of magnitude are consistent with the presence of andalusite in leucosomes, as discussed previously. The higher P estimate for the syn-F_{1b} or syn-F_{1c} assemblage of sample 1112m (3.92 ± 0.62 kbar) and the even higher P estimate for the syn-S_2 assemblage of sample 2064 (4.30 ± 0.60 kbar) are consistent with increasing-pressure P–T–t paths inferred from petrogenetic grids and with metamorphism-folding relationships, as discussed previously. Thus, the quantitative pressure estimates confirm that the Mount Stafford area followed an anticlockwise P–T–t path.

11.14 Discussion

In the Mount Stafford area, metamorphic isograds cut across shallow-dipping metasedimentary rocks. All fold generations appear to have been accompanied by metamorphism, implying grossly compressional deformation during heating. Although we cannot determine whether apparently equivalent folding episodes were synchronous in all metamorphic zones, it appears that the higher-grade rocks began to be heated earlier than the lower-grade rocks, implying proximity to a thermal source. Moreover, the metamorphic grade varies markedly over a small lateral distance, which implies grossly perturbed thermal conditions and a local geotherm of 75°C per km. The pressure was uniformly low (around 2.5 kbar initially, rising to about 4 kbar later). These conditions, occurring at a depth of only 10 km or so, suggest that the heat source was local and yet external to the observed metasedimentary succession, and most probably external to the crust. The observation that melt leucosomes migrated very little (no more than a metre or so, and commonly much less) implies that heat was transferred from the higher- to the lower-grade rocks in the Mount Stafford area by conduction or aqueous fluids or both, rather than by melts. This is consistent with the apparently diachronous relationship between metamorphism and folding.

A satisfactory model for the Mount Stafford metamorphism must involve abnormal access of mantle heat during crustal thickening, while maintaining pressures low enough and temperatures high enough to produce amphibolite to granulite facies assemblages at the appropriately high crustal level. A relatively local magmatic heat source is required, to account for the very high local thermal gradient, but a more regional rise in temperature is also necessary, to account for the more extensive nature of the metamorphism at Mount Stafford than the very local contact metamorphism that typically occurs around high-level plutons in areas that have not undergone regional heating.

We have no direct information on the nature of the local heat source. No clear evidence of syn-metamorphic mafic magmatism has yet been found at Mount Stafford, all mafic rocks having been subjected to the same granulite and amphibolite facies metamorphism as the metasedimentary rocks, although they have to yet been dated radiometrically. Nevertheless, pre-metamorphic mafic intrusions may reflect the early stages of heating of the region.

Furthermore, though they are broadly associated with the highest-grade rocks, granitoids in the north and east of the area cut the metamorphic zones (Fig. 11.1), contain enclaves of deformed and metamorphosed sedimentary rocks (Fig. 11.2), and have been subjected only to the later episodes of deformation that affected the metasediments. Nevertheless, granitic magmas are capable of transferring heat to the upper crust, especially if they are rich in heat-producing elements (e.g. Loosveld & Etheridge 1990), and can produce relatively localized, low-pressure regional metamorphism of the 'regional aureole' type (White *et al.* 1974, Vernon 1982) if they are superimposed on regional pre-heating. A simple model involving rise of geotherms above a rising body of granitic magma, followed by the eventual breakthrough of some of the zone boundaries by the intruding granitoid, has been presented by den Tex (1963). This could account for the present distribution of granitic rocks and metamorphic zones at Mount Stafford. Metamorphism produced by rise of granitic magma is likely to be not only local, but relatively short-lived, which is consistent with the relatively fine grain size of some of the Mount Stafford metasedimentary and meta-igneous granofelses. The model also predicts that higher-grade rocks should heat up earlier, with respect to the folding sequence, than lower-grade rocks, as inferred for Mount Stafford.

The replacement of cordierite by symplectic aggregates of andalusite, biotite, and quartz, and the numerous other overprinting assemblages, are consistent with an anticlockwise *P–T–t* path for the M_1 event. Conditions passed from the andalusite field to the sillimanite field during increase of pressure, and then back to the andalusite field while leucosome melts cooled. The very low pressure, together with the open style of most of the folds, indicate that crustal thickening during metamorphism M_1 was relatively minor, namely around 5 km – certainly insufficient to produce an over-thickened crust. The Mount Stafford area was then overthrust, during ductile deformation (F_2), by rocks of the Mount Weldon area immediately to the east (Collins *et al.* 1989a,b), resulting in the growth of syn-S_2 aggregates of sillimanite, garnet, and biotite after cordierite in the eastern Mount Stafford area. This appears to represent a temporally and spatially distinct metamorphism (M_2), and its effects are described elsewhere (Clarke *et al.* 1990). Both M_1 and M_2 have areally restricted effects in the Reynolds Range region, as do the later M_3 and M_4 events (Clarke *et al.* 1990, Clarke & Powell 1990). Each of the M_{1-4} events was synchronous with compressive deformation, but they have not destroyed the coherency of the regional geology (Clarke & Powell 1990).

Models that can account for the *regional* heating during compression of the terrain are those of Bird & Baumgardner (1981), which involves crust/mantle delamination accompanied by crustal thickening, and Loosveld & Etheridge (1990), based on the work of Houseman *et al.* (1981), which involves crustal thickening and convective lithospheric thinning. However, although these models are relevant to low-pressure metamorphism during thickening on an appropriately broad scale, they not directly applicable to Mount Stafford, where metamorphism is relatively local and occurred at a very high crustal level (around 8–14 km depth). As pointed out by Loosveld & Etheridge (1990), asthenospheric upwelling is likely to cause melting of the lower crust, so that crustal, as well as mantle-derived magmas should rise through the abnormally heated crust. We see no evidence of mafic magmas synchronous with the Mount Stafford metamorphism, but we suggest that granitic magma produced by the partial melting of deep crust by mafic magma is commonly responsible for the transfer of heat to relatively local, high-level terrains, such as the Mount Stafford area. We also suggest that the sequence of folding may be an equally local response to this heating and the consequent release of fluids. However, some broader-scale regional heating (possibly related to unexposed mafic intrusions resulting from a mantle perturbation) is also necessary, to account for the regional extent of the metamorphism, compared with much more localized granite-induced contact metamorphism that is typical of areas not subjected to regional heating accompanying the granite intrusion.

Continental rifting has been proposed as a tectonic setting to explain the high-temperature, low-pressure metamorphic rocks of the Pyrenees (Wickham & Oxburgh 1985). However, only compressional deformation accompanied metamorphism in the Mount Stafford area, which precludes such a setting as being the cause of M_1, with the cautionary note that the observed M_1 zoning is essentially one-dimensional, and direct evidence of the cause of the metamorphism has apparently been destroyed by later intrusions. Moreover, the fact that successive metamorphic and compressional events have not destroyed the coherency of the regional geology of the entire Anmatjira–Reynolds Range region suggests that the area behaved almost passively to asthenospheric thermal perturbations (Clarke & Powell 1990).

11.15 Conclusions

Granulite facies metamorphism in the Mount Stafford area occurred at increasing, but remarkably low pressure (around 2.5–4 kbar). Metamorphism accompanied folding, and P–T conditions followed an anticlockwise path. Post-peak assemblages enable post-peak cooling during relatively minor crustal thickening to be monitored in terms of reactions in the KFMASHTO system. The ultimate cause of the metamorphism was external to the crust, most probably being an asthenospheric thermal perturbation. This may also be applicable to other granulite facies terrains. However, the immediate cause of the regional metamorphism in the Mount Stafford area probably involved transfer of heat to the middle and upper crust by granitic magma, producing the local, relatively short-lived, extremely low-pressure metamorphic event. In effect, this superimposed a local heat source on a region pre-heated by mafic magma resulting from a mantle disturbance.

Acknowledgements

This project was supported by Australian Research Council Grant No. A38415716 and a Macquarie University research grant. T. J B. Holland, A. R. Norman, and D. Waters are thanked for critical comments.

References

Bell, T. H. & M. J. Rubenach 1983. Sequential porphyroblast growth and crenulation cleavage development during progressive deformation. *Tectonophysics* **92**, 171–94.

Bird, P. & J. Baumgardner 1981. Steady propagation of delamination events. *Journal of Geophysical Research* **86**, 4891–903.

Blake, D. H. & R. W. Page 1988. The Proterozoic Davenport Province, central Australia: regional geology and geochronology. *Precambrian Research* **40/41**, 329–40.

Bohlen, S. R. 1987. Pressure–temperature–time paths and a tectonic model for the evolution of granulites. *Journal of Geology* **95**, 617–32.

Bosworth, T. O. 1910. Metamorphism around the Ross of Mull Granite. *Quarterly Journal of the Geological Society of London* **66**, 376–96.

Clarke, D. B., C. B. McKenzie, G. K. Muecke & S. W. Richardson 1976. Magmatic andalusite from the South Mountain batholith, Nova Scotia. *Contributions to Mineralogy and Petrology* **56**, 279–87.

Clarke, G. L., W. J. Collins & R. H. Vernon 1990. Successive Early Proterozoic metamorphic events in the Anmatjira Range, central Australia. *Journal of Metamorphic Geology*, in press.

Clarke, G. L., M. Guiraud, R. Powell & J.-P. Burg 1987. Metamorphism in the Olary Block, South Australia: compression with cooling in a Proterozoic fold belt. *Journal of Metamorphic Geology* **5**, 291–306.

Clarke, G. L. & R. Powell 1990. Proterozoic granulite facies metamorphism in the southeastern Reynolds Range, central Australia: geological context, *P–T* path, and overprinting relationships. *Journal of Metamorphic Geology*, submitted.

Clarke, G. L., R. Powell & M. Guiraud 1989. Low-pressure granulite facies metapelitic assemblages and corona textures from MacRobertson Land, east Antarctica: the importance of Fe_2O_3 and TiO_2 in accounting for spinel-bearing assemblages. *Journal of Metamorphic Geology* **7**, 323–35.

Clemens, J. D. & V. J. Wall 1981. Origin and crystallization of some peraluminous (S-type) granitic magmas. *Canadian Mineralogist* **19**, 111–31.

Collins, W. J. & C. Teyssier 1989. Crustal-scale ductile fault systems in the Arunta Inlier, central Australia. *Tectonophysics* **158**, 49–66.

Collins, W. J., R. H. Vernon & G. L. Clarke 1989a. Two terranes of contrasting structural evolution in the Anmatjira Range, central Australia. *Journal of Structural Geology*, submitted.

Collins, W. J., I. S. Williams & W. Compston 1989b. Three short-lived granulite facies events in the Arunta Block, central Australia. *Geology*, submitted.

den Tex, E. 1963. A commentary on the correlation of metamorphism and deformation in space and time. *Geologie en Mijnbouw* **42**, 170–6.

Ellis, D. J. 1987. Origin and evolution of granulites in normal and thickened crusts. *Geology* **15**, 167–70.

England, P. C. 1987. Diffuse continental deformation: length scales, rates and metamorphic evolution. *Philosophical Transactions of the Royal Society of London* **A321**, 3–22.

England, P. C. & S. W. Richardson 1977. The influence of erosion upon the mineral facies of rocks from different metamorphic environments. *Journal of the Geological Society of London* **134**, 210–13.

Grant, J. A. 1985. Phase equilibria in partial melting of pelitic rocks. In *Migmatites*, J. R. Ashworth (ed.), 86–144. Glasgow: Blackie.

Hensen, B. J. 1986. Theoretical phase relations involving cordierite and garnet revisited: the influence of oxygen fugacity on the stability of sapphirine and spinel in the system Mg–Fe–Al–Si–O. *Contributions to Mineralogy and Petrology* **92**, 362–7.

Hoffer, E. 1976. The reaction sillimanite + biotite + quartz $\rightleftharpoons$ cordierite + K-feldspar + H_2O and partial melting in the system K_2O–FeO–MgO–Al_2O_3–SiO_2–H_2O. *Contributions to Mineralogy and Petrology* **55**, 127–30.

Holdaway, M. J. 1971. Stability of andalusite and the aluminum silicate phase diagram. *American Journal of Science* **271**, 97–131.

Holdaway, M. J. & S. M. Lee 1977. Fe–Mg cordierite stability in high-grade pelitic rocks based on experimental, theoretical and natural observations. *Contributions to Mineralogy and Petrology* **63**, 175–98.

Holland, T. J. B. & R. Powell 1990. An enlarged and updated internally consistent thermodynamic dataset with uncertainties and correlations: the system K_2O–Na_2O–CaO–MgO–MnO–FeO–Fe_2O_3–Al_2O_3–TiO_2–SiO_2–C–H_2–O_2. *Journal of Metamorphic Geology* **8**, in press.

Houseman, G. A., D. P. McKenzie & P. Molnar 1981. Convective instability of a thickened boundary layer and its relevance for the thermal evolution of continental convergence belts. *Journal of Geophysical Research* **86**, 6115–32.

Johannes, W. 1985. The significance of experimental studies for the formation of migmatites. In *Migmatites*, J. R. Ashworth (ed.), 36–85. Glasgow: Blackie.

Kenah, C. & L. S. Hollister 1983. Anatexis in the Central Gneiss Complex, British Columbia. In *Migmatites, melting and metamorphism*, M. P. Atherton & C. D. Gribble (eds), 142–62. Nantwich, Cheshire: Shiva.

Kerrick, D. M. & J. A. Speer 1988. The role of minor element solid solution on the andalusite–sillimanite equilibrium in metapelites and peraluminous granitoids. *American Journal of Science* **288**, 152–92.

Kretz, R. 1966. Interpretation of the shape of mineral grains in metamorphic rocks. *Journal of Petrology* **7**, 68–94.

Kretz, R. 1983. Symbols for rock-forming minerals. *American Mineralogist* **68**, 277–9.

Lister, G. S. & P. F. Williams 1983. The partitioning of deformation in flowing rock masses. *Tectonophysics* **92**, 1–33.

Loosveld, R. J. H. & M. A. Etheridge 1990. A model for low-pressure facies metamorphism during crustal thickening. *Journal of Metamorphic Geology*, submitted.

McLellan, E. L. 1983. Contrasting textures in metamorphic and anatectic migmatites: an example from the Scottish Caledonides. *Journal of Metamorphic Geology* **1**, 241–62.

Manning, D. A. C. & M. Pichavant 1983. The role of fluorine and boron in the generation of granitic melts. In *Migmatites, melting and metamorphism*, M. P. Atherton & C. D. Gribble (eds), 94–109. Nantwich, Cheshire: Shiva.

Mawer, C. K., D. C. Rubie & A. J. Brearley 1988. A model for rapid melting in crustal shear zones: implications for mechanisms of melt migration (abstract). *EOS, Transactions, American Geophysical Union* **69**, 1411.

Mehnert, K. R. 1968. *Migmatites and the origin of granitic rocks*. Amsterdam: Elsevier.

Merrill, R. B., J. K. Robertson & P. J. Wyllie 1970. Melting reactions in the system $NaAlSi_3O_8$–$KAlSi_3O_8$–SiO_2–H_2O to 20 kilobars compared with results for other feldspar–quartz–H_2O and rock–H_2O systems. *Journal of Geology* **78**, 558–69.

Noakes, L. C. 1953. The structure of the Northern Territory in relation to mineralization. In *Fifth Empire Mineral Congress: geology of Australian ore deposits*, 284–96.

Phillips, G. N. 1980. Water activity changes across an amphibolite–granulite facies transition, Broken Hill, Australia. *Contributions to Mineralogy and Petrology* **75**, 377–86.

Phillips, G. N. & V. J. Wall 1981. Evaluation of prograde metamorphic conditions: their implications for the heat source and water activity during metamorphism in the Willyama Complex, Broken Hill, Australia. *Bulletin de Minéralogie* **104**, 801–10.

Powell, R. & T. J. B. Holland 1988. An internally consistent dataset with uncertainties and correlations: 3. Applications to geobarometry, worked examples and a computer program. *Journal of Metamorphic Geology* **6**, 173–204.

Richardson, S. W., M. C. Gilbert & P. M. Bell 1969. Experimental determination of kyanite–andalusite and andalusite–sillimanite equilibria: the aluminum silicate triple point. *American Journal of Science* **267**, 259–72.

Sandiford, M. A. 1985. The metamorphic evolution of granulites at Fyfe Hills: implications for Archaean crustal thickness in Enderby Land, Antarctica. *Journal of Metamorphic Geology* **3**, 155–78.

Selverstone, J., F. S. Spear, G. Franz & G. Morteani 1984. High-pressure metamorphism in the SW Tauern Window, Austria: *P–T* paths form hornblende–kyanite–staurolite schists. *Journal of Petrology* **25**, 501–31.

Sheraton, J. W., L. A. Offe, R. J. Tingey & D. J. Ellis 1980. Enderby Land, Antarctica–an unusual Precambrian high-grade terrain. *Journal of the Geological Society of Australia* **27**, 1–18.

Stewart, A. J. 1981. *Reynolds Range region*. 1 : 100 000 Geological Map Series. Canberra: Australian Bureau of Mineral Resources, Geology & Geophysics.

Stewart, A. J., L. A. Offe, A. J. Glikson, R. G. Warren & L. P. Black 1980. Geology of the northern Arunta Block, Northern Territory. *Australian Bureau of Mineral Resources, Geology & Geophysics Record* 1980/83, 292 pp.

Stewart, A. J., R. D. Shaw & L. P. Black 1984. The Arunta Inlier: a complex ensialic mobile belt in central Australia. Part 1. Stratigraphy, correlations and origin. *Australian Journal of Earth Sciences* **31**, 445–55.

Thompson, A. B. & P. C. England 1984. Pressure–temperature–time paths of regional metamorphism II. Their inference and interpretation using mineral assemblages in metamorphic rocks. *Journal of Petrology* **25**, 929–55.

Turner, F. J. 1981. *Metamorphic petrology*, 2nd edn. New York: McGraw-Hill.

Tuttle, O. F. & N. L. Bowen 1958. Origin of granite in the light of experimental studies in the system $NaAlSi_3O_8$–$KAlSi_3O_8$–SiO_2–H_2O. *Geological Society of America, Memoir* 74, 153 pp.

Vernon, R. H. 1968. Microstructures of high-grade metamorphic rocks at Broken Hill, Australia. *Journal of Petrology* **9**, 1–22.

Vernon, R. H. 1976. *Metamorphic processes*. London: Allen & Unwin.

Vernon, R. H. 1978. Pseudomorphous replacement of cordierite by symplectic intergrowths of andalusite, biotite and quartz. *Lithos* **11**, 283–9.

Vernon, R. H. 1979. Formation of late sillimanite by hydrogen metasomatism (base-leaching) in some high-grade gneisses. *Lithos* **12**, 143–52.

Vernon, R. H. 1982. Isobaric cooling of two regional metamorphic complexes related to igneous intrusions in southeastern Australia. *Geology* **10**, 76–81.

Vernon, R. H. 1986. K-feldspar megacrysts in granites – phenocrysts, not porphyroblasts. *Earth-Science Reviews* **23**, 1–63.

Vernon, R. H. 1987. Oriented growth of sillimanite in andalusite, Placitas – Juan Tabo area, New Mexico, U.S.A. *Canadian Journal of Earth Sciences* **24**, 580–90.

Vernon, R. H. 1988. Sequential growth of cordierite and andalusite porphyroblasts, Cooma Complex, Australia: microstructural evidence of a prograde reaction. *Journal of Metamorphic Geology* **6**, 255–69.

Vernon, R. H. & W. J. Collins 1988. Igneous microstructures in migmatites. *Geology* **16**, 1126–9.

Vernon, R. H. & G. D. Pooley 1981. SEM/microprobe study of some symplectic intergrowths replacing cordierite. *Lithos* **14**, 75–82.

Vielzeuf, D. & J. R. Holloway 1988. Experimental determination of the fluid-absent melting relations in the pelitic system. Consequences for crustal differentiation. *Contributions to Mineralogy and Petrology* **98**, 257–76.

Warren, R. G. 1983. Metamorphic and tectonic evolution of granulites, Arunta Block, central Australia. *Nature* **305**, 300–3.

Warren, R. G. & A. J. Stewart 1988. Isobaric cooling of Proterozoic high-temperature metamorphites in the northern Arunta Block, central Australia. *Precambrian Research* **40/41**, 175–98.

Weiss, L. E. & D. B. McIntyre 1957. Structural geometry of Dalradian rocks at Loch Leven, Scottish Highlands. *Journal of Geology* **65**, 575–602.
Wells, P. R. A. 1980. Thermal models for the magmatic accretion and subsequent metamorphism of continental crust. *Earth and Planetary Science Letters* **46**, 253–65.
White, A. J. R., B. W. Chappell & J. R. Cleary 1974. Geologic setting and emplacement of some Australian Paleozoic batholiths and implications for intrusion mechanisms. *Pacific Geology* **8**, 159–71.
Wickham, S. M. & E. R. Oxburgh 1985. Continental rifts as a setting for regional metamorphism. *Nature* **318**, 330–3.
Windrum, D. P. & M. T. McCulloch 1986. Nd and Sr isotopic systematics of central Australian granulites: chronology of crustal development and constraints on the evolution of lower continental crust. *Contributions to Mineralogy and Petrology* **94**, 289–303.
Wyborn, L. A. I. 1977. *Aspects of the geology of the Snowy Mountains region and their implications for the tectonic evolution of the Lachlan Fold Belt*. Unpubl. PhD thesis, Australian National University, Canberra.

Zwart, H. J. 1962. On the determination of polymetamorphic mineral associations and its application to the Bosost area (Central Pyrenees). *Geologische Rundschau* **52**, 38–65.

CHAPTER TWELVE

Archaean and Proterozoic high-grade terranes of East Antarctica (40–80°E): a case study of diversity in granulite facies metamorphism

S. L. Harley & B. J. Hensen

12.1 Introduction and general considerations

In recent years there has been a tendency for workers in high-grade terranes to seek and identify unified and somewhat generalized models for the origin of granulites, usually based on observations in classic terranes such as the Adirondacks (Bohlen 1987) and southern India (e.g. Newton *et al.* 1980, Hansen *et al.* 1984). In particular, some studies have emphasized a perceived uniformity in granulite $P-T$ conditions and $P-T-t$ (pressure–temperature–time) paths (Newton & Perkins 1982, Bohlen 1987), and hence have modelled granulite terranes in terms of one particular tectonic setting such as a magmatic arc. In contrast, Harley (1989a) has emphasized *diversity* in the important features of granulite terranes, most notably in their $P-T$ conditions of formation and $P-T-t$ paths but also in their lithological constitution and age structures, and has considered a spectrum of possible tectonic settings and mechanisms for granulite formation.

The pressure–temperature ($P-T$) realm and $P-T-t$ paths of a large number of granulite terranes or occurrences are summarized in Figure 12.1, based on the data tabled and discussed by Harley (1989a). A general $P-T$ condition of 7.5 ± 1 kbar and 800 ± 50°C (Bohlen 1987) encompasses nearly 50% of $P-T$ data for granulites worldwide, but it is clear that a great number of granulite occurrences *do not* fit into this 'best-fit' $P-T$ box. A significant group of low-P ($\leqslant$5 kbar) and high-P ($\geqslant$9 kbar) granulites are apparent in Figure 12.1A. The former are often associated with magmatism (e.g. anorthosite massifs) and the latter have been metamorphosed at $P-T$ conditions which

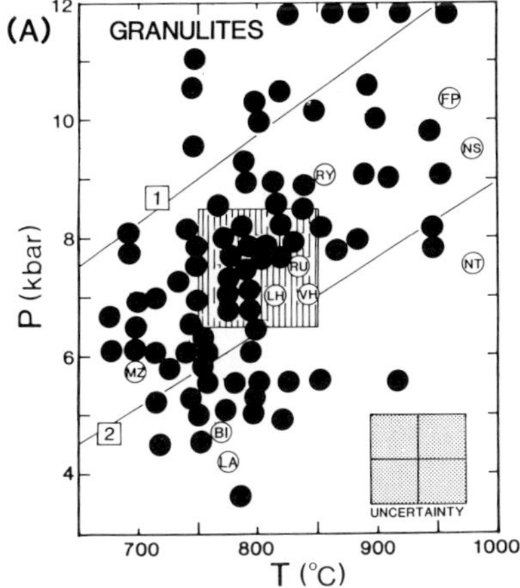

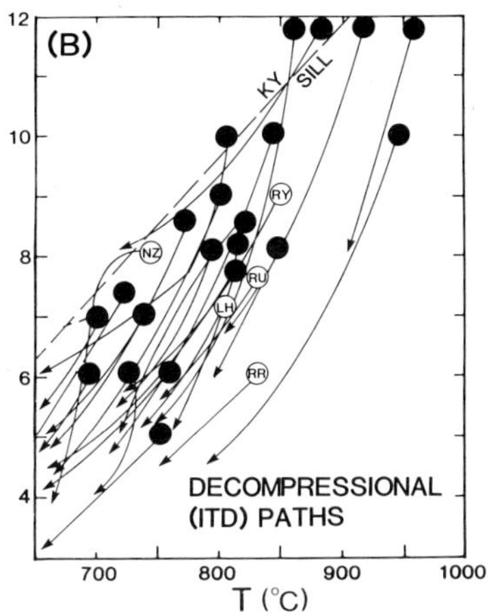

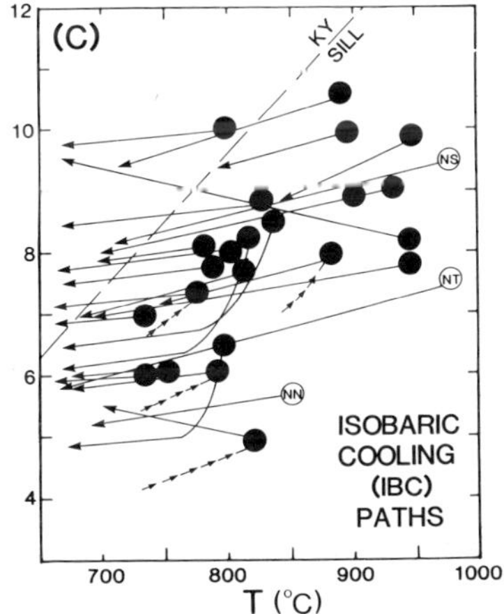

Figure 12.1 Pressure–temperature (*P–T*) conditions and *P–T–t* paths for granulites worldwide (modified from Harley 1989a, Figs 1, 10 and 11, by permission of Cambridge University Press). (A) *P–T* estimates for peak conditions. The vertical ruled box is the 'best-fit' *P–T* estimate (Bohlen 1987) for granulites. The shaded box indicates typical uncertainties on *P–T* estimates. Reaction curve (1) is the incoming of garnet in a typical quartz–tholeiite composition (garnet on high-*P* side), after Green & Ringwood (1967). Curve (2) represents the incoming of garnet at high *P* in undersaturated basalt (Ito & Kennedy 1971). (1) and (2) essentially separate high-, medium-, and low-*P* granulites (Green & Ringwood 1967). (B) Examples of decompressional (ITD) *P–T–t* paths. Temperature uncertainties on each path are in the range ±30–60°C. (C) Examples of isobaric cooling (IBC) *P–T–t* paths. Pressure uncertainties on each path are in the range ±0.5–1 kbar. Short arrowed prograde segments for some counterclockwise *P–T–t* path terranes are also shown. Abbreviations of Antarctic terranes or areas: BI, Bolingen Islands; FP, Forefinger Point; LA, Larsemann Hills; LH, Lützow – Holm Bay; MZ, Molodezhnaya; NS, Napier Complex, Scott Mts; NT, Napier Complex, Tula Mts; NN, Napier Complex, Napier Mts; NZ, Napier Complex, shear zones; RY, Rayner Complex; RU, Rauer Group (south); RR, Rauer Group (north); VH, Vestfold Hills.

overlap with those determined for crustal xenoliths entrained in basaltic volcanics (Griffin & O'Reilly 1986). There is also a group of important granulite terranes which exhibit *very high-T* (VHT) metamorphism with peak temperatures in the range 900–1000°C.

The large range in peak *P–T* conditions shown in Figure 12.1A is defined through both mineral assemblage constraints and quantitative geothermobarometry of granulites (Harley 1989a). Certain key *mineral assemblages* may be indicators of specific *P–T* fields. As discussed by Harley (1989a) and earlier workers (e.g. Ellis *et al.* 1980, Ellis 1983, Hensen & Green 1973), VHT granulite metamorphism can be recognized from the occurrence of sapphirine + quartz in metapelites, relatively Mg-rich pigeonite or aluminous pyroxenes in metabasites and meta-ironstones, and wollastonite–scapolite assemblages in metacalc-silicates. *Low-P* granulite metamorphism is suggested by the occurrence of spinel + quartz in metapelites (e.g. Waters, 1986) and fayalite + quartz in meta-ironstones or felsic orthogneisses, whereas *high-P* granulite metamorphism may be characterized by garnet-bearing parageneses in quartz-bearing metabasites. The threefold baric division of the granulite facies on the basis of metabasite mineral assemblages (Green & Ringwood 1967) (Fig. 12.1A) is still fundamental to our understanding of the *P–T* realm of the granulite facies.

P–T estimates given in Figure 12.1A are mostly based on geothermobarometry of peak mineral assemblages. Barometers are based mainly on continuous net-transfer equilibria which occur within specific mineral assemblages (e.g. garnet–cordierite–sillimanite–quartz, Hensen & Green 1973; garnet–rutile–sillimanite–ilmenite–quartz, Bohlen 1987) and in which the compositions (e.g. Fe–Mg, Ca–Mg) of the solid solutions vary only as functions of *P* and *T* (see Hensen & Harley, Ch. 2). Barometer calibrations based either on direct experimental data within the chemical system of interest, or on reversals of end-member analogue reactions coupled with activity-composition relationships for the solid solutions, are now available for many granulite assemblages. Pressures calculated using such calibrations are generally consistent with, and refine somewhat, the broader assemblage constraints mentioned above.

Temperatures of granulite metamorphism are often calculated using Fe–Mg exchange reactions between coexisting phases (e.g. garnet–biotite, garnet–orthopyroxene, garnet–clinopyroxene) or Ca–Mg transfer between coexisting pyroxenes. However, *peak T* conditions are often not recorded by these techniques because of re-equilibration between coexisting phases on cooling towards blocking temperatures for cation exchange (e.g. Ellis & Green 1985, Harley 1985a, Bohlen 1987), and an underestimate of peak temperature results. This problem is particularly important in the cases of VHT granulites where mineral assemblages often indicate maximum temperatures higher than those calculated using cation exchange thermometers.

Granulite terranes also exhibit a spectrum of *P–T–t paths*, which provide us with some constraints on the type of tectonic or magmatic settings in which the metamorphism took place (England & Thompson 1986, Bohlen 1987, Ellis 1987, Sandiford & Powell 1986b, Harley 1989a). *P–T–t* paths which involve significant decompression at high *T* (isothermal-decompression paths, ITD) are recognized from many granulite terranes (Fig. 12.1B) and indicate that peak metamorphism occurred in overthickened crust. In contrast, *P–T–t* paths in which cooling from high *T* occurred at near-constant pressure (isobaric-cooling paths, IBC) (Fig. 12.1C) suggest that peak granulite conditions were attained in crust of near-normal thickness (e.g. Ellis 1980, Bohlen 1987). The *P–T–t* paths depicted in Figures 12.1B & C are constrained by the following:

(a) The geothermobarometry of texturally earlier and later mineral assemblages or generations, or of zoned minerals in specific assemblages responsive to *P–T* changes.
(b) Textural features, such as the development of garnet lamellae and grains from earlier aluminous phases (IBC) or the replacement of garnet by symplectites of lower-*P* assemblages (e.g. orthopyroxene–plagioclase, orthopyroxene–cordierite) (ITD), which can be interpreted in terms of mineral reactions and hence define a *P–T–t* trajectory. The combination of a variety of reaction textures and hence inferred mineral reactions from a number of related rocks can lead to very well defined *P–T–t* paths (Harley 1989a).

Some important features of granulite *P–T–t* paths include the following:

(a) Some cooling generally attends decompression in the case of ITD paths, and most ITD-type granulites record decompression intervals of 2–4 kbar. Decompressional *P–T–t* paths are preserved in granulites which show a wide range in 'peak' *P–T* conditions, including some VHT granulites and low-*P* types.
(b) Granulites exhibiting isobaric cooling (IBC) *P–T–t* paths also preserve a wide range in initial or peak *P–T* conditions. IBC paths are obtained from deep-crustal (7–10 kbar) and mid-crustal or lower-*P* granulites (4–7 kbar).
(c) 'Isobaric' cooling usually does involve some decrease in pressure (0.5–2 kbar), but typical $\mathrm{d}P/\mathrm{d}T$ gradients are only 0.3–0.5 kbar per 100°C.

Other features and implications of these paths are discussed in more detail by Harley (1989a), and will be considered here in relation to specific East Antarctic examples in subsequent sections. It is clear that great diversity exists in granulite metamorphism, based on the *P–T* and related data. In this chapter the granulite facies terranes of East Antarctica (40–80°E) will be described and analysed as a demonstration of this diversity.

Coastal exposures of the East Antarctic Shield are dominated by granulite facies terranes which range in age from over 3000 Ma to perhaps 750 Ma (Tingey, unpublished data; James & Tingey 1983). These terranes show variety and diversity in their geological histories and *P–T–t* evolutions which reflect differences in their mechanisms of formation. Much of this diversity can be related to the relative importance of, and interplay between, tectonic (thickening, thinning) and magmatic processes in the formation of granulites. In the region to be discussed in this chapter, between 40°E and 80°E, diversity in granulite metamorphism is well illustrated by the occurrence of several very distinctive terranes:

(a) The *c.* 3070–3000 Ma Napier Complex, which preserves good evidence for regional granulite metamorphism at extreme temperatures (1000°C) of rock units perhaps as old as 3950 Ma. This terrane also shows evidence for near-isobaric cooling (IBC) subsequent to the metamorphic maximum, and for continued lower-crustal residence for a period of some 2000 Ma. It is cut by several generations of Proterozoic dolerite dykes.
(b) The 3000–2500 Ma Vestfold Block, which preserves a polymetamorphic history culminating in a late phase of extensive magmatic accretion at 2500 Ma. This terrane is dissected spectacularly by several generations of Proterozoic dyke swarms.
(c) A Late Proterozoic Complex in which a *c.* 1000 Ma tectonothermal event variably affected basement and cover rocks of Proterozoic age, and also locally reworked Archaean granulite precursors. Near-isothermal decompressional *P–T–t* paths (ITD) characterize the different parts of this belt.
(d) The Lützow – Holm Bay Complex, which preserves an apparently prograde and single-cycle amphibolite to granulite transition and definite evidence for a clockwise *P–T–t* path terminated by near-isothermal decompression.

The details of the geological and metamorphic histories of these terranes and sub-areas within them will be considered here, and their bearing on models for the origins of granulites in general, and the development of the East Antarctic Shield in particular, will be assessed. The reader is referred to Figures 12.2, 3, 7 & 8 for the location of various place names used in the text.

Figure 12.2 Location map of East Antarctica between 40°E and 80°E, showing the diversity of pressure–temperature data and paths obtained in the region. The shaded region is the Antarctic coastline. The small boxes are pressure (*P*, vertical axes in kbar) – temperature (*T*, horizontal axes in °C × 100) diagrams for the indicated areas or localities. Solid arrows in these boxes define continuous *P–T–t* paths; dashed lines indicate inferred links between earlier and later *P–T* conditions in overprinted, polymetamorphic, granulites. Circled numbers 1, 2, and 3 refer to the amphibolite-, transitional-, and granulite-grade zones, respectively, in the Lützow–Holm Complex. Sources of data are given in the text.

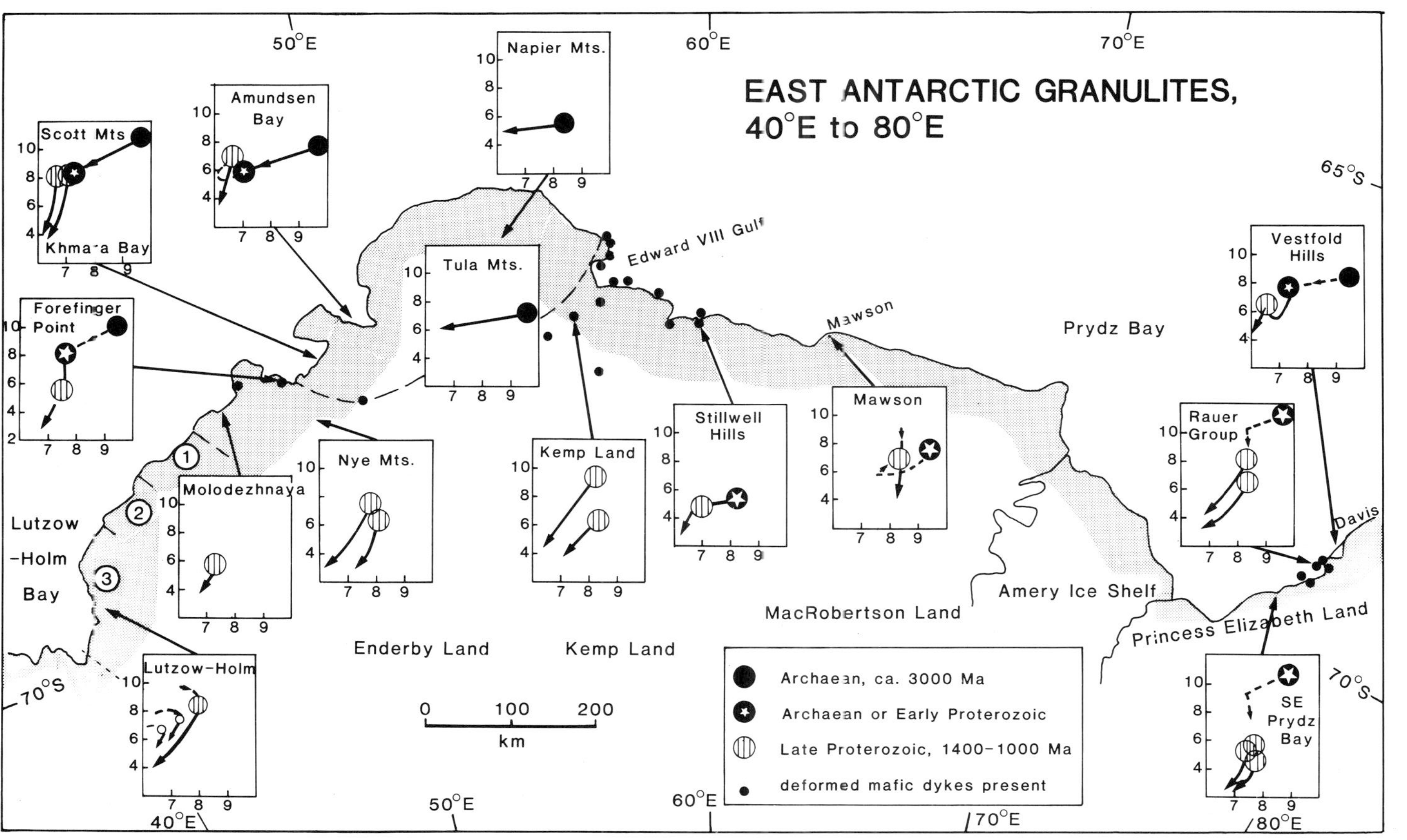
EAST ANTARCTIC GRANULITES,
40°E to 80°E
Scott Mts
Khma-a Bay
Amundsen Bay
Napier Mts.
Tula Mts.
Forefinger Point
Molodezhnaya
Nye Mts.
Kemp Land
Stillwell Hills
Mawson
Rauer Group
Vestfold Hills
SE Prydz Bay
Lutzow-Holm
Edward VIII Gulf
Mawson
Prydz Bay
Davis
Lutzow -Holm Bay
Enderby Land
Kemp Land
MacRobertson Land
Amery Ice Shelf
Princess Elizabeth Land
0 100 200
km
Archaean, ca. 3000 Ma
Archaean or Early Proterozoic
Late Proterozoic, 1400–1000 Ma
deformed mafic dykes present
40°E
50°E
60°E
70°E
80°E
65°S
70°S

12.2 The Napier Complex

12.2.1 Geological background and general features

The geology and geological history of the Napier Complex (Fig. 12.2) have been described extensively (Sheraton *et al.* 1980, 1987; James & Black 1981; Black & James 1983; Black *et al.* 1983a,b; Sandiford & Wilson 1983, 1984; Harley & Black 1987) and will not be discussed here except within the context of the metamorphic history. Important features are summarized in Table 12.1 and presented in Figures 12.2 and 12.3. Although layered paragneiss sequences of mainly supracrustal origin (Sheraton *et al.* 1980, 1987; Sandiford & Wilson 1986) are important components of the complex, the terrane is dominated by felsic to intermediate orthogneisses. Paragneiss protoliths include quartzites, magnesian quartzites, ironstones (including Mn- and Ba-rich types), aluminous pelites, and quartzofeldspathic rock types. Mafic and ultramafic granulites occur both within layered gneiss sequences and in massive orthogneisses. Felsic orthogneisses, which are particularly dominant in the southern and northern parts of the complex, include two broad types (Fig. 12.3):

(a) Orthopyroxene–mesoperthite–quartz granitic orthogneisses (charnockites) which are relatively undepleted in Y and HREE (Sheraton & Black 1983, Sheraton *et al.* 1985). An early- to syn-metamorphic (D_1) 3070 Ma orthogneiss from Proclamation Island is of this type, as are many orthogneisses which intrude paragneisses in the Khmara Bay and Tula Mountains areas.
(b) Orthopyroxene–plagioclase–quartz tonalitic orthogneisses which dominate in the Southern Scott Mountains (Sheraton *et al.* 1987).

These orthogneisses show depletion in Y and HREE, and hence have been interpreted to be derived from melting of mafic precursors containing garnet or amphibole or both. Although they may have been emplaced too at the time of the main metamorphism, substantial crustal prehistories are inferred for examples of such orthogneisses from Fyfe Hills and Mount Sones, where zircon age populations of up to 3800 Ma and 3930 Ma respectively (Compston & Williams 1982, Black *et al.* 1986a) have been recognized. The first recognized granulite event in this terrane therefore post-dated the formation of some crustal precursors by as long as 900 Ma. The relative volume of such very old crust within the Napier Complex compared with crust produced coevally with the major tectonism at 3070–2900 Ma remains a major outstanding question.

Subsequent to the major tectonism the only significant felsic magmatism

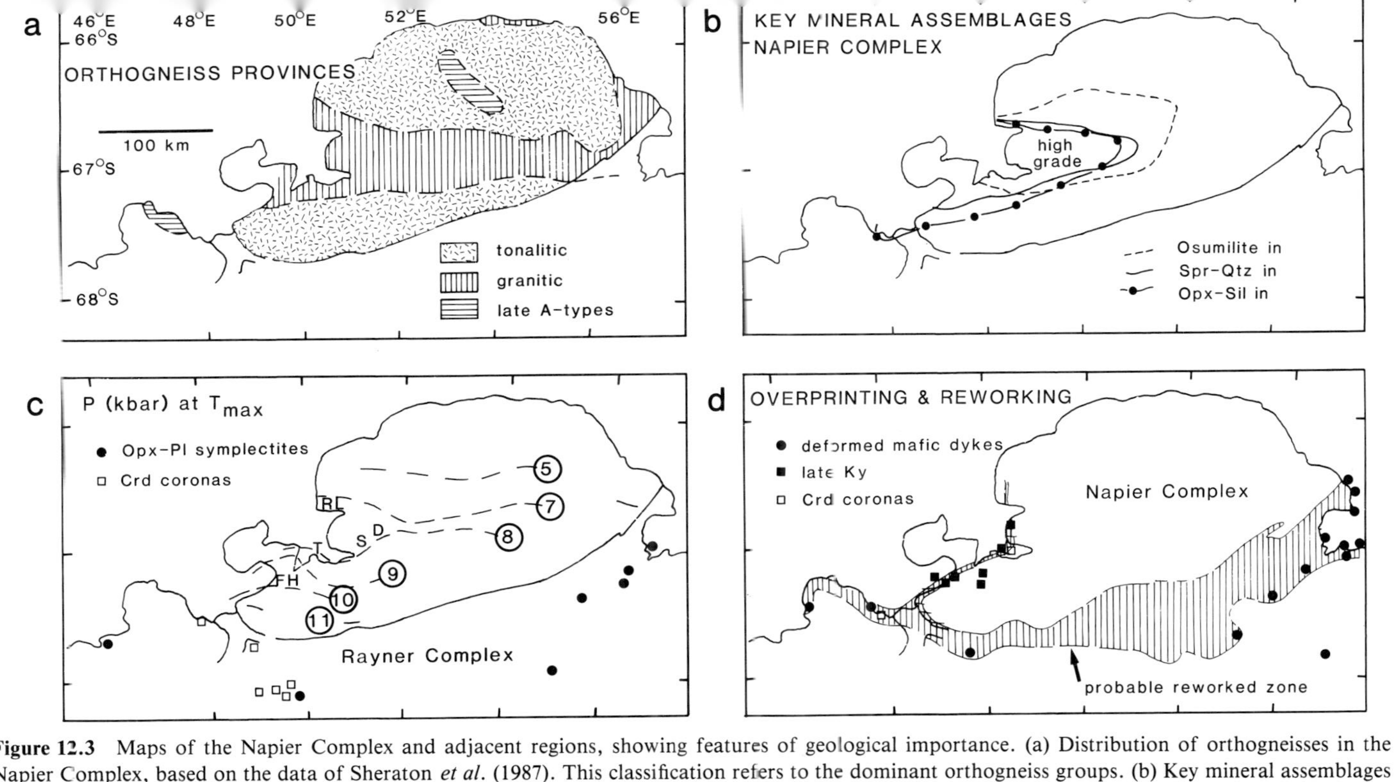

Figure 12.3 Maps of the Napier Complex and adjacent regions, showing features of geological importance. (a) Distribution of orthogneisses in the Napier Complex, based on the data of Sheraton *et al.* (1987). This classification refers to the dominant orthogneiss groups. (b) Key mineral assemblages in the Napier Complex. The indicated assemblages occur in geographical regions within the curved 'isograds'. Data from Ellis (1980), Grew (1980, 1982), Sheraton *et al.* (1980, 1987), Harley (1985a), Sandiford (1985a), and Harley & Black (1987). (c) Distribution of pressures at maximum temperatures recorded in the Napier Complex. Circled numbers are pressures in kbar. Data from Ellis (1980), Harley (1983, 1985a), Ellis & Green (1985), and Sandiford (1985a). Also shown is the distribution of distinctive post-deformational reaction textures in granulites of the Rayner Complex (Sheraton *et al.* 1980, Ellis 1983, Black *et al.* 1987). Place names are as follows: D, Dallwitz Nunatak (945); FH, Fyfe Hills; RL, Mt Riiser-Larsen; S, Mt Sones; T, Tonagh Island. (d) Distribution of indicators of Proterozoic (and younger) overprinting or reworking of older, mainly Archaean, crust. Data from Sheraton *et al.* (1980, 1987), Sandiford & Wilson (1984), Harley (1985b and unpublished data), and Sandiford (1985b).

involved the emplacement of intrusives similar to A-type granites (Collins *et al.* 1982) in the Napier Mountains at 2840 Ma and 2480 Ma (Black *et al.* 1986b). These granites were deformed in an upright folding episode (D_3, James & Black 1981) at *c.* 2460 Ma. High-grade pegmatites were also locally emplaced into the granulites or derived from them by local anatexis (Sheraton *et al.* 1980, 1987; Grew & Manton 1979; Grew 1981a). 2500 Ma U–Pb ages obtained on accessory minerals (e.g. zircon, perrierite) in these pegmatites have been interpreted (Grew & Manton 1979, Grew 1981a, Sandiford & Wilson 1984) to date the time of the most intense metamorphic event correlated to D_1 and D_2, in contrast to the 3070–2900 Ma age noted above. The presence of sapphirine, sillimanite, orthopyroxene, and rare beryllian surinamite enclaves within some pegmatites (Grew 1981a) is consistent with their formation under granulite conditions (Hölscher *et al.* 1986). At present we prefer the older age for the highest-T metamorphism as it is based on a variety of independent isotopic techniques (Rb–Sr, Sm–Nd, conventional and ion-microprobe zircon U–Pb) applied to the granulites themselves, and ascribe the younger pegmatite-based ages to the D_3 granulite event outlined below.

Several generations of mafic dykes were emplaced in the Proterozoic, including a high-Mg suite intruded at 2500–2300 Ma, and the much younger (1190 ± 200 Ma) Amundsen Dyke suite (Sheraton & Black 1981). These dykes form important structural/time markers as they are usually undeformed in the Napier Complex but are deformed and metamorphosed in adjacent terranes affected by late Proterozoic tectonism (Sheraton *et al.* 1980).

The extreme regional very high-temperature (VHT) metamorphism, which is such an outstanding feature of the Napier Complex, occurred throughout two early deformation episodes (D_1 and D_2), probably at and after 3070 Ma (Black & James 1983). D_1 is notable for its intense fabrics, including strong down-plunge stretching lineations, developed coevally with high-grade mineralogies. The high shear strains indicated by such features (Sandiford & Wilson 1984) can, however, be interpreted to result from either compression or extension in the absence of consistent sense-of-shear indicators and marker units. In addition, there is very little evidence as to whether the granulites were undergoing burial or unroofing during this event. A minimum age for D_2 (2840 ± 200 Ma) has been obtained from a biotite granite (Black *et al.* 1986b), and it is apparent that limited isotopic resetting occurred at high grades until *c.* 2900 Ma (Harley & Black 1987). The timespan of D_1 and D_2, and hence of the highest-T metamorphism, remains poorly defined, but the 200–300 Ma timespan suggested by the isotopic constraints is not consistent with the *maximum* timescales for the thermal relaxation of lithosphere perturbed by important thermal processes (Harley & Black 1987). The very-high-grade metamorphic conditions ($>900^\circ$C) inferred from textural evidence to have continued throughout the D_1–D_2 interval (Black & James 1983) could not be sustained for 200 Ma in crust of reasonable conductivity.

12.2.2 High-temperature metamorphism in the Napier Complex

Mineral assemblages developed in a wide range of rock types together with conventional geothermobarometry show conclusively that the earliest high-grade metamorphism occurred at temperatures (*T*) in excess of 950°C, and even up to 1020°C in the southwestern parts of the Napier Complex. Very high-*T* mineral assemblages in magnesian metapelites have been documented by several workers (Dallwitz 1968; Sheraton *et al.* 1980, 1987; Ellis 1980, 1983; Ellis *et al.* 1980; Grew 1980, 1982; Harley 1983, 1985a, 1986; Sandiford 1985a; Motoyoshi & Matsueda 1984), whose combined studies show that the FMAS ($FeO-MgO-Al_2O_3-SiO_2$) assemblages Spr + Qtz, Spr + Opx + Qtz, Opx + Sil + Qtz, and Spl + Qtz are of *regional* occurrence (Fig. 12.3). Such assemblages are only stable at *P–T* conditions beyond the stability of Grt + Crd (Hensen & Green 1973, Ellis *et al.* 1980) (Figs 12.4 & 6).

Metapelite assemblages can be considered using the FMAS *P–T* grid of Figure 12.4, modified from Hensen & Green (1973) in the light of recent experiments and thermodynamic calculations (Bertrand *et al.* 1989, Aranovich & Podlesskii 1989). The relevance of this grid, considered to be applicable for low f_{O_2} conditions (<QFM) (Hensen 1986, see Hensen & Harley, Ch. 2), is confirmed by the occurrence of a (Spl, Opx, Qtz) assemblage, Spr + Grt + Sil + Crd, which is unique to the grid (Harley 1986). *P–T* fields relevant to the reported assemblages, shown in Figure 12.4, indicate maximum *P–T* conditions near the [Spl] point (11 kbar, 1040°C) in the Khmara Bay – Scott Mountains area. Somewhat lower pressures are indicated by assemblages in the Tula Mountains, and prograde Crd + Sil + Qtz assemblages (± Grt) in the Napier Mountains indicate both lower *P* and *T* conditions there (Sheraton *et al.* 1980) (Fig. 12.2). It should be noted that the FMAS grid presented here (Fig. 12.4) maximizes the *P–T* stability field of cordierite (+ Grt) as it is based on data pertaining to volatile-bearing, and H_2O-rich, cordierite. The *P*-positions of the [Spl] and other invariant points, and hence the related univariant curves, will be displaced to 1 kbar lower pressures if CO_2–cordierite data are applicable in these granulites (Aranovich & Podlesskii 1989). The recent theoretical and experimental studies of cordierite stability are at variance with the lower *P–T* position of the FMAS grid presented by Grew (1980, 1982) and with the lower-*P* (7 kbar) position of the $MgO-Al_2O_3-SiO_2$ (MAS) invariant point [Grt, Spl] derived through extrapolation of high-*T* MAS-system experiments (see discussion in Hensen & Harley, Ch. 2). This MAS point, which lies at higher *P* than [Spl], is now inferred to occur at 12–13 kbar (Aranovich & Podlesskii 1989, Bertrand *et al.* 1989).

The KFMAS ($K_2O-FeO-MgO-Al_2O_3-SiO_2$) assemblages osumilite (Osu) + Spr + Opx + Qtz and Osu + Grt + Opx + Qtz, indicative of both high *T* (near [Spl]) and very low a_{H_2O} (Ellis *et al.* 1980, Grew 1982), also occur on a regional scale (Fig. 12.3). Very low inferred a_{H_2O} conditions ($\leqslant 0.05$) are

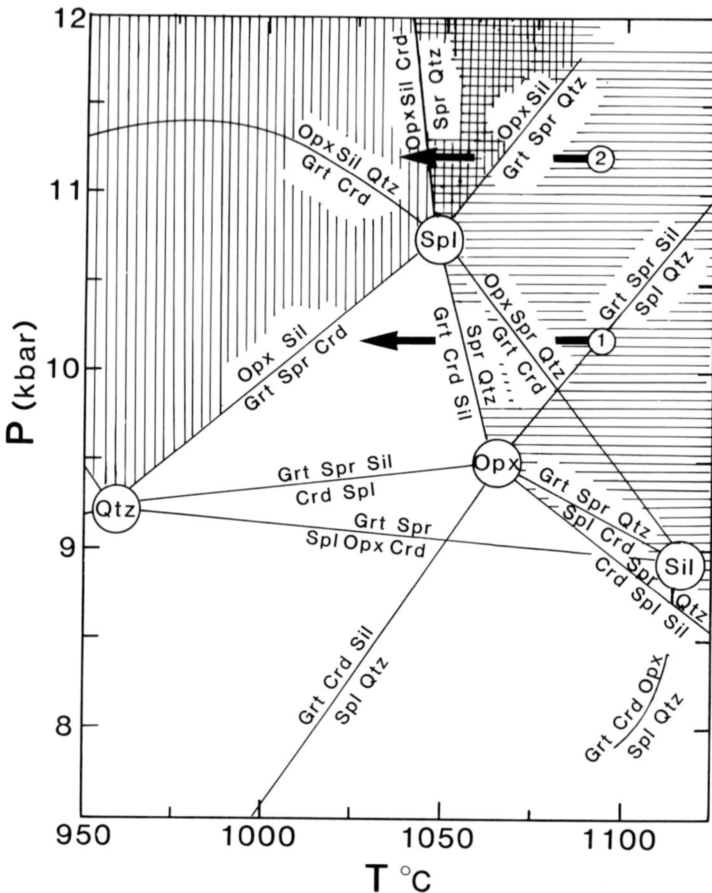

Figure 12.4 Pressure (P) – temperature (T) diagram depicting the mineral reaction relations in model FMAS metapelitic granulites at high temperatures. The grid is modified from Hensen & Green (1973) in the light of new experimental (Bertrand *et al.* 1989) and theoretical (Hensen 1986, Aranovich & Podlesskii 1989) constraints (modified from Harley 1989a, Fig. 5 by permission of Cambridge University Press). Invariant points are denoted by large circles labelled by the non-participating phase. Ruled fields show the relative $P–T$ conditions of formation of regionally important assemblages in the Napier Complex: vertical ruled area, orthopyroxene + sillimanite stable; horizontal ruled area, sapphirine + quartz stable. Isobaric cooling paths 1 and 2 are discussed in the text.

supported by compositional data on relatively calcic (up to An_{20}) alkali-feldspar mesoperthites (Sheraton *et al.* 1980, 1987; Harley 1985a), which formed initially as homogeneous subsolidus phases, and by the stability of Opx + Kfs + Qtz with respect to melt at 1000°C.

High metamorphic temperatures (970–1020°C) are also indicated by the occurrence of exsolved and inverted pigeonites (Mg_{42}–Mg_{47}) in meta-ironstones (Grew 1982, Sandiford & Powell 1986a, Harley 1987a). Some metapyroxenites and garnet–pyroxenites contain clinopyroxenes which have exsolved multiple phases (Opx, Pl, Grt, Spl, Mag, Ilm); re-integrated pre-exsolution compositions of such pyroxenes also indicate metamorphic temperatures in excess of 950°C (Ellis & Green 1985).

Geothermometry of coexisting Grt + Opx pairs yields calculated temperatures above 900°C and sometimes nearer 1000°C when minimum K_D (Fe–Mg) values are used (Harley 1985a); these T conditions are also consistent with the high initial alumina contents (10–12 wt% Al_2O_3) of orthopyroxenes coexisting with low-Ca garnet or with sillimanite (see Hensen & Harley, Ch. 2). Geobarometry of both metapelitic Grt + Sil or Grt + Opx rock types and felsic to mafic orthogneisses (Harley 1981, 1983, 1985a; Grew 1980; Sandiford 1985a; Motoyoshi & Matsueda 1984) yields reasonably consistent results and confirms the inference that pressures at T_{max} were greater in the Scott Mountains – Khmara Bay area (8–11 kbar) than in the areas near Amundsen Bay (6–8 kbar). Unpublished results for parts of the east Tula Mountains and Amundsen Bay have been added to previously published data to produce a geo-isobar map (Fig. 12.3) illustrating this spatial variation. It is not clear whether these isobars are discordant to or concentric with the boundary of the Napier Complex against the Rayner Complex to the south and east.

12.2.3 P–T–t *path: near-isobaric cooling and prolonged lower-crustal residence*

Near-isobaric cooling (IBC) to temperatures of 700–750°C occurred subsequent to D_2 (i.e. considered here to be later than 2900 Ma) and prior to the imposition of the third deformation (D_3) and associated high-pressure granulite to amphibolite facies metamorphism at 2460 Ma (Black *et al.* 1983b). Locally developed reaction textures in granulites affected by D_3 and associated shearing also indicate IBC subsequent to those events (Harley 1985a; Sandiford 1985a; Black *et al.* 1983b; Sheraton *et al.* 1980, 1987).

Evidence for IBC in high-grade metapelites is provided by the occasional development of coronas where one or more uni- and divariant equilibria are crossed. Reference to Figure 12.4 shows that Grt + Crd assemblages will eventually result from IBC at pressures up to 10–11 kbar, whereas secondary Opx + Sil + Grt assemblage would be developed under IBC at the highest pressures or in quartz-deficient rock types. Coronas of Grt + Crd ± Sil which separate Spr + Qtz, coronas of Crd ± Grt on sapphirine in original Spr + Opx + Qtz assemblages, and Spr + Sil (± Grt) on Spl + Qtz, are common in the Tula Mountains and Amundsen Bay. These imply retrograde reactions along an IBC path similar to path 1 in Figure 12.4, although the pressures may be somewhat lower when the rôles of minor components (Fe^{3+}, Ca) and a_{H_2O} are taken into account. Retrograde coronas in the Scott Mountains and other southerly areas often involve the development of Opx + Sil coronas on Spr + Qtz via the reaction (Spl, Crd, Grt): Spr + Qtz = Opx + Sil, and the formation of garnet in previous Opx + Sil ± Qtz assemblages. These textures are consistent with IBC along a path such as path 2 in Figure 12.4, at pressures near or greater than the [Spl] point.

Post-D_2 IBC is also indicated by textures in felsic and mafic to ultramafic pyroxene-bearing rock types, where secondary garnet is sometimes developed as euhedral neoblasts on earlier garnet and pyroxene (Fig. 12.5a), fine trails

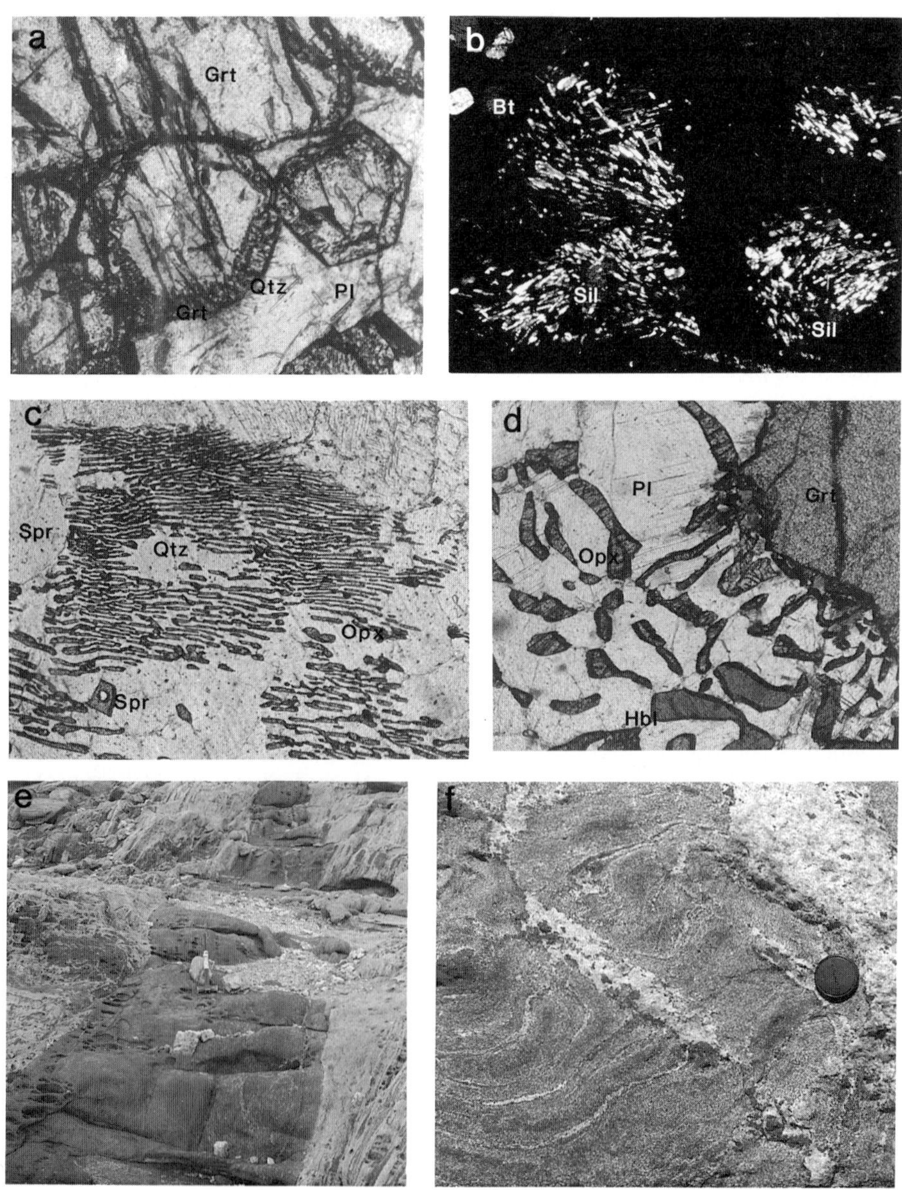

and rims between pyroxenes and plagioclase, or lamellae and rims in and on earlier aluminous pyroxenes. Increased Grt–Pyx Fe–Mg K_D values for neoblast pairs compared with earlier coarser phases, and decreases in the alumina contents of pyroxenes near garnet neoblasts, support the textural inference of IBC and furthermore indicate the operation of reactions such as

high-Al pyroxene (Opx,Cpx) = lower-Al pyroxene + garnet
and Pyx + Pl = Grt + Qtz

(Harley 1983, 1985a; Ellis & Green 1985). These textures post-date pyroxene exsolution lamellae and in some cases pre-date D_3 (Harley 1985a). However, it is also clear that that many garnet-forming reactions were initiated during and following D_3 (e.g. Sheraton *et al.* 1987, Sandiford 1985a), as secondary garnet occurs in seams of S_3 (and later) recrystallization and nucleated on fractured or internally deformed pyroxene porphyroclasts.

The P–T conditions of post-D_2 to -D_3 isobaric cooling (Fig. 12.2) have been established both on local (Ellis 1980, Black *et al.* 1983a, Ellis & Green 1985, Sandiford 1985a) and regional (Harley 1983, 1985a; Harley & Black 1987) scales. Post-D_2 but pre-D_3 cooling through 300–350°C from initial peak temperatures of 950–1020°C occurred with only minor decreases in pressure to 5–7 kbar in the Tula Mountains and 6–9 kbar in areas further south. Although the timescale of this cooling is not well constrained by isotopic data, it is believed from consideration of thermal models for the cooling of thickened crust and lithosphere (Harley & Black 1987) that thermal re-equilibration to *c.* 700–600°C occurred within 50–100 Ma of the D_2 event. The Napier Complex was probably isostatically and thermally equilibrated (i.e. stable) well before reactivation in the D_3 tectonothermal episode, even though the

---

Figure 12.5 (a) Example of secondary garnet regrowth, in this case on primary garnet, in felsic granulites of the Napier Complex. Euhedral rims consist of Grt + Qtz intergrowths. The initial assemblage is Grt + Opx + Pl + Qtz (±Sil). Sample 49891, Mt Hollingsworth; field width = 1.3 mm. (b) Possible evidence for a low-P prograde path in the Napier Complex? Bundles of sillimanite needles and rare biotite (Bt) enclosed in garnet (Mg_{52}) in a Grt + Opx + Pl + Qtz granulite. Sample 49657, McIntyre Island; field of view 2.2 mm wide. (c) Oriented symplectite consisting of sapphirine (Spr) and quartz (Qtz) with minor orthopyroxene (Opx), similar to examples suggested by Hensen & Motoyoshi (1988) to be pseudomorphs after primary cordierite. Sample 49834, Tonagh Island; field of view 1.3 mm wide. (d) Example of Opx + Pl intergrowth replacing Grt ± Hbl. Sample 65630, Dragon Pt, Rauer Group; field of view 1.3 mm wide. Such textures are typical of Grt-bearing mafic granulites within the Proterozoic Complexes. (e) Example of metamorphosed and deformed but clearly discordant mafic dyke, Scherbinina Island, Rauer Group. Dyke cuts felsic Opx–Pl gneisses and mafic layers. Note the development of steep L-fabric in the dyke (foreground). Sledgehammer shaft 70 cm long. (f) Concordant (stromatic) and discordant (agmatic) migmatite structures developed in mafic granulite, Filla Island, Rauer Group. Neosomes contain coarse sub- to euhedral Opx with plagioclase. Dark palaeosomes contain Opx + Cpx + Grt + Pl ± Hbl. Coarse Hbl is sometimes developed on the neosome margins. Camera lens-cap 5 cm in diameter.

P–T estimates for the latter event (5–8 kbar and 650–700°C) (Black *et al.* 1983a,b; Harley 1985a; Sandiford 1985a) lie on the lower-*T* segment of the overall IBC path.

Further independent evidence which confirms the IBC paths of Figures 12.1 and 12.2 is provided by liquidus studies of the early (*c* 2400 Ma) and late (*c* 1200 Ma) mafic dyke suites (Kuehner & Green 1987), which demonstrate that emplacement pressures for both suites were near 8 kbar. These results imply that the granulites presently exposed in the Napier Complex were buried at 20–28 km throughout the 2500–1200 Ma time interval as well as from 3000 Ma to 2500 Ma. Thus 2000 Ma of mid- to lower-crustal residence is inferred for this Archaean granulite terrane, a true 'craton', at least until its local and marginal remobilization and partial exhumation in late Proterozoic and younger tectonic events (see later sections).

12.2.4 The prograde path of the Napier Complex

The prograde *P–T–t* path of a granulite terrane is an important clue to the origin of that terrane, and yet is seldom well constrained. Ellis (1987) and Harley (1989a,b) have discussed the origin of the Napier Complex in terms of crustal thickening models which imply a clockwise *P–T–t* trajectory and significant decompression *prior* to the attainment of the thermal maximum and subsequent IBC. On the other hand, some granulite terranes characterized by IBC (e.g. Adirondacks, Namaqualand, and Broken Hill) have been shown to have evolved to their thermal maxima along a counterclockwise *P–T–t* path where burial accompanies heating (e.g. Bohlen 1987, Waters 1986). Such counterclockwise *P–T–t* paths would be consistent with magmatic (e.g. Wells 1980) or extensional (Sandiford & Powell 1986b) settings for the granulite metamorphism. Little direct evidence for a pre-peak higher-pressure history, such as inclusions of kyanite or pseudomorphs after very magnesian garnet (e.g. Opx + Spr + Qtz or Opx + Spr + Sil in appropriate proportions), has been found in the Napier Complex. The presence of Cr-bearing kyanite porphyroblasts in a fuchsite–quartzite from Mount Charles (S. L. Harley, unpublished data) may, however, suggest a prograde higher-*P* evolution. The Y and HREE-depleted characteristics of tonalitic orthogneisses have been cited by Ellis (1987) as evidence for their generation in overthickened crust, but it is not yet apparent whether these orthogneisses were formed contemporaneously with D_1.

Possible evidence in favour of a counterclockwise *P–T–t* evolution includes the following:

(a) The presence of fine needles and trails of sillimanite, often in several separate bundles, within coarse D_1 garnet ($X_{Mg} = 52$) porphyroblasts (Fig. 12.5b) (Black & James 1983, Harley 1985a).
(b) The occurrence of early spinel-bearing assemblages.

(c) The presence of the lower-*P* (5 kbar) and *T* granulites in the Napier Mountains.

Textural relations in a Spr + Grt + Crd + Sil gneiss from Mt Sones (Harley 1986) *could* also be interpreted in terms of a counterclockwise path. In this rock rare *spinel* occurs included in Spr + Grt + Sil, which suggests an up-pressure crossing of the FMAS reaction

$$\text{Spl} + \text{Crd} = \text{Spr} + \text{Grt} + \text{Sil}\ (\text{Qtz, Opx})$$

Although Harley (1986) suggested that this reaction did indeed occur, he concluded that it occurred as a continuous and isobaric transition in the Zn-bearing natural-rock system where spinel stability is extended to higher *P* compared with the FMAS grid.

A new and most important contribution to the identification of the prograde path has been the recognition (Sheraton *et al.* 1987, Hensen & Motoyoshi 1988) of distinctive Spr + Opx + Qtz symplectites in many Spr + Qtz metapelites in the Tula Mountains (Fig. 12.5c). Hensen & Motoyoshi (1988) have demonstrated through modal analysis that such symplectites often approximate closely cordierite (Mg_{87}) in composition and hence may be pseudomorphs after earlier-formed cordierite. If this interpretation is correct, it would imply that the granulites of the Napier Complex as a whole evolved along counterclockwise *P–T–t* paths, or at least along paths which were within the stability limits of cordierite for most of their prograde segments, and would therefore argue against collisional or tectonic thickening models for the origin of the Napier Complex. Possible models for the production of these IBC-type granulites will be discussed later in Section 12.10. At present, the pre-peak *P–T* history of the Napier Complex remains a rather open and controversial question.

12.3 The Vestfold Hills Block

12.3.1 General geological framework

The Vestfold Hills (Figs 12.2 & 8) comprise the second Archaean granulite facies terrane in the sector under discussion. This *c.* 400 km^2 block of Archaean gneisses is densely dissected by several generations of post-tectonic mafic dykes with different orientations, ranging in age from *c.* 2400 Ma to 1300 Ma (Collerson & Sheraton 1986b). The granulites consist of a variety of gneisses of igneous and sedimentary parentage, deformed and metamorphosed in a series of events between *c.* 3000 Ma and 2450 Ma (Collerson & Sheraton 1986a, Collerson *et al.* 1983). Although superficially similar to the Napier Complex in this age range for the major tectonothermal events, there are considerable differences in the intensity and grades of the *c.* 2450 Ma events, in the timing and extent of orthogneiss emplacement, and in the *P–T*–fluid conditions of pre-2500 Ma metamorphism.

Two major Archaean crustal episodes can be distinguished. The first of these includes several poorly resolved events which culminated in the metamorphism, partial melting, and deformation of pre-existing mafic volcanics and pelitic, psammitic, and calcareous sediments, together with the emplacement of strongly Y- and HREE-depleted tonalitic orthogneisses, the Mossel Gneiss (Collerson *et al.* 1983). The complexities of this *c.* 3000–2800 Ma episode can only be seen in regions unaffected by the intense deformational effects associated with the second episode.

The second episode, dated at 2450 Ma (Collerson *et al.* 1983), involved granulite facies metamorphism accompanied by an often intense deformation (D_2) which produced strong stretching lineations. A volumetrically important suite of tonalitic and dioritic to monzonitic and granitic orthogneisses, the Crooked Lake Gneisses (Collerson *et al.* 1983) were emplaced at this time. These orthogneisses are strongly depleted in Y and HREE and, like the older Mossel Gneisses, exhibit geochemical signatures suggestive of a derivation from a garnet ± amphibole-bearing mafic source (Sheraton & Collerson 1984). The presence of these juvenile contributions to the crust at *c.* 2500 Ma and the intensity of the coeval deformation provide a marked contrast with the relatively gentle 2500 Ma events in the Napier Complex. If the alternative (2500 Ma) age relations for the Napier Complex (Grew & Manton 1979) are adopted, the Vestfold Hills would still be distinctive in the partial preservation of a pre-2500 Ma granulite history.

12.3.2 P–T *conditions and metamorphic assemblages*

Metamorphic conditions associated with the 2500 Ma episode have been estimated at about 800 ± 50°C and 6–8 kbar, based mainly on assemblages developed in Fe-rich orthogneisses, mafic pyroxene–hornblende granulites, and Grt + Sil + Bt or Grt + Bt + Opx metapelites (Collerson & Sheraton 1986a). No more thorough *P–T* study of these granulites has yet been completed because of the general lack of garnet in mafic granulites and the dominance of Fe-rich quartzite and semipelitic rock types in the paragneisses.

Metamorphic conditions associated with events preceding D_2, and probably preceding or associated with partial melting and the generation of the Mossel Gneisses, have been estimated from sapphirine-bearing assemblages developed in Mg–Al-rich, quartz-deficient lithologies (emeries) which occur as rare boudins in the Mossel Gneiss (Harley 1987c & unpublished data). These boudins are similar to Spr + Spl + Opx xenoliths reported from the Mawson Charnockite at Mawson station (Segnit 1957, Sheraton *et al.* 1982), but in addition contain the assemblage Spr (Mg_{87}) + Sil + Crn + Crd (Mg_{91}), which indicates initial *T* conditions of 900–950°C at 5–10 kbar by comparison with simple (MAS) system reaction grids (Hensen 1987).

12.3.3 Rôle of partial melting in D_1–M_1

In areas of low D_2 strain, earlier migmatitic structures can be recognized in pelitic and, more spectacularly, in mafic granulites. Cpx + Opx + Hbl + Pl + Ilm granulites in Long Peninsula (Tryne Metavolcanics) preserve agmatitic and patch or bleb migmatite structures in areas with low percentages of leucosome, and nebulitic and schollen structure where either local melting has been more extensive or where migrating melts of external derivation contributed more to the total melt budget. In all cases the leucosomes are tonalitic to trondhjemitic in composition, although they may be more mafic and enriched in orthopyroxene where segregation has been incomplete. These migmatite features suggest that the early tonalitic orthogneisses, the Mossel Gneisses, were derived at least *in part* by *in-situ* anatexis of local mafic pyroxene–hornblende granulites.

12.3.4 P–T–t *path and lower-mid-crustal residence*

The Archaean and younger P–T–t path deduced for the Vestfold Hills is summarized in Figures 12.2 and 12.8. The relationship of the older higher P–T estimates, obtained from the pre-D_2 boudins, to the *c.* 2450 Ma granulite event is not constrained and the sense of P–T change between these two episodes is not known. The dashed path (Fig. 12.8) is constructed for the simplest case, which assumes that the Vestfold crust was not exposed between the events and that some thickening of the crust and burial of the earlier sequence occurred in D_2. The post-D_2 history is well constrained by liquidus studies of the post-deformational dyke swarms (Kuehner & Green 1987), which indicate near-isobaric cooling at least until after the emplacement at 7.5 kbar or more of a group of high-Mg dykes and associated stocks and sheets at *c.* 2400 Ma. Experimental duplication of the phenocryst assemblage of a younger (1380 Ma) tholeiitic dyke suggests its emplacement at less than 5.5 kbar (Kuehner & Green 1987). These results indicate that the erosion or removal of some 9 km thickness of overlying crust occurred in the Vestfold block in the early to mid-Proterozoic. The causes of this partial exhumation could be either of the following:

(a) The post-2500 Ma crust was overthickened to some extent by tectonic thickening (D_2) and by the accretion of the Crooked Lake Gneisses.
(b) The crustal column was overthickened at 2400 Ma and 1700 Ma by the cumulative accretion of mafic magmas at the base of the crust.

Explanation (a) is preferred here because the potential thermal effects of underaccreting *c.* 9 km of magma are not observed in the gneisses.

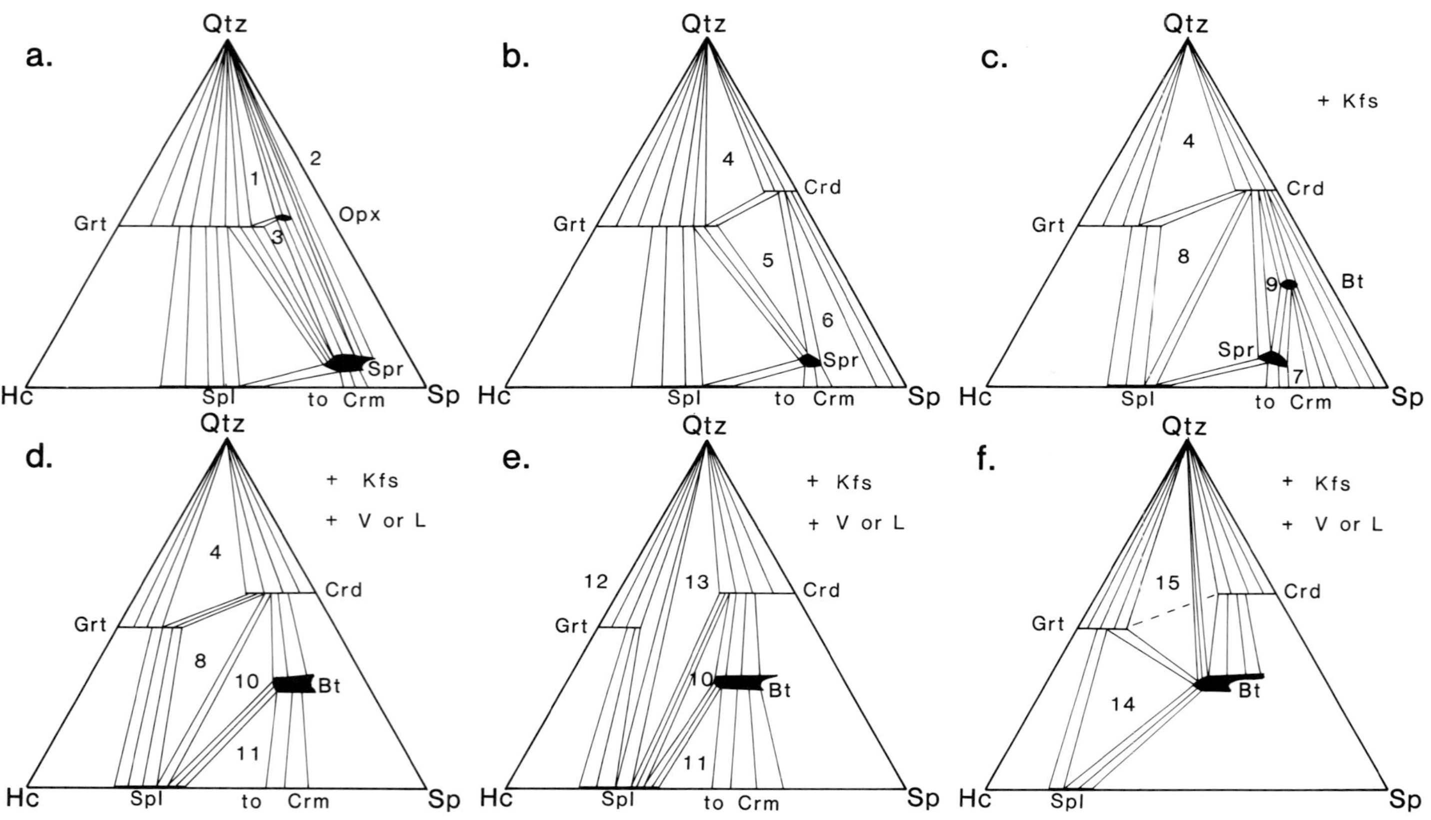

a.
Qtz
1
2
3
Grt
Opx
Spr
Hc
Spl
to Crm
Sp
b.
Qtz
4
Crd
Grt
5
6
Spr
Hc
Spl
to Crm
Sp
c.
Qtz
+ Kfs
4
Crd
Grt
8
9
Bt
Spr
7
Hc
Spl
to Crm
Sp
d.
Qtz
+ Kfs
+ V or L
4
Crd
Grt
8
10
Bt
11
Hc
Spl
to Crm
Sp
e.
Qtz
+ Kfs
+ V or L
12
13
Crd
Grt
10
Bt
11
Hc
Spl
to Crm
Sp
f.
Qtz
+ Kfs
+ V or L
15
Crd
Grt
Bt
14
Hc
Spl
Sp

12.4 The Lützow – Holm Bay Complex

The granulites of the Lützow – Holm Bay Complex form the highest-grade zone of a medium-pressure progressive metamorphic belt (Hiroi *et al.* 1983a), which increases in grade from east to west. West of an inferred tectonic break with the Rayner Complex granulite and upper-amphibolite facies gneisses around Molodezhnaya (Figs 12.2 & 7), the metamorphic grade increases from an amphibolite facies St–Ky zone, through a 50 km wide transitional zone where orthopyroxene appears in metabasites at the expense of orthoamphibole, to an extensive granulite facies belt characterized by Grt + Bt + Sil pelites (Fig. 12.7), Grt ± Spl-bearing mafic and ultramafic hornblende–pyroxene granulites, and Wo + Scp bearing calc-silicates. Details of this zonal sequence are illustrated in Figure 12.7. This apparent example of progressive metamorphism is supported by studies of the granulites themselves, in which a *clockwise* prograde *P–T* history through the high-*P* amphibolite facies is indicated by:

(a) Ky ± St inclusions within garnet porphyroblasts in metapelites which contain sillimanite as the only matrix aluminosilicate (Hiroi *et al.* 1983b, Shiraishi *et al.* 1987, Motoyoshi *et al.* 1987); and
(b) the bulk-rock compositions of scapolite-bearing calc-silicates, which correspond closely to epidote (Hiroi *et al.* 1987).

Peak-metamorphic conditions in the granulite facies were 8–10 kbar and 800–850°C (Motoyoshi *et al.* 1987, Shiraishi *et al.* 1987), based mainly on geothermobarometry of Grt + Opx + Cpx + Pl + Hbl mafic and ultramafic granulites and compatible mineral assemblages in metapelites (Grt + Sil + Bt + Pl + Qtz) and calc-silicates (Scp + Cpx + Pl ± Wo ± Cal ± Qtz). The post-peak *P–T–t* path has been shown (Shiraishi *et al.* 1987, Hiroi *et al.* 1983a) to have involved significant decompression to 4 kbar at temperatures greater than 700°C (Fig. 12.1). Evidence for this *P–T–t* path includes:

(a) symplectic Opx + Spl + Pl pseudomorphs after garnet in metabasic and ultrabasic rock types (Hiroi *et al.* 1986);

Figure 12.6 Projections from sillimanite (and other phases as indicated) onto the plane SiO_2–$FeAl_2O_4$–$MgAl_2O_4$ (Qtz–Hc–Sp) for metapelitic granulites from East Antarctica. Observed stable associations indicated by numbered three-phase fields. Data sources given in the text. V and L are vapour and liquid (melt) respectively; either may also be present in Bt-bearing assemblages. (a) Assemblages from the Napier Complex (1, 2, 3) and Long Point, Rauer Group (3). (b) Assemblages from the Napier Complex (4, 5), Long Point (4, 5), and the Northern Vestfold Hills (6). (c) Assemblages from the Rayner Complex (4, 8), Molodezhnaya (4), the Rauer Group (4, 8), and the Vestfold Hills (7, 9). (d–f) Assemblages reported from the Larsemann Hills (4, 8, 10, 11, 12, 13) and Brattstrand Bluffs (4, 8, 10, 14, 15) in the Prydz Bay area, and from Molodezhnaya (15).

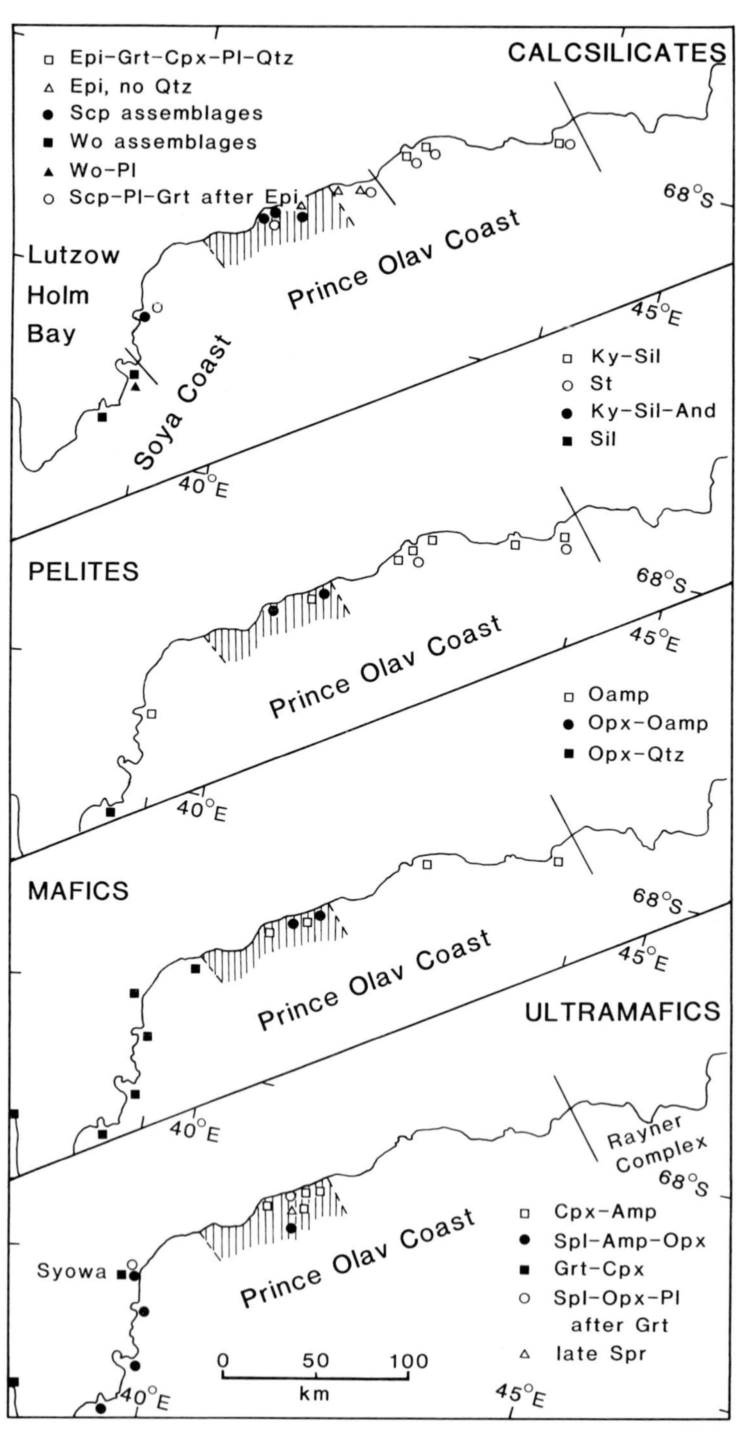

CALCSILICATES
Epi-Grt-Cpx-Pl-Qtz
Epi, no Qtz
Scp assemblages
Wo assemblages
Wo-Pl
Scp-Pl-Grt after Epi
Lutzow
Holm
Bay
Soya Coast
Prince Olav Coast
68°S
45°E
40°E
Ky-Sil
St
Ky-Sil-And
Sil
PELITES
Prince Olav Coast
68°S
45°E
40°E
Oamp
Opx-Oamp
Opx-Qtz
MAFICS
Prince Olav Coast
68°S
45°E
40°E
ULTRAMAFICS
Rayner
Complex
68°S
Cpx-Amp
Spl-Amp-Opx
Grt-Cpx
Spl-Opx-Pl
after Grt
late Spr
Syowa
Prince Olav Coast
0
50
100
km
40°E
45°E

(b) Opx + Pl coronas between garnet and clinopyroxene in quartz-bearing metabasites;
(c) rims of grossular on plagioclase which had previously developed from scapolite in Wo + Scp calcsilicates (Hiroi *et al.* 1987) – this indicates that cooling below 700°C occurred at pressures substantially *less* than the initial pressures of 8–10 kbar (Harley 1989a);
(d) rims and moats of Crd + Opx between Grt + Qtz in metapelites (Hiroi *et al.* 1983b).

This *P–T–t* path is one of the most completely defined of the paths reported herein, as the prograde evolution is better constrained. However, there is some disagreement over the *age* of the main metamorphism and decompression. Whereas earlier Rb–Sr data suggested an age of *c.* 1000 Ma, similar to metamorphic ages from the Rayner Complex and other late Proterozoic granulite areas, recent Rb–Sr isochron results (Shibata *et al.* 1986) indicate a metamorphic age of *c.* 700 Ma. This younger age is unique for granulite metamorphism in this part of Antarctica, but is similar to ages obtained from undeformed, post-tectonic, granitoids and pegmatites from areas to the east (Black *et al.* 1987), and to the age of calc-alkaline magmatism in the Yamato–Belgica low-*P*–high-*T* terrane to the west (Shiraishi *et al.* 1987).

12.5 Proterozoic overprinting in the Napier Complex and Napier relics within the Rayner Complex, Enderby Land

12.5.1 Shear zones within the Napier Complex

The deformational and metamorphic features of mainly east–west-trending, steep, ductile shear zones which dissect the Napier Complex in the Khmara Bay region have been described by Black *et al.* (1983b), Sandiford (1985b), and Harley (1985b). Initial movements on these appear to have been strike-slip in nature (Black *et al.* 1983b, Sandiford 1985b), and are inferred to have occurred at *c.* 1100–1000 Ma (Black *et al.* 1983b, 1984). Dominantly dextral strike-slip with some dip-slip component or transpression in these zones would be consistent with west-directed thrusting of the Rayner Complex, interpreted from deformation patterns observed in areas further east (Clarke 1987; see Section 12.7) and with the recorded *P–T–t* paths in the shear zones. These

Figure 12.7 Metamorphic features of the Lützow – Holm Bay region, based on data from Hiroi *et al.* (1983a,b, 1986, 1987), Motoyoshi *et al.* (1987), and Shiraishi *et al.* (1987), for metamorphosed calc-silicates, pelites, mafics, and ultramafics. The vertical ruled field is the 'transition zone' defined by the above authors and occurring between amphibolite facies gneisses to the east and granulites to the west. Mineral abbreviations are as given in the text. Additional names are as follows: Epi, epidote; Oamp, orthoamphibole. The long solid line on the east side marks the boundary between the Rayner Complex and the Lützow–Holm Complex.

ductile zones were reworked by lower-grade mylonite zones preserving steeply plunging lineations, which are thought to be related to north-directed thrusting and local pegmatite emplacement at *c*. 500–550 Ma (Grew 1981a, Black *et al*. 1983b, Sandiford 1985b, Clarke 1987).

The Proterozoic metamorphic histories of the shear zones are well defined by mineral and reaction textures developed within overprinted metapelites. Initial Ky + Grt + anthophyllite + Qtz assemblages are successively overprinted by Ky + Grt + gedrite, Crd + anthophyllite, and Crd + St + anthophyllite assemblages produced in reactions such as:

$$\text{orthoamphibole} + \text{Ky} = \text{Crd} + \text{St}$$

$$\text{Grt} + \text{Ky} + \text{Qtz} = \text{Crd} + \text{Pl} \pm \text{St}$$

and

$$\text{gedrite} \pm \text{Ky} + \text{Qtz} = \text{anthophyllite} + \text{Crd} \pm \text{St}$$

consistent with ITD from *c*. 8 kbar and 680–720°C to 3–5 kbar and 600–650°C (Harley 1985b, Sandiford 1985b) (Figs 12.2 & 3).

Further examples of near-east–west-trending shear zones of probable Proterozoic age occur in Amundsen Bay and the Tula Mountains (Figs 12.2 & 3). At Tonagh Island (Harley, unpublished data), initial Archaean Grt (Mg_{45}) + Sil + Qtz + Pl + Rt gneisses are overprinted, first, by Grt–Ky–Rt– Qtz assemblages. These are in turn overprinted by post-shearing coronas and textures which involve the replacement of fractured garnet by aggregates of Crd + anthophyllite coexisting with Mg_{32-35} garnet; the replacement of rutile by ilmenite; the formation of fine sillimanite and fibrolite needles embaying earlier Sil and Ky; and finally the formation of coarse staurolite xenoblasts and poikiloblasts overprinting and replacing anthophyllite + Crd ± Bt. These textural features and calculated *P–T* data indicate decompression from 5–6 kbar and 600–650°C to only 3 kbar and 600°C. Kyanite at this locality is an index of the late Proterozoic overprint, as appears to be the case in most of the occurrences reported previously (Sandiford & Wilson 1984, Black *et al*. 1983b, Sheraton *et al*. 1987) and depicted in Figure 12.3.

Shear zones of probable late Proterozoic age also cut *c*. 1190 Ma Amundsen Dykes within the Napier Complex, leading to the development of secondary garnet in the dykes at *P–T* conditions of 6–8 kbar and 650–700°C (Sheraton *et al*. 1987, Kuehner 1986). These *P–T* conditions are consistent with those relevant to the development of Ky in the metapelites considered above.

An important feature of these shear zones and the coeval or later magmatism localized within them is the geochemical (Sandiford 1985b) and isotopic (Black *et al*. 1983b) evidence that some of the crust underlying the Napier Complex granulites must be relatively hydrous and undepleted in heat-producing and LIL elements, in contrast to the Napier Complex granulites themselves. Grew (1981a) reports the occurrence of boron-bearing and beryllian phases (beryl, tourmaline, and dumortierite) in late pegmatites (520 Ma) cutting the shear zones. The mineral assemblages developed within the shear zones are interpreted to record some 10–14 km uplift induced in an overriding crustal block (the Napier Complex) underplated from the south or south-east by younger and lower-grade crust of the Rayner Complex during late Proterozoic thrusting or transpression (Harley 1985b, 1989b; Sandiford 1985b).

12.5.2 Napier Complex relics in the Rayner Complex

Within the late Proterozoic Rayner Complex of Enderby Land, probable reworked relics of the Napier Complex have been identified at Forefinger Point (Fig. 12.2), where a series of magnesian metapelites contain both quartz-bearing and quartz-free Opx + Sil ± Grt assemblages, variably overprinted by textures indicative of metamorphism and decompression (ITD) that are at least partly related to the late Proterozoic (*c.* 1000 Ma) tectonism. These overprints include: the abundant development of biotite, later partially rimmed by Opx + Kfs + Pl; Crd moats between Opx and Sil ± Qtz; Spr or Spl + Crd symplectite replacing Sil; Spr + Crd + Pl symplectites replacing Sil and possibly formed by the reaction

$$\text{Opx} + \text{Sil} + \text{mesoperthite} + \text{V (or liquid)} = \text{Spr} + \text{Pl} + \text{Crd} + \text{Bt};$$

and Spr + Opx ± Spl lamellar intergrowths after Mg_{55}–Mg_{58} garnet. Preliminary mineral zoning data and *P–T* calculations (S. L. Harley and others, unpublished data) suggest the path depicted in Figure 12.2, wherein Archaean granulites equilibrated at 9–11 kbar and 950°C were overprinted in the late Proterozoic at *c.* 8 kbar and 750–850°C and subsequently involved in ITD to 5–7 kbar and 700°C.

12.6 The Rayner Complex in Enderby Land

12.6.1 Geological framework, lithology, and protolith ages

The Rayner Complex (Ravich & Kamenev 1975) is a large region of late Proterozoic granulite to upper amphibolite facies metamorphism (Sheraton *et al.* 1980) which occurred at 1000 ± 50 Ma (Grew 1978, Black *et al.* 1983b, 1987). In Enderby Land it has been distinguished from the Napier Complex by the absence of undeformed tholeiitic dykes (Sheraton *et al.* 1980, 1987). The relatively sharp boundary between the Rayner Complex and the Napier Complex in this region has probably been reactivated by *c.* 500 Ma faulting associated with north-directed thrusting in the Cambrian (Clarke 1988), so that relics of the Archaean Napier Complex are *not* precluded from occurring within the body of the younger belt, as noted in the previous section.

At Molodezhnaya and in the Nye Mountains paragneiss and orthogneiss components are both abundant. Paragneisses include Fe-rich pelites and semi-pelites, calcsilicates, ultramafics, and mafic granulites which may in part be deformed equivalents of Amundsen dykes. The paragneisses differ from those present in the Napier Complex in the rarity of magnesian pelites and the presence of significant calc-silicates and marbles within the Proterozoic successions. Model age (Sr and Nd) calculations on one paragneiss indicate its derivation from crust with an average age of only 1300 Ma, and hence suggest that at least some of the paragneiss precursors were deposited after the generation and emplacement of many orthogneisses in the mid- to late Proterozoic (Black *et al.* 1987).

Orthogneisses are dominated by K_2O-rich, I-type (Chappell & White 1974) granitoids. At Molodezhnaya some orthogneisses may have been emplaced near the time of peak metamorphism and deformation at 1022 ± 62 Ma (Grew 1978). However, it is clear from a combination of isotopic techniques (Black *et al.* 1987) that many felsic to intermediate orthogneisses were emplaced at 1500–1400 Ma, and that most of these were derived from the remelting of 1800–2000 Ma crust rather than the much older Napier Complex. Whether *all* of the extensive supracrustal sequence represents a *c.* 1300 Ma or younger cover to these orthogneisses remains as an open question.

These composite-aged sequences were deformed in at least two folding episodes (Grew 1978; Sheraton *et al.* 1980, 1987; Black *et al.* 1987), the first of which is only manifested in intrafolial folds and stretching lineations. The second deformation (D_2) produced large-scale upright folds which typically trend east–west, and occurred contemporaneously with the main metamorphism (Grew 1978, Black *et al.* 1987).

12.6.2 Mineral assemblages and P–T *conditions*

Mineral assemblages are generally those typical of moderate-grade medium-*P* granulites: Opx + Cpx + Hbl + Pl ± Qtz in mafic rock-types; Opx + Kfs + Pl + Qtz in felsic and intermediate orthogneisses; and various pelitic assemblages consistent with the stability of either the Bt + Sil or Grt + Crd tie lines in different areas (Fig. 12.6). Very-high-*T* indicators such as mesoperthite, pigeonite, or Spr + Qtz are generally absent. Some spatial variations in peak *P–T* conditions are indicated by variations in diagnostic mineral composition parameters. For example, at Molodezhnaya Grt–Crd–Sil–Qtz assemblages contain only Mg_{22} garnet (Grew 1981b), whereas higher-*P* conditions are inferred for areas further east where Mg_{39} garnet occurs in the same assemblage (Black *et al.* 1987).

Largely on the basis of metapelite assemblages, Grew (1981b) calculated *P–T* conditions of 5–6 kbar and 700°C for granulites at Molodezhnaya

(Fig. 12.2). Geothermobarometry of a varied selection of metapelites and felsic to mafic orthogneisses (Ellis 1983, Black *et al.* 1987) indicates peak *P–T* conditions of 6–8 kbar and 750–800°C in the Nye Mountains area (Fig. 12.2). Calculated fluid activities (at 750°C) based on Grt + Bt + Sil + Kfs + Qtz assemblages from these areas yield a_{H_2O} values of 0.3–0.4, significantly higher than those inferred for the Napier Complex (S. L. Harley, unpublished data).

Garnet porphyroblasts in metapelites frequently contain inclusions of lower and similar grade phases (e.g. Bt, Sil, Opx) and more rarely early muscovite (Ellis 1983). Based on inclusion and matrix phases in a garnet metapelite from the Nye Mountains, the prograde dehydration reaction

$$\mathrm{Bt + Sil + Qtz = Grt + Crd + Kfs + V}$$

was inferred by Ellis (1983), although it is quite likely that melting was also involved in these migmatitic paragneisses:

$$\mathrm{Bt + Sil + Qtz\ (\pm\ V) = Grt + Crd + Kfs + L}$$

In either case, the prograde reaction could have occurred at pressures significantly less than 10 kbar.

12.6.3 Retrograde P–T–t *path*

Post-peak near-isothermal decompression (ITD) to 3–4 kbar at *T* in the range 600–700°C (Fig. 12.1) is characteristic of the Rayner Complex in Enderby Land and most other late Proterozoic granulite sequences over the whole 40–80°E sector. Important textures indicative of ITD in the Rayner Complex include:

(a) Opx + Crd symplectites or simple Crd coronas on resorbed and embayed metapelite garnets (Ellis 1983) (Fig. 12.3); and
(b) Opx + Pl symplectites and moats on resorbed Grt porphyroblasts and on Grt + Cpx or Grt + Hbl contacts in mafic granulites (Ellis 1983, Sheraton *et al.* 1987) (e.g. Fig. 12.5d).

The spatial extent of such textures is indicated in Figure 12.3c. So far, quantitative geothermobarometry has only been attempted on a limited number of these examples (Ellis 1983, Black *et al.* 1987).

12.6.4 Later events superimposed on granulites in Enderby Land

Late, upright, greenschist facies shear zones are a common feature of the Proterozoic and Archaean complexes in Enderby Land. These may be synchronous with, but usually post-date, the emplacement of post-deformational pegmatites and granitic intrusives thought to be derived from the remelting of the 1800–2000 Ma crust of the Proterozoic belts (Black *et al.* 1987). Rb–Sr whole-rock ages of *c.* 540 Ma obtained for some granitic intrusives are interpreted (Black *et al.* 1987) to date a metamorphic resetting event probably associated with the shearing. There is convincing U–Pb isotopic evidence (Black *et al.* 1987) that many of the granites and pegmatites were emplaced at 760–770 Ma, an age comparable to the age of magmatism in the Yamato–Belgica Complex further west (Shiraishi *et al.* 1987). This age for post-deformational magmatism may also be relevant to areas as far east as Prydz Bay (Clarke 1988, Harley 1987c, Sheraton *et al.* 1987).

12.7 The Rayner Complex: Kemp Land and MacRobertson Land

12.7.1 Geological framework, lithology, and protolith ages

The boundary between the Rayner and Napier Complexes in Kemp Land (Fig. 12.2) is gradational. Proterozoic overprinting of Archaean granulites occurs in a region perhaps as large as 5000 km^2, as identified by the occurrence of deformed metatholeiite dykes (Figs 12.2 & 3) and limited Rb–Sr isotopic data (Sheraton & Black 1983, Clarke 1988) consistent with Archaean protolith ages. Archaean relics and reworked equivalents appear to be of importance in the Edward VIII Gulf area (Sheraton *et al.* 1987) and as far east as Fold Island and Stillwell Hills, where Clarke (1988) reports the occurrence of a *c.* 2700 Ma felsic orthogneiss within an area dominated by paragneisses rich in metapelites, metaquartzites, and other supracrustals. Supracrustal suites also containing mafic granulites and calc-silicates are common in the region as a whole, whereas felsic orthogneisses are less abundant. The proportion of reworked Archaean material compared with metamorphosed equivalents of younger intrusives and sedimentary cover sequences is not yet established.

Late Proterozoic tectonism in this region involved up to three phases of post-dyke deformation, which are locally superimposed on earlier fabrics of probable Archaean age (James *et al.* 1987). An early recumbent phase (D_1), typified by strong east–west stretching lineations and west-verging folds, is overprinted by two upright (D_2, D_3) folding phases (Clarke 1988). The later upright phases include a variably developed D_2 event which locally produced east–west-trending structures and a later, more pervasive, D_3 event which produced north–south-trending folds and downdip stretching lineations (Clarke 1987). The Proterozoic D_1 and D_3 events have been interpreted in terms of a west-directed shear (Clarke 1988) involving the thrusting of Archaean precursors and younger sequences westward over the Napier Complex.

An important syn- to late-tectonic intrusive, the Mawson Charnockite, outcrops over an area of 5000 km^2 in MacRobertson Land. This 964 Ma orthogneiss (Black *et al.* 1987) is only affected by the D_3 event defined by

Clarke (1987, 1988) and thus provides a maximum age constraint for this deformation. Xenoliths and rafts of country-rock paragneiss which occur within the Mawson Charnockite share the metamorphic and deformational features observed within the paragneiss successions further west. Geochemical studies (Young & Ellis 1987) show that the Mawson Charnockite is composite and polygenetic in that two distinctive chemical suites occur. A high-TiO_2 suite, with high K_2O and Sr initial ratios, is interpreted as a melt derived from crustal sources at 7 kbar, while a low-TiO_2 suite, characterized by Y and HREE depletion, was produced at somewhat deeper levels (>10 kbar) as a high-temperature melt from a garnet-bearing mafic source (Young & Ellis 1987, Ellis 1987).

12.7.2 Metamorphic P–T *conditions*

High-pressure granulite assemblages (e.g. Grt + Opx + Cpx + Pl) are developed in mafic to intermediate orthogneisses, including metamorphosed Amundsen Dykes, in the zone of reworking referred to earlier within Kemp Land (Sheraton *et al.* 1987). A typical mafic granulite from this region yields peak P–T conditions of 8–10 kbar and 700–800°C (Ellis 1983; and see Fig. 12.2). Lower pressures of 5–7 kbar at 800–850°C are inferred, however, for most parts of the Rayner Complex away from this transition zone (Black *et al.* 1987, Clarke 1987) and in areas where the Proterozoic deformations are most intense. Sheraton *et al.* (1987) have suggested that the lower calculated pressures only reflect the *late* stages of the Rayner metamorphic evolution, whereas the high-P zone records P–T conditions established earlier in the metamorphic evolution in areas only partially affected by overprinting at lower pressures. It is possible that the high-P zone was uplifted by overriding the adjacent, relatively cold, Napier Complex at an early stage in the late Proterozoic evolution, and hence escaped further thermal perturbation associated with D_3, which affected areas south and east where syn-D_3 magmatism may also have been important.

Clarke (1987) has inferred an early phase of apparent IBC in Grt + Crd + Bt + Sil gneisses from the Stillwell Hills, where a Proterozoic fabric defined by Bt + Sil overprints earlier Grt + Crd consistent with the reaction

$$\text{Grt + Crd + Kfs + V = Bt + Sil + Qtz}$$

However, this apparent cooling event (Fig. 12.2) may be the consequence of lower-T reworking of *Archaean* assemblages by Proterozoic metamorphism and hydration imposed at 4–6 kbar and 700–750°C. Other evidence for the prior existence of relatively high-T granulites (>850°C) overprinted at somewhat lower T is provided by the occurrence of a Spr + Opx (enstatite) + Spl +

Table 12.1 Summary and correlation of geological features of East Antarctic granulites, 40°E–80°E.

FEATURE	NAPIER COMPLEX	VESTFOLD HILLS	LUTZOW HOLM BAY
OLDEST PROTOLITHS	3930-3700 Ma	3100 Ma (?)	Not known; probably mid Proterozoic
FELSIC ORTHOGNEISSES	Ages 3900-3000 Ma Some symmetamorphic. HREE-depleted tonalitic and non-depleted granitic suites.	HREE-depleted older tonalites (Mossel gneisses) ca. 2800-3100 Ma Varied later Syn-M(D)2 group 2500 Ma.	Common
PARAGNEISS, SUPRACRUSTALS & OTHER LITHOLOGIES	Mg- and Fe-rich aluminous pelites; quartzites; ironstones; ultramafics. Carbonate protoliths rare.	Fe-rich aluminous and quartzitic types; carbonates; ironstones; mafic volcanics and ultramafics. Rare Mg-rich evaporites (?)	Fe-rich aluminous pelites; quartzites; marbles and calcsilicates; ultramafics and mafics.
EARLY M-D EVENTS			
AGE	3100-3000 Ma (M1-D1) to 2800 Ma (?) (M2-D2)	3100-2800 Ma (M1-D1)	1100 Ma or 750 Ma (M1-D1, M2-D2)
P-T CONDITIONS	7-11 kbar, 950-1000°C (SW parts) 5-7 kbar, 800-850°C (NE parts)	7-8 kbar, 900-950°C (recorded in rare boudins)	8-10 kbar, 700-800°C in highest grade zone; lower P and T conditions in zones in E parts.
P-T PATHS	IBC to 700°C at 5-9 kbar	poorly known	ITD to 3-4 kbar at 600-650°C.
DEFORMATION OR SETTING	Intense; flat lying folds and strong L fabrics; later reclined folds	Often intense, but locally pre-D2 intrusive and sedimentary structures preserved	Isoclinal folds; stretching lineations Not well documented

SUBSEQUENT M-D EVENTS			
AGE	2500 Ma (M3-D3)	2500 Ma (M2-D2)	
P-T CONDITIONS	5-8 kbar, 650-750°C	608 kbar, 800°C	
P-T PATH	Further IBC at 5-8 kbar	IBC at 6-8 kbar, based on younger dyke data.	
DEFORMATION	Open upright large-scale folds; local S-fabrics.	Close to isoclinical folds, now with steep axes.	
AGE	1100 Ma (?) Localised (M4-D4, M5-D5)	1100 Ma (Localised) (M3-D3)	
P-T CONDITIONS	5-7 kbar, 650-720°C	5-6 kbar, 650-700°C	
P-T PATH	ITD to 3-4 kbar at 600-650°C.	Poorly known	
DEFORMATION	Localised retrograde shear zones, subhorizontal L fabrics.	Very minor. Main effects are garnet growth in dykes from SW Vestfolds.	
POST-TECTONIC FELSIC MAGMATISM?	Granitoids 2800-2500 Ma. Pegmatites at 2500 Ma, 770 Ma and possibly 550-500Ma.	Pegmatites.	Pegmatites.
MAFIC DYKES	2400 and ca. 1200 Ma undeformed except in shear zones (D4-D5)	2400 Ma, 1800 Ma, 1400 Ma Undeformed.	None.
SHEAR ZONES AND MLYONITES	Proterozoic shear zones (D4-D5). Upright amphibolite facies; mylonites (greenschist facies) at ca. 500 Ma (?)	Several orientations and ages greenschist mylonite zones Proterozoic to 500 Ma.	Low grade (greenschist) narrow mylonite zones.

Table 12.1 (*Continued*)

RAYNER COMPLEX	KEMP & MACROBERTSON LAND	SE PRYDZ BAY	RAUER GROUP
Rare 3000 Ma reworked Napier Complex; otherwise 2000-1700 Ma felsics	Reworked Napier Complex (3000 Ma?)	Not known. Possible rare Archaean or Mid Proterozoic	Possible 3500 to > 3000 Ma relics in much of E Rauers.
Age range; 1500-1400 Ma I-types, anorthosites; some syn-M2 charnockites 1000 Ma.	Relics from reworked Archaean; probable 1500-1400 Ma I-types; syn-M2 Mawson Charnockite and related.	Minor in most parts. Some late tectonic granites. S-type leucogneisses common.	Abundant ca. 1400 Ma or older HREE-depleted tonalitic types; K-rich granitic suites; S-type leucogneisses.
Fe-rich pelites, semipelites. Carbonates. Common layered gneiss sequences.	Fe-rich pelites; quartzites; Carbonates; Ironstones.	Dominant Fe-rich aluminous pelites; quartzites. Minor carbonates, rare mafics.	Minor. Fe-rich pelite, quartzite, and calcsilicate sequences. Mg-rich pelite and marble boudins, rafts.
3000 Ma (very local relics)	3000-2500 Ma (local relics) (M1-D1)	Pre-1000 Ma (Archaean?)	Pre-1400 Ma (Archaean?) (M1-D1, M2-D2)
9-10 kbar, 900-1000°C (Forefinger Point)	5-8 kbar, 800-900°C (mainly Kemp Land)	9-11 kbar, 900-950°C (Sostrene Is. mafic)	9-12 kbar, 950-1000°C (Mg-pelites). Long Point.
IBC or ITD followed by IBC	IBC?	ITD to 4-5 kbar, 750°C in M2 overprint	ITD to 5-7 kbar, 700-800°C in M3 overprint
		Local relic body.	Local folded rafts, boudins

1000-900 Ma (M2-D2, M3-D3)	$\geq$ 964 Ma (M2-D2, M3-D3)	1100-1000 Ma (M1-D1, M2-D2)	1400 Ma (maximum) (M3-D3), 1000 Ma (M4-D4)
7-9 kbar, 750-850°C to 5 kbar, 700-750°C	7-9 kbar, 800-900°C (W) and 5-7 kbar, 750-850°C (E)	3.5-5 kbar, 700-750°C	5-8 kbar, 820-860°C
ITD to 3-5 kbar at 600-650°C	ITD to 4-5 kbar in W parts, and to 5 kbar in E.	minor ITD to 3-4 kbar at 600°C-700°C; then some cooling	ITD to 3-5 kbar at 700°C
Early flat-lying isoclines, L-fabrics variable. Later upright EW folds	Early flat lying isoclines, strong L-fabrics. Later upright structures.	Flat-lying structures with strong L-fabrics refolded about more upright to reclined structures. Variable intensity & local high strain zones.	Pre-dyke isoclinal reclined folds; steep areas, stretching L-fabrics. Post-dyke upright folding and ductile high-strain zones, variable.
Granites and Pegmatites at 770 Ma	Pegmatites	A-type granitoids at 500 Ma Pegmatites.	Pegmatites
Deformed equivalents of 1200 Ma dykes; rare.	Deformed 1200 Ma dykes common. Contain Grt + Cpx assemblages.	Proterozoic ones probably absent.	Several cross-cutting groups, deformed and metamorphosed. Rare undeformed alnoitic dykes.
Low grade mylonites 500 Ma.	Greenschist facies mylonites, N directed thrust movement.	Narrow mylonite zones in some areas. 500 Ma.	Common mylonite zones, several orientations; probably 500 Ma or younger

Crd boudin within the Mawson Charnockite at Mawson Station (Sheraton *et al.* 1982).

12.7.3 Retrograde P–T–t *paths*

Near-isothermal decompression (ITD) to 4–5 kbar at 600–700°C is inferred from the widespread occurrence of Opx + Pl ± Mag symplectites on Grt + Cpx + Hbl assemblages in Kemp Land (Sheraton *et al.* 1987, Ellis 1983). Decompression from *c.* 8 kbar to 4 kbar is considered to have occurred subsequent to dyke emplacement and the main deformation events. In MacRobertson Land the amount of ITD is not well constrained, but some is inferred from the occurrence of Crd coronas in Grt + Bt + Sil gneisses.

12.8 The south-east Prydz Bay region

12.8.1 Geological framework and general features

The geology of the south-east Prydz Bay area has been described by Sheraton & Collerson (1983), Collerson & Sheraton (1986a), Stüwe *et al.* (1989), Thost *et al.* (1988), and I. C. W. Fitzsimons & S. L. Harley (unpublished data). South of the Rauer Group the granulite terrane is dominated by paragneisses; orthogneisses are only important in the Munro Kerr Mountains (Fig. 12.8) and in some small isolated outcrops. The paragneisses are often migmatitic, and are dominated by pelites and semipelites. Garnet-bearing leucogneisses, interpreted as well segregated partial melts derived from the metapelties (Harley 1987c, Stüwe *et al.* 1989), are also common.

The apparent lack of deformed mafic dykes (Sheraton & Collerson 1983) and isotopic data which indicate that most of the paragneisss precursors were derived from mid-Proterozoic crust (Sheraton *et al.* 1984) have led to the interpretation that the paragneiss precursors were deposited subsequent to *c.* 1300 Ma (Collerson & Sheraton 1983, Stüwe *et al.* 1989). Much of the Prydz Bay granulites may represent metamorphosed cover sequences, with older basement relics exposed locally (e.g. Harley 1987c, Thost *et al.* 1988).

The 'cover sequences' were affected by at least two major deformation events (D_1, D_2) associated with high-*T* metamorphism and partial melting at *c.* 1100–1000 Ma, and by a younger and probably unrelated large-scale open folding which is responsible for regional variations in orientation. The earliest fold phase (D_1) produced stretching lineations, and flat-lying isoclinal folds, and was in part coeval with the generation of partial melts and their segregation into sheet-like leucogneiss bodies (Stüwe *et al.* 1989; Fitzsimons & Harley, unpublished data). D_2 produced east-west trending and steeply plunging folds, caused sheathing of D_1 fold structures, and led to the progressive reorientation and boudinage of leucosomes and leucogneisses (Stüwe *et al.* 1989).

D_2 in the Bolingen Islands and Brattstrand Bluffs to Steinnes Peninsula area produced asymmetric zones of high shear strain, separated by blocks or zones preserving D_1 fabrics and discordant leucosome–restite relationships with relatively little overprint (Thost *et al.* 1988; Fitzsimons & Harley, unpublished data). The D_1 and D_2 events may correlate with D_3 and D_4 described by Harley (1987c) in the Rauer Group, or they may be a progression related solely to D_4 in the chronology of Harley (1987c).

12.8.2 Metamorphic assemblages, P–T *conditions, and partial melting*

Most paragneisses record relatively low-P and high-T metamorphism at 3.5–5.5 kbar and 750–800°C (Fig. 12.8) (Stüwe & Powell 1989, Thost *et al.* 1988; Fitzsimons & Harley, unpublished data). Typical mineral assemblages (Fig. 12.6) include Grt + Bt + Sil + Qtz + Kfs + Ti-phases and Grt + Crd + Sil + Spl + Kfs ± Qtz; (Grt + Opx)-bearing metapelites are rare. Early (S_1) fabrics defined by Sil + Spl ± Bt indicate that low-P conditions were prevalent on the prograde P–T–t path prior to the onset of partial melting (Stüwe & Powell 1989).

Partial melting was of particular importance in the evolution of these paragneisses, many of which can be interpreted as being composed of varying proportions of partially extracted former melt and complementary restite components (Stüwe & Powell 1989). Pelites frequently exhibit migmatitic features and good field evidence for the generation and extraction of leucocratic melts, in equilibrium with Grt + Crd, during D_1 and in the early stages of continuing D_2 shear (Fitzsimons & Harley, unpublished data). Migmatitic features include the following:

(a) Palaeosomes or pelitic restites occurring as schlieren, screens, pods, and rafts in or adjacent to granitic (Kfs + Qtz + Grt) leucosomes. These quartz-absent restites contain oxidized Spl + Sil + Crd + Mag assemblages.
(b) Leucosomes containing minor Grt and sometimes Crd euhedra. Crd is less common in well segregated leucogneiss sheets (1.3 m thickness) than in lensoidal or fine leucosomes, suggesting its origin as a restite or melanosome phase.
(c) Localized, discordant Crd and Crd–Grt-rich pegmatities, interpreted as *melanosomes* retained after the extraction of Kfs–Qtz leucocratic melts, i.e. the solid phases produced in incongruent melting reactions.

Melting occurred at low pressures (*c.* 4–5 kbar) and at 750°C through reactions involving the breakdown of biotite:

$$\text{Bt} + \text{Sil} + \text{Qtz} + \text{Mag} = \text{Crd} + \text{Spl} + \text{Kfs} + \text{Ilm} + \text{L}$$

and

$$Bt + Sil + Qtz + Mag = Grt + Crd + Kfs + Ilm + L$$

Considerably higher initial pressure estimates are obtained from localities where orthogneisses occur. Thost *et al.* (1988) calculate peak *P–T* conditions of at least 9 kbar and 900°C for a Grt (Mg_{47}) + Cpx + Opx + Hbl mafic granulite from Sostrene Island (Fig. 12.8), and Stüwe & Powell (1989) report average pressures of 7–8 kbar at 800°C for a pelite from the Munro–Kerr Mountains. It is proposed that these high-*P* estimates relate to basement gneiss zones exhumed within the cover sequence, and that the pressures may refer to metamorphic events preceding the recognized late Proterozoic evolution. It is, for example, possible that the high-*P* granulites are relics of Archaean or mid-Proterozoic metamorphic events.

12.8.3 Retrograde P–T–t *paths*

Isothermal decompression (ITD) through 0.5–1.5 kbar at *c.* 700°C occurred subsequent to D_2 in the Larsemann Hills (Stüwe & Powell 1989), as evidenced

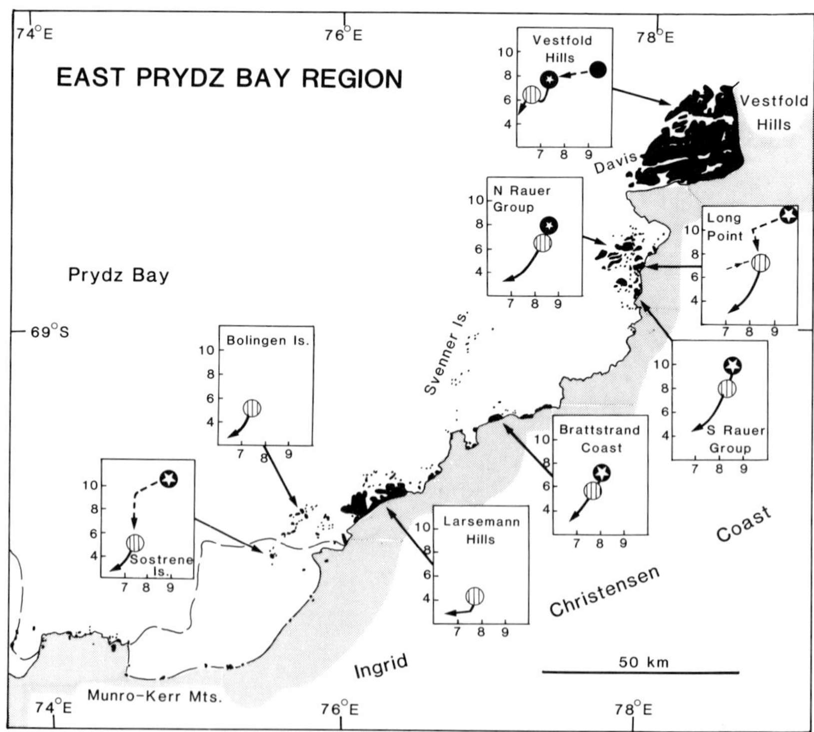

Figure 12.8 Location map of the East Prydz Bay region. See Figure 12.2 for details of symbols and *P–T–t* path data.

by the formation of Crd rims on Grt + opaques, and Spl + Ilm on magnetite, in metapelites. Similar extents of ITD occurred elsewhere in Prydz Bay, but are generally not resolvable using conventional geothermobarometry. The high-pressure mafic granulite from Sostrene Island (Thost *et al.* 1988) records two stages of decompression, the first of which led to the replacement of Grt + Cpx (± Hbl) by Opx + Pl + Ilm symplectites at 5 kbar and 750°C, and was probably related to late Proterozoic overprinting of an older granulite. The second decompressional effect is the isochemical breakdown of garnet alone to Opx + Spl + Pl + Mag, which occurred subsequent to D_2.

Late garnet regrowth occurred locally (Stüwe & Powell 1989) and subsequent to the ITD history. This regrowth might be related to post-tectonic pegmatite emplacement or to the D_3 event of Stüwe *et al.* (1989). The precise timing of these events is not known, but they may be either 770 Ma or 550–500 Ma in age by comparison with the ages obtained for intrusive activity and low-grade metamorphism in the Rayner Complex (Black *et al.* 1987).

12.9 The Rauer Group

12.9.1 Geological framework, lithological constitution, and general features

The Rauer Group represents the northernmost exposures of the late Proterozoic metamorphic complex on the eastern side of Prydz Bay, separated from the Archaean Vestfold Block by the Sorsdal Glacier (Collerson *et al.* 1983, Sheraton & Collerson 1983). The regional geology and structural history of the Rauer Group have been described by Harley (1987c, 1989c), who recognized three broad lithological groups:

(a) Well layered gneisses comprising mafic pyroxene granulite, migmatitic Bt–Grt–Sil–Qtz-feldspar metapelites and quartzitic garnet gneisses, garnet-bearing leucogneisses; and rare boudinaged ultramafics, ironstones, and Wo + Scp calc-silicates. These gneisses also occur as xenoliths or rafts within the orthogneisses which dominate the outcrop, and often may preserve early fabrics cut by intrusive contacts.

(b) Composite layered mafic–felsic gneisses comprising migmatitic mafic granulite interleaved with or veined by orthogneisses of mainly tonalitic composition. Areas of localized partial melting of mafic granulite have given rise to HREE-depleted anhydrous leucosomes in these lithologies (Tait & Harley 1988). This unit is in many ways comparable to the Mossel Gneisses of the Vestfold Hills except that garnet is common both in the mafic and felsic components.

(c) Felsic to intermediate orthogneisses, some of which preserve intrusive contacts with units (a) and (b). Recent fieldwork (S. L. Harley, unpublished data), however, indicates that at least two suites of orthogneiss occur:

(i) an early HREE-depleted group of granodioritic to tonalitic composition which may be derived from early Proterozoic or Archaean precursors (Sheraton *et al.* 1984) and which are extensively cut by several generations of deformed mafic dykes;
(ii) a suite of dioritic and granitic orthogneisses which often preserve relict igneous structures (Kfs phenocrysts, graded and rhythmic layering) and which were emplaced during a later Proterozoic deformation event (D_3, see below).

These major groups, along with garnet–leucogneiss sheets, were interleaved about major north-west–south-east trending reclined to recumbent isoclines in a deformation (D_3, Harley 1987c) which *pre-dates* the emplacement of most metatholeiite dykes (Fig. 12.5e) but is constrained to be of late Proterozoic age (<1400 Ma) by model age data (Sheraton *et al.*1984). A considerable time gap may have occurred between this deformation and associated high-grade metamorphism and the next deformation (D_4) if the metamorphosed mafic dykes are equivalent to the *c.* 1370 ± 125 Ma tholeiitic dyke swarms of the Vestfold Block (Collerson & Sheraton 1986b) and if D_4 is similar in age to the late deformation events in the Rayner Complex (1000–960 Ma, Black *et al.* 1987). D_4 resulted in open to tight upright folds and upright ductile shear zones, and produced biotite foliations or spaced fracture cleavages in deformed dykes and orthogneisses. Although rather variable, the intensity of D_4 structures appears to increase southwards. It is this deformation which probably correlates with the D_2 event recognized for south-east Prydz Bay (see Section 12.8).

Later greenschist facies shear zones dissect the Rauer Group granulites and also cut planar (post-D_4) pegmatites. The shear zones and pegmatites are thought to be 550–500 Ma old, by comparison with isotopic data on pegmatites from the Rayner Complex (Black *et al.* 1987, Harley 1987c) and the granite at Landing Bluff in south-east Prydz Bay (Sheraton & Black 1988).

12.9.2 Metamorphic assemblages and P–T *conditions*

It is evident that some polymetamorphic lithologies may preserve pre-D_3, perhaps even Archaean, metamorphic assemblages. In the north-east Rauer Group, Archaean model ages have been obtained on felsic orthogneisses (Sheraton *et al.* 1984) which themselves host distinctive Fo–Di–Spl marbles, and Mg-rich aluminous metapelite rafts which contain anomalous mineral assemblages (Fig. 12.6) interpreted (S. L. Harley, unpublished data) as Archaean or early Proterozoic relics (Fig. 12.8). At Long Point (Fig. 12.8), quartz-bearing Grt + Sil + Bt gneisses enclosed within felsic orthogneiss contain magnesian garnets (Mg_{57}) indicating high equilibration pressures (>9 kbar). More spectacularly, a quartz-deficient metapelite raft (SH218) contains the assemblages Opx + Grt + Sil and Opx + Grt + Sil + Spr, implying *P–T* conditions bounded by the (Spl, Qtz) and (Spl, Crd) reactions of Figure 12.4. Garnet porphyroblasts coexisting with Opx + Sil are the *most magnesian* yet found in granulite facies metapelites worldwide (Mg_{72}). *P–T* conditions estimated through Grt–Opx thermobarometry are 10–12 kbar and 1000°C (S. L. Harley, unpublished data), clearly outside of the range of *P–T* conditions pertinent to the late Proterozoic D_3–D_4 evolution. This metapelite also exhibits reaction textures similar to those reported from Forefinger Point in the Rayner Complex and previously (Section 12.5) suggested to be a response to the incorporation of earlier-formed granulites into the late Proterozoic ITD régime.

Diagnostic granulite assemblages of probable late Proterozoic age (pre-D_3 to D_3) include Grt + Cpx + Opx + Pl ± Hbl (mafic granulites), Grt + Opx + Pl + Qtz ± Bt (orthogneisses), Grt + Sil + Rt + Qtz + feldspars ± Ilm ± Bt (Fig. 12.6) (metapelites), and Wo + Scp + Di + Grs ± Pl ± Cal ± Qtz (calc-silicates) (Harley 1987c). Cordierite-bearing metapelites are rare, in contrast to south-east Prydz Bay granulites, and restricted to Mg-rich and often quartz-deficient horizons (Harley 1987c, 1989c). Detailed geothermobarometry of mafic to felsic orthogneisses (Harley 1988) indicates peak (D_3) *P–T* conditions of 7–8.5 kbar and 860 ± 40°C in the south and southeastern parts of the Rauers, and 6–7.5 kbar at similar temperatures in the north-west Rauers. These *P* estimates are corroborated by barometry of Grt + Sil + Pl + Ilm + Rt + Qtz assemblages (7.5–8.5 kbar) in southern localities, and of Grt + Crd + Opx + Qtz assemblages (6.4 kbar) from a locality nearer the Vestfold Hills (Harley 1989c). Temperatures of 820°C or more are also implied by the occurrence of calcic scapolite (Mei_{90}) and wollastonite with anorthite (An_{97-99}) in some calc-silicates. The presence of grossular (Grs_{70-85}) as zoned early grains and as secondary rims in such calcsilicates also constrains a_{CO_2} to be low (<0.2; Harley 1987c), whereas mineral-fluid calculations for adjacent metapelite assemblages generally indicate low a_{H_2O} conditions (<0.4). Taken together, these results suggest that dominantly fluid-absent granulite metamorphism accompanied D_3.

12.9.3 Partial melting in mafic granulites

Mafic granulites from the composite layered gneiss group (b) often preserve pre-D_3 migmatitic structures ranging from stromatic to schollen and agmatic (Tait & Harley 1988) (Fig. 12.5f). Leucosomes contain euhedral orthopyroxene, are dioritic in bulk composition, and are essentially dry. Whether these leucosomes solely represent melt, or melt *and* the solid products of peritectic melting reactions (Waters 1988) is not clear from mass-balance calculations because of later (D_3, D_4) re-equilibration and the introduction of biotite in younger events (Tait & Harley 1988). Petrogenetic modelling of the melting relations of the pyroxene (±Hbl) granulite palaeosomes, and liquidus studies of the HREE-depleted leucosomes, indicate that water-undersaturated or vapour-absent melting at >10 kbar and $>960^\circ$C could have produced these high-T migmatites. These $P-T$ conditions overlap with those of the Mg–pelite relic considered earlier, and may be related to the same pre-D_3 event.

12.9.4 Retrograde P–T–t *path*

ITD through at least 2–3 kbar and at temperatures greater than 700–750°C (Fig. 12.8) is indicated both by geothermobarometry of late mineral assemblages in mafic and felsic orthogneisses (Harley 1988) and metapelites (Harley 1989c). Textural evidence in support of such a decompressional history includes:

(a) Opx + Pl symplectites and mantles on earlier Grt + Cpx + Hbl (Fig. 12.5d);
(b) Pl moats and rims replacing Grt in felsic orthogneisses;
(c) Crd ± Spl moats and coronas on earlier Grt + Sil + Bt;
(d) Crd + Opx symplectites on Grt + Opx + Qtz in magnesian metapelites; and
(e) growth of Ilm + Sil aggregates at the expense of Grt + Rt.

While these textures post-date D_3, their relation to D_4 is sometimes ambiguous. Harley (1987c) considered that decompression pre-dated D_4, which was therefore superimposed at pressures of 3–5 kbar. However, recent fieldwork which shows that some metatholeiite dykes contain embayed garnet in felsic segregations, suggests that this period of decompression outlasted D_4.

Post-D_3 and perhaps -D_4 decompression to *c*. 5–6 kbar and 700–750°C, subsequent to a phase of localized partial melting, is indicated also by multistage reaction textures in the metapelite SH218:

(a) Crd + Spr intergrowths replacing Sil, formed by the continuous reaction

$$\text{Sil} + \text{Opx} + = \text{Crd} + \text{Spr} \qquad (\text{Grt, Spl, Qtz})$$

(b) pseudomorphous replacement of garnet by lamellar intergrowths of Spr + Opx ± Crd, formed through the reaction

$$\text{Grt} = \text{Spr} + \text{Opx} + \text{Crd}\ (\pm\ \text{Spl}) \qquad (\text{Spl, Sil, Qtz}),$$

and

(c) later embayment of garnet by Opx + Crd symplectites which post-date biotite rinds.

12.10 Discussion: tectonic models for granulite terranes and the evolution of the East Antarctic Shield

12.10.1 Tectonic models for the Archaean IBC-type terranes

In the absence of good constraints on the prograde metamorphic path (see Section 12.2) it is possible to suggest a spectrum of models for the generation of IBC granulites in terranes such as the Napier Complex or Vestfold Block. As noted by several workers (Ellis 1980, 1987; Harley 1985a, 1989a,b; Sandiford 1985a; Bohlen 1987) the occurrence of near-isobaric cooling in granulites implies that the thermal peak of metamorphism occurred in crust of near-normal thickness, and hence subsequent to isostatic readjustment after any earlier thickening either by tectonic processes (collision) or by magmatic accretion.

Tectonic thickening caused by collision and terminating in IBC has been proposed by Ellis (1987) and Harley (1987d, 1989a). Ellis (1987) argued that IBC would *normally* occur in rocks within the lower plate or lower part of doubly thickened crust, subsequent to a decompressional phase caused by uplift and erosion. The decompressional phase in such erosion-controlled models, however, typically occurs over a long time interval (50–100 Ma) (England & Thompson 1984, 1986). It is therefore likely, given such slow uplift rates, that the lower parts of the thickened crust will have approached steady-state thermal conditions by the end of the decompressive phase, and hence will not retain any initial anomalously high-temperature conditions. To retain the high temperatures found in the Napier Complex much more rapid exhumation of the hot lower crust is required.

Based largely on recent geophysical and dynamic models for diffusely thickened collisional zones such as the Tibetan Plateau (England 1987), and large-scale post-collisional extensional terranes exemplified by the Basin and Range Province (Sonder *et al.* 1987), Harley (1987d, 1989a,b) has suggested that IBC from high temperatures in mid- to lower-crustal régimes could occur where crustal thickening through collision is followed by *very* rapid extensional thinning at rates of 2–5 mm per annum. Rapid, isothermal or even increasing-T decompression would occur during the extensional phase, followed by post-extensional IBC from high temperatures (Figs. 12.9a & b). The temperatures reached at depth depend on the pre-extensional temperature profile, the extent of accretion of syn-extensional magmas generated from decompressing asthenosphere, and the rate and time span of extension (Sonder *et al.* 1987). As discussed in detail by Harley (1989b), a spectrum of recorded retrograde paths ranging from IBC to rapid, high-T, ITD (followed perhaps by some IBC) are possible depending upon the amount of overthickening of the crust and the magnitude of extensional strain rates (Fig. 12.9b). Such a collision-dominated model for the origin of the Napier Complex would be favoured if high-P relics (Ky, Mg–Grt) were discovered *and* if the dominant orthogneiss

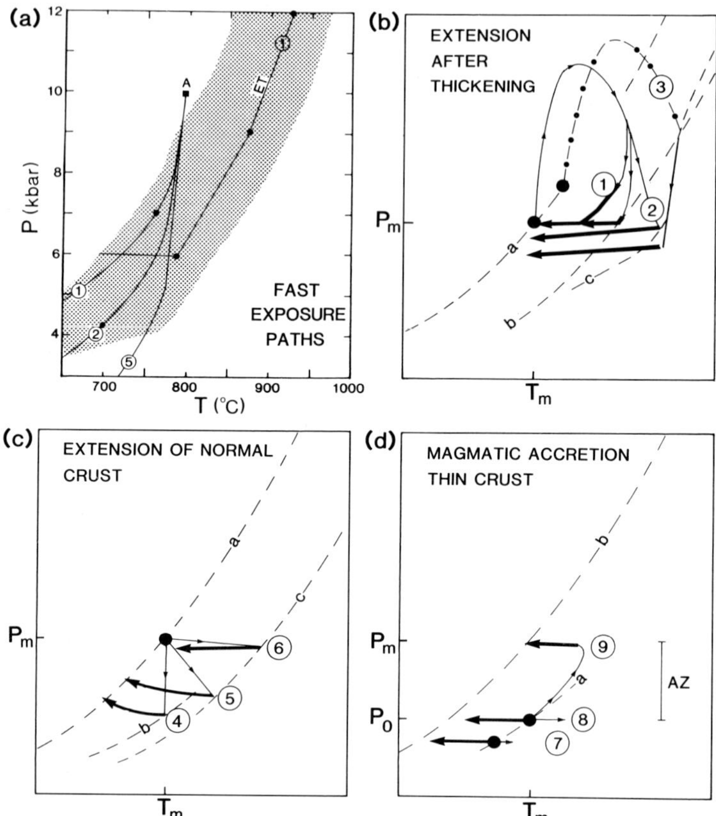

Figure 12.9 Some models for granulite P–T–t paths and genesis (modified from Harley 1989a, Figs 13b and 14a,c,d, by permission of Cambridge University Press). (a) Model exhumation paths for fast rates of 'tectonic' stripping. Paths are labelled for different exposure rates (in mm/yr); solid circles represent 10 Ma time intervals. Paths A (1, 2, 5) based on the model of Albarède (1976); Path ET based on the models of England & Thompson (1986), with initial heat flow of 75 mW m^{-2}, conductivity K of 2.5 W m^{-1}K^{-1}, basal Q of 3 mW m^{-2}, and internal crustal heat production of 1.3 μW m^{-3}. Rock originally buried to 50 km in doubly thickened crust. The shaded field is the generalized P–T–t path régime for ITD-type granulites (Fig. 12.1b). Note the final phase of isobaric cooling from 800°C in the ET path. (b) Schematic P–T–t paths for continental thickening followed by rapid tectonic thinning (extension), based on Sonder *et al.* (1987). Initial geotherm a; late-extension geotherm b; late-extension plus magmatism triggered by the detachment of the lithospheric thermal boundary layer indicated by geotherm c. Paths 1 and 2 correspond to increasing rates of extension and hence decompression. Path 3 includes decompression-induced melting of lithosphere followed by isobaric cooling of magmas emplaced into the crust. *Recorded* P–T–t path segments indicated by heavy lines. (c) Extension of crust of normal thickness (geotherm a) through either symmetrical (to geotherm b) or asymmetric extension (to geotherm c), after Sandiford & Powell (1986b). Path 4, lower-crustal rock (at P_m, base of crust) in symmetrical pure-shear model; path 5, lower crust in or below the detachment zone near the region of subcrustal thinning, simple-shear (asymmetric) model (Wernicke 1985); path 6, path for rocks at the base of the crust in or beyond the 'discrepant zones' of Wernicke (1985). Preserved P–T–t paths indicated by heavy lines. (d) Magmatic accretion into thinned crust. Initial geotherm in thinned crust a; final, post-magmatic geotherm b; accretion zone AZ thickens the crust to normal thickness, P_m. Paths 7 and 8 are for rocks *above* the accretion zone; path 9 is relevant to rocks initially at P_0 and subsequently buried beneath the accretion zone. Similar P–T–t paths (heavy lines) would be recorded if the initial crust was thicker, but cooling would not be isobaric (Harley 1989a).

sequences were proven to be substantially older than, and hence unrelated to, the major metamorphic event.

IBC-type granulites may also be generated through the extension of normal-thickness crust (Sandiford & Powell 1986b). In symmetrical extensional models, deep-level (7–11 kbar) IBC would not be possible for rocks which had previously been at the base of a normal-thickness crust, but near-isobaric heating and cooling paths could occur in and above the low-angle detachment zones proposed in the asymmetric simple-shear extensional models (Sandiford & Powell 1986b) (Fig. 12.9c). In most extensional settings, however, magmatic accretion also plays an important rôle, and may in reality dominate the thermal and hence *P–T–t* histories in the deep crust.

Magmatic accretion, either in magmatic arc or continental extensional settings, has been proposed by many workers as an effective means of producing IBC-type granulites in crust of near-normal thickness (Wells 1980, Phillips & Wall 1981, Waters 1986, Bohlen 1987). Counterclockwise *P–T–t* paths, possibly relevant to the Napier Complex (Hensen & Motoyoshi 1988), are generated in rocks within and below the accretion zone. IBC, or shallow dP/dT cooling and decompression, *P–T–t* paths follow the metamorphic peak, depending upon the post-accretion crustal thickness (Harley 1989a) (Fig. 12.9d).

The viability of magmatic accretion as a mechanism of high-*T* granulite formation depends on the proportion of newly accreted magmatic material compared to pre-existing crust, and the temperatures of the accreted magmas. Accretion or 'hot-spot' models previously proposed for the generation of the Napier Complex granulites (Ellis 1980, Harley 1985a) invoke the existence of a deeper, unexposed heat source. However, the similarity of peak-metamorphic temperatures (*c*. 1000°C) in granulites metamorphosed at 7–11 kbar suggests that these granulites are *within*, and perhaps below, any magmatic accretion zone. Detailed geochronological studies of orthogneisses in the Tula Mountains and Scott Mountains are required to show the significance of syn-D_1 magmatic accretion in the genesis of the unusual Napier Complex granulites.

In the case of the Vestfold Hills, the post-2450 Ma near-isobaric cooling history and prolonged lower- to mid-crustal residence is consistent with the accretion of a large volume of high-*T* magmas (the Crooked Lake Gneisses), perhaps within a magmatic arc or extending continental margin. Little metamorphic evidence exists to constrain the prograde path in this granulite facies event, but it is apparent from the relics of previously metamorphosed and deformed gneisses that crustal thicknesses of at least 20–25 km had been attained at some time prior to the 2450 Ma accretion and deformation events. Folding and accretion of the Crooked Lake Gneisses thickened this crust to perhaps 40–45 km prior to early Proterozoic dyke emplacement (see Section 12.3). The tectonic implications of the pre-2450 Ma (*c*. 3000–2800 Ma) granulite event are not well understood, although Collerson & Sheraton

(1986a) have suggested a magmatic arc setting for this metamorphism, mainly from a consideration of the geochemical characteristics of the early orthogneisses.

12.10.2 Models for the late Proterozoic granulite terranes

From the foregoing sections it is clear that *P–T–t* paths characterized by decompression through 1–4 kbar (uplift of 3–12 km) are typical of the late Proterozoic (1100–1000 Ma) granulite terranes throughout the region, and of the possibly late Proterozoic or younger Lützow – Holm Bay Complex. Regional variations do occur in the temperatures at which decompression took place (650–850°C), the maximum recorded pressures (4–6 to 8–10 kbar), and possibly the timing of uplift (Harley 1988, 1989b).

This general late Proterozoic decompressional history has been related to a major phase of collisional orogeny (Harley 1988, 1989b) which locally involved thrusting and differential movements within the Archaean terranes (Clarke 1987, Sandiford 1985b, Harley 1985b) as well as strong deformation within Proterozoic domains. The interpretation of Sheraton *et al.* (1984) and Stüwe & Powell (1989) that certain paragneiss-dominated cover sequences (e.g. Prydz Bay) were mainly deposited subsequent to mafic dyke emplacement suggests the possibility (Harley 1988) that the late Proterozoic tectonothermal event involved thickening of a previously extended and 'hot' continental margin. In detail (Harley 1988, 1989b; Stüwe & Powell 1989), simple collisional models in which erosion terminates the heating of the crust require quite extreme initial thermal conditions, and basal heat flux plus crustal heat productivity combinations, in order to generate the high-*T* decompressional paths, particularly those at pressures of only around 5 kbar. Such ITD paths could be generated through relatively rapid (1–2 mm per annum) uplift and exposure of overthickened crust in a collisional–extensional model (Harley 1989b) (Fig. 12.9a) in which magmatic accretion during the late-collisional and rapid exhumation phases might also be important. The timescales for the observed metamorphism and decompression implied by this type of model are rather short (<20 Ma) compared with simple collisional models (20–60 Ma). The applicability of these generalized models needs, therefore, to be tested by detailed isotope geochronology aimed at determining the timespans of the deformational–metamorphic events and uplift rates.

A significant feature of the late Proterozoic terranes is that the paragneiss-dominated, or 'cover sequence', areas preserve evidence of peak metamorphism at substantially lower pressures (3.5–6 kbar) than those recorded in orthogneiss-dominated or mixed, probable 'basement' terranes (Figs. 12.2 & 8). In the light of this observation, Stüwe & Powell (1989) argue against collisional models for the late Proterozoic metamorphism in Prydz Bay and invoke a large amount of magmatic accretion into thinned crust (an 'asthenospheric perturbation'), centred beneath the exposed granulites in the

Larsemann Hills. Although the emplacement of large volumes of magma could indeed give rise to the pronounced mid-crustal thermal anomaly indicated by these granulites, and could lead to some decompression if the eventually accreted crust was overthick (*c.* 40 km), there is a lack of field evidence (e.g. charnockites or mafic intrusives) for such magmatism. Models invoking magmatic accretion alone do not account for the deformational histories of the terranes. Magmatic accretion alone is also an inefficient means of producing very high temperatures in the 'basement' granulites where a large amount of older crust was already present. However, as noted previously, magmatic accretion may have played an important rôle in controlling metamorphic grade within a collisional orogenic setting. Relatively low-*P* granulite metamorphism and subsequent decompression could occur in a broadly collisional context if, after initial burial in early thrusting, the young cover sequences became resident in an 'upper-plate' (England & Thompson 1984) above a décollement or detachment at the modified basement–cover contact, beneath which further tectonic thickening occurred. It is clear that much more structural, metamorphic, and particularly geochronological data are needed before detailed models quantifying the relative rôles of tectonic and magmatic thickening can be applied to the late Proterozoic terranes.

12.10.3 Stabilization of the East Antarctic Shield and the exhumation of the granulite terranes

It has been demonstrated that the Archaean granulite terranes described herein remained stable essentially cratonic elements for at least 1500–2000 Ma subsequent to their major tectonothermal, granulite-forming, episodes. The presently exposed granulites in these areas remained resident in the lower or middle crust at least until reworking in the *c.* 1000 Ma Proterozoic events (Sections 12.3–5). The relationship between the Napier Complex and the Rayner Complex in Enderby Land (Fig. 12.10, modified after Harley 1989b) is suggested to involve *c.* 1100–1000 Ma underplating of the older Napier Complex by Rayner Complex crust in a transpressional régime related to thrusting further east, so that en-masse block uplift of the south-west Napier Complex through 10–14 km occurred at that time (Harley 1989b). Subsequent to the late Proterozoic tectonism, therefore, most of the granulites now exposed throughout the 40–80°E sector were buried at depths equivalent to 3–4 kbar (10–15 km) in what can be regarded as an essentially unified shield area in which the various granulite terranes shared a common post-1000 Ma evolution, including pegmatite and granite emplacement at 770 Ma and 550–500 Ma, shearing and local low-grade metamorphism at 550–500 Ma, and the post-500 Ma emplacement of alkaline dyke suites (Sheraton *et al.* 1987).

The eventual exhumation of these Antarctic granulites did not occur until after a widespread but often weak greenschist facies overprint, often associated with *c.* 550 Ma shearing in mylonite zones, which took place at *c.* 3 kbar

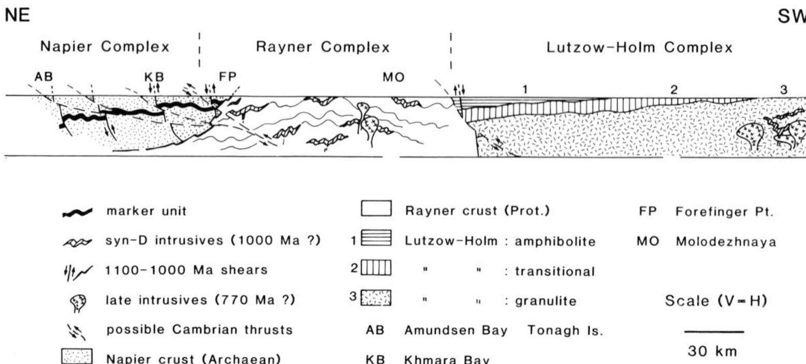

Figure 12.10 Schematic NE–SW cross-section illustrating an interpretation of the relationships between the Napier Complex, the Rayner Complex, and the Lützow–Holm Complex in western Enderby Land and eastern Dronning Maud Land. Late Proterozoic structures are interpreted to have caused initial uplift and exhumation of the Napier Complex; Cambrian and perhaps more recent faulting are believed to have resulted in sufficient thickening to cause later exposure through erosion.

and 500°C (Bt-zone assemblages; Harley 1987c). This Cambrian event, suggested to result from north- or north-west-directed thrusting (Clarke 1987), is probably responsible for some of the major discontinuities between terranes and the eventual exhumation of the granulites to within 6 km of the Earth's surface through erosion of the post-thrusting thickened crust. At least some of the granulite terranes were exposed and providing sedimentary detritus by the early Permian (James & Tingey 1983).

Acknowledgements

We thank Mike Brown for inviting us to write this review, and John Sheraton and Ed Grew for their careful comments. The assistance and support of ANARE for fieldwork in Antarctica, and ASAC grants to S.L.H. (1987–9) and B.J.H. (1988–9) are gratefully acknowledged. The support of British Council (B.J.H.) and TransAntarctic Association (S.L.H.) travel grants enabled the conception and preparation of this work. We thank our students, Ian Fitzsimons, Rhoda Tait, and Doug Thost, for their inputs and use of initial data, and G. Clarke, C. Delor, D. J. Ellis, P. England, P. Kinny, Y. Motoyoshi, J. Platt, and D. Young for discussions.

References

Albarède, F. 1976. Thermal models of post-tectonic decompression as exemplified by the Haut-Allier granulites (Massif Central, France). *Bulletin de la Société Géologique de France* **18**, 1023–32.

Aranovich, L. Ya. & K. K. Podlesskii 1989. Geothermobarometry of high grade metapelites: simultaneously operating reactions. In *Evolution of metamorphic belts*, J. S. Daly, R. A. Cliff & B. W. D. Yardley (eds), (in press). Geological Society of London, Special Publication.

Bertrand, P., D. J. Ellis & D. H. Green 1989. Stabilité des assemblages Sa–Qz et Hy–Sil–Qz dans le système FMAS, sous faible P_{H_2O} et f_{O_2}. *Comptes Rendus de l'Académie des Sciences, Série II* **308**, 1437–42.

Black, L. P., J. D. Fitzgerald & S. L. Harley 1984. Pb isotopic composition, colour, and microstructure of monazites from a polymetamorphic rock in Antarctica. *Contributions to Mineralogy and Petrology* **85**, 141–8.

Black, L. P., S. L. Harley, S. S. Sun & M. T. McCulloch 1987. The Rayner Complex of East Antarctica: complex isotopic systematics within a Proterozoic mobile belt. *Journal of Metamorphic Geology* **5**, 1–26.

Black, L. P. & P. R. James 1983. Geological history of the Archaean Napier Complex of Enderby Land. In *Antarctic Earth science*, R. L. Oliver, P. R. James & J. B. Jago (eds), 11–15. Canberra: Australian Academy of Science.

Black, L. P., P. R. James & S. L. Harley 1983a. The geochronology, structure, and metamorphism of early Archaean rocks at Fyfe Hills, Enderby Land, Antarctica. *Precambrian Research* **21**, 197–222.

Black, L. P., P. R. James & S. L. Harley 1983b. Geochronology and geological evolution of metamorphic rocks in the Field Islands area, East Antarctica. *Journal of Metamorphic Geology* **1**, 277–303.

Black, L. P., J. W. Sheraton & P. R. James 1986a. Late Archaean granites of the Napier Complex, Enderby Land, Antarctica: a comparison of Rb–Sr, Sm–Nd, and U–Pb isotopic systematics in a complex terrain. *Precambrian Research* **32**, 343–68.

Black, L. P., I. S. Williams & W. Compston 1986b. Four zircon ages from one rock: the history of a 3930 Ma-old granulite from Mount Sones, Enderby Land, Antarctica. *Contributions to Mineralogy and Petrology* **94**, 427–37.

Bohlen, S. R. 1987. Pressure–temperature–time paths and a tectonic model for the evolution of granulites. *Journal of Geology* **95**, 617–32.

Chappell, B. W. & A. J. R. White 1974. Two contrasting granite types. *Pacific Geology* **8**, 173–4.

Clarke, G. L. 1987. Structural constraints on the Proterozoic reworking of Archaean crust in the Rayner Complex, MacRobertson and Kemp Land Coast, East Antarctica. *Abstracts, 5th International Symposium on Antarctic Earth Science*, Cambridge, p. 29.

Clarke, G. L. 1988. Structural constraints on the Proterozoic reworking of Archaean crust in the Rayner Complex, MacRobertson and Kemp Land Coast, East Antarctica. *Precambrian Research* **40–41**, 137–56.

Collerson, K. D., E. Reid, D. Millar & M. T. McCulloch 1983. Lithological and Sr–Nd isotopic relationships in the Vestfold block: implications for Archaean and Proterozoic crustal evolution in the east Antarctic. In *Antarctic Earth science*, R. L. Oliver, P. R. James & J. B. Jago (eds), 77–84. Canberra: Australian Academy of Science.

Collerson, K. D. & J. W. Sheraton 1986a. Bedrock geology and crustal evolution of the Vestfold Hills. In *Antarctic oasis: terrestrial environments and history of the Vestfold Hills*, J. Pickard (ed.), 21–62. Sydney: Academic Press.

Collerson, K. D. & J. W. Sheraton 1986b. Age and geochemical characteristics of a mafic dyke swarm in the Archaean Vestfold Block, Antarctica: inferences about Proterozoic dyke emplacement in Gondwana. *Journal of Petrology* **27**, 853–86.

Collins, W. J., S. D. Beams, A. J. R. White & B. W. Chappell 1982. Nature and origin of A-type granites with particular reference to southeastern Australia. *Contributions to Mineralogy & Petrology* **80**, 189–200.

Compston, W. & I. S. Williams 1982. Protolith ages from inherited zircon cores measured by a high mass resolution ion microprobe. *Fifth International Conference on Geochronology, Cosmochronology and Isotope Geology*, Nikko, Japan, 63–4 (abstract).

Dallwitz, W. B. 1968. Coexisting sapphirine and quartz in granulite from Enderby Land, Antarctica. *Nature* **219**, 476–7.

Ellis, D. J. 1980. Osumilite–sapphirine–quartz granulites from Enderby Land, Antarctica: *P–T* conditions of metamorphism, implications for garnet–cordierite equilibria and the evolution of the deep crust. *Contributions to Mineralogy and Petrology* **74**, 201–10.

Ellis, D. J. 1983. The Napier and Rayner Complexes of Enderby Land, Antarctica: Contrasting styles of metamorphism and tectonism. In *Antarctic Earth science*, R. L. Oliver, P. R. James & J. B. Jago (eds), 20–4. Canberra: Australian Academy of Science.

Ellis, D. J. 1987. Origin and evolution of granulites in normal and thickened crusts. *Geology* **15**, 167–70.

Ellis, D. J. & D. H. Green 1985. Garnet-forming reactions in mafic granulites from Enderby Land, Antarctica – implications for geothermometry and geobarometry. *Journal of Petrology* **26**, 633–62.

Ellis, D. J., J. W. Sheraton, R. N. England & W. B. Dallwitz 1980. Osumilite–sapphirine–quartz granulites from Enderby Land, Antarctica – mineral assemblages and reactions. *Contributions to Mineralogy and Petrology* **72**, 123–43.

England, P. C. 1987. Diffuse continental deformation: length scales, rates and metamorphic evolution. *Philosophical Transactions of the Royal Society of London* **A321**, 3–22.

England, P. C. & A. B. Thompson 1984. Pressure–temperature–time paths of regional metamorphism, I. Heat transfer during the evolution of regions of thickened continental crust. *Journal of Petrology* **25**, 894–928.

England, P. C. & A. B. Thompson 1986. Some thermal and tectonic models for crustal melting in continental collision zones. In *Collisional tectonics*, M. P. Coward & A. C. Ries (eds), 83–94. Geological Society of London, Special Publication 19.

Green, D. H. & A. E. Ringwood 1967. An experimental investigation of the gabbro to eclogite transformation and its petrological applications. *Geochimica et Cosmochimica Acta* **31**, 767–833.

Grew, E. S. 1978. Precambrian basement at Molodezhnaya station, East Antarctica. *Geological Society of America Bulletin* **89**, 801–13.

Grew, E. S. 1980. Sapphirine + quartz association from Archean rocks in Enderby Land, Antarctica. *American Mineralogist* **65**, 821–36.

Grew, E. S. 1981a. Surinamite, taaffeite, and beryllian sapphirine from pegmatites in granulite-facies rocks of Casey Bay, Enderby Land, Antarctica. *American Mineralogist* **66**, 1022–33.

Grew, E. S. 1981b. Granulite facies metamorphism at Molodezhnaya Station, East Antarctica. *Journal of Petrology* **22**, 297–336.

Grew, E. S. 1982. Osumilite in the sapphirine–quartz terrane of Enderby Land, Antarctica: implications for osumilite petrogenesis in the granulite facies. *American Mineralogist* **67**, 762–87.

Grew, E. S. & W. I. Manton 1979. Archean rocks in Antarctica: 2.5-billion-year uranium–lead ages of pegmatites in Enderby Land. *Science* **206**, 443–5.

Griffin, W. L. & S. Y. O'Reilly 1986. The lower crust in eastern Australia: xenolith evidence. In *The nature of the lower continental crust*, J. B. Dawson, D. A. Carswell, J. Hall & K. H. Wedepohl (eds), 363–74. Geological Society of London, Special Publication 24.

Hansen, E. C., R. C. Newton & A. S. Janardhan 1984. Fluid inclusions in rocks from the amphibolite-facies gneiss to charnockite progression in southern Karnataka, India: direct evidence concerning the fluids of granulite metamorphism. *Journal of Metamorphic Geology* **2**, 249–64.

Harley, S. L. 1981. *Garnet–orthopyroxene assemblages as pressure–temperature indicators.* Unpubl. PhD thesis, University of Tasmania, 365 pp.

Harley, S. L. 1983. Regional geobarometry – geothermometry and metamorphic evolution of Enderby Land, Antarctica. In *Antarctic Earth science*, R. L. Oliver, P. R. James & J. B. Jago (eds), 25–30. Canberra: Australian Academy of Science.

Harley, S. L. 1985a. Garnet–orthopyroxene bearing granulites from Enderby Land, Antarctica: metamorphic pressure–temperature–time evolution of the Archaean Napier Complex. *Journal of Petrology* **26**, 819–56.

Harley, S. L. 1985b. Paragenetic and mineral–chemical relationships in orthoamphibole-bearing gneisses from Enderby Land, East Antarctica: a record of Proterozoic uplift. *Journal of Metamorphic Geology* **3**, 179–200.

Harley, S. L. 1986. A sapphirine–cordierite–garnet–sillimanite granulite from Enderby Land, Antarctica: implications for FMAS petrogenetic grids in the granulite facies. *Contributions to Mineralogy and Petrology* **94**, 452–60.

Harley, S. L. 1987a. A pyroxene-bearing meta-ironstone and other pyroxene–granulites from Tonagh Island, Enderby Land, Antarctica: further evidence for very high temperature (>980°C) Archaean regional metamorphism in the Napier Complex. *Journal of Metamorphic Geology* **5**, 341–56.

Harley, S. L. 1987b. Archaean sapphirine granulites from the Vestfold Hills. In *Abstracts, 5th International Symposium on Antarctic Earth Science*, Cambridge, p. 150.

Harley, S. L. 1987c. Precambrian geological relationships in high-grade gneisses of the Rauer Islands, East Antarctica. *Australian Journal of Earth Sciences* **34**, 175–207.

Harley, S. L. 1987d. Metamorphic evolution of granulites from the Rauer Group, East Antarctica: decompression following Proterozoic collision. In *Abstracts, 5th International Symposium on Antarctic Earth Science*, Cambridge, p. 61.

Harley, S. L. 1988. Proterozoic granulites from the Rauer Group, East Antarctica. I. Decompressional pressure–temperature paths deducted from mafic and felsic gneisses. *Journal of Petrology* **29**, 1059–95.

Harley, S. L. 1989a. The origins of granulites: a metamorphic perspective. *Geological Magazine* **126**, 215–47.

Harley, S. L. 1989b. The crustal evolution of some East Antarctic granulites. In *Antarctic Earth science*, M. R. A. Thompson (ed.), In press. Cambridge: Cambridge University Press.

Harley, S. L. 1989c. Metamorphic evolution of granulites from the Rauer Islands, East Antarctica: decompression following Proterozoic collision. In *Antarctic Earth science*, M. R. A. Thompson (ed.), in press. Cambridge: Cambridge University Press.

Harley, S. L. & L. P. Black 1987. The Archaean geological evolution of Enderby Land, Antarctica. In *Evolution of the Lewisian and comparable Precambrian high-grade terrains*, R. G. Park & J. Tarney (eds), 285–96. Geological Society of London, Special Publication 27.

Hensen, B. J. 1986. Theoretical phase relations involving garnet and cordierite revisited: the influence of oxygen fugacity on the stability of sapphirine and spinel in the system Mg–Fe–Al–Si–O. *Contributions to Mineralogy and Petrology* **92**, 362–7.

Hensen, B. J. 1987. *P–T* grids for silica-undersaturated granulites in the system MAS ($n + 4$) and FMAS ($n + 3$) – tools for the derivation of *P–T–t* paths of metamorphism. *Journal of Metamorphic Geology* **5**, 255–71.

Hensen, B. J. & D. H. Green 1973. Experimental study of the stability of cordierite and garnet in pelitic compositions at high pressures and temperatures. III. Synthesis of experimental data and geological applications. *Contributions to Mineralogy and Petrology* **38**, 151–66.

Hensen, B. J. & Y. Motoyoshi 1988. Sapphirine–quartz–orthopyroxene symplectites after cordierite in granulites from the Napier Complex, Antarctica: evidence for a counterclockwise *P–T* path? (Abstract) *Terra Cognita* **8**, 263.

Hiroi, Y., K. Shiraishi, Y. Nakai, T. Kano & S. Yoshikura 1983a. Geology and petrology of the Prince Olav Coast, East Antarctica. In *Antarctic Earth science*, R. L. Oliver, P. R. James & J. B. Jago (eds), 32–5. Canberra: Australian Academy of Sciences.

Hiroi, Y., K. Shiraishi, K. Yanai & K. Kizaki 1983b. Aluminum silicates in the Prince Olav and Soya Coasts, East Antarctica. *Memoirs of the National Institute of Polar Research (Japan)*, special issue **28**, 115–31.

Hiroi, Y., K. Shiraishi, Y. Motoyoshi, S. Kanisawa, K. Yanai & K. Kizaki 1986. Mode of occurrence, bulk chemical compositions, and mineral textures of ultramafic rocks in the Lutzow–Holm Complex, East Antarctica. *Memoirs of the National Institute of Polar Research* (*Japan*), Special Issue 43, 62–84.

Hiroi, Y., K. Shiraishi, Y. Motoyoshi & T. Katsushima 1987. Progressive metamorphism of calc-silicate rocks from the Prince Olav and Soya Coasts, East Antarctica. *Proceedings of the National Institute of Polar Research* (*Japan*), *Symposium on Antarctic Geoscience* **1**, 73–97.

Hölscher, A., W. Schreyer & D. Lattard 1986. High-pressure, high-temperature stability of surinamite in the system $MgO–BeO–Al_2O_3–SiO_2–H_2O$. *Contributions to Mineralogy and Petrology* **92**, 113–27.

Ito, K. & G. C. Kennedy 1971. An experimental study of the basalt–garnet granulite–eclogite transition. In *The structure and physical properties of the Earth's crust*, J. G. Heacock (ed.), 303–14. American Geophysical Union Monograph 14.

James, P. R. & L. P. Black 1981. A review of the structural evolution and geochronology of the Archaean Napier Complex of Enderby Land, Australian Antarctic Territory. *Geological Society of Australia Special Publication* **7**, 71–83.

James, P. R., P. Ding & L. Rankin 1987. Structural geology of the early Precambrian gneisses of northern Fold Island, Mawson Coast, Kemp Land, East Antarctica. *Abstracts, 5th International Symposium on Antarctic Earth Science*, Cambridge, p. 76.

James, P. R. & R. Tingey 1983. The Precambrian geological evolution of the East Antarctic metamorphic shield – a review. In *Antarctic Earth science*, R. L. Oliver, P. R. James & J. B. Jago (eds), 5–10. Canberra: Australian Academy of Sciences.

Kuehner, S. 1986. *Mafic dykes of the East Antarctic shield; experimental, geochemical and petrological studies focussing on the Proterozoic evolution of the crust and mantle*. Unpubl. PhD thesis, University of Tasmania.

Kuehner, S. M. & D. H. Green 1987. Uplift history of the East Antarctic Shield: constraints imposed by high-pressure experimental studies of Proterozoic mafic dykes. *Abstracts, 5th International Symposium on Antarctic Earth Science*, Cambridge, p. 84.

Motoyoshi, Y. & H. Matsueda 1984. Archaean granulites from Mt Riiser-Larsen in Enderby Land, East Antarctica. *Memoirs of the National Institute of Polar Research, Japan*, special issue **33**, 103–25.

Motoyoshi, Y., S. Matsubara & H. Matsueda 1987. Progressive prograde metamorphism of the Lutzow – Holm Bay region, East Antarctica. *Abstracts, 5th International Symposium on Antarctic Earth Science*. Cambridge, p. 100.

Newton, R. C. & D. Perkins III 1982. Thermodynamic calibration of geobarometers based on the assemblages garnet–plagioclase–orthopyroxene–(clinopyroxene)–quartz. *American Mineralogist* **67**, 203–22.

Newton, R. C., J. V. Smith & B. F. Windley 1980. Carbonic metamorphism, granulites, and crustal growth. *Nature* **288**, 45–50.

Phillips, G. N. & V. J. Wall 1981. Evaluation of prograde regional metamorphic conditions: their implications for the heat source and water activity during metamorphism in the Willyama Complex, Broken Hill, Australia. *Bulletin de Minéralogie* **104**, 801–10.

Ravich, M. G. & E. N. Kamenev 1975. *Crystalline basement of the Antarctic Platform*. New York: Wiley.

Sandiford, M. A. 1985a. The metamorphic evolution of granulites at Fyfe Hills: implications for Archaean crustal thickness in Enderby Land, Antarctica. *Journal of Metamorphic Geology* **3**, 155–78.

Sandiford, M. A. 1985b. The origin of retrograde shear zones in the Napier Complex: implications for the tectonic evolution of Enderby Land, Antarctica. *Journal of Structural Geology* **7**, 477–88.

Sandiford, M. & R. Powell 1986a. Pyroxene exsolution in granulites from Fyfe Hills, Enderby Land, Antarctica: evidence for 1000°C metamorphic temperatures in Archean continental crust. *American Mineralogist* **71**, 946–54.

Sandiford, M. A. & R. Powell 1986b. Deep crustal metamorphism during continental extension: modern and ancient examples. *Earth and Planetary Science Letters* **79**, 151–8.

Sandiford, M. A. & C. J. L. Wilson 1983. The geology of the Fyfe Hills – Khmara Bay region, Enderby Land. In *Antarctic Earth science*, R. L. Oliver, P. R. James & J. B. Jago (eds), 16–19. Canberra: Australian Academy of Science.

Sandiford, M. A. & C. J. L. Wilson 1984. The structural evolution of the Fyfe Hills – Khmara Bay region, Enderby Land, East Antarctica. *Australian Journal of Earth Sciences* **31**, 403–26.

Sandiford, M. A. & C. J. L. Wilson 1986. The origin of Archaean gneisses in the Fyfe Hills region, Enderby Land: field occurrence, petrography and geochemistry. *Precambrian Research* **31**, 37–68.

Segnit, E. R. 1957. Sapphirine-bearing rocks from MacRobertson Land, Antarctica. *Mineralogical Magazine* **31**, 690–7.

Sheraton, J. W. & L. P. Black 1981. Geochemistry and geochronology of Proterozoic tholeiite dykes of East Antarctica: evidence for mantle metasomatism. *Contributions to Mineralogy and Petrology* **78**, 305–17.

Sheraton, J. W. & L. P. Black 1983. Geochemistry of Precambrian gneisses: relevance for the evolution of the East Antarctic shield. *Lithos* **16**, 273–96.

Sheraton, J. W. & L. P. Black 1988. Chemical evolution of granitic rocks of the East Antarctic shield, with particular reference to post-orogenic granites. *Lithos* **21**, 37–52.

Sheraton, J. W., L. P. Black & M. T. McCulloch 1984. Regional geochemical and isotopic characteristics of high-grade metamorphics of the Prydz Bay area: the extent of Proterozoic reworking of Archaean continental crust in east Antarctica. *Precambrian Research* **26**, 169–98.

Sheraton, J. W. & K. D. Collerson 1983. Archaean and Proterozoic geological relationships in the Vestfold Hills – Prydz Bay area, Antarctica. *BMR Journal of Australian Geology and Geophysics* **8**, 119–28.

Sheraton, J. W. & K. D. Collerson 1984. Geochemical evolution of Archaean granulite-facies gneisses of the Vestfold Block and comparisons with other Archaean gneiss complexes in the East Antarctic Shield. *Contributions to Mineralogy and Petrology* **87**, 51–64.

Sheraton, J. W., R. N. England & D. J. Ellis 1982. Metasomatic zoning in sapphirine-bearing granulites from Antarctica. *BMR Journal of Australian Geology and Geophysics* **7**, 269–73.

Sheraton, J. W., D. J. Ellis & S. M. Kuehner 1985. Rare-earth element geochemistry of Archaean orthogneisses and evolution of the East Antarctic shield. *BMR Journal of Australian Geology and Geophysics* **9**, 207–18.

Sheraton, J. W., L. A. Offe, R. J. Tingey & D. J. Ellis 1980. Enderby Land, Antarctica – an unusual Precambrian high grade metamorphic terrain. *Journal of the Geological Society of Australia* **27**, 1–18.

Sheraton, J. W., R. J. Tingey, L. P. Black, L. A. Offe & D. J. Ellis 1987. Geology of Enderby Land and Western Kemp Land, Antarctica. *Australian Bureau of Mineral Resources Bulletin* 223, 51 pp.

Shibata, K., K. Yanai & K. Shiraishi 1986. Rb–Sr whole-rock ages of metamorphic rocks from eastern Queen Maud Land, East Antarctica. *Memoirs of the National Institute of Polar Research* (*Japan*), special issue **43**, 133–48.

Shiraishi, K., Y. Hiroi, Y. Motoyoshi & K. Yanai 1987. Plate tectonic development of Late Proterozoic paired metamorphic complexes in Eastern Queen Maud Land, East Antarctica. In

Gondwana six: structure, tectonics, and geophysics, G. W. McKenzie (ed.), 309–18. Washington, DC: American Geophysical Union.

Sonder, L. J., P. C. England, B. P. Wernicke & R. L. Christiansen 1987. A physical model for Cenozoic extension of western North America. In *Continental extensional tectonics*, M. P. Coward, J. F. Dewey & P. L. Hancock (eds), 187–201. Geological Society of London, Special Publication 28.

Stüwe, K., H.-M. Braun & H. Peer 1989. Geology and structure of the Larsemann Hills area, Prydz Bay, East Antarctica. *Australian Journal of Earth Sciences* **36**, 219–41.

Stüwe, K. & R. Powell 1989. Low pressure granulite facies metamorphism in the Larsemann Hills area, East Antarctica: petrology and tectonic implications for the evolution of the Prydz Bay arca. *Journal of Metamorphic Geology* **7**, 465–83.

Tait, R. E. & S. L. Harley 1988. Local processes involved in the generation of migmatites within mafic granulites. *Transactions of the Royal Society of Edinburgh: Earth Sciences* **79**, 209–22.

Tingey, R. J. 1981. Geological investigations in Antarctica 1968–1969: the Prydz Bay – Amery Ice Shelf – Prince Charles Mountains area. *Bureau of Mineral Resources, Australia, Record*, 1981/34.

Thost, D., Y. Motoyoshi & B. J. Hensen 1988. Low pressure granulite metamorphism in the Bolingen Islands, East Antarctica. (Abstract) *Terra Cognita* **8**, 247.

Waters, D. J. 1986. Metamorphic zonation and thermal history of pelitic gneisses from Western Namaqualand, South Africa. *Transactions of the Geological Society of South Africa* **89**, 97–102.

Waters, D. J. 1988. Partial melting and the formation of granulite facies assemblages in Namaqualand, South Africa. *Journal of Metamorphic Geology* **6**, 387–404.

Wells, P. R. A. 1980. Thermal models for the magmatic accretion and subsequent metamorphism of continental crust. *Earth and Planetary Science Letters* **46**, 253–65.

Wernicke, B. 1985. Uniform-sense normal simple shear of the continental lithosphere. *Canadian Journal of Earth Sciences*, **22**, 108–25.

Young, D. N. & D. J. Ellis 1987. Geochemical constraints on the tectonic setting of the Proterozoic mobile belt of East Antarctica. *Abstracts, 5th International Symposium on Antarctic Earth Science*, Cambridge, p. 173.

CHAPTER THIRTEEN

Prograde metamorphism, anatexis, and retrogression of the Scourian Complex, north-west Scotland

Ian Cartwright

13.1 Introduction

The Lewisian of north west Scotland is an eastern extension of the North Atlantic Craton and is probably the most intensively studied Precambrian gneiss terrain in the world. Recent reviews of mainland Lewisian geology include summaries of gneiss geochemistry (Tarney & Weaver 1987), metamorphic conditions and causes of metamorphism (Sills & Rollinson 1987; Barnicoat 1987; Cartwright & Barnicoat 1987, 1989; Barnicoat *et al.* 1987), structural evolution (Jensen 1984, Coward & Park 1987, Park *et al.* 1987), and geophysical properties (Hall 1987). The mainland Lewisian may be subdivided into two regions (Fig. 13.1). The central Scourian Complex largely preserves metamorphic mineralogies and tectonic fabrics of Archaean age; the adjacent Laxfordian areas underwent extensive Proterozoic reworking. This contribution discusses the metamorphic evolution of the Scourian Complex, which is summarized in Table 13.1.

13.1.1 Geological history of the Scourian Complex

The *P–T*–time path for the Scourian Complex, based on estimates from Tables 13.2 and 13.3, is shown in Figure 13.2. The rocks of the Scourian Complex were separated from the mantle at 2.95 Ga and metamorphosed in the 2.7 Ga Badcallian granulite facies episode, with peak conditions estimated at between 750–800°C at 0.7–0.8 GPa or 1150–1250°C at 1.5–1.8 GPa (Table 13.2). Highest-grade conditions occurred contemporaneous with, or slightly after, the formation of pervasive flat-lying fabrics in a regional subhorizontal shear event which affected much of the Scourian Complex (e.g. Sheraton *et al.*

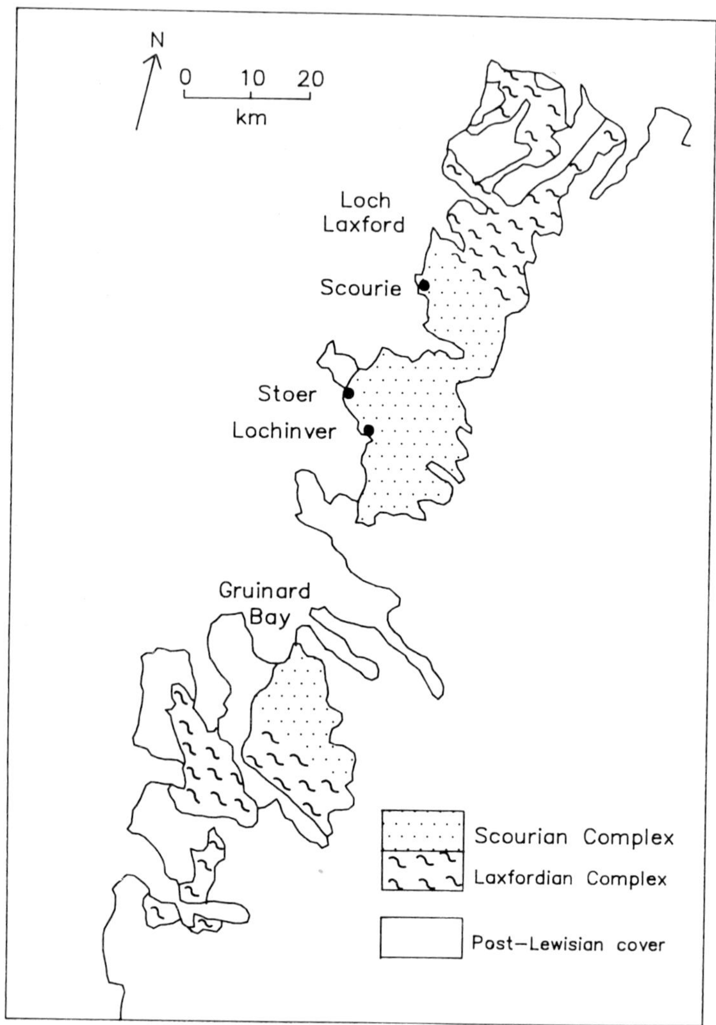

Figure 13.1 General geological map of the mainland Lewisian, showing the extent of the Scourian and Laxfordian complexes and localities discussed in the text.

1973b, Cartwright *et al.* 1985). Suites of granitic to trondhjemitic veins and sheets were also emplaced close to the climax of the Badcallian event (O'Hara & Yarwood 1978, Rollinson & Windley 1980b, Cartwright *et al.* 1985). Subsequent to the Badcallian metamorphism, the Scourian Complex underwent a period of locally intense hydrous retrogression accompanied by pegmatite intrusion and the formation of numerous dip-slip shear zones during the 2.5 Ga Inverian episode, when pressures and temperatures had fallen to 500–625°C and 0.3–0.6 GPa. The Scourie dykes were intruded at 2.0 Ga at

Table 13.1 Geological history of the Scourian Complex.

	Age (Ga)	Event
F	2.95	Formation of supracrustals and orthogneisses
B	2.7	Badcallian granulite facies metamorphism and intense subhorizontal deformation
I	2.5-2.4	Inverian retrogression and shearing, intrusion of late-Scourian pegmatites
D	2.0	Intrusion of Scourie dyke swarm
L	1.9-1.7	Laxfordian retrogression and deformation, intrusion of Laxfordian granites
E	1.0	Exposure at pre-Torridonian land surface

Letters refer to Figure 13.2

Data sources: F - Hamilton *et al.* (1979), Whitehouse & Moorbath (1986)
B - Pidgeon & Bowes (1972), Chapman & Moorbath (1977), Humphries & Cliff (1982), Cohen *et al.* (1988a)
I - Evans (1965), Cohen *et al.* (1988a)
D - Cohen *et al.* (1988b)
L - Moorbath & Park (1972), Lambert & Holland (1972), Taylor *et al.* (1984)
E - Moorbath (1969)

ambient conditions of 450–550°C and 0.5–0.7 GPa. Further minor retrogression accompanied by dominantly strike-slip shearing occurred at 500–650°C and 0.4–0.7 GPa in the 1.9–1.7 Ga Laxfordian episode, and the Complex was exposed at the sub-Torridonian land surface at 1.0 Ga.

13.1.2 Petrology of Scourian gneisses

The Scourian Complex is dominated by tonalitic gneisses (Peach *et al.* 1907; Sutton & Watson 1951, Sheraton *et al.* 1973a) which constitute up to 80% of the total outcrop. In addition, there are minor volumes of basic to ultramafic orthogneisses (O'Hara 1961; Bowes *et al.* 1964; Davies 1974, 1977) and supracrustal lithologies, including semipelites, calc-silicates, and ironstones (Peach *et al.* 1907; Sutton & Watson 1951; Barooah 1970; Okeke *et al.* 1983; Davies 1974; Barnicoat & O'Hara 1979; Rollinson 1980; Cartwright *et al.* 1985;

Table 13.2 *P–T* estimates for the peak of Badcallian metamorphism.

Estimate	Author	*P* (GPa)	*T* (°C)	Method	Loc.
1	O'Hara (1967)	1.5-1.8	1000-1100	Px parageneses	S
2	Muecke (1969)	0.7-0.9	700-800		S
3	O'Hara & Yarwood (1978)	1.5±0.3	1150-1250 1000-1100	Reintegration of highly exsolved fsp, two-fsp, px parageneses, gnt stability in metabasic rocks	S
4	Pride & Muecke (1980)	>1.1	>900	Cpx-opx temperature, gnt stability in metabasic rocks	S
5	Rollinson (1980)	<1.1	915±130	Scap-cpx-plg temperature, cpx-plg-qtz pressure	S
6	Savage & Sills (1980)	1.1-1.5	1000±100	Estimation of pre-equilibration compositions in metabasics	D
7	Rollinson (1981)	<1.1	820±50	Gnt-cpx-opx-plg-qtz	S
8	Newton & Perkins (1982)	0.7-0.85	840	Plg-opx-gnt-qtz	S
9	Barnicoat (1983)	>0.85	>1000	Cpx-gnt temperature, gnt-cpx = opx-plg-sp pressure	S
10	Cartwright & Barnicoat (1987)	>1.1	900-990	Phase equilibria, gnt-cpx temperature, ky stability	St
11	Sills & Rollinson (1987)	0.7-0.8	750-800	Gnt-cpx, gnt-opx, gnt-hbl temperatures, gnt-opx-plg-qtz pressure	S,D,A
12	Barnicoat & O'Hara (1979)	-	~1000	Reintegration of ferro-pigeonite in metaironstone	S

Localities: A = Achiltibuie; D = Drumbeg; S = Scourie; St = Stoer (Fig. 13.1)

Mineral abbreviations: amp - amphibole, ap - apatite, bt - biotite, cc - calcite, chl - chlorite, cpx - clinopyroxene, ep - epidote, fsp - feldspar, gnt - garnet, hbl - hornblende, ilm - ilmenite, ksp - alkali feldspar, ky - kyanite, mar - margarite, mt - magnetite, mus - muscovite, olv - olivine, opx - orthopyroxene, par - paragonite, plg - plagioclase, px - pyroxene, py - pyrite, qtz - quartz, scap - scapolite, sil - sillimanite, sp - spinel, sta - staurolite, zir - zircon.

Figure 13.2 *P–T–t* path for the Scourian Complex. The *P–T* conditions for the Badcallian event (B) are the favoured conditions from Table 13.2 (see text). Conditions during re-equilibration of garnet-clinopyroxene assemblages (G), Inverian retrogression (I), intrusion of the Scourie dykes (D), and Laxfordian retrogression (L) are from Table 13.3. The ages assigned to these events (in Ga) are from Table 13.1 and imply extremely slow cooling and uplift rates during cooling from the peak of metamorphism. A period of isobaric cooling during G is implied by the breakdown and subsequent regrowth of garnet-clinopyroxene assemblages in Scourian metabasic lithologies (O'Hara & Yarwood 1978, Barnicoat 1983). Heating and/or reburial could have happened between Inverian times and dyke intrusion, although the data do not require this.

Table 13.3 *P–T* estimates for post-Badcallian events.

Estimate	Author	*P* (GPa)	*T* (°C)	Method
a) Estimates recording re-equilibration of peak metamorphic mineralogies				
12	Savage & Sills (1980)	0.8-1.2	800-900	Gnt-cpx = opx-plg-sp
13	Barnicoat (1983)	0.8	800-900	Gnt-cpx = opx-plg-sp
b) Inverian retrogression				
14	Cartwright & Barnicoat (1986)	0.3-0.6	500-625	Mar-par-chl-sil stability
15	Sills (1982, 1983)	0.5-0.7	575-625	Gnt-bt, gnt-hbl, two fsp, hbl-plg, mus-chl, ep stability
c) Scourie dyke intrusion				
16	O'Hara (1977)	0.5-0.9	350-550	Cpx-gnt, two fsp, olv-plg
17	Tarney (1963)	0.5-0.6	500	Cpx-gnt
d) Laxfordian conditions				
18	Beach (1973)	0.4-0.6	500-650	Sta stability
19	Sills (1983)	0.4-0.7	>500	Par content of mus

Abbreviations as for Table 13.2

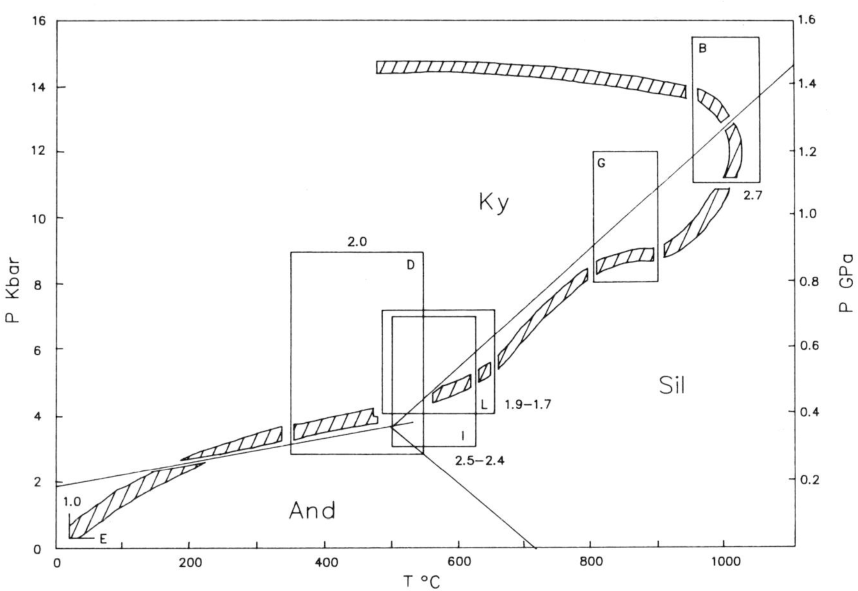

Cartwright 1986; Cartwright & Barnicoat 1986, 1987). Representative granulite facies mineral assemblages are listed in Table 13.4. Retrogressed tonalite and basic gneisses show the replacement of clino- and orthopyroxene by hornblende and locally biotite accompanied by an increase in the albite content of plagioclase; ultramafic rocks were converted to talc–tremolite-bearing assemblages during retrogression.

Granitic to trondhjemitic sheets and veins are found within the tonalitic and basic orthogneisses and several of the supracrustal layers (Rollinson & Windley 1980b; O'Hara & Yarwood 1978; Barnicoat 1983; Cartwright *et al.* 1985; Cartwright & Barnicoat 1986, 1987; Cartwright 1988). The sheets and veins that occur within the tonalites were termed collectively 'acid leucogneisses' by Cartwright *et al.* (1985). The leucogneisses and the other suites of acidic veins comprise quartz and plagioclase with K-feldspar in the granitic lithologies. Mafic minerals usually constitute <10 modal% of these lithologies and are dominantly hornblende and biotite, but rarely clinopyroxene, orthopyroxene, or garnet are present; muscovite is a common late phase. The sheets and veins range from a few millimetres to tens of metres in thickness and the larger sheets may be laterally extensive over several kilometres. They are locally discordant to layering in the host gneisses and mafic selvages are occasionally present. From relationships with tectonic fabrics these lithologies are inferred to have been emplaced during the prograde stages of the Badcallian tectonometamorphic event (O'Hara & Yarwood 1978, Rollinson &

Table 13.4 Representative granulite facies mineral assemblages.

	Meta-igneous Lithologies
Ultramafic Gneisses	olv-cpx-opx-sp-amp
Metabasic Gneisses	plg-cpx-opx ± gnt ± mt
Tonalitic Gneisses	qtz-plg-cpx-opx ± ap ± mt ± ilm ± zir
	Metasediments
Ironstones	qtz-opx-cpx-mt ± gnt ± plg ± py
Calcsilicates	qtz-plg-cpx ± ksp ± ap ± scap ± cc
Metapelites	qtz-plg ± ky ± sta ± hbl ± ap ± zir

Data from Barnicoat & O'Hara (1979), Rollinson (1980), Barnicoat (1983), Cartwright (1986). Mineral abbreviations as for Table 13.2

Windley 1980b, Cartwright & Barnicoat 1987). The origin of the acidic veins and sheets is discussed below.

Granulite facies Scourian gneisses are extremely depleted in U, Th, Rb, Pb, and K relative to petrologically similar rocks in the Laxfordian areas and to lithologies in amphibolite facies gneiss terrains in general (Weaver & Tarney 1981, Tarney & Weaver 1987). Geochemical depletion has commonly been assumed to have occurred during the granulite facies metamorphism as a response to either the ingress of CO_2-rich fluids (e.g. Weaver & Tarney 1983) or the extraction of partial melts (e.g. Pride & Muecke 1980, 1982). However, Tarney & Weaver (1987) envisage that depletion occurred prior to the granulite facies metamorphism as part of the igneous evolution of the terrain.

This contribution discusses the following topics, all of which have been the subject of recent controversy (see discussion in Park & Tarney 1987): (a) the P–T conditions of the peak of metamorphism; (b) the causes of the high-grade Badcallian metamorphism; (c) the rôle of anatexis in the formation of depleted granulite facies assemblages; and (d) the causes of the Inverian retrogression.

13.2 Conditions of Badcallian metamorphism

As summarized in Table 13.2 and Figure 13.3, estimates of Badcallian peak-metamorphic conditions for the area between Stoer and Scourie (Fig. 13.1) vary widely from 700–800°C at 0.7–0.8 GPa (estimate 11) to 1250°C at up to 1.8 GPa (estimate 3). A reasonably consistent group of P–T estimates from geothermometry, geobarometry, pyroxene exsolution, and phase equilibria studies place peak conditions at 1000 ± 50°C and >1.1 GPa (estimates 1, 4, 5, 6, 9, 10, and 12). Porphyroblasts of kyanite are found within metasedimentary gneisses in the Stoer and Scourie areas which, from textural evidence, are inferred to be peak-metamorphic (Cartwright & Barnicoat 1986, Barnicoat *et al.* 1987). No peak-metamorphic sillimanite is reported from the Scourian Complex; occurrences reported from the Scourie–Loch Laxford areas (Beach 1973, Davies 1974, Okeke *et al.* 1983), near Stoer (Cartwright & Barnicoat 1986), and north of Loch Maree (Crane 1978) occur in Inverian or Laxfordian assemblages, often replacing earlier kyanite. The presence of kyanite as the stable aluminosilicate phase at the peak of metamorphism in the Scourie–Stoer area provides an important constraint on the pressure and temperature of the Badcallian metamorphism (Barnicoat *et al.* 1987, Cartwright & Barnicoat 1989). The group of P–T studies yielding 1000°C and >1.1 GPa do in the main lie within the kyanite field and are consistent with the observed aluminosilicate stability at the peak of metamorphism. Maximum metamorphic pressures are difficult to assess; however, the stability of spinel peridotite assemblages in ultramafic bodies constrains maximum pressures at 1000°C to <1.8 GPa (O'Hara 1977).

In a review of Badcallian P–T conditions, Sills & Rollinson (1987) suggested

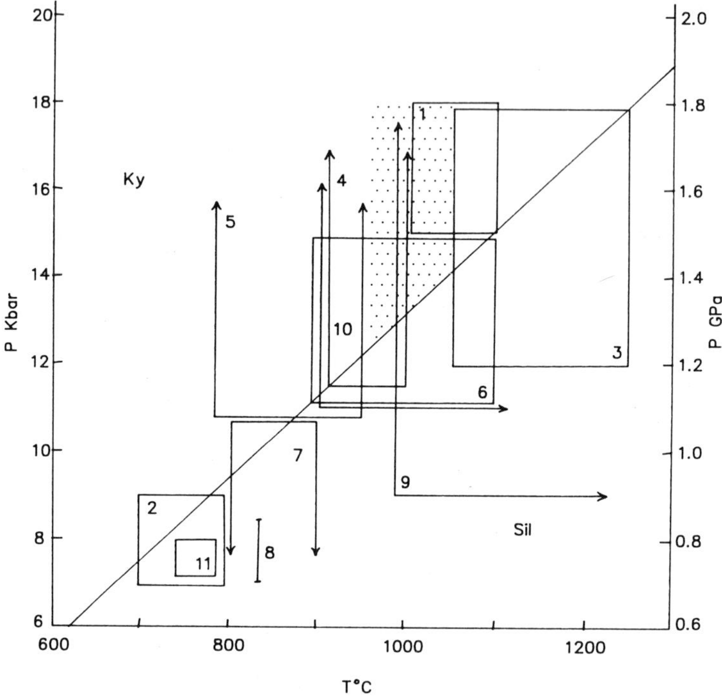

Figure 13.3 Summary of Badcallian *P–T* estimates (data from Table 13.2). The favoured conditions are shown by the shaded area. Lower *P–T* estimates are probably the result of down-temperature re-equilibration of mineral compositions and conflict with the stability of kyanite at this stage. The higher *P–T* estimates are from two-feldspar geothermometry applied to inappropriate assemblages. From Barnicoat *et al.* (1987, Fig. 1), by permission of Scottish Academic Press (Journals) Ltd.

that high temperatures (> 1000°C) from two-feldspar geothermometry in rare orthopyroxene-bearing granitic sheets that contain mesoperthitic alkali feldspars (e.g. Table 13.2, estimate 3 and Rollinson 1982) may be igneous and that peak-metamorphic *P–T* conditions throughout the Scourian Complex were possibly much lower, 750–800°C at 0.7–0.8 GPa (estimate 11). This conclusion concurs with earlier studies by Muecke (1969), Newton & Perkins (1982), and Rollinson (1981) who also proposed relatively low peak-metamorphic *P–T* conditions (estimates 2, 7, and 8). However, these estimates lie within the sillimanite field and are inconsistent with the presence of kyanite at the peak of metamorphism. The Scourian Complex underwent slow cooling from the peak of metamorphism (the data of Table 13.2 imply that cooling rates may have been as slow as 2–3°C Ma^{-1} for the period between 2.7 and 2.5 Ga) which would cause re-equilibration of mineral compositions in all but the coarsest-grained lithologies. Barnicoat (1987) estimated that only rocks with garnets and clinopyroxenes of > 5–10 mm diameter would preserve garnet–clinopyroxene temperatures of 1000°C during cooling. It is almost certain that the lower *P–T* values underestimate peak-metamorphic conditions due to resetting of mineral compositions during cooling. Furthermore, the 900–1000°C temperature estimates 10 and 12 (Table 13.2) are from metasedimentary lithologies and hence must record metamorphic not magmatic conditions.

Much higher *P–T* estimates of 1100–1250°C and 1.2–1.8 GPa (estimate 3) were proposed by O'Hara & Yarwood (1978), on the basis of highly exsolved feldspars from a granulite facies granite sheet. However, as noted by Ghiorso (1984), the re-integrated feldspar compositions used by O'Hara & Yarwood (1978) lie on the same side of the consolute curve in the ternary feldspar system and cannot represent equilibrium compositions. A second pair of re-integrated feldspars from the study of O'Hara & Yarwood (1978) do probably represent equilibrium compositions (Ghiorso 1984) and yield temperatures of 1000–1100°C at the pressure of 1.5 GPa proposed in that study.

It is concluded that the Badcallian metamorphism in the Stoer–Scourie area took place at 1000 ± 50°C and >1.1 GPa. It has been tacitly assumed that peak-metamorphic conditions were uniform across the Scourian Complex (e.g. Sills & Rollinson 1987; Cartwright & Barnicoat 1987, 1989; Barnicoat *et al.* 1987). Estimating Badcallian conditions in areas that underwent intense Inverian retrogression (such as the Loch Laxford and Gruinard Bay areas; Fig. 13.1) is difficult. However, it is possible that determination of peak-metamorphic pressures and temperatures in those areas will reveal differences in Badcallian conditions within the complex. In particular, the Gruinard Bay area, which is petrologically and geochemically distinct from the northern portions of the Scourian Complex (Davies 1977, Rollinson & Fowler 1987) may represent a different crustal level.

13.3 Partial melting during granulite facies metamorphism

In common with many metamorphic terrains, there has been much debate as to the relative rôles of anatexis and the action of CO_2-rich fluids in the formation of the Scourian granulite facies lithologies and in effecting geochemical depletions. Several studies (e.g. O'Hara 1977; O'Hara & Yarwood 1978; Pride & Muecke 1980; Barnicoat 1983, 1987; Cartwright & Barnicoat 1986, 1987; Cartwright 1988) have regarded the granulites as the restites of partial melting (cf. Fyfe 1973). By contrast, largely on the basis of geochemical data, several other authors (e.g. Rollinson & Windley 1980a, 1980b; Weaver & Tarney 1981, 1983; Tarney & Weaver 1987; Park & Tarney 1987) have argued against large-scale anatexis occurring during the Badcallian metamorphism, and consider that the dehydration of the gneisses was probably due to the action of CO_2-rich fluids (cf. Newton *et al.* 1980). Evidence for anatexis in the metabasic and supracrustal gneisses will be outlined first, followed by discussion of partial melting in the tonalitic gneisses.

13.4 Partial melting in metabasic and metasedimentary lithologies

There is general agreement (even amongst those who discount partial melting as a major process in the Scourian Complex as a whole) that the metasediments and basic gneisses do preserve evidence for undergoing anatexis during Badcallian metamorphism (e.g. Park & Tarney 1987, Tarney & Weaver 1987). Barnicoat (1983) and Cartwright & Barnicoat (1986, 1987) proposed that locally cross-cutting acidic veins in metasedimentary and metabasic gneisses represented aggregated partial melts. These veins lie close to the cotectic surfaces in the albite–orthoclase–quartz system (Fig. 13.4) and, as noted above, are inferred to have been emplaced close to the peak of metamorphism. The restriction of veins of distinct petrology and major-element geochemistry to one lithological unit and the presence of mafic selvages and small (mm–cm) patches of leucosome material imply that these melts were locally derived (Cartwright & Barnicoat 1986, 1987).

13.5 Partial melting in the tonalitic gneisses

Although it is accepted that the metasedimentary and metabasic gneisses probably underwent anatexis during Badcallian metamorphism, there is less agreement as to whether the tonalites melted at the same stage. As these lithologies dominate the Scourian Complex, this argument bears on the overall rôle of partial melting in the shaping of the geochemical, isotopic, and mineralogical character of the terrain. During prograde metamorphism the tonalitic gneisses would have melted had they contained hydrous minerals, even in the absence of a vapour phase (Wyllie 1983). Only if the tonalites were anhydrous during the later stages of prograde metamorphism could partial melting have been prevented. Several studies (e.g. Sheraton *et al.* 1973a; Pride & Muecke 1980; Barnicoat 1983, 1987; Cartwright & Barnicoat 1986, 1987; Cartwright 1988) concluded that the acid leucogneisses represent the neosomes of local anatexis of the tonalites. However, Rollinson & Windley (1980b) and Rollinson (1982) proposed that these lithologies were derived from the fractional crystallization of the igneous precursors to the tonalites. These two contrasting theories are discussed below.

13.5.1 Initial hydration state of the tonalites

The initial hydration state of the tonalites is crucial to the discussion of anatexis in these lithologies. The tonalites are almost certainly metamorphosed plutonic igneous lithologies, possibly formed in an ocean–continent collision zone (e.g. Rollinson & Fowler 1987, Tarney & Weaver 1987). Modern equivalents to the precursors of these gneisses may be the deeper-level rocks of the Andean Cordillera of southern America (Bartholomew & Tarney 1984). Partial melting in subduction-zone environments occurs under hydrous conditions forming H_2O-bearing melts that give rise to amphibole- or biotite-bearing lithologies (e.g. Wyllie 1984). Rollinson & Windley (1980b) and Weaver & Tarney (1980) suggested that amphibole was an important early phase in the igneous precursors to the Scourian tonalites, and Tarney & Weaver (1987) proposed that these lithologies were emplaced as hydrous magmas.

If the tonalites were initially amphibole- or biotite-bearing, partial melting during Badcallian metamorphism would have been inevitable unless dehydration occurred. The most commonly invoked cause of dehydration is as a response to the ingress of CO_2-rich fluids which reduce a_{H_2O} below that defined by amphibole- or biotite-bearing assemblages.

13.5.2 CO_2 infiltration during metamorphism

By analogy with studies of 'charnockitization' in southern India (Newton *et al.* 1980) and due to the common occurrence of high-density CO_2 inclusions in granulite facies terrains, Weaver & Tarney (1983) proposed that granulite facies lithologies in the Scourian Complex were the result of CO_2 infiltration. However, there are several drawbacks with this model:

(a) The Scourian lithologies contain ilmenite–magnetite assemblages that buffer f_{O_2} at values below the upper stability of graphite. Infiltration of CO_2 would readily overcome their buffering capacity and result in the precipitation of graphite (Lamb & Valley 1984, 1985). As graphite is not reported from the Scourian gneisses, it is implausible that these lithologies have interacted with significant volumes of CO_2-rich fluids during metamorphism.

(b) No arguments to explain the timing of CO_2 influx at 2.7 Ga coincident with the highest-grade conditions have yet been advanced.

(c) The occurrence of CO_2-fluid inclusions in Adirondack granulites was discussed by Lamb *et al.* (1987), who showed that peak-metamorphic assemblages in that terrain buffered a_{CO_2} at lower values than those of the fluid in the inclusions. That study concluded that high-density CO_2 fluid inclusions may be trapped after the peak of metamorphism. Even if the CO_2-rich fluid inclusions were peak-metamorphic, their presence does not allow discrimination between CO_2 introduced from an external source and small volumes of CO_2-rich fluids which may remain after partial melting of crustal lithologies (Powell 1983a,b). The available data do not support CO_2-flushing as a viable mechanism of generating the Scourian granulites.

Partial melting during the granulite facies event may have been prevented if the tonalites were dehydrated during earlier metamorphic episodes (occurring in the period between 2.95 and 2.7 Ga). However, no evidence exists for such episodes and this possibility cannot be considered further.

It is considered that the Scourian tonalite gneisses contained hydrous minerals during prograde Badcallian metamorphism and underwent anatexis at that stage. Melting was probably mainly by vapour-absent reactions (Powell 1983a,b) which, assuming that typical amphibolite facies tonalites contained ~20% hornblende (Cartwright 1988), could form ~20% melt with a water content of 2% at 1000°C; biotite-bearing tonalites would undergo higher degrees of melting. If melting has occurred, evidence for it should be apparent, and the major and trace-element geochemistry of the Scourian Complex tonalites must be reconciled with anatectic models.

13.5.3 Origin of the acid leucogneisses

The acid leucogneisses are suites of granitic to trondhjemitic veins and sheets which occur within the tonalites. As discussed earlier, these lithologies comprise variable quantities of quartz, alkali feldspar, and plagioclase with mafic minerals (dominantly hornblende and biotite) constituting <10 modal% of these lithologies. The granulite facies granitic sheets of O'Hara & Yarwood (1979) and Rollinson & Windley (1980b) are probably part of this suite of rocks; however, such anhydrous mineralogies are extremely rare (only reported locally from the Scourie area). Sheraton *et al.* (1973a), Pride & Muecke (1982), O'Hara & Yarwood (1978), Cartwright & Barnicoat (1987), and Cartwright (1988) proposed that these leucogneisses represented the aggregated products of partial melting of the tonalites. This proposition is consistent with the major-element geochemistry of the leucogneisses which, in common with the sets of veins from the metabasic and metasedimentary lithologies, plot close to the cotectic surfaces in the Ab–Or–Qtz system (Fig. 13.4). Additionally, relations between the leucogneisses and tectonic fabrics in the Stoer and Scourie areas indicate that the sheets were emplaced close to the peak of the Badcallian event (Cartwright & Barnicoat 1987, O'Hara & Yarwood 1978, Rollinson & Windley 1980b).

Figure 13.4 (a) Projection of normative (CIPW) compositions of veins into the Ab–Or–Qtz system. Veins from the basic gneisses (×), plagioclase amphibolite supracrustals (●), and the white mica paragneisses (vertical lines) plot close to cotectic surfaces in this projection as do the acid leucogneisses (diagonal ornament). (b) Projection of host gneisses into the Ab–Or–Qtz system: basic gneisses (vertical lines); tonalites (diagonal pattern); and plagioclase amphibolites (●). These data are consistent with the veins and sheets representing the aggregated products of crustal anatexis. Data from Rollinson & Windley (1980b), Cartwright & Barnicoat (1986, 1987), Cartwright (1986), and Table 13.5. Cotectic lines are in the water-saturated (solid lines) and anhydrous (dashed lines) systems at 0.1, 0.5, and 1.0 GPa (Luth 1969, Huang & Wyllie 1975).

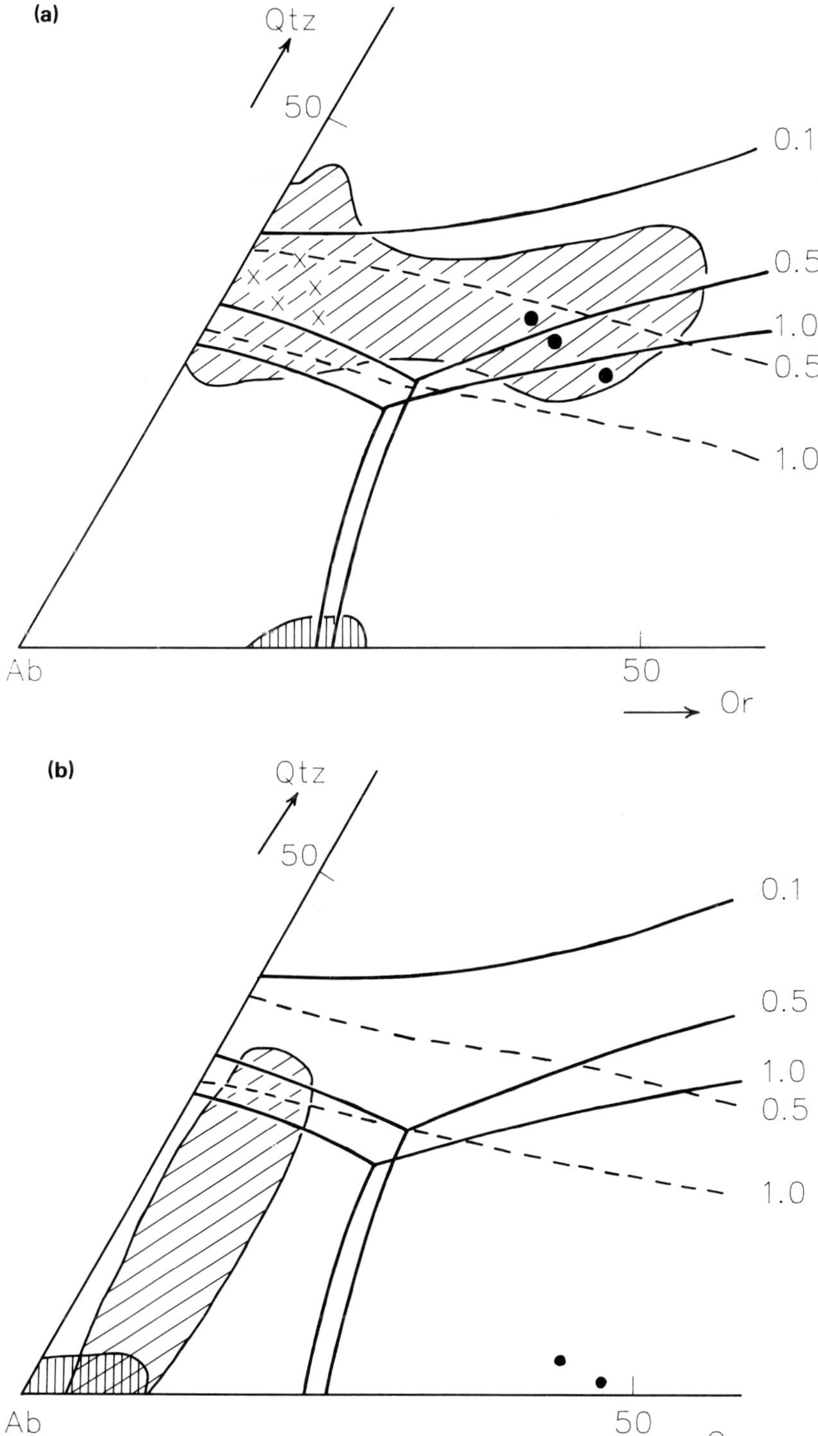
(a)
Qtz
50
0.1
0.5
1.0
0.5
1.0
Ab
50
Or
(b)
Qtz
50
0.1
0.5
1.0
0.5
1.0
Ab
50
Or

Table 13.5 Analyses and Ab : Or : Qtz ratios (from CIPW norm) of plagioclase amphibolites.

Wt%	Host amphibolites		Felsic Veins		
Oxide	C85047	C84218	C84237	C84383	C84234
SiO_2	65.01	60.85	72.12	71.13	70.94
TiO_2	0.60	1.90	0.02	0.31	0.23
Al_2O_3	17.97	18.04	15.39	15.30	14.95
Fe_2O_3	2.24	2.92	0.11	1.94	0.87
MgO	0.46	0.60	0.01	0.62	1.38
MnO	0.03	0.20	0.00	0.04	0.05
CaO	2.18	3.49	1.05	1.66	2.91
Na_2O	5.90	5.81	4.05	4.15	4.07
K_2O	4.70	4.77	5.83	3.24	3.57
P_2O_5	0.01	0.02	0.00	0.07	0.08
LOI	0.71	0.72	0.23	1.05	0.88
Total	99.81	99.32	98.81	99.51	99.93
% Ab-Or-Qtz					
Ab	58.3	61.8	37.9	41.3	40.5
Or	32.5	35.4	37.1	23.0	28.2
Qtz	9.2	2.8	25.0	35.7	31.3

By contrast, Rollinson & Windley (1980b) considered that the leucogneisses were formed by fractional crystallization from the igneous precursors to the tonalitic gneisses. Rollinson (1982) furthered this idea by suggesting that the leucogneisses were formed at low (0.05 GPa, 0.5 kbar) pressure and represented melts with <1 wt% H_2O. However, this model has several drawbacks (as discussed in Cartwright 1988). First, Rollinson & Windley (1980b) consider, in common with most other studies, that the leucogneisses were intruded at the peak of Badcallian metamorphism. If this is so, the generation and emplacement of the leucogneisses post-dates the formation of the tonalites by ~200 Ma. Secondly, a granitic melt formed at 0.05 GPa and 1000°C with a water content of 1 wt% would crystallize if temperatures dropped by a few tens of degrees, or if the ambient pressure was raised (changes which almost certainly happened between 2.95 and 2.7 Ga). Additionally, in the fractional crystallization model, the tonalitic liquid gives rise first to trondhjemitic leucogneisses and then to the granitic compositions. However, trondhjemitic melts at 1000°C and 0.05 GPa must have H_2O contents of >1 wt% and are unlikely to evolve to granitic liquids with lower H_2O contents.

It is concluded that the leucogneisses are not related to the tonalites by fractional crystallization, and that the timing and composition of these lithologies are most consistent with their representing aggregated partial melts of the tonalitic gneisses formed during prograde Badcallian metamorphism. However, it is not possible to determine unambiguously whether these lithologies are entirely locally derived, and it is likely that at least some of the leucogneisses were derived from melting of rocks below the current exposure level. Unlike the metasediments and metabasic gneisses where the neosomes are easily distinguished from the host gneisses, the similarity in appearance of the leucogneisses and tonalites makes identification of millimetre- or centimetre-sized segregations and veins difficult. Small diffuse granitic segregations and veins with mafic selvages are occasionally observed in the tonalites; these may be equivalents of the larger leucogneiss sheets and may represent the products of very local anatexis.

13.5.4 Origins of anhydrous leucogneiss sheets

The rare anyhdrous leucogneisses discussed by Rollinson & Windley (1980b) and O'Hara & Yarwood (1979) are clearly anomalous. Regardless of the origin of the leucogneisses, if these sheets were emplaced at 1000°C and >1.1 GPa (as proposed in both models) they must represent melts with initial H_2O contents of at least 2 wt% (Whitney 1975, Luth 1976, Wyllie 1983). Water-undersaturated granitic melts may crystallize pyroxene, but exsolution of H_2O during the later stages of cooling (e.g. Cartwright 1988, Waters 1988) would result in retrogression of that pyroxene. These anhydrous assemblages may represent sheets from which residual liquids have escaped prior to becoming H_2O-saturated.

13.6 Major and trace element constraints on melting

Melting of the Scourian Complex would have affected both the major and trace-element geochemistry of the gneisses. Weaver & Tarney (1981, 1983), and Park & Tarney (1987) consider that the Laxfordian areas of the Lewisian (Fig. 13.1) represent lower-grade equivalents of the Scourian Complex that did not undergo granulite facies metamorphism. The equivalence of the Laxfordian and Scourian areas is difficult to prove, and Badcallian metamorphic conditions in the Laxfordian zones are not known with any certainty due to the later intense Laxfordian reworking of these areas. Nevertheless, relative to Laxfordian rocks, the Scourian granulites are depleted in U, Th, Rb, and Pb and also have lower K_2O, Na_2O, and SiO_2 contents (e.g. Weaver & Tarney 1983, Table 1). Typical Scourian tonalites are also depleted in these elements relative to possible modern-day analogues. These broad geochemical differences between the Scourian and Laxfordian areas are compatible with the extraction of melts from the Scourian Complex (Pride & Muecke 1980, 1982;

Weaver & Tarney 1983). Pride & Muecke (1980, 1982) estimate that extraction of 10–20% granitic partial melt from the Laxfordian crust would produce a restite with broadly the major element composition of the Scourian lithologies, and also explain the lower Rb, Th, and U concentration of the granulites with respect to the Laxfordian rocks and upper-crustal lithologies in general. Radiogenic isotope data are also consistent with anatexis occurring during Badcallian metamorphism. Scourian tonalites show significant fractionation of Sm/Nd on a centimetre scale, resulting in a variation in model ages between 3.25 and 2.95 Ga; however, no large-scale homogenization of Nd is apparent. These results are consistent with Sm-Nd fractionation occurring due to amphibole breakdown during melting (Burton *et al.* 1988). Cohen *et al.* (1988a) interpret depletions in Rb, Sr, U, Th, and Pb isotopes at 2.67 Ga as being due to anatexis and the removal of a melt phase. The leucogneisses have higher Th, U, and Rb concentrations than the tonalites (Pride & Muecke 1980) which is consistent with an anatectic origin for these lithologies.

The conclusion that Scourian gneisses, including the tonalites, are restites agrees with much of the chemical and isotopic data. However, there are problems reconciling the rare earth element (REE) and trace-element geochemistry of the gneisses with such a model.

13.6.1 REE chemistry of the leucogneisses and tonalites

REE data from leucogneisses and granulite facies tonalites from Scourie and the calculated REE pattern of a granitic melt (M) that is in equilibrium with the tonalite (assuming a typical granulite facies modal mineralogy of 15% quartz, 65% plagioclase, 12% clinopyroxene, 5% orthopyroxene, and 3% magnetite) are summarized in Figure 13.5. The theoretical melt composition has higher total REE than the granulite facies tonalites (the proposed restites) and a negative Eu anomaly. By contrast, the leucogneisses have lower total REEs than either the amphibolite or granulite facies lithology and have a marked positive Eu anomaly (although the Eu concentration is close to that of the model melt). Clearly, there are problems relating the leucogneisses and tonalites by any simple model of melting.

Granitic and trondhjemitic sheets with similar REE patterns are found in a number of basement terrains (Condie *et al.* 1986). Several explanations have been advanced to explain the REE profiles, including: disequilibrium melting (Pride & Muecke 1982); the presence of accessory minerals such as monazite and zircon in the restites; a cumulate origin for the sheets (Condie *et al.* 1986); and entrainment of residual crystals (Condie *et al.* 1986). These proposals may explain anomalous REE profiles in individual sheets, but are unlikely to account for the ubiquitious REE geochemistry of this suite of rocks as a whole. It is possible that assumptions made in calculating the model melt composition are in error. The mineral–melt distribution coefficients (K_D) used in Figure 13.5 are based on experimental studies utilizing low-pressure water-

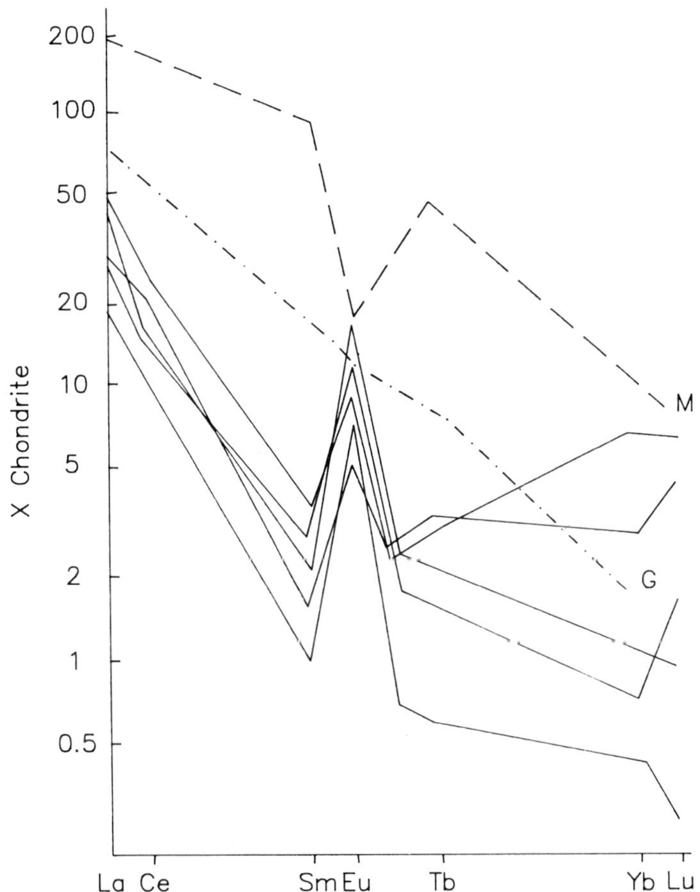

Figure 13.5 Chondrite-normalized REE patterns for acid leucogneisses and a typical granulite facies Scourian tonalite (G); data from Weaver & Tarney (1980, 1983) and Pride & Muecke (1980). The dashed line shows theoretical REE profile of a granitic melt (M) in equilibrium with the granulite facies tonalite (distribution coefficients from Hanson 1978), which is similar to REE profiles of high-level granites (e.g. Condie *et al.* 1986) but contrasts markedly with those observed in the leucogneisses.

saturated melts (Hanson 1978). However, melts formed by crustal anatexis during granulite facies metamorphism are almost certain to be water-under-saturated (Powell 1983a,b; Cartwright 1988). Such melts are highly polymerized and, as the number of octahedral sites in which the trivalent REE are located is reduced as melt polymerization increases, probably have lower K_D values for the REE (Flynn & Burnham 1978). Increased melt polymerization may explain the anomalously low trivalent REE concentrations in the leucogneisses, while Eu as a divalent ion is present at closer to the expected levels.

13.6.2 Other trace-element problems

Other trace-element abundances (e.g. Ba concentrations or Ce/Y ratios) have also been used to discount anatexis as an important process in the Scourian Complex (e.g. Rollinson & Windley 1980a, Weaver & Tarney 1983, Tarney & Weaver 1987). However, these studies were based on mass balance calculations utilizing models of single-stage equilibrium melting. It is probable that crustal anatexis involves the formation of batches of melt at discrete intervals as fluid-absent melting reactions are intersected (Powell 1983a,b; Cartwright & Barnicoat 1986, 1987). Considerable fractionation and decoupling of trace elements are possible under such a melting régime (Cartwright & Barnicoat 1986, Fig. 7). Other likely processes, such as zone refining, changes in the restite mineralogy as melting proceeds, or fractional melting, add further uncertainties to these calculations. Additionally, the possible effects of crustal melts having different structures to those which are commonly used to determine mineral–melt partition coefficients need to be evaluated. The trace-element data may provide an important constraint on anatexis; however, there is a need for more detailed small-scale studies than have hitherto been attempted.

Account of possible variable trace-element and REE geochemistry of the melts must be made in modelling the effects of melt extraction on the geochemistry of the complex as a whole (e.g. Weaver & Tarney 1983). The proposition that the geochemical differences between amphibolite and granulite facies terrains may be partially primary (Tarney & Weaver 1987) must also be assessed.

13.7 Causes of the Badcallian metamorphism

The elevated temperatures of granulite facies metamorphism are usually considered to be due either to the emplacement of large bodies of magma into the crust (Wells 1980, Bohlen 1987), or to tectonic thickening (England & Richardson 1977, England & Thompson 1984). In the Scourian Complex, the only lithologies that are sufficiently voluminous to cause granulite facies metamorphism by the input of magmatic heat are the tonalites. Peak-metamorphic conditions resulting from magmatic heating would occur within 35 Ma of the emplacement of the igneous lithologies (Wells 1980, Barnicoat 1987). However, Badcallian metamorphism occurred at 2.7 Ga, some 200–250 Ma after the emplacement of the tonalites. This large age gap, and the absence of evidence for any other major igneous activity just prior to the Badcallian metamorphism, renders magmatic heat models untenable in explaining the causes of Badcallian metamorphism. The magmatic heat input models also predict almost isobaric cooling from the peak of metamorphism and anticlockwise P–T–t paths. The P–T–t path defined by the Scourian Complex is broadly clockwise (Fig. 13.2), although a period of isobaric cooling soon after the peak of metamorphism is implied by garnet–clinopyroxene reaction textures (Barnicoat 1983).

Barnicoat (1987) presented the results of one-dimensional heat-flow modelling, which showed that tectonic thickening with reasonable crustal and mantle heat flows could generate the conditions of Badcallian metamorphism. This model is also consistent with the clockwise *P–T–t* path shown in Figure 13.2. The subhorizontal Badcallian fabrics plausibly record deformation associated with crustal thickening, and the common observation that peak-metamorphic minerals overgrow these fabrics accords with the predictions that the thermal peak of metamorphism post-dates the crustal thickening.

13.8 Causes of Inverian retrogression

The Inverian retrogression occurred when the Scourian Complex had cooled to 500–625°C and involved hydration of the Badcallian granulite facies assemblages to amphibolite facies mineralogies (Sills 1982, 1983; Cartwright & Barnicoat 1986, 1987). The intensity of Inverian retrogression varies markedly (Fig. 13.6). The Gruinard Bay, Stoer, and Laxford Bridge regions underwent extensive retrogression and preserve few relic granulite facies assemblages. By contrast, the Scourie area comprises largely unaltered granulite facies lithologies with retrogressed assemblages found only within, or immediately adjacent to, Inverian shear zones. Cartwright (1988) proposed that fluids exsolved during latter stages of crystallization of the leucogneiss melts were the cause of the Inverian retrogression. Granitic or trondhjemitic melts formed during prograde metamorphism at temperatures of 1000°C must have had H_2O contents of >2 wt% (Whitney 1975, Luth 1976, Wyllie 1983). During cooling, the melts would have crystallized, with the residual liquids becoming more water-rich, until eventually water saturation occurred and a fluid phase was exsolved (Fig. 13.7). Comparison with experimental data indicates that fluids would be exsolved from these crystallizing melts at 600–625°C (Fig. 13.7). These fluids were probably the cause of the Inverian retrogression, a proposition supported by: (a) the large-scale correlation between the areas that contain abundant leucogneisses and those which have undergone the most intense Inverian retrogression (Fig. 13.6); (b) the coincidence of the temperature of retrogression and that at which fluids are predicted to have been exsolved from the crystallizing melts; and (c) small-scale retrogression observed around leucogneiss veins in otherwise unaltered granulite facies lithologies (Cartwright 1988, Fig. 6c). Additionally, Cohen *et al.* (1988a) suggested that final textural equilibration within the Scourian Complex granulites at 2.4–2.5 Ga may have been related to the final crystallization of melts in the complex.

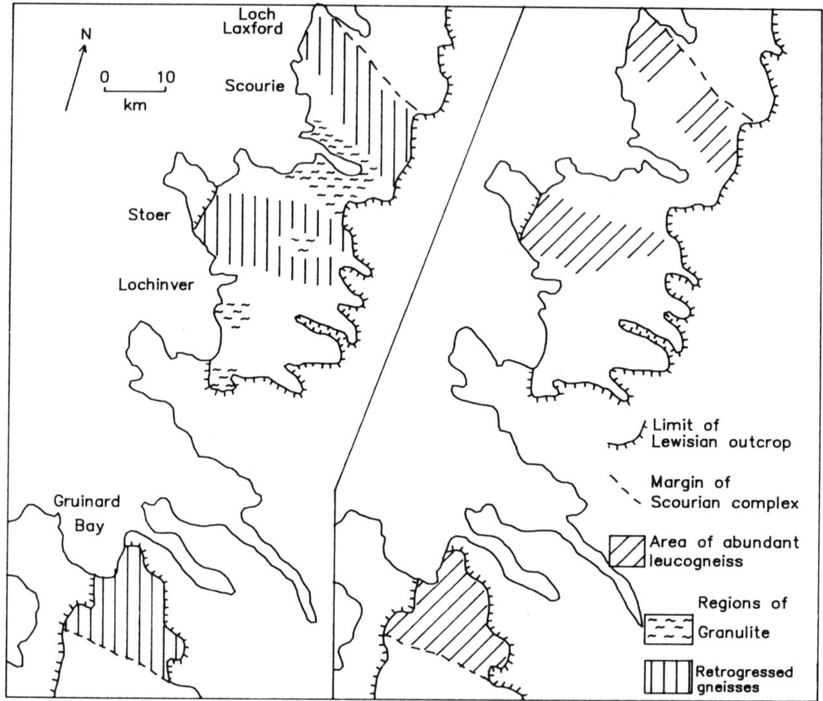

Figure 13.6 Comparative maps of the Scourian Complex (data sources as for Cartwright 1988, Fig. 7) showing: (left) the regions which preserve largely unaltered granulite facies mineralogies and those which have undergone extensive Inverian retrogression; (right) the areas containing abundant leucogneisses. There is a marked correlation between the areas containing abundant leucogneisses and those which underwent the most intense retrogression, which is consistent with fluids exsolved from the crystallizing leucogneisses being the cause of the retrogression.

Even if each leucogneiss sheet represented a melt with the minimum 2 wt% H_2O, sufficient quantities of melts are present in the retrogressed areas to provide the volumes of fluid required to hydrate the surrounding lithologies. Retrogression in regions which contain few leucogneisses is limited to relatively minor volumes of fluid channelled along the Inverian shears (possibly from underlying leucogneiss-rich zones). Amphibolite facies retrogression of the Scourian Complex was almost inevitable, regardless of the origin of the leucogneisses. If the leucogneisses were derived from fractional crystallization (Rollinson & Windley 1980b), or even originated from a source outside the complex, they would still represent melts that existed at the peak of metamorphism (O'Hara & Yarwood 1978, Rollinson & Windley 1980b, Rollinson 1982, Sills & Rollinson 1987). Hence, they would have initially contained >2 wt% H_2O, and their crystallization history would be identical to that discussed above, which assumed that the leucogneisses represented melts of the tonalites. Water saturation accompanied by the exsolution of a fluid phase at

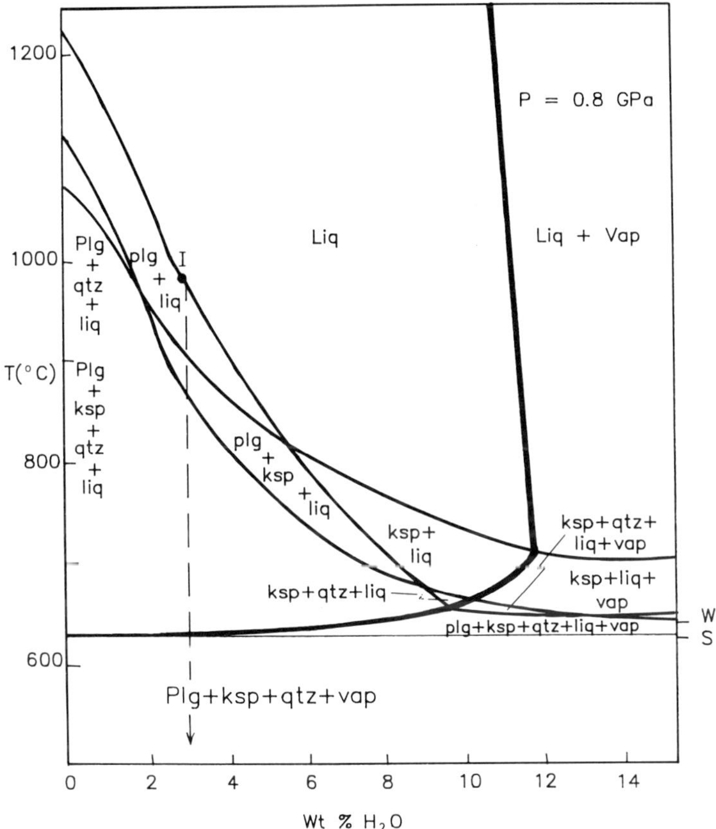

Figure 13.7 Pseudobinary isobaric T–wt% H_2O phase diagram for a synthetic granitic composition at 0.8 GPa, the average pressure between the Badcallian and Inverian events (data from Whitney 1975, Luth 1976). Melt I originally on the liquidus at ~1000°C contains 3 wt% H_2O. During cooling (represented by the dashed line), crystallization of plg, ksp, and qtz occurs and the residual liquid becomes more H_2O-rich until water saturation occurs at ~625°C (W). The melt finally crystallizes at the solidus (S) at ~620°C. This general model predicts that final melt crystallization will not occur until temperatures of ~620°C, and that a fluid phase is evolved from initially water-undersaturated melts close to their solidus temperatures. Polybaric crystallization is closely similar to the isobaric case considered here (Cartwright 1988). This model is applicable to crystallization of the leucogneiss melts regardless of their origin.

600–625°C would still have occurred during cooling from the peak of metamorphism.

Other explanations of the sources of the Inverian fluids include: (a) volatilization from underlying rocks undergoing prograde metamorphism; (b) a mantle source; and (c) volatilization from underlying subducting oceanic crust (e.g. Sills 1983). None of these proposals satisfactorily explains either the timing or distribution of the Inverian retrogression. Models (a) and (c) require the presence of unobserved lithologies which are undergoing contrasting tectonometamorphic processes to the uplift and cooling that was occurring in the rocks of the Scourian Complex.

A problem with the model discussed above is that there is a time lag of 200 Ma between the formation of the leucogneisses and the Inverian retrogression (Rollinson, pers. comm.). However, if it is accepted that the leucogneisses were emplaced at the peak of metamorphism (which most authors do, regardless of their beliefs on the ultimate origins of these rocks), this problem is not specific to the view advanced here that the leucogneisses were formed by crustal anatexis. Whatever the origin of these lithologies, if they were melts at 1000°C, they would not have completed crystallization until temperatures had fallen to ~600–650°C, and exsolution of water during cooling would cause retrogression, as described above. Water-undersaturated granitic melts have extremely high viscosities. A granite melt at 1000°C with 2 wt% H_2O has a viscosity of $\sim 10^{10}$ Pa s (calculated using the method of Shaw 1972); a similar liquid with 5 wt% H_2O has a viscosity of $\sim 10^5$ Pa s. It is difficult for such highly viscous liquids to migrate within the crust (McKenzie 1984) and this may explain why these liquids have not escaped. The slow cooling rate of the Scourian complex is itself problematic (as discussed below) and the relative ages of metamorphic events may require revision.

Similar causes for retrogression following granulite facies metamorphism have been invoked for the Broken Hill area of Australia (Corbett & Phillips 1981) and the Namaqualand granulites of southern Africa (Waters 1988). The correlation between potential aggregated melts and amphibolite facies retrogression may provide a useful test for assessing the rôle of anatexis during granulite facies metamorphism.

13.9 Discussion and conclusions

The Scourian Complex underwent granulite facies metamorphism in the Badcallian event at 2.7 Ga at peak conditions of 1000°C and >1.1. GPa. The granulite facies assemblages are most plausibly restites of anatexis, and the suites of granitic to trondhjemitic veins and segregations emplaced close to the climax of the Badcallian event probably represent the aggregated products of partial melting. The melts would have almost certainly been water-undersaturated and highly polymerized, which may account for their anomalous REE geochemistry. During cooling from the peak of metamorphism, melt crystallization occurred and exsolution of water-rich fluids during the later stages of crystallization (when temperatures had reached 625–600°C) caused the Inverian retrogression.

The formation and retrogression of granulite facies assemblages occurred as a response to anatexis and melt crystallization. The granulite facies lithologies were formed as restites of melting due to the partitioning of H_2O into the melts; during cooling, H_2O is returned from the melts to the gneisses. This

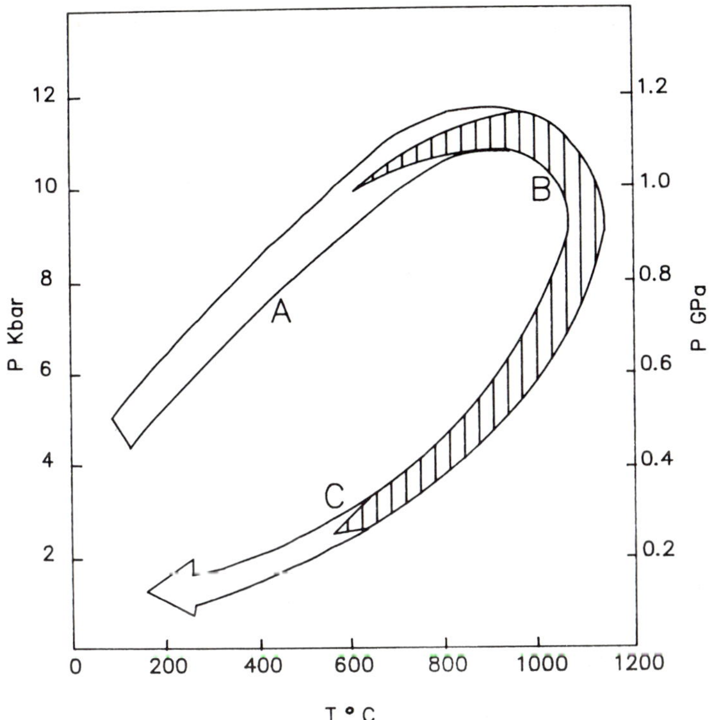

Figure 13.8 Schematic *P–T–t* loop for lower-crustal terrains showing the location of H_2O. In zone A, H_2O is contained within hydrous phases and grain-boundary fluids. The transition between A and B occurs in the range 600–1000°C and represents periods of anatexis in successively more refractory compositions, partitioning H_2O into the melts. During cooling, the melts crystallize and become more water-saturated until H_2O is exsolved, allowing rehydration of the granulite facies assemblages. Melts formed by crustal anatexis are likely to be initially markedly water-undersaturated, and the transition from B to C will occur over a short time interval at ~625–600°C (Cartwright 1988).

'cycling' of volatiles during pro- and retrograde metamorphism is schematically illustrated in Figure 13.8. Many other granulite facies terrains, such as Broken Hill, Australia (Corbett & Phillips 1981); Namaqualand, South Africa (Waters 1988); and West Greenland (Bridgwater *et al.* 1973, Korstgård *et al.* 1987) show a period of amphibolite facies retrogression during cooling from the peak of granulite facies metamorphism which may have been caused by fluids derived from crystallizing melts. This view of prograde and retrograde granulite facies metamorphism has the advantage of simplicity by considering fluids as being 'internal' to the terrain and not being introduced from external sources, for which little evidence exists. In the Scourian Complex, the model also accounts for the timing of the prograde and retrograde metamorphism and the distribution of granulite and retrogressed assemblages; such explanations are not inherent to models which invoke the influx of externally derived fluids.

13.9.1 Cooling from the peak of metamorphism

The geochronology data (Table 13.1) imply that cooling and uplift rates between the peak of Badcallian metamorphism at 2.7 Ga and the Inverian retrogression at 2.5 Ga were only 2–3°C Ma^{-1} and 4–5 MPa Ma^{-1} (~0.15 mm per annum); average rates between 2.7 and 1.0 Ga are <1°C Ma^{-1} and <1 MPa Ma^{-1} (~0.03 mm per annum). These rates are at least an order of magnitude slower than cooling and uplift rates recorded from most recent orogenic belts and pose problems for tectonic models of the Scourian Complex. The Sm–Nd study of Humphries & Cliff (1982) and the radiogenic isotope data of Cohen *et al.* (1988a) are consistent with steady cooling occurring between the Badcallian and Inverian events, and imply that the complex was not exhumed prior to 2.5 Ga. A possible tectonic explanation of the slow cooling rates is that the Lewisian Complex formed part of an extensive tectonically thickened block (much of the North Atlantic Craton underwent granulite facies metamorphism at 2.9–2.7 Ga, e.g. Korstgård *et al.* 1987). By contrast to relatively small orogenic belts (e.g. the Alps) large tectonic blocks are eroded inwards from their margins and are probably exhumed more slowly (Newton, pers. comm.); a modern-day analogue may be the Tibetan Plateau. A consequence of slow cooling is that extensive re-equilibration of isotopic and mineralogical systems occurs, and this is evident in the Scourian Complex lithologies from the Sm–Nd data (Humphries & Cliff 1982) and the coarse-grained exsolution and symplectite textures (O'Hara & Yarwood 1978, Barnicoat and O'Hara 1979, Savage and Sills 1980, Barnicoat 1983). However, it is possible that the rather simplistic interpretation of the geochronology data in Table 13.1 is in error, and that future studies may either revise the age of individual events or reveal a more complex *P–T–t* history. In either case the basic story outlined in this chapter will probably not be affected.

Acknowledgements

The ideas discussed in this chapter have benefitted from discussions with A. C. Barnicoat, M. J. O'Hara, J. W. Valley, and W. R. Fitches. Reviews of the manuscript by G. T. R. Droop, W. Perkins, and A. C. Barnicoat, and of an earlier version by H. R. Rollinson, W. R. Fitches, and an anonymous referee helped clarify the discussion. P. K. Harvey and B. P. Atkin of Nottingham helped with the XRF analyses. A. Thawley and W. Edwards helped produce the figures. I. C. gratefully acknowledges receipt of NATO postdoctoral fellowship GT8/F/86/GS/1.

References

Barnicoat, A. C. 1983. Metamorphism of the Scourian complex, NW Scotland. *Journal of Metamorphic Geology*, **1**, 163–82.
Barnicoat, A. C. 1987. The causes of the high grade metamorphism of the Scourie complex, NW Scotland. In *The evolution of the Lewisian and comparable Precambrian high grade terrains*, R. G. Park & J. Tarney (eds) 73–9. Geological Society of London, Special Publication 27.
Barnicoat, A. C., I. Cartwright & M. J. O'Hara 1987. Kyanite in the mainland Lewisian complex. *Scottish Journal of Geology* **23**, 209–13.
Barnicoat, A. C. & M. J. O'Hara 1979. High temperature pyroxenes from an ironstone at Scourie, Sutherland. *Mineralogical Magazine* **43**, 371–5.
Barooah, B. C. 1970. Significance of calc-silicate rocks and meta-arkoses in the Lewisian complex southeast of Scourie. *Scottish Journal of Geology* **6**, 221–5.
Bartholomew, D. S. & J. Tarney 1984. Geochemical characteristics of magmatism in the southern Andes (45–46°S). In *Andean magmatism: chemical and isotopic constraints*, R. S. Harmon & B. Barreiro (eds), 220–9. Nantwich, UK: Shiva.
Beach, A. 1973. The mineralogy of high-temperature shear zones at Scourie, NW Scotland. *Journal of Petrology* **14**, 231–48.
Bohlen, S. R. 1987. Pressure–temperature–time paths and a tectonic model for the evolution of granulites. *Journal of Geology* **95**, 617–32.
Bowes, D. R., A. E. Wright & R. G. Park 1964. Layered intrusive rocks in the Lewisian of the northwest Highlands of Scotland. *Quarterly Journal of the Geological Society of London* **120**, 153–92.
Bridgwater, D., J. V. Watson & B. F. Windley 1973. The Archaean craton of the North Atlantic region. *Philosophical Transactions of the Royal Society of London* **A273**, 493–512.
Burton, K. W., A. S. Cohen & R. K. O'Nions 1988. Investigation of dehydration and melt loss in the lower crust. *Chemical Geology* **70**, 13.

Cartwright, I. 1986. *The geological history of the Lewisian complex at Stoer, NW Scotland.* Unpubl. PhD thesis, University College of Wales.
Cartwright, I. 1988. Crystallization of melts, pegmatite intrusion and the Inverian retrogression of the Scourian complex, north-west Scotland. *Journal of Metamorphic Geology* **6**, 77–93.
Cartwright, I. & A. C. Barnicoat 1986. The generation of quartz-normative melts and corundum-bearing restites by crustal anatexis: petrogenetic modelling based on an example from the Lewisian of north-west Scotland. *Journal of Metamorphic Geology* **4**, 79–99.
Cartwright, I. & A. C. Barnicoat 1987. Petrology of Scourian supracrustal rocks and orthogneisses from Stoer, NW Scotland: implications for the geological evolution of the Lewisian complex. In *The evolution of the Lewisian and comparable Precambrian high grade terrains*, R. G. Park & J. Tarney (eds), 93–107. Geological Society of London, Special Publication 27.
Cartwright, I. & A. C. Barnicoat 1989. Geological history of the Scourian complex. In *Evolution of metamorphic belts*, J. S. Daly, R. A. Cliff & B. W. D. Yardley (eds), 297–301. Geological Society of London, Special Publication 42.
Cartwright, I., W. R. Fitches, M. J. O'Hara, A. C. Barnicoat & S. O'Hara 1985. Archaean supracrustal rocks from the Lewisian near Stoer, Sutherland. *Scottish Journal of Geology* **21**, 187–96.
Chapman, H. J. & S. Moorbath 1977. Lead isotope measurements from the oldest recognized Lewisian gneiss of north-west Scotland. *Nature* **268**, 41–2.
Cohen, A. S., R. K. O'Nions & M. J. O'Hara 1988a. Melting, depletion and textural equilibration of Lewisian granulites. *Chemical Geology* **70**, 6.

Cohen, A. S., F. G. Waters, R. K. O'Nions & M. J. O'Hara 1988b. A precise crystallization age for the Scourie dykes and a new chronology for crustal development in north-west Scotland. *Chemical Geology* **70**, 19.

Condie, K. C., G. P. Bowling & P. Allen 1986. Origin of granulites in an Archean high grade terrane, southern India. *Contributions to Mineralogy and Petrology* **92**, 93–103.

Corbett, G. J. & G. N. Phillips 1981. Regional retrograde metamorphism of a high grade terrain: the Willyama complex, Broken Hill, Australia. *Lithos* **14**, 59–73.

Coward, M. P. & R. G. Park 1987. The role of mid-crustal shear zones in the Early Proterozoic evolution of the Lewisian. In *The evolution of the Lewisian and comparable Precambrian high grade terrains*, R. G. Park & J. Tarney (eds), 127–38. Geological Society of London, Special Publication 27.

Crane, A. 1978. Correlation of metamorphic fabrics and the age of Lewisian metasediments near Loch Maree. *Scottish Journal of Geology*, **14**, 225–46.

Davies, F. B. 1974. A layered basic complex in the Lewisian south of Loch Laxford, Sutherland. *Journal of the Geological Society of London* **130**, 279–84.

Davies, F. B. 1977. The Archaean evolution of the Lewisian complex of Gruinard Bay. *Scottish Journal of Geology* **13**, 189–96.

England, P. C. & S. W. Richardson 1977. The influence of erosion upon the mineral facies of rocks from different metamorphic environments. *Journal of the Geological Society of London* **134**, 201–13.

England, P. C. & A. B. Thompson 1984. Pressure–temperature–time paths of regional metamorphism, I. Heat transfer during the evolution of regions of thickened continental crust. *Journal of Petrology* **25**, 894–928.

Evans, C. R. 1965. Geochronology of the Lewisian basement near Lochinver, Sutherland. *Nature* **207**, 54–6.

Flynn, R. T. & C. W. Burnham. 1978. An experimental determination of rare earth partition coefficients between a chloride containing vapor phase and silicate melts. *Geochimica et Cosmochimica Acta* **42**, 685–701.

Fyfe, W. S. 1973. The granulite facies, partial melting, and the Archaean crust. *Philosophical Transactions of the Royal Society of London* **A273**, 457–62.

Ghiorso, M. S. 1984. Activity–composition relations in the ternary feldspars. *Contributions to Mineralogy and Petrology* **87**, 282–96.

Hall, J. 1987. Physical properties of Lewisian rocks: implications for deep crustal structure. In *The evolution of the Lewisian and comparable Precambrian high grade terrains*, R. G. Park & J. Tarney (eds) 185–92. Geological Society of London, Special Publication 27.

Hamilton, P. J., N. M. Evensen, R. K. O'Nions & J. Tarney 1979. Sm–Nd systematics of Lewisian gneisses: implications for the origin of granulites. *Nature* **277**, 25–8.

Hanson, G. N. 1978. The application of trace elements to the petrogenesis of igneous rocks of granitic composition. *Earth and Planetary Science Letters* **38**, 26–43.

Huang, W. L. & P. J. Wyllie 1975. Melting reactions in the system $NaAlSi_3O_8$–$KAlSi_3O_8$–SiO_2 to 35 kilobars, dry and with excess water. *Journal of Geology* **83**, 737–48.

Humphries, F. J. & R. A. Cliff 1982. Sm–Nd dating and cooling history of Scourie granulites, Sutherland. *Nature* **295**, 515–7.

Jensen, L. N. 1984. Quartz microfabric of the Laxfordian Canisp shear zone, NW Scotland. *Journal of Structural Geology* **6**, 293–302.

Korstgård, J., B. Ryan & R. Wardle 1987. The boundary between Proterozoic and Archaean crustal blocks in central West Greenland and northern Labrador. In *The evolution of the Lewisian and comparable Precambrian high grade terrains*, R. G. Park & J. Tarney (eds), 247–59. Geological Society of London, Special Publication 27.

Lamb, W. & J. W. Valley 1984. Metamorphism of reduced granulites in low CO_2 vapour-free environment. *Nature* **312**, 56–8.

Lamb, W. & J. W. Valley 1985. C–O–H fluid calculations and granulite genesis. In *The deep Proterozoic crust in the North Atlantic provinces*, A. C. Tobi & J. L. R. Touret (eds), 119–31. Dordrecht: Reidel.

Lamb, W., J. W. Valley & P. E. Brown 1987. Post-metamorphic CO_2-rich fluid inclusions in granulites. *Contributions to Mineralogy and Petrology* **96**, 485–95.

Lambert, R. St. J. & J. G. Holland 1972. A geochronological study of the Lewisian from Loch Laxford to Durness, Sutherland, N. W. Scotland. *Journal of the Geological Society of London* **128**, 3–19.

Luth, W. C. 1969. The systems $NaAlSi_3O_8$–SiO_2 and $KAlSi_3O_8$–SiO_2 to 20 kb and the relationship between H_2O content, P_{H_2O} and P_{Total} in granitic magmas. *American Journal of Science* **267-A**, 325–41.

Luth, W. C. 1976. Granitic rocks. In *The evolution of the crystalline rocks*, D. K. Bailey & R. Macdonald (eds), 335–417. London: Academic Press.

McKenzie, D. 1984. The generation and compaction of partially molten rock. *Journal of Petrology* **25**, 713–65.

Moorbath, S. 1969. Evidence for the age of deposition of the Torridonian sediments of NW Scotland. *Scottish Journal of Geology* **5**, 154–70.

Moorbath, S. & R. G. Park 1972. The Lewisian chronology of the southern region of the Scottish mainland. *Scottish Journal of Geology* **8**, 51–74.

Muecke, G. K. 1969. *Petrogenesis of granulite facies rocks from the Lewisian of Scotland.* Unpubl. DPhil thesis, University of Oxford.

Newton, R. C. & D. Perkins 1982. Thermodynamic calibration of geobarometers based on the assemblages garnet–plagioclase–orthopyroxene–(clinopyroxene)–quartz. *American Mineralogist* **67**, 203–22.

Newton, R. C., J. V. Smith & B. F. Windley 1980. Carbonic metamorphism, granulites, and crustal growth. *Nature* **288**, 45–50.

O'Hara, M. J. 1961. Zoned ultrabasic and basic gneiss masses in the early Lewisian metamorphic complex at Scourie, Sutherland. *Journal of Petrology* **2**, 248–76.

O'Hara, M. J. 1967. Mineral parageneses in ultramafic rocks. In *Ultramafic and related rocks* P. J. Wyllie (ed.), 393–403. Chichester: Wiley.

O'Hara, M. J. 1977. Thermal history of excavation of Archaean gneisses from the base of the continental crust. *Journal of the Geological Society of London* **134**, 185–200.

O'Hara, M. J. & G. Yarwood 1978. High pressure–temperature point on an Archaean geotherm, implied magma genesis by crustal anatexis and consequences for garnet–pyroxene thermometry and barometry. *Philosophical Transactions of the Royal Society of London* **A288**, 441–56.

Okeke, P. O., G. D. Borley & J. Watson 1983. A geochemical study of Lewisian metasedimentary granulites and gneisses in the Scourie–Laxford area of NW Scotland. *Mineralogical Magazine* **47**, 1–9.

Park, R. G., A. Crane & M. Niamatullah 1987. Early Proterozoic structure and kinematic evolution the southern mainland Lewisian. In *The evolution of the Lewisian and comparable Precambrian high grade terrains*, R. G. Park & J. Tarney (eds), 139–51. Geological Society of London, Special Publication 27.

Park, R. G. & J. Tarney 1987. The Lewisian complex: a typical Precambrian high-grade terrain?

In *The evolution of the Lewisian and comparable Precambrian high grade terrains*, R. G. Park & J. Tarney (eds), 13–25. Geological Society of London, Special Publication 27.

Peach, B. N., J. Horne, W. Gunn, C. T. Clough, L. W. Hinxman & J. J. H. Teall, 1907. The geological structure of the north-west Highlands of Scotland. *Geological of Great Britain, Memoir.*

Pidgeon, R. T. & D. R. Bowes 1972. Zircon U–Pb ages of granulites from the central region of the Lewisian, NW Scotland. *Geological Magazine* **109**, 247–58.

Powell, R. 1983a. Fluids and melting under upper amphibolite facies conditions. *Journal of the Geological Society of London* **140**, 629–33.

Powell, R. 1983b. Processes in granulite facies metamorphism. In *Migmatites, melting, and metamorphism*, M. P. Atherton & C. D. Gribble (eds), 127–39. Nantwich, UK: Shiva.

Pride, C. & G. K. Muecke 1980. Rare earth element geochemistry of the Scourian complex, NW Scotland – evidence for the granite–granulite link. *Contributions to Mineralogy and Petrology* **73**, 403–12.

Pride, C. & G. K. Muecke 1982. Geochemistry and origin of granitic rocks, Scourian complex, NW Scotland. *Contributions to Mineralogy and Petrology* **80**, 379–85.

Rollinson, H. R. 1980. Mineral reactions in a granulite-facies calc-silicate rock from Scourie. *Scottish Journal of Geology* **163**, 153–64.

Rollinson, H. R. 1981. Garnet–clinopyroxene thermometry and barometry in the Scourian complex, NW Scotland. *Lithos* **14**, 225–38.

Rollinson, H. R. 1982. Evidence from feldspar compositions of high temperatures in granite sheets in the Scourie complex, NW Scotland. *Mineralogical Magazine* **46**, 73–6.

Rollinson, H. R. & M. B. Fowler 1987. The magmatic evolution of the Scourian complex at Gruinard Bay. In *The evolution of the Lewisian and comparable Precambrian high grade terrains*, R. G. Park & J. Tarney (eds), 51–71. Geological Society of London, Special Publication 27.

Rollinson, H. R. & B. F. Windley 1980a. Selective elemental depletion during metamorphism of Archaean granulites, Scourie, NW Scotland. *Contributions to Mineralogy and Petrology* **72**, 257–63.

Rollinson, H. R. & B. F. Windley 1980b. An Archaean granulite-grade tonalite–trondhjemite–granite suite from Scourie, NW Scotland. *Contributions to Mineralogy and Petrology* **72**, 265–81.

Savage, D. & J. D. Sills 1980. High pressure metamorphism in the Scourian of NW Scotland: evidence from garnet granulites. *Contributions to Mineralogy and Petrology* **74**, 153–63.

Shaw, H. R. 1972. Viscosities of magmatic silicate liquids: an empirical method of prediction. *American Journal of Science* **272**, 870–93.

Sheraton, J. W. A. C. Skinner & J. Tarney 1973a. The geochemistry of Scourian gneisses of the Assynt distinct. In *The early Precambrian of Scotland and related rocks of Greenland*, R. G. Park & J. Tarney (eds), 13–30. University of Keele.

Sheraton, J. W., A. C. Skinner & J. Tarney 1973a. The geochemistry of Scourian gneisses of the Assynt district. In *The early Precambrian of Scotland and related rocks of Greenland*, R. G. Park & J. Tarney (eds), 13–30. University of Keele.

Sills, J. D. 1982. The retrogression of ultramafic granulites from the Scourian of NW Scotland. *Mineralogical Magazine* **46**, 55–61.

Sills, J. D. 1983. Mineralogical changes occurring during the retrogression of Archaean gneisses from the Lewisian complex of NW Scotland. *Lithos* **16**, 113–24.

Sills, J. D. & H. R. Rollinson 1987. Metamorphic evolution of the mainland Lewisian complex. In *The evolution of the Lewisian and comparable Precambrian high grade terrains*, R. G. Park & J. Tarney (eds), 81–92. Geological Society of London, Special Publication 27.

Sutton, J. & J. Watson 1951. The Pre-Torridonian metamorphic history of the Loch Torridon and Scourie areas in the NW Highlands, and its bearing on the chronological classification of the Lewisian. *Quarterly Journal of the Geological Society of London* **106**, 241–307.

Tarney, J. 1963. Assynt dykes and their metamorphism. *Nature* **199**, 672–4.

Tarney, J. & B. L. Weaver 1987. Geochemistry of the Scourian complex: petrogenesis and tectonic models. In *The evolution of the Lewisian and comparable Precambrian high grade terrains*, R. G. Park & J. Tarney (eds), 45–56. Geological Society of London, Special Publication 27.

Taylor, P. N., N. W. Jones & S. Moorbath 1984. Isotopic assessment of relative contributions of crust and mantle sources to the magma genesis of Precambrian granitoid rocks. *Philosophical Transactions of the Royal Society of London* **A310**, 605–25.

Waters, D. J. 1988. Partial melting and the formation of granulite facies assemblages in Namaqualand, South Africa. *Journal of Metamorphic Geology* **6**, 387–404.

Weaver, B. L. & J. Tarney 1980. Rare earth geochemisty of Lewisian granulite-facies gneisses, northwest Scotland: implications for the petrogenesis of the Archaean lower continental crust. *Earth and Planetary Science Letters* **51**, 279–96.

Weaver, B. L. & J. Tarney 1981. Lewisian gneiss geochemistry and Archaean crustal development models. *Earth and Planetary Science Letters* **55**, 171–80.

Weaver, B. L. & J. Tarney 1983. Elemental depletion in Archaean granulite-facies rocks. In *Migmatites, melting, and metamorphism*, M. P. Atherton & C. D. Gribble (eds), 250–63. Nantwich, UK: Shiva.

Wells, P. R. A. 1980. Thermal models for the magmatic accretion and subsequent metamorphism of continental crust. *Earth and Planetary Science Letters* **46**, 253–65.

Whitehouse, M. J. & S. Moorbath 1986. Pb–Pb systematics of Lewisian gneisses – implications for crustal differentiation. *Nature* **319**, 488–9.

Whitney, J. A. 1975. The effects of pressure, temperature, and X_{H_2O} on phase assemblage in four synthetic rock compositions. *Journal of Geology* **83**, 1–31.

Wyllie, P. J. 1983. Experimental studies on biotite- and muscovite-granites and some crustal magmatic sources. In *Migmatites, melting, and metamorphism*, M. P. Atherton & C. D. Gribble (eds), 12–26. Nantwich, UK: Shiva.

Wyllie, P. J. 1984. Constraints imposed by experimental petrology on possible and impossible magma sources and products. *Philosophical Transactions of the Royal Society of London* **A310**, 439–56.

Index

DURHAM UNIVERSITY LIBRARY
3 0104 00512701 8